T0135056

Genetic and Evolutionary Computation

Series Editors:
Wolfgang Banzhaf, Michigan State University, East Lansing, MI, USA
Kalyanmoy Deb, Michigan State University, East Lansing, MI, USA

More information about this series at http://www.springer.com/series/7373

Wolfgang Banzhaf • Betty H.C. Cheng
Kalyanmoy Deb • Kay E. Holekamp
Richard E. Lenski • Charles Ofria
Robert T. Pennock • William F. Punch
Danielle J. Whittaker
Editors

Evolution in Action: Past, Present and Future

A Festschrift in Honor of Erik D. Goodman

 Springer

Editors
Wolfgang Banzhaf
Michigan State University
East Lansing, MI, USA

Betty H.C. Cheng
Michigan State University
East Lansing, MI, USA

Kalyanmoy Deb
Michigan State University
East Lansing, MI, USA

Kay E. Holekamp
Michigan State University
East Lansing, MI, USA

Richard E. Lenski
Michigan State University
East Lansing, MI, USA

Charles Ofria
Michigan State University
East Lansing, MI, USA

Robert T. Pennock
Michigan State University
East Lansing, MI, USA

William F. Punch
Michigan State University
East Lansing, MI, USA

Danielle J. Whittaker
Michigan State University
East Lansing, MI, USA

ISSN 1932-0167   ISSN 1932-0175 (electronic)
Genetic and Evolutionary Computation
ISBN 978-3-030-39833-0   ISBN 978-3-030-39831-6 (eBook)
https://doi.org/10.1007/978-3-030-39831-6

This Springer imprint is published by the registered company Springer Nature Switzerland AG
The registered company address is: Gewerbestrasse 11, 6330 Cham, Switzerland

Dr. Erik D. Goodman, 2016

Preface

The BEACON Congress in August 2018 provided an opportunity to celebrate its founding director Professor Erik Goodman's remarkable achievements on the occasion of his 75th birthday. A number of the presentations at that 2018 BEACON Congress were specifically dedicated to him. This volume *Evolution in Action — Past, Present, and Future* combines some of these with additionally solicited contributions from colleagues and members of BEACON who could not attend the 2018 Congress.

The contributions published here range from refreshingly personal stories of encounters with Dr. Goodman, the history and achievements of the BEACON Center for the Study of Evolution in Action under his leadership, to research topics presented along the lines of the research thrust groups in the BEACON Center. These topics dominate Parts I to V of the book. An important focus of BEACON has always been education about evolution, and this is reflected in the contributions in Part VI. At our request, Dr. Goodman graciously provided us with an account of his own life story, which forms the last part of this volume.

All chapters published in this volume have undergone editorial review. Additionally, some chapters (10, 11, 17, 18, 20, 21, 23) have undergone additional external anonymous peer review at the request of their authors.

We would like to thank all contributors and reviewers for their diligent work in bringing together this exciting volume. We would also like to thank those who have helped the editors, from organizing our group meetings to bringing contributions together in a manageable LaTeX document. In particular, we thank Honglin Bao, Connie James, Stephen Kelly, Iliya Miralavy, Ian Whalen, and Yuan Yuan for their invaluable help.

East Lansing, Michigan
February 12, 2019

Wolfgang Banzhaf
Betty H.C. Cheng
Kalyanmoy Deb
Kay E. Holekamp
Richard E. Lenski
Charles Ofria
Robert T. Pennock
Bill F. Punch
Danielle J. Whittaker

List of Contributors

Christoph Adami[1]
Department of Microbiology & Molecular Genetics, Michigan State University, East Lansing, MI, USA

Hawlader A. Al Mamun
CSIRO Data61, Commonwealth Scientific and Industrial Research Organisation, Canberra, ACT, AUSTRALIA

Marilyn J. Amey
College of Education, Michigan State University, East Lansing, MI, USA

Wolfgang Banzhaf[1]
Department of Computer Science & Engineering, Michigan State University, East Lansing, MI, USA

Jeffrey E. Barrick[1]
Department of Molecular Biosciences, The University of Texas at Austin, Austin, TX, USA

Eric Berling[1]
Center for Interdisciplinarity, Michigan State University, East Lansing, MI USA

Zachary D. Blount[1]
Department of Microbiology & Molecular Genetics, Michigan State University, East Lansing, MI, USA; and Department of Biology, Gambier, OH, USA

Sada Boyd[1]
Energy and Environment Program, North Carolina A&T State University, Greensboro, NC, USA

Alita Burmeister[1]
Department of Ecology & Evolutionary Biology, Yale University, New Haven, CT, USA

Judi Brown Clarke[1]
BEACON Center for Evolution in Action, Michigan State University, East Lansing, MI, USA

Daniel E. Deatherage[1]
Department of Molecular Biosciences, The University of Texas at Austin, Austin, TX, USA

Kalyanmoy Deb[1]
Department of Electrical & Computer Engineering, Michigan State University, East Lansing, MI, USA

Meghan A. Duffy[2]
Department of Ecology & Evolutionary Biology, University of Michigan, Ann Arbor, MI, USA

Subhrajit Dutta
Department of Civil Engineering, National Institute of Technology Silchar, Cachar, Assam, INDIA

Ian Dworkin[1]
Department of Biology, McMaster University, 1280 Main St. West, Hamilton, Ontario, CANADA

Rasheena Edmundson[1]
Department of Biology, Bennett College, Greensboro, NC, USA

Akamu J. Ewunkem[1]
Department of Nanoengineering, Joint School of Nanoscience & Nanoengineering, North Carolina A&T State University & UNC Greensboro, Greensboro, NC, USA

Zhun Fan[1]
Shantou University, Shantou, Guangdong, CHINA

Patricia L. Farrell-Cole[1]
Van Andel Institute, Grand Rapids, MI, USA

Francisco Fernández de Vega
University of Extremadura, Mérida, SPAIN

Ellen F. Foxman
Department of Laboratory Medicine; and Department of Immunology, Yale School of Medicine, New Haven, CT, USA

Amir H. Gandomi[1]
Faculty of Engineering and Information Technology, University of Technology, Sydney, NSW, AUSTRALIA

Eben Gering[1]
Department of Integrative Biology; and Ecology, Evolutionary Biology, & Behavior Program, Michigan State University, East Lansing, MI, USA

Thomas Getty[1]
Department of Integrative Biology; and Ecology, Evolutionary Biology, & Behavior Program, Michigan State University, East Lansing, MI, USA

Heather Goldsby[1]
Department of Computer Science & Engineering, Michigan State University, East Lansing, MI, USA

Cedric Gondro[1]
Department of Animal Science, Michigan State University, East Lansing, MI, USA

Joseph L. Graves Jr[1]
Department of Nanoengineering, Joint School of Nanoscience & Nanoengineering, North Carolina A&T State University & UNC Greensboro, Greensboro, NC, USA

Jian Han[1]
Department of Biology, North Carolina A&T State University, Greensboro, NC, USA

Scott H. Harrison[1]
Department of Biology, North Carolina A&T State University, Greensboro, NC, USA

Faraz Hasan
Department of Computer Applications, B.S. Abdur Rahman Crescent Institute of Science & Technology, Chennai, Tamil Nadu, INDIA

Erik Hemberg
Computer Science & Artificial Intelligence Lab, Massachusetts Institute of Technology, Cambridge, MA, USA

Arend Hintze[1]
Department of Computer Science & Engineering; and Department of Integrative Biology, Michigan State University, East Lansing, MI, USA

Kay Holekamp[1]
Department of Integrative Biology, and Ecology, Evolutionary Biology, & Behavior Program, Michigan State University, East Lansing, MI, USA

Gisela Hussey
College of Veterinary Medicine, Michigan State University, East Lansing, MI, USA

Akiko Iwasaki
Department of Immunology, Yale School of Medicine; and Department of Molecular, Cellular & Developmental Biology, Yale University, New Haven, CT, USA; and Howard Hughes Medical Institute, Chevy Chase, MD, USA

Julius H. Jackson[1]
Department of Microbiology & Molecular Genetics, Michigan State University, East Lansing, MI USA

Tanush Jagdish[1]
Department of Microbiology & Molecular Genetics, Michigan State University, East Lansing, MI, USA; and Department of Physics and of Biology, Kalamazoo College, Kalamazoo, MI, USA. Currently at: Program for Systems Biology, Harvard University, Cambridge, MA, USA

Constance James[1]
BEACON Center for the Study of Evolution in Action, Michigan State University, East Lansing, MI, USA

Liesel Jeffers-Francis[1]
Department of Biology, North Carolina A&T State University, Greensboro, NC, USA

Zaki Ahmad Khan
Department of Zoology College of Life Sciences, Nanjing Agricultural University, Weigang, Nanjing, Jiangsu, CHINA

Benjamin Kerr[1]
Department of Biology, University of Washington, Seattle, WA, USA

Thomas LaBar[1]
Department of Microbiology & Molecular Genetics, Michigan State University, East Lansing, MI, USA

Amy M. Lark[1]
Department of Cognitive & Learning Sciences, Michigan Tech University, Houghton, MI, USA

Zachary Laubach[1]
Department of Integrative Biology; and Ecology, Evolutionary Biology, & Behavior Program, Michigan State University, East Lansing, MI, USA

Kenna Lehmann[1]
Department of Integrative Biology; and Ecology, Evolutionary Biology, & Behavior Program, Michigan State University, East Lansing, MI, USA

Richard E. Lenski[1]
Department of Microbiology & Molecular Genetics, Michigan State University, East Lansing, MI, USA

Wenji Li
Shantou University, Shantou, Guangdong, CHINA

Barbara Lundrigan[1]
Department of Integrative Biology; Program in Ecology, Evolutionary Biology & Behavior; and Michigan State University Museum, Michigan State University, East Lansing, MI, USA

Tian-tong Luo
Department of Zoology College of Life Sciences, Nanjing Agricultural University, Weigang, Nanjing, Jiangsu, CHINA

Teresa L. McElhinny[1]
Department of Integrative Biology, Michigan State University, East Lansing, MI, USA

Katherine D. McLean
Department of Ecology & Evolutionary Biology, University of Michigan, Ann Arbor, MI, USA

Chet McLeskey[1]
Center for Interdisciplinarity, Michigan State University, East Lansing, MI, USA

Louise S. Mead[1]
Department of Integrative Biology, Michigan State University, East Lansing, MI, USA

Paul Meek
NSW Department of Primary Industries, National Marine Science Centre, Coffs Harbour, NSW; and University of New England, School of Environmental and Rural Science, AUSTRALIA

Justin R. Meyer[1]
Division of Biological Sciences, University of California, San Diego, La Jolla, CA, USA

Risto Miikkulainen[1]
The University of Texas at Austin, Austin, TX, USA; and Cognizant, Inc., USA

Tracy Montgomery[1]
Department of Integrative Biology; and Ecology, Evolutionary Biology, & Behavior Program, Michigan State University, East Lansing, MI, USA

J. Jeffrey Morris[1]
Department of Biology, University of Alabama at Birmingham, Birmingham, AL, USA

Andrea Morrow[1]
Department of Integrative Biology; Program in Ecology, Evolutionary Biology & Behavior; and Michigan State University Museum, Michigan State University, East Lansing, MI, USA

Charles Ofria[1]
Department of Computer Science & Engineering, Michigan State University, East Lansing, MI, USA

Una-May O'Reilly[2]
Computer Science & Artificial Intelligence Lab, Massachusetts Institute of Technology, Cambridge, MA, USA

Michael O'Rourke[1]
Center for Interdisciplinarity, Department of Philosophy; and AgBioResearch, Michigan State University, East Lansing, MI, USA

Robert T. Pennock[1]
Lyman Briggs College; Department of Philosophy; Department of Computer Science & Engineering, Michigan State University, East Lansing, MI, USA

Percy Pierre[1]
Department of Electrical and Computer Engineering, Michigan State University, East Lansing, MI, USA

Kristen L. Rhinehardt[1]
Department of Nanoengineering, Joint School of Nanoscience & Nanoengineering, North Carolina A&T State University & UNC Greensboro, Greensboro, NC, USA

Gail Richmond
Department of Teacher Education, Michigan State University, East Lansing, MI, USA

Connie A. Rojas[1]
Department of Integrative Biology, Michigan State University, East Lansing, MI, USA

Matthew R. Rupp[1]
BEACON Center for the Study of Evolution in Action, Michigan State University, East Lansing, Michigan, USA

Jory Schossau[1]
Department of Integrative Biology, Michigan State University, East Lansing, MI, USA

Laura Smale[1]
Department of Psychology; Department of Integrative Biology; Neuroscience Program; Program in Ecology, Evolutionary Biology and Behavior, Michigan State University, East Lansing, MI, USA

Joan E. Strassmann[2]
Department of Biology, Washington University in St. Louis, St. Louis, MO, USA, and Wissenschaftskolleg zu Berlin, GERMANY

Rachel Sullivan[1]
Department of Microbiology & Molecular Genetics, Michigan State University, East Lansing, MI; present address: Waisman Center, University of Wisconsin-Madison, Madison, WI, USA

Kevin R. Theis[1]
Department of Biochemistry, Microbiology and Immunology, Wayne State University, Detroit, MI, USA

Misty D. Thomas[1]
Department of Biology, North Carolina A&T State University, Greensboro, NC, USA

Julie Turner[1]
Department of Integrative Biology; and Ecology, Evolutionary Biology, & Behavior Program, Michigan State University, East Lansing, MI, USA

Paul E. Turner[1]
Department of Ecology & Evolutionary Biology, Yale University, New Haven, CT, USA

Brian D. Wade[1]
Department of Plant, Soil & Microbial Sciences, Michigan State University, East Lansing, MI, USA

Aaron P. Wagner[1]
Metron Inc, Reston, VA, USA

Bethany R. Wasik
Department of Ecology & Evolutionary Biology, Yale University, New Haven, CT, USA, and Department of Ecology & Evolutionary Biology, Cornell University, Ithaca, NY, USA

Brian R. Wasik
Department of Ecology & Evolutionary Biology, Yale University, New Haven, CT, USA and Baker Institute for Animal Health, Department of Microbiology & Immunology, College of Veterinary Medicine, Cornell University, Ithaca, NY, USA

Patricia Weber
College of Veterinary Medicine, Michigan State University, East Lansing, MI, USA

Emily G. Weigel[1]
Department of Biological Sciences, Georgia Institute of Technology, Atlanta, GA, USA

Ian Whalen[1]
Department of Computer Science & Engineering, Michigan State University, East Lansing, MI, USA

Danielle J. Whittaker[1]
BEACON Center for the Study of Evolution in Action, Michigan State University, East Lansing, MI, USA

Michael Wiser[1]
BEACON Center for the Study of Evolution in Action, Michigan State University, East Lansing, MI, USA

Gabriel Yedid
Department of Zoology College of Life Sciences, Nanjing Agricultural University, Weigang, Nanjing, Jiangsu, CHINA

Yuan Yuan[1]
Department of Computer Science & Engineering, Michigan State University, East
Lansing, MI, USA

Luis Zaman[1]
Department of Ecology & Evolutionary Biology, University of Michigan, Ann
Arbor, MI, USA

Guijie Zhu
Shantou University, Shantou, Guangdong, CHINA

Jian-long Zhu
Department of Zoology College of Life Sciences, Nanjing Agricultural University,
Weigang, Nanjing, Jiangsu, CHINA

[1] BEACON Member
[2] BEACON External Advisory Committee Member

Contents

Part VI Evolution Education

Part I
The BEACON Center for Evolution in Action

Chapter 1
2010: A BEACON Odyssey

Richard E. Lenski

Abstract Life often follows a peculiar and winding path. This paper connects some events in my life with how I got to know Erik Goodman and how the BEACON Center for the Study of Evolution in Action came into being. Some of these events are from memory, while others were recorded in emails. Of course, BEACON has had many participants and so there are many narrative threads, which together are woven into the tapestry of BEACON and our lives.

Key words: Avida, BEACON, chance, contingency, Erik Goodman, experimental evolution, science funding, synergy

1.1 Life's Winding Path

Life often follows a peculiar and winding path. That's true in our individual lives, in the rise and fall of civilizations, and in the history of life on Earth. Understanding what propels an evolving population along a particular path has been a major focus of my group's research. In particular, we study the repeatability of evolution, the roles of chance and necessity, and how outcomes are sometimes contingent on prior events that seemed inconsequential when they occurred [4, 5, 8, 10, 13, 16, 17, 18, 20].

This paper presents a related story, albeit a personal one that connects some events in my academic life with how I got to know Erik Goodman, and how the BEACON Center for the Study of Evolution in Action came into being. Some of these events are from memory, while others were recorded in emails. I'll quote from some of those emails for posterity's sake. I must emphasize that this is *my* story, not *the* story. An important aspect of any complex organization like BEACON is that

Richard E. Lenski
BEACON Center for the Study of Evolution in Action, Michigan State University, East Lansing, MI, USA e-mail: lenski@msu.edu

© Springer Nature Switzerland AG 2020
W. Banzhaf et al. (eds.), *Evolution in Action: Past, Present and Future*,
Genetic and Evolutionary Computation, https://doi.org/10.1007/978-3-030-39831-6_1

so many people have been instrumental in its genesis and success. Thus, there are many unique and winding narrative threads, and together they are woven into the tapestry of BEACON and our lives.

1.2 A Squash Match

My thread begins on a squash court in 1994. I was playing in a recreational league and had a match with Wolfgang Bauer, a colleague in the Department of Physics and Astronomy here at MSU. I forget who won our match, but no matter—it turned out to be a huge win *off* the court. We chatted about science after playing, and Wolfgang said that he had a visitor coming to his lab who would give a talk on "Self-organized criticality in living systems" or something like that. Self-organized criticality was a fairly new concept then [3], but I had read about it in the popular-science press. An example of a system that exhibits self-organized criticality is a sand pile. Drop one grain after another on a pile and not much happens until, suddenly, one more grain triggers an avalanche that changes the shape of the entire pile. The subtle internal structuring that leads to the sudden transition is what physicists had dubbed self-organization.

When Wolfgang said the talk would be on self-organization and evolution, I might have rolled my eyes. Or at least I thought to myself that this talk might be a physicist trying to explain how evolution works, without understanding natural selection. But I said I'd attend, and I did.

The speaker was Christoph Adami, who was then a postdoc at Caltech. Chris gave a lively and interesting talk about his experiments with Tierra—an artificial living system in which computer programs replicate, mutate, and compete inside a virtual world. And he certainly understood natural selection, so my worry was unfounded. A couple of years earlier I'd read about Tierra in a News and Views piece by the evolutionary biologist John Maynard Smith [15], and I had been intrigued by the idea of such a fast-evolving and novel system.

Back to Chris's talk on self-organized criticality: in one slide that especially caught my attention, he showed a step-like increase in the fitness of an evolving population (figure 1 in [1]). Chris interpreted these dynamics as mutations raining down on the programs, with each mutation having little effect until one of them triggered an adaptive transition—much like grains of sand falling inconsequentially onto a pile until one of them triggers an avalanche. The analogy was intriguing, but I thought it was wrong, or at least incomplete—I thought there was a simpler explanation, which I'll get to in a moment. I was excited by this slide because in our work with bacteria we saw similar dynamics—sudden jumps in the bacteria's fitness punctuating periods of relative stasis (figure 5 in [13]).

My interpretation was that these dynamics were caused by the rise of lineages with new beneficial mutations. Each mutant lineage begins as a single individual, and it spreads more or less exponentially according to its fitness advantage (if it survives loss by random drift). But despite its exponential increase, for a long time the mutants remain a minority and thus have little effect on the *average* fitness of the population. Only when the mutants become sufficiently common does one see an appreciable rise in mean fitness; and then the mean changes so quickly that the overall trajectory has a step-like appearance. Besides his intelligence and enthusiasm, one thing that I immediately liked about Chris was that, as I explained my interpretation after his talk, he was not defensive and, instead, was receptive to this alternative explanation.

Chris and I corresponded occasionally after his talk, and I gave him feedback about some projects he planned to pursue using Avida, a new program that he was developing to study evolution in digital organisms. Chris also told me that he was working on a book about artificial life, based on a class he was teaching on that subject. He said that he would send me a copy when his book was published.

1.3 The French Connection

My family and I packed our bags and moved to Montpellier, France, after the summer of 1997. My wife Madeleine had spent half a year as a child in Paris when her father took a sabbatical, and she wanted a similar experience for our kids. I knew some evolutionary biologists in Montpellier, but I had no specific plans for new research. Instead, I hoped that some time away from MSU would allow me to catch up on a backlog of writing.

Chris Adami's book, *Introduction to Artificial Life*, was published early in 1998, and as promised he sent me a copy [1]. I remember finding it in our mailbox one afternoon. Along with the book, there was a disc in a sleeve at the back that contained the Avida program. I found an old email from February 24 of that year:

> Bonjour Chris, I just got the book and CD, which look very nice!!

(Only now, while writing this piece, did I realize that the same day I received Chris's book happened to be the tenth anniversary of the start of the LTEE, my long-term evolution experiment with *E. coli*.) Once I started reading the book, I couldn't put it down. Besides being absorbed by the information in the text, I was eager to give Avida a spin. The software ran flawlessly, and soon I was running little experiments and seeing artificial life evolve before my eyes.

In short, I was hooked. I remember telling Madeleine, only half in jest, that maybe I should call my group back at MSU and tell them to close the lab down, that I had switched from evolving microbes to evolving programs. On Madeleine's advice, that didn't happen, and my lab continued its exciting and productive research on the microbial side. However, I think that more than half of my own attention over the next several years was on Avida. I wrote Chris again on March 2, just six days after receiving his book:

> I spent this weekend reading your book from cover to cover and beginning to play around with the avida program. In short, I think the book and program are quite fantastic. You

should be very proud indeed! ... I have 101 ideas for cool experiments – so many, in fact, that my head is spinning.

The next day, I wrote the book-reviews editor at the journal *Science*:

I know that Science does not publish unsolicited book reviews. Therefore, I am writing to ask whether you would consider asking me to write a review of a fascinating book and companion CD. The book is titled "Introduction to Artificial Life" ... After I picked it up, it was hard to put the book down, and I did so only to explore the computer program that comes on the CD. That program quickly and simply demonstrates the real-time evolution of programs that compete for CPU time inside a personal computer.

The editor invited my review, and 11 days later I sent a draft to Chris for his feedback. (My speedy writing about something so interesting and fun as Avida was balm for my soul. In the intervening period, a dear young friend had died, and I flew with one of my children to California for the funeral.) Chris responded that same day:

I am flattered. In fact, I don't feel the need to change anything in your review (except maybe correct that the official abbreviation of the California Institute is Caltech, not Cal Tech). Also, I was wondering whether there is any way we can give some credit to my graduate student Charles Ofria, who wrote all of avida ...

Those changes were made, and my book review [9] was published on May 8, not long before it was time to return to Michigan. I ended my review with a question about generalization, a nod to a great French biologist, and a look to the future:

[One] may reasonably ask whether results can be extended to real organisms. In extrapolating from the genetics of bacteria to animals, Jacques Monod is said to have quipped that "What is true for *E. coli* is also true for elephants, only more so." Is what is true for avidians also true for real organisms, or is it less so? It will be interesting to see.

1.4 Strange Creatures and Fun Science

While I was writing the book review, I also ran experiments with Avida. We often think of scientific experiments as carefully planned exercises with explicit hypotheses and precise methods, data collection, and analyses. These are important parts of science, of course, but the process often begins another way—namely, informal experiments that involve *playing* with a system that has caught one's fancy.

In my early experiments, I evolved some strange creatures whose behavior I could not understand. On March 15, I wrote Chris:

I also saw some strange, persistent genotypes ... They cannot replicate themselves, but are evidently replicated by (or, alternatively, repeatedly spun off from) closely related programs ... Moreover, when I seeded these 'parasites' into an empty lattice – and having set all mutation rates to zero! – some (but not all) of them spun off their self-replicating relatives spontaneously. Weird, huh?

Chris replied that same day:

I think we've seen that. We have a pretty good idea about what's going on there, and it is definitely 'normal' behavior. I'll let Charles comment on that.

Charles Ofria understood the inner workings of Avida better than anyone. I sent Charles the genomes of what I called a "gallery of weird creatures" to investigate.

Over the following weeks, months, and years, my role grew from discovering weird creatures. Although these curiosities were interesting, and some of them might even have relevance for biology, I thought that they would impede understanding and more rigorous experimentation. Therefore, I suggested to Chris and Charles that the user should be able to turn off these complications "in order to have simpler scenarios to analyze and test."

1.5 Synergistic Science

I also pondered what experiments I most wanted to perform using Avida. Santiago Elena (then a postdoc in my lab) and I had recently published a paper in *Nature* titled "Test of synergistic interactions among deleterious mutations in bacteria" [7]. The point of that paper was to test one of the main hypotheses for the evolution of sexual recombination. According to that hypothesis, sex could be advantageous for eliminating deleterious mutations, provided that genotypes with multiple deleterious mutations tend to be less fit than expected from their individual effects. Santi performed experiments to test the assumption that harmful mutations tend to interact synergistically by generating 225 *E. coli* mutants that had one, two, or three random insertion mutations, and then competing each mutant against the unmutated progenitor to estimate its relative fitness. Although the average fitness declined as the number of mutations increased, there was no tendency for synergistic interactions.

What might a similar experiment with Avida reveal? Would it support our findings with bacteria, or would it instead provide evidence in favor of the hypothesis of synergistic interactions? The experiments and analyses seemed straightforward. And with Avida, one could have much greater statistical power, with more mutants in each class and more mutations in the most highly mutated classes. On April 13, I wrote Charles and asked for tools that would help us perform these experiments in Avida:

> I would like to be able to take a given creature and generate any number of mutants derived from it. Each mutant genotype should have exactly N (e.g., 1, 2, or 10) random mutations ... I would like to be able to take a specific pair of mutants derived from the same progenitor (as above) and introduce all of their mutations together into the same creature ... I would like a new routine that allows me to take a given genotype, inoculate it into an empty grid, and obtain [its] fitness [when] mutation [is] "turned off" ...

Charles responded the same day. He had already implemented a "landscaping" tool (as in fitness landscapes), one

> where you can either ask it to try ALL possible mutations up to n steps away, or to sample them ...

He thought everything else I wanted could be done as well. He also had a request:

I'm not sure when you'll be back in the US, but if you have time I wouldn't mind taking a few days at some point this summer to come to Michigan and work more directly with you.

We continued to correspond as Charles developed the necessary tools, and together we worked on details of the experimental plan. On May 12, I emailed Charles an ambitious plan that involved examining over half a million mutant Avidians. Charles replied the next day, making it all sound so easy:

No problem ... Sounds good ... Currently it takes about 1 second to run about 500 genotypes through a test CPU.

On July 3 he sent an update, saying things were proceeding well and he was expanding the scope of a key analysis by 10-fold. On July 13, I received by fax (remember those things?) a set of figures with preliminary data. As I'm prone to do, I replied with questions and asking for additional analyses. And as Charles is prone to do, he proposed to increase the scale of the experiment even further.

We also arranged for Charles, Chris, and Travis Collier (a Caltech undergrad working with them) to visit MSU in early August to work on this, our first, paper. I don't have emails from those days because they were here visiting, but as I recall we spent about 16 hours a day for several days holed up in a small office with a whiteboard and a computer—Charles and Travis tweaking the Avida software and running experiments, and Chris and me hashing out how to parameterize the effects we were trying to measure. Soon Chris was sending me various models fit to the data. I replied by suggesting an alternative model, and Chris wrote back triumphantly—and with his customary enthusiasm—on August 26:

The new function fits the data perfectly! I remember now that I had suggested this form earlier (in the car when we were driving to the restaurant) but for some reason decided that it would not change anything from the quadratic fit. Obviously that is wrong ... There is very little (actually, none at all) doubt that this is the way to fit.

Things slowed down a bit after that, as I settled back into microbial research, faced deadlines that had accumulated while I was on sabbatical, and both Chris and I began a semester of teaching. But on March 10, 1999, I wrote Chris and Charles, after I had begun to convert our results into a paper:

I've gotten so excited thinking about Avida that I'm back to working on the "epistasis" paper even though it is only about 10th on my must-do list ... Maybe I really should close my bacterial lab!

I had some questions about details of methods and requests for more polished figures, and I proposed we submit the paper to *Nature* "because it echoes issues from our E. coli paper there." On March 20, I sent them a draft of the paper. After a few rounds of editing, we submitted it in early April.

The reviews arrived on May 21, and I wrote with considerable excitement to Chris and Charles:

The cover letter and two reviews are on their way. As you read over them, you'll see how favorable the reviewers were! There really isn't that much to do ... My hope is that we can get a quick acceptance and be able to list this manuscript as "in press" on our proposal.

That would lend some extra credibility to the Avida work for microbiologists who might not fully understand or sympathize with digital microbes! So far, I've been working on re-organizing the abstract and next two paragraphs to fulfill [the editor's] instructions. I've also changed the title and elsewhere to avoid the jargon-ish "epistasis" as much as possible, instead talking about the form of interaction more generally.

The paper was revised and accepted for publication on May 27, and our first collaborative project—"Genome complexity, robustness and genetic interactions in digital organisms"—appeared in *Nature* that August [11]. We had evolved 174 Avidian progenitors; then, starting from each of those progenitors, we obtained the relative fitness of every possible mutant with exactly one point mutation and a million or more mutants with two through ten mutations; and we analyzed tens of thousands of two-mutation recombinants derived from each progenitor. We observed frequent interactions between mutations (epistasis) including many cases of synergistic epistasis, but we saw no excess synergism relative to the alternative. In short, the results were qualitatively consistent with what Santiago and I had reported two years earlier for *E. coli*—but with the far greater replication and precision that Avida provided.

In the meantime, Charles, Chris, and I were discussing possible experiments to address other questions, and as implied by the reference above to "our proposal" we were thinking about how to support an expanding research program. I had also been working with Charles (and behind the scenes) to see whether we could get him to MSU after he finished his Ph.D. at Caltech.

1.6 Dollars and Sense

On January 14, 1999, the NSF—under the leadership of Rita Colwell, a distinguished microbial ecologist—issued a call for proposals:

> As a first step in a longer-term effort to understand the nature and dynamics of biocomplexity, NSF announces a special competition to support integrated research on the functional interrelationships between microorganisms ... and the biological, chemical, geological, physical, and/or social systems that jointly comprise complex environmental systems.

So that afternoon, I wrote Chris and Charles:

> There might be something here for us, say a series of expts comparing avida and ecoli under a variety of evolutionary scenarios ... Notice that multi-institutional collaborations are favored, with a focus on "biocomplexity" and microorganisms.

A pre-proposal would be due in mid-March. Charles wrote back on February 6:

> I finally had time to sit down (and grab Chris!) and really go through the Biocomplexity program announcement. As Chris put it, they've only come one step away from actually writing our names in there! ... Obviously we should play off of the three approaches we take. You do experimental work in actual biological systems, I model those systems allowing much easier statistical study, and Chris applies a theoretical approach using statistical mechanics and the like. So we can all come at the same problem from very different perspectives and really work off each other. Especially nice is the fact that they're looking for

collaborations from more than one university, and from more than one field ... I can't afford
to take off too much time at this point though. Speaking of which, FYI, my thesis defense
is set for May 13th ...

As a graduate student, Charles wasn't eligible to be co-PI on the proposal, though
he was listed as a key participant. Also, I thought we needed more microbiological
expertise. So I contacted Margaret (Peg) Riley, a talented molecular evolutionist
(then at Yale, now at the University of Massachusetts, Amherst), about teaming up
with Chris and me as a third co-PI. Her lab would sequence some genes from the
LTEE bacteria to get a handle on the rate of genome evolution (this was many years
before it was affordable to sequence a set of complete genomes). Peg was interested,
and Chris sent around a "draft of a draft" of the pre-proposal on March 1. After an
intense week of back and forth about the science, we submitted the pre-proposal on
March 8, a full week early. On April 19, we heard that the NSF invited us to submit
a full proposal, which would be due in mid-June. In the meantime, our paper using
Avida to study genetic interactions had been accepted by *Nature*, so we could cite it
as "in press" in the full proposal, and Charles defended his dissertation at Caltech.

After overcoming a bug or two in NSF's online submission system, we submitted
the full proposal on June 11. Charles moved to MSU in early August to begin a postdoc, and on September 8 we got the news that our biocomplexity project would be
funded. One of the first orders of business for Charles was to purchase and config-
ure a Beowulf cluster, so that we could scale up our Avida experiments. That cluster
was placed in the basement of the Plant and Soil Sciences Building, which is the
building where my lab was then. I look back on the biocomplexity grant as a sort
of dry run for BEACON, albeit on a smaller scale. It had interdisciplinary research,
multiple institutions, proven collaborations, and lots of ideas for moving forward
along with intense competition to get there. Many of the ideas panned out, too, as
evidenced by exciting papers including, on the Avida side, demonstrating the survival of the flattest [19] and the evolutionary origin of complex features [12], and on
the bacterial front, identifying beneficial mutations [6] and starting to characterize
the rate of genome evolution [14] in the LTEE. Two other important developments
during that period were Rob Pennock (who I had just met at the ALife VII meeting
in Portland) moving to MSU in 2000, and Charles becoming a tenure-track faculty
member in computer science in 2002.

1.7 Best Lunch Deal Ever

In 2003, the NSF issued a call for Science and Technology Centers (STCs), which
they do only every few years. I had moved from UC-Irvine to MSU to join one of
the first STCs: the Center for Microbial Ecology (CME), directed by the amazing
Jim Tiedje. On March 15, I wrote Charles:

I think you should consider going after an NSF Science & Technology Center – that's what
got the CME going at MSU and thus scientifically why I came here. With you [and others]
defining a field of, say, digital evolution or genetic programming – and by also pulling

some additional players on the biology side including me [and others] – I think you'd have a reasonable shot at a very important center-level grant ... The tie-in between the basic science and industry would be very appealing to both MSU and NSF. I suggest you think seriously about this, and pull together a group of your colleagues to see if they are interested. You'd probably want someone a bit more senior to be the director – Erik G strikes me as one possibility.

Charles was too early in his career to direct an STC, and I didn't like the idea of being an STC director myself—how Jim Tiedje could simultaneously direct a successful STC and run an extraordinarily productive lab was beyond my understanding.

So Charles and I began to make plans. He responded with a proposed center name of "Experimental and Applied Evolution." I was concerned that the proposed center be seen as sufficiently distinct from the CME:

I think you should consider an explicit focus on the computational side of things, rather than experimental evolution in the broader context. My reasoning is that NSF likes to spread things around and things that come too close to the CME – such as including microbial evolution – might actually get in the way of success. So definitely have ties to biologists, and to biological systems, but make computer science the focus. "Genetic programming and digital evolution" might be the title ...

But Charles wisely pushed back against my suggestion to narrow the center's focus:

Erik sounds quite keen on the STC idea and would like to meet (possibly for lunch) to talk more about it. Would you be interested/available to join us some time this week? Also, I do see your point about not wanting to choose something too close to the CME, but I think that as long as microbes aren't the focal point ... the rewards will be worth it.

So Erik, Charles, and I went to Sindhu's restaurant for lunch, and we had a great discussion. Charles and I proposed to Erik that he would be the director of the hoped-for STC. Erik was interested, but he said he needed some time to think about it. After just a few days, Charles told me Erik had said yes, so that was the best lunch deal ever!

1.8 Sunken Treasure

Before he could firmly commit to being director, Erik had to meet with MSU administrators about disclosures and recusals related to his company, Red Cedar Technology. It took a few weeks to get meetings with all of the higher-ups, but everything worked out. Meanwhile, it was time to make plans and build a team. On April 11, Charles wrote with a proposed set of thrust areas for the center:

1. The Universal Foundations of Evolution. The goal of this area is to understand general principles that govern all evolving systems, primarily through controlled experiments. Much of the work being done in Avida or E. coli would fall here ...
2. Tools and Methods for Directed Evolution. In this area, we would be harnessing the principles that we discover in the first area in order to create tools that will make things evolve in directions that we want ... A lot of Erik Goodman's work on multi-level GAs would fit here ...

3. Applications of Evolution to Real World Problems. Here, we take the tools constructed in area two, and actually do something useful with them. Be it building parts of cars as Erik does, or evolving proteins ...

4. Bringing Evolution to the Masses. Okay, this one needs a better title, but basically it's the education component. I think that the education portion of this project is so strong, that it warrants being brought front and center with its own thrust area ... I think we could put together multiple courses, but in a modular fashion where they could be mixed and matched as needed ...

In late April, though, we ran into a problem. The NSF deadline for submitting the pre-proposal was June 3, but MSU announced they would need a draft by May 7 for an internal review, because each institution could submit a maximum of five pre-proposals to the NSF and the number of interested groups at MSU exceeded that limit. The days that followed were hectic but productive: writing text that would dovetail our ideas with the NSF goals, drafting budgets, identifying partners and participants, collecting bio-sketches, preparing an organizational chart, and assembling it all. I realized what a superb leader Erik was—dedicated, organized, even-handed, timely, and a funny and nice guy on top of all that! In the wee morning hours of May 7, Erik sent a revised org chart and text, along with these words:

Going to bed ... coffee has worn off and my tail is draggin'.

That afternoon Erik submitted our pre-pre-proposal to the MSU administration.

On Thursday, May 15, we were told that our pre-pre-proposal was selected to move forward, so we had to get back to work in order to submit a pre-proposal to NSF in a little over two weeks. By Sunday, Erik was sending around a section on industry involvement, having already lined up several companies interested in becoming affiliates. The next couple of weeks were a blur of meetings, emails, writing, editing, and polishing. Decisions had to be made about font size, and what points would have to be cut in order to squeeze into the allotted page limit. In the evening of Saturday, May 31, Erik sent an email with the subject "Shrunken summary attached." We were coming down to the wire. Emails and attachments with names reflecting the various sections of the proposal were flying among us every day and almost around the clock, including an edited summary from Rob Pennock on the morning of June 3, the due date at NSF. That afternoon, it was submitted. We would have to wait until October to hear whether we were invited to submit a full proposal for our center, which we'd named the Center for Applied Evolutionary Dynamics and Computation.

The NSF's decision arrived on October 17:

Review panels of experts met recently at the National Science Foundation to evaluate pre-proposals submitted to the above referenced Program Solicitation. I regret to inform you that your preproposal was not among those invited to submit a full proposal.

I wrote Erik some words of solace:

Well, the good news is that we don't need to do all that additional work! Reviewer #2 got it right, by the way, so read that one to feel better ... I'm sure there was terrific competition, and we probably didn't fit into any easily definable area ... I want to thank you personally for your tremendous leadership on this project ... And I look forward to working with you in other ways to develop our vision further!

Wolfgang Bauer, who years earlier had introduced me to Chris Adami, also weighed in:

> Let's not get discouraged by this. The work on hammering out the STC pre-proposal was worth the effort. I had a similar experience [in another area] ... But we regrouped and ... the community-building exercise across our campus and at other universities is still paying big dividends ... Bummer (to use the technical term), but no reason to hang our heads.

Wolfgang's remarks proved to be prophetic.

1.9 A BEACON Is Born

Things hummed along nicely over the next few years. The Avida team of students and collaborators continued to expand and thrive under Charles' direction, even after the biocomplexity grant had ended. In late 2004, Kay Gross called my attention to the NSF program for Long-Term Research in Environmental Biology, or LTREB, and she suggested I should apply to provide core support for the LTEE. My application was successful, so in 2005—over 17 years after it began—the LTEE had its own funding. That same year, Simon Levin, a theoretical ecologist at Princeton, pulled me (and I pulled Charles and Chris) into a multi-institution, multidisciplinary DARPA project focused on mathematical biology. That allowed us to sustain our momentum from the biocomplexity grant, and it provided critical funds as the LTEE entered the genomics era. So we remained primed and alert for another chance to compete for an STC.

In 2008, the NSF issued its next call for STC proposals, and we sprang into action, once again under Erik's leadership. I suggested we call ourselves the Center for the Study of Evolution in Action, or CSEA (to be pronounced "cc"), and we organized under that banner for a while. At some point, though, Erik, Charles, and Rob came up with a far better name in BEACON. The acronym was forced—Bio/computational Evolution in Action CONsortium—but it captured the idea of shining light on evolution as an on-going process. It also nicely echoed Dobzhansky's famous assertion that "Nothing in biology makes sense except in the light of evolution," and it offered possibilities for an attractive logo.

We organized ourselves locally and chose four partner institutions—North Carolina A&T State University, University of Idaho, University of Texas, and University of Washington—where we knew excellent colleagues with interests in evolution on both the computational and biological sides. (We had proposed a different set of partner institutions in our previous application for an STC. One was Caltech, where Frances Arnold, a 2018 Nobel Laureate, would have been part of the center.) On the local front, Kay Holekamp joined the team as a co-PI, adding a talented biologist whose fieldwork complemented my laboratory-based biology, while Erik and Charles represented engineering and computation, and Rob added valuable breadth as a philosopher of science whose work highlighted evolution as an exemplar of the scientific method. In August, we once again submitted a pre-pre-proposal for MSU's internal review and competition. After that was approved, we polished the text, pulled together budgets and bio-sketches, and prepared our pre-proposal for the NSF. On October 15, Charles emailed all the participants:

> The pre-proposal was sent out today around 3pm, a couple of hours before it was due.

Then we waited. On February 2, 2009, Erik sent an email with the subject line "Good news! BEACON: STC for Study of Evolution in Action is invited to submit full proposal!" He wrote:

> I just received notification from NSF that our preproposal has passed the screening and we are invited to submit a full proposal ... I'm including below the Panel Summary ... They REALLY liked it! ... Now the next stage of work begins ... The full proposal will be due on April 30, 2009, so we have only three months to get the whole thing done ... Regardless of whatever the Groundhog Day prediction was, I'm seeing spring starting to emerge!

Now we had to prepare not a pre-proposal, but the real thing. Back to work! The number of meetings and emails was almost overwhelming, though they were also productive and even fun, such as coming up with a mission statement for BEACON and contemplating members of an external advisory board. Of course, much of the work was drafting and crafting the text that described our proposed activities in research, training, education, diversity, and outreach. But a lot of effort also went into collecting all of the required elements for a full proposal: detailed budgets, identifying possible conflicts of interest according to NSF rules (including hundreds of former students and recent collaborators who would be excluded as potential reviewers), formulating a management plan to cover the diverse activities of our proposed center, and obtaining institutional commitments. The MSU administration also formed an internal "red team" (a mock review panel) to critique our draft proposal. On April 30, Charles wrote to all the participants that the full proposal had been submitted to NSF, and thanking everyone "for an incredible amount of help in pulling this monster together." Back to waiting ...

1.10 Bringing Home the BEACON

On August 13, Erik sent an email with the subject line "WOW!!! BEACON gets a SITE VISIT from NSF!!! Congratulations!" He went on to say:

> The BEACON Center has made it to the next step in the STC review process. We have been selected as one of the (expected to be 12) proposals to be scheduled for a site visit. Of that number, about 5-7 are to be funded, with the decision coming in February. So WE MADE THE CUT! We've made it from 450 down to 12 it's pretty rarefied air up here! ... We have a lot of work to do very quickly ... We'll need to do a lot of talking before the site visit, and will get started shortly. In the meantime, congratulations to the whole team on making it to this point, and let's do our best to take the last step as well!

The next week, we heard that our site visit was scheduled for October 22-23. And even before then, we had to respond in writing to the NSF reviews by September 2. While the reviews were very positive overall, we nevertheless had to address any and all concerns expressed by the 13 (yes, 13) reviewers and in the panel summary.

(We also had to seek guidance from the NSF about how to respond to one review that was unprofessional and even offensive.) The set of reviews collectively ran to more than 70 pages, and our reply was 10 single-spaced pages. These things require a lot of work, but Erik led the team with his customary energy, commitment, and skill.

The preparation for the site visit was intense, as we planned who would talk on each topic in the precisely orchestrated NSF-mandated schedule. For the research portion, we developed a matrix of speakers who would collectively showcase the scope, diversity, and connectedness of our questions, study systems, partner institutions, and participants. We held rehearsals to make sure that all the presenters stuck to their allotted time.

At the end of the long and exhausting but also exhilarating first day of the site visit, we were then given a homework assignment, and a nontrivial one at that. The site visitors met in closed session for an hour or so, discussing our proposal and presentations. From their discussion, they came up with a set of questions we were charged with answering—in writing—by first thing the next morning! All of the BEACON presenters from MSU and our partners were invited to dinner at the home of Titus Brown and Tracy Teal. After grabbing a quick bite, those of us on the leadership team found a quiet room and went back to work: discussing the questions, considering our answers, and writing our responses. We worked to the wee hours of the morning, and a bottle of whiskey appeared when it was clear that we were just about done.

The next morning, we presented our report and answered follow-up questions. The site visitors also met with Erik and administrators from MSU and our partner institutions, before they secluded themselves to write their final report. Our work was done: we had the satisfying feeling that we had done all we could do, and that the site visitors and NSF staff were impressed with what they had seen. Still, the site visitors took it as their charge to highlight in their final report not only the center's strengths, but also to identify possible weaknesses. And so we spent the next two weeks carefully crafting our response to their report. On November 10, Erik sent it to Joan Frye, who directed the STC program at NSF. He ended his email with this:

We are, needless to say, eager to hear the results of the next stage of decisions!

Just an hour later, Erik wrote the four co-PIs a message with the subject line "BEACON – burning bright!"

Joan just confirmed that she received our report ... "Fantastic" was the one word confirmation ... So, my thanks especially to this group, who have worked so hard in putting this together at every stage! Morning, noon and night, always crunching away! Awesome! It'll be such a relief to HAVE the center – that has to be easier than trying to get it ...

Still, don't count your chickens before they hatch, and so it was back to waiting.

The year 2009 marked the 150th anniversary of Charles Darwin's book The Origin of Species, and November was the month of its publication. While we waited for word from the NSF, Wolfgang Bauer sent an email to the BEACON team:

As I went to lunch today, I walked past a couple of guys who were passing out free copies of the "150th Anniversary Edition" of Darwin's work on the origin of species. Cool, a free book! I took one. I actually made a special detour to snag a copy, because I smelled a rat. Sure enough, this is a "special" edition with creationist (sorry, I mean "intelligent design") commentary. You know, all of the usual pseudo-scientific drivel ... Sorry for adding one more item to your email inbox, but I thought that this was interesting enough to share.

If nothing else, it was a clear sign that BEACON was needed.

On January 28, 2010, Erik was told that decisions about the STC competition would be announced the week of February 8. He drafted a press release, in case the news was good, and sent it around to the co-PIs for feedback. On February 8, we heard nothing—a blizzard had hit Washington and shut down the Federal government.

The next day, however, Joan Frye wrote Erik with some news:

> It is my pleasure to inform you that we are recommending an award for your STC proposal. Congratulations! I am snowed in and technically, the office is "closed" today, but I thought you would want to get the word.

Moments later, Erik wrote all the participants with the subject line "BEACON IS FUNDED!!!" He signed off as "Director, BEACON Center for the Study of Evolution in Action"—and just below his new signature he added:

> (Ooh, THAT was fun!)

Meanwhile, East Lansing was experiencing a heavy snowstorm as well. But that didn't stop a bunch of us from going out for dinner that evening to celebrate!

So that's the story of one of life's peculiar and winding paths: how a squash match in 1994—along with years of hard work by many people—led to the founding of the BEACON Center for the Study of Evolution in Action in 2010. Let me close my reminiscences with some thanks. Thank you, Erik, for your dedication, energy, communication, generosity, and skill that brought BEACON into being and ensured its success. Also, thanks to Wolfgang Bauer for inviting me to Chris Adami's talk long ago, and to Chris, Charles, and all the talented and dedicated people who have worked with me on both microbes and digital organisms. Last but not least, thanks to Brian Baer, Judi Brown Clarke, Connie James, Louise Mead, Danielle Whittaker, and Darcie Zubek, all of whom have made Erik's job manageable and allowed the BEACON Center to achieve its goals with efficiency and enjoyment.

References

1. Adami, C.: *Self-organized criticality in living systems.* Physics Letters A **203**, 29–32 (1995)
2. Adami, C.: *Introduction to Artificial Life.* Springer, New York (1998)
3. Bak, P., Tang, C., Wiesenfeld, K.: *Self-organized criticality: An explanation of 1/f noise.* Physical Review Letters **59**, 381–384 (1987)
4. Blount, Z.D., Borland, C.Z., Lenski, R.E.: *Historical contingency and the evolution of a key innovation in an experimental population of* Escherichia coli. Proc Natl Acad Sci USA **105**, 7899–7906 (2008)
5. Blount, Z.D., Lenski, R.E., Losos, J.B.: *Contingency and determinism in evolution: Replaying life's tape.* Science **362**, eaam5979 (2018)

6. Cooper, T.F., Rozen, D.E., Lenski, R.E.: *Parallel changes in gene expression after 20,000 generations of evolution in* Escherichia coli. Proc Natl Acad Sci USA **100**, 1072–1077 (2003)
7. Elena, S.F., Lenski, R.E.: *Test of synergistic interactions among deleterious mutations in bacteria.* Nature **390**, 395–398 (1997)
8. Good, B.H., McDonald, M.J., Barrick, J.E., Lenski, R.E., Desai, M.M.: *The dynamics of molecular evolution over 60,000 generations.* Nature **551**, 45–50 (2017)
9. Lenski, R.E.: *Get a life.* Science **280**, 849–850 (1998)
10. Lenski, R.E.: *Convergence and divergence in a long-term experiment with bacteria.* American Naturalist **190**, S57–S68 (2017)
11. Lenski, R.E., Ofria, C., Collier, T.C., Adami, C.: *Genome complexity, robustness and genetic interactions in digital organisms.* Nature **400**, 661–664 (1999)
12. Lenski, R.E., Ofria, C., Pennock, R.T., Adami, C.: *The evolutionary origin of complex features.* Nature **423**, 139–144 (2003)
13. Lenski, R.E., Travisano, M.: *Dynamics of adaptation and diversification: a 10,000-generation experiment with bacterial populations.* Proc Natl Acad Sci USA **91**, 6808–6814 (1994)
14. Lenski, R.E., Winkworth, C.L., Riley, M.A.: *Rates of DNA sequence evolution in experimental populations of* Escherichia coli *during 20,000 generations.* J Mol Evol **56**, 498–508 (2003)
15. Maynard Smith, J.: *Byte-sized evolution.* Nature **335**, 772–773 (1992)
16. Meyer, J.R., Dobias, D.T., Weitz, J.S., Barrick, J.E., Quick, R.T., Lenski, R.E.: *Repeatability and contingency in the evolution of a key innovation in phage lambda.* Science **335**, 428–432 (2012)
17. Ostrowski, E.A., Ofria, C., Lenski, R.E.: *Genetically integrated traits and rugged adaptive landscapes in digital organisms.* BMC Evolutionary Biology **15**, 83 (2015)
18. Travisano, M., Mongold, J.A., Bennett, A.F., Lenski, R.E.: *Experimental tests of the roles of adaptation, chance, and history in evolution.* Science **267**, 87–90 (1995)
19. Wilke, C.O., Wang, J., Ofria, C., Lenski, R.E., Adami, C.: *Evolution of digital organisms at high mutation rates leads to survival of the flattest.* Nature **412**, 331–333 (2001)
20. Woods, R.J., Barrick, J.E., Cooper, T.F., Shrestha, U., Kauth, M.R., Lenski, R.E.: *Second-order selection for evolvability in a large* Escherichia coli *population.* Science **331**, 1433–1436 (2011)

Chapter 2
A Strong Director Facilitates the Successes of All BEACON Members: A Personal Example

Kay E. Holekamp

Abstract Under the direction of Erik Goodman, BEACON has been a remarkable success. It has been a fabulous boon to those of us whose research interests are constantly evolving. Here I explain how Erik's guidance of BEACON facilitated the professional development of many graduate students and post-docs in my lab, and allowed me to reinvent myself as a scientist multiple times during the past decade.

Key words: Evolution of intelligence, communication, cooperation, collective action, spotted hyenas

2.1 Introduction

Although the BEACON Center was born of collective action by several different individuals at five different American universities, the effective oversight and enormous long-term success of the Center have both been largely attributable to the efforts of the Center's Director, Erik Goodman. Throughout the past several years NSF officials have repeatedly lauded BEACON as an exemplary STC, prompting an inquiry as to why this is the case. I suggest that BEACON's extraordinary success, in the eyes of both its participants and those of the oversight people at NSF, is largely due to the unique and wonderful traits exhibited by Erik in his role as BEACON's Director. A successful STC Director must possess the intellectual and leadership skills required to earn the respect of a diverse group of over-confident, maverick scientists, and herd them toward achievement of a common goal. A successful director must also be patient, kind, good-natured and forgiving, and must ensure that all decision-making processes are completely transparent. A successful director must put the needs of people ahead of other needs, and in the case of

Kay E. Holekamp

Department of Integrative Biology and Ecology, Evolutionary Biology, and Behavior Program, Michigan State University, East Lansing, MI, 48824, USA e-mail: holekamp@msu.edu

© Springer Nature Switzerland AG 2020

W. Banzhaf et al. (eds.), *Evolution in Action: Past, Present and Future,*

Genetic and Evolutionary Computation, https://doi.org/10.1007/978-3-030-39831-6_2

BEACON, be particularly attentive to the needs of budding scientists in the next generation who hope to work at the interface between computer science and biology. Perhaps most importantly, a successful STC Director must have a clear vision of what his or her errant group of scientists might ultimately be capable. Erik possesses all these traits in spades.

Erik's vision has had life-changing effects on many of my BEACON colleagues and many of my own PhD and postdoctoral students. BEACON support of projects designed to collect pilot data for larger, externally funded research has already enabled folks in my lab to acquire approximately $2,500,000 in external support since 2010, with proposals submitted to acquire over $2,500,000 in additional support still pending. These awards and proposals are listed in the supplement to this chapter, as are the five post-docs and 19 PhD students in my lab who have benefitted directly from BEACON support of one form or another. But these 24 post-docs and graduate students were not the only people in my lab to benefit from Erik's vision. Erik's support and guidance allowed me personally to stretch in several new intellectual directions, and allowed me to reinvent myself multiple times while BEACON was in force at MSU. For this I will be forever grateful.

2.2 Re-invention Number One

My first re-invention involved transforming myself from a run-of-the-mill ethologist focused on the behavioral development of spotted hyenas into an evolutionary ecologist able to use both real and digital hyenas as model organisms for testing hypotheses about the evolution of cooperation and the mediation of collective action. With BEACON support of two pilot initiatives, Risto Miikkulainen and I were able to secure a large R01 grant from NIH that supported Padmini Rajagopalan and Aditya Rawal in Risto's lab, and supported both field and endocrine benchwork in my lab to elucidate relationships among cooperative collective action, motivational variables and the hormones that might predispose some individuals to be more cooperative than others, and also more responsive than others to the communication signals emitted by groupmates. This work is still in progress, but it, in turn, has sparked new projects by several different graduate students in my lab, and those have since generated an impressive number of publications [29, 47, 49, 50, 68, 69, 70, 71, 82, 91, 92, 93, 94, 95, 96, 97].

As another aspect of this re-invention, we were able to collect pilot data using extraordinary recent innovations in telemetry to inquire about how hyenas decide whether and with whom to cooperate. BEACON post-docs Frants Jensen, Ariana Strandburg-Peshkin, and Andrew Gersick designed collars for spotted hyenas that give us a hyena's location every second of every day for several weeks. Even better, is that accelerometers and magnetometers on the collars tell us exactly how the hyena is moving in every instant. But the real coup here was adding microphones and sound boards to the collars such that we are able to hear everything the hyena hears for several weeks, and see from the accelerometer and magnetometer data how it responds to what it hears. The pilot data we collected with BEACON seed funds from six hyenas fitted with these fancy collars have allowed us to develop and submit major proposals to the European Human Frontiers in Science Program and various other funding agencies to fit every hyena in an entire clan with these collars to study their decision-making in the context of collective action.

2.3 Re-invention Number Two

My second re-invention involved shifting my research focus to the evolution of intelligence in animals and machines. I have become particularly fascinated by the evolution of domain-general intelligence, which cannot be accounted for by most hypotheses forwarded to suggest why intelligence evolves [38]. The cognitive demands imposed on animals in their natural habitats vary considerably among species, so behavioral ecologists and evolutionary psychologists have often proposed that intelligence is comprised of an aggregate of special abilities that have evolved in response to specific environmental challenges (e.g., [2, 20, 21]). For example, spatial memory is very well developed in squirrels and seedcaching birds (e.g., [8, 41]). These domain-specific cognitive mechanisms or "modules" are activated under particular circumstances, enhancing fitness by improving the animal's ability to solve specific types of problems posed by the environment. At the neurological level, modules are often conceptualized as dedicated brain areas serving domain-specific cognitive functions that can be selectively activated or inhibited; for example, the hippocampus mediates much spatial memory in mammals and the suprachiasmatic nucleus mediates time-keeping. Abundant evidence shows that certain species are exceptionally good at solving some types of ecological or social problems, but not others (e.g, [19, 79]), and that these specialized abilities enhance fitness (e.g, [39, 78]). Thus, there is a great deal of support for the evolution and maintenance of domain-specific cognitive modules in both humans and other animals.

Interestingly, as noted by Burkart et al [14], there are also several lines of evidence incompatible with a strictly modular view of intelligence, suggesting that domain-general processes might evolve in animals as well as domain-specific modules. First, much interspecific variation in both brain size [37, 38, 74] and problem-solving ability [9, 52] remains unexplained by domain-specific selection pressures. Second, brains exhibit remarkable developmental plasticity and they change in response to experience even in adulthood (e.g., [73]); both these forms of plasticity are incompatible with a purely domain-specific view of intelligence [66, 67].

Third, there are robust differences among species in their ability to learn and to solve novel problems e.g., [10, 53]). If intelligence involved nothing more than co-evolving modules, we would expect different species to rely on different suites of modules based on the adaptive problems each faces, but instead some species consistently outperform others across all tested cognitive abilities [27].

Fourth, animals have executive functions, which appear unlinked to any specific cognitive adaptations [24]. Executive functions are often conceptualized as 'conductors' of a symphonic assemblage of more elementary cognitive functions. Executive functions make possible phenomena such as taking time to think before acting, coping with novel or unexpected challenges, resisting temptations and staying focused. Core executive functions include inhibitory control, working memory, selective attention and the ability to adapt rapidly and flexibly to altered circumstances [24]. These executive functions have been well documented in several animal species, and are known to affect their decision-making, as also occurs in humans [24, 34, 45, 52].

Fifth, in humans the existence is well-documented of both modules and domain-general processes. Performance across tasks in various cognitive domains is positively correlated, and factor-analytical procedures applied to data sets documenting individual performance across tasks consistently reveal a single 'general factor,' called 'g,' that loads positively overall and can explain a significant amount of variation (e.g., [17, 90]). Interestingly, g better predicts life outcomes in humans than does any specific cognitive ability [23]. Most current models of human cognition involve a hierarchical structure, with g and executive functions being closely related, and both situated at a higher level in the hierarchy than domain-specific cognitive abilities [14, 59]; g is closely correlated with measures of executive function in rodents [45, 57, 58] and primates (reviewed by [14]). For example, Kolata et al [45] found a significant relationship in mice between their general learning abilities and selective attention, as indicated by their ability to ignore salient distracting stimuli while they performed a discrimination task.

Finally, g has now been calculated in a substantial number of animal species using psychometric factor-analytical approaches like those used in humans, and as in humans, it accounts for 17-48% of the variance in performance on multiple tasks (e.g., rodents, [58]; primates, [72]; dogs, [5, 63]; birds, [44, 77]). Importantly, g does not simply capture anxiety, personality traits, stress, motivational state or other non-cognitive processes, nor is it merely a statistical artifact [14, 28, 57]. With recent NSF support, awarded thanks to pilot data collected with BEACON funds, we are now assessing selection on general intelligence in response to evolutionary novelty, using spotted hyenas as our subjects. Our null hypothesis, as suggested by Burkart et al. [14], is that evolving intelligence is strictly modular. We define general intelligence as the suite of cognitive mechanisms that appear to enhance the animal's ability to engage in flexible, innovative behaviors when confronted with a problem [26], and we are quantifying it as g, calculated using a psychometric factor-analytical approach.

Although there is now considerable evidence that g exists in non-human animals, the origins of general intelligence remain a fascinating mystery. Most hypotheses suggesting explanations for the evolution of animal intelligence posit that general intelligence evolves as a by-product of selection for domain-specific cognitive abilities such as foraging efficiency [60, 98, 100] or social agility [11, 15, 16, 25, 35, 72, 74, 99]. In contrast, however, the "Cognitive Buffer" hypothesis [1, 22, 48, 84, 85] suggests that domain-general intelligence is favored directly by natural selection to help animals cope with novel or unpredictable environments, where it is adaptive because it enables individuals to exhibit flexible behavior, and thus find innovative solutions to problems threatening their survival and reproduction.

In support of the Cognitive Buffer hypothesis, innovation rates and brain size, and thus presumably general intelligence, predict colonization success in novel environments by birds [87], mammals [86], fishes [80], amphibians and reptiles [3]. These correlations suggest that general intelligence, indicated by brain size and innovative behavior as its proxies, may indeed enhance fitness in new environments. Kotrschal & Taborsky [46] have shown experimentally that even small environmental changes enhance cognitive abilities in evolving fish. However, no experimental work has yet been conducted in the wild to test predictions of the cognitive buffer hypothesis, nor have selection gradients been assessed on cognitive traits in different environments.

Sol et al. [88] suggested that behavioral flexibility, a hallmark of general intelligence, permits animals to cope with the human-induced rapid environmental change currently taking place around the globe. Behavioral flexibility might be expected to have particularly strong effects on fitness in cities, which represent some of the most extreme novel environments confronted by animals today [83]. In cities, animals may need to exploit new food resources, cope with a new suite of potential predators and competitors, develop new navigation strategies, and adjust their communication to cope with new noise [7, 33, 83]. Evidence is quickly accumulating that adaptation to city life alters traits ranging from vigilance and immuno-competence to life-history patterns (e.g., [18, 51, 61]). The relationship between animal cognition and urbanization is also of enormous contemporary interest. Therefore, we are currently conducting experiments to test predictions of the Cognitive Buffer hypothesis in the context of urbanization [42]. We are assessing general intelligence and its fitness consequences across a range of human-modified landscapes, some of which expose animals to novel conditions that differ dramatically from those under which they evolved [31, 32, 43].

Some recent work consistent with the Cognitive Buffer hypothesis suggests that urban animals might be better at problem-solving than conspecifics inhabiting rural areas (e.g., [6, 65]), but other data suggest a more complicated story. For example, Snell-Rood & Wick [83] found partial support for the hypothesis that urban environments select for increased behavioral plasticity, but this selection appeared to be most pronounced early during the urban colonization process. Their data also suggest that behavioral plasticity may be favored in rural environments that are changing because of human activity. In our current work we are evaluating the relationship between degree of urbanization and seven cognitive abilities in multiple clans of free-living spotted hyenas living in three different habitats, each at a different stage of the urbanization process. Administration of our test battery will eventually allow us to assess performance on each cognitive task within and among clans. Our test battery includes five tests of basic cognitive abilities and two tests assessing the executive functions of inhibitory control and selective attention.

The Cognitive Buffer hypothesis is only one of the two leading direct explanations for the evolution of general intelligence. Burkart et al. [14] instead prefer its primary competitor, the Cultural Intelligence hypothesis, which stresses the critical importance of social inputs during the ontogenetic construction of survival-relevant skills. They argue that existing data from vertebrates, and especially from great apes, support the Cultural Intelligence hypothesis. However, the general intelligence explained by the Cultural Intelligence hypothesis is actually quite limited, so I believe we must seek a more robust explanation for its evolution. I believe that the Cognitive Buffer hypothesis offers a better alternative because it can account for phenomena the Cultural Intelligence hypothesis leaves unexplained. Burkart et al. [14] argue that fundamental preconditions for the evolution of large brains and great general intelligence include a slow life history and high survivorship, which are possible only in species not subject to unavoidable extrinsic mortality such as high predation pressure [74]. However, much can be learned by considering apparent exceptions to "rules" like these, so I offer the octopus as one such exception [38].

Most octopuses are strictly solitary except when copulating, have very short lives, have countless predators, and produce thousands of offspring, most of which die. Nevertheless, they have some of the largest brains known among invertebrates [36, 101], they exhibit a great deal of curiosity about their environments [62], they recognize individual humans [4], they exhibit pronounced individual differences [56, 81], they use tools, and they play [54, 55]. Octopuses thus appear to exhibit a considerable amount of general intelligence without any opportunity whatsoever for social learning. Clearly the cultural intelligence hypothesis cannot account for the general intelligence apparent in creatures like these.

Similarly, the cultural intelligence hypothesis offers little promise with respect to evolving general intelligence in machines [40]. Computer scientists and robotic engineers have understood for decades that the embodiment of intelligent machines affects their ability to adapt and learn via feedback obtained during their interactions with the environment, mediated by sensors and activators [12, 13, 30, 76]. Most hypotheses forwarded to explain the evolution of intelligence in animals, including the Cultural Intelligence hypothesis, fail to address the question of how morphological traits outside the nervous system might have shaped intelligence. In creatures such as octopuses and primates, mutations affecting nervous system structure or function, which might generate less-stereotyped and more-flexible behavior, are visible to selective forces in the environment because they can be embodied in the limbs. Thus, greater intelligence is more likely to evolve in these animals than in those whose interactions with their environments are more highly constrained.

Roboticists have also realized that logic alone cannot generate much intelligent behavior in their machines, and that to achieve better performance, their robots must also want things. The skills discovered by evolutionary algorithms are diverse, and many such skills may occur within a single population of digital organisms or robots, but individual agents are rarely motivated to acquire a large array of skills. As a result, most current evolutionary algorithms produce domain-specific intelligence in machines that rarely possess more than a small set of skills, and are thus suited to performing only tasks that demand that particular skill set. Although an intrinsic motivation to explore the environment has been imitated in artificial agents via machine learning [64, 75], the production of generalist learners within an evolutionary context remains highly problematic [89].

Any selection pressure that promotes behavioral diversity or flexibility within the organism's lifetime, including the ability to learn from experience, should theoretically result in enhanced general intelligence. Novel or changing environments should select for individuals who can learn as much as possible in their lifetimes, as suggested by the Cognitive Buffer hypothesis. Indeed, Stanton & Clune [89] developed an evolutionary algorithm that produces agents who explore their environments and acquire as many skills as possible within their life-times while also retaining their existing skills. This algorithm encourages evolution to select for curious agents motivated to interact with things in the environment that they do not yet understand, and engage in behaviors they have not yet mastered. This algorithm has two main components, a fitness function that rewards individuals for expressing as many unique behaviors as possible, and an intralife novelty score that quantifies the types of behaviors rewarded by the algorithm. Agents are also provided with an intra-life novelty compass that indicates which behaviors are considered novel within the environment. The intra-life novelty compass may simply identify and direct agents toward areas of high expected learning, because new knowledge often promotes the ability to perform new skills. Aligned with these results, we suggest that the primary value of the Cultural Intelligence hypothesis is to offer social learning as an intra-life novelty compass, but that this hypothesis provides neither the requisite fitness function nor anything analogous to an intra-life novelty score. A viable hypothesis explaining the evolution of large brains and general intelligence should be able to account for general intelligence in virtually any species where it is known to occur, and should be able to predict the conditions under which we can develop machines with general intelligence as well. The Cultural Intelligence hypothesis appears unable to do these things, but the Cognitive Buffer Hypothesis can.

So, for allowing me to re-invent myself in these multiple ways over the past decade, I owe Erik Goodman, along with the wonderful people he assembled as BEACON colleagues, an enormous debt of gratitude. BEACON has been a fabulously enriching experience for me, my students and postdocs, and on behalf of all of us, I thank Erik for his extraordinary leadership, which has made BEACON such an exemplary success story.

Acknowledgements This material is based in part upon work supported by the National Science Foundation under Cooperative Agreement No. DBI-0939454. Any opinions, findings, and conclusions or recommendations expressed in this material are those of the author(s) and do not necessarily reflect the views of the National Science Foundation. The work described here was also made possible in part by NSF grants IOS1755089 and DEB1353110.

References

1. Allman, J., McLaughlin, T., Hakeem, A.: *Brain weight and life-span in primate species.* Proceedings of the National Academy of Sciences of the United States of America **90**, 118–122 (1993)

2. Amici, F., Barney, B., Johnson, V.E., Call, J., Aureli, F.: *A modular mind? A test using individual data from seven primate species.* PLOS ONE **7**, e51,918 (2012)

3. Amiel, J.J., Tingley, R., Shine, R.: *Smart moves: Effects of relative brain size on establishment success of invasive amphibians and reptiles.* PLoS One **6**, e18,277 (2011)

4. Anderson, R.C., Mather, J.A., Monette, M.Q., Zimsen, S.R.M.: *Octopuses (Enteroctopus dofleneni) recognize individual humans.* Journal of Applied Animal Welfare Science **13**, 261–272 (2010)

5. Arden, R., Adams, M.J.: *A general intelligence factor in dogs.* Intelligence **5**, 79–85 (2016)

6. Audet, J.N., Ducatez, S., Lefebvre, L.: *The town bird and the country bird: Problem solving and immunocompetence vary with urbanization.* Behav. Ecol. **27**, 637–644 (2016)

7. Bateman, P.W., Fleming, P.A.: *Big city life: Carnivores in urban environments.* J. Zool. Lond. **287**, 1–23 (2012)

8. Bednekoff, P., Balda, R., Hile, A.: *Long-term spatial memory in fur species of seed-caching corvids.* Anim. Behav. **53**, 335–341 (1997)

9. Benson-Amram, S., Dantzer, B., Stricker, G., Swanson, E.M., Holekamp, K.E.: *Brain size predicts problem-solving ability in mammalian carnivores.* Proceedings of the National Academy of Sciences of the United States of America **113**, 2532–2537 (2016)

10. Bitterman, M.E.: *The comparative analysis of learning.* Science **188**, 699–709 (1975)

11. Borrego, N., Gaines, M.: *Social carnivores outperform asocial carnivores on an innovative problem.* Anim. Behav. **114**, 21–26 (2016)

12. Brooks, R.A.: *Elephants don't play chess.* Robotics and Autonomous Systems **6**, 1–16 (1990)

13. Brooks, R.A.: *Intelligence without reason.* In: Proceedings of the 12th International Joint Conference on Artificial Intelligence (IJCAI-91), pp. 569–595. Morgan Kaufmann, San Mateo, CA (1991)

14. Burkart, J.M., Schubiger, M.N., van Schaik, C.P.: *The evolution of general intelligence.* Behav. Brain Sci. **40**, e224 (2017)

15. Byrne, R.W.: The evolution of intelligence. In: P. Slater, T. Halliday (eds.) Behaviour and evolution, pp. 223–265. Cambridge University Press, Cambridge, UK (1994)

16. Byrne, R.W., Whiten, A. (eds.): *Machiavellian Intelligence: Social expertise and the evolution of intellect in monkeys, apes and humans.* Oxford University Press (1988)

17. Carroll, J.B.: *Human cognitive abilities: A survey of factor-analytic studies.* Cambridge University Press, New York (1993)

18. Charmantier, A., Demeyrier, V., Lambrechts, M., Perret, S., Grégoire, A.: *Urbanization is associated with divergence in pace-of-life in great tits.* Front. Ecol. Evol. **5**, 53 (2017)

19. Cheney, D.L., Seyfarth, R.M.: Social and non-social knowledge in vervet monkeys. In: R. Byrne, A. Whiten (eds.) Machiavellian Intelligence: Social expertise and the evolution of intellect in monkeys, apes and humans, pp. 255–270. Oxford University Press, UK (1988)

20. Cosmides, L., Tooby, J.: Unraveling the enigma of human intelligence: Evolutionary psychology and the multimodular mind. In: R. Sternberg, J. Kaufman (eds.) The evolution of intelligence, pp. 145–198. Erlbaum (2001)

21. Cosmides, L., Tooby, J.: *Evolutionary psychology: New perspectives on cognition and motivation.* Ann. Rev. Psych. **64**, 201–229 (2013)

22. Deaner, R.O., Barton, R.A., van Schaik, C.P.: Primate brains and life history: Renewing the connection. In: P.M. Kappeler, M.E. Pereira (eds.) Primate Life Histories and Socioecology, pp. 233–265. University of Chicago Press, Chicago, IL (2003)

23. Deary, I.J., Penke, L., Johnson, W.: *The neuroscience of human intelligence differences.* Nat. Rev. Neurosci. **11**, 201–211 (2010)

24. Diamond, A.: *Executive functions.* Ann. Rev. Psychol. **64**, 135–168 (2013)

25. Dunbar, R.I.M.: *The social brain: Mind, language and society in evolutionary perspective.* Ann. Rev. Anthropol. **325**, 163–181 (2003)

26. Farris, S.: *Evolution of brain elaboration.* Phil. Trans. Roy Soc. B **370**, 20150,054 (2015)

27. Fernandes, H.B.F., Woodley, M.A., te Nijenhuis, J.: *Differences in cognitive abilities among primates are concentrated on g: Phenotypic and phylogenetic comparisons with two meta-analytical databases.* Intelligence **46**, 311–322 (2014)

28. Galsworthy, M.J., Paya-Cano, J.L., Monleo, S., Plomin, R.: *Evidence for general cognitive ability (g) in heterogeneous stock mice and an analysis of potential confounds*. Genes, Brain Behav. **1**, 88–95 (2002)
29. Gersick, A.S., Cheney, D.L., Schneider, J.M., Seyfarth, R.M., Holekamp, K.E.: *Long-distance communication facilitates cooperation among spotted hyaenas (crocuta crocuta)*. Animal Behaviour **103**, 107–116 (2015)
30. Goldman, A., de Vignemont, F.: *Is social cognition embodied?* Trends in Cognitive Sciences **13**, 154–159 (2009)
31. Green, D.S., Johnson-Ulrich, L., Couraud, H.E., Holekamp, K.E.: *Anthropogenic disturbance induces opposing population trends in spotted hyenas and African lions*. Biodiversity and Conservation **27**, 1–19 (2018)
32. Greenberg, J.R., Holekamp, K.E.: *Human disturbance affects personality development in a wild carnivore*. Animal Behaviour **132**, 303–312 (2017)
33. Gross, K., Pasinelli, G., Kunc, H.P.: *Behavioral plasticity allows short-term adjustment to a novel environment*. Am. Nat. **176**, 456–464 (2010)
34. Hauser, M.D.: *Perseveration, inhibition and the prefrontal cortex: A new look*. Curr. Opin. Neurobiol. **9**, 214–222 (1999)
35. Herrmann, E., Call, J., Hernandez-Lloreda, M.V., Hare, B., Tomasello, M.: *Humans have evolved specialized skills of social cognition: The cultural intelligence hypothesis*. Science **317**, 1360–1366 (2007)
36. Hochner, B., Shormrat, T., Fiorito, G.: *The octopus: A model for comparative analysis of the evolution of learning and memory mechanisms*. Biological Bulletin **210**, 308–317 (2006)
37. Holekamp, K.E.: *Questioning the social intelligence hypothesis*. Trends Cog. Sci. **11**, 65–69 (2007)
38. Holekamp, K.E., Benson-Amram, S.: *The evolution of intelligence in mammalian carnivores*. Interface Focus p. 20160108 (2017)
39. Holekamp, K.E., Dantzer, B., Stricker, G., Yoshida, K.C.S., Benson-Amram, S.: *Brains, brawn and sociality: A hyaena's tale*. Animal Behaviour **103**, 237–248 (2015)
40. Holekamp, K.E., Van Meter, P.E., Swanson, E.M.: *Developmental constraints on behavioral flexibility*. Philosophical Transactions of the Royal Society B. **368**, 20120,350 (2013)
41. Jacobs, L., Liman, E.: *Grey squirrels remember the locations of buried nuts*. Anim. Behav. **41**, 103–110 (1991)
42. Johnson-Ulrich, L., Johnson-Ulrich, Z., Holekamp, K.E.: *Proactive behavior, but not inhibitory control, predicts repeated innovation in spotted hyenas tested with a multi-access box*. Animal Cognition **21**(3), 379–392 (2018)
43. Johnson-Ulrich, L., Lehmann, K.D.S., Turner, J.W., E., H.K.: Testing cognition in the umwelt of the spotted hyena. In: N. Bueno-Guerra, F. Amici (eds.) The experimental umwelt: A practical guide to animal cognition, pp. 244–265. Cambridge University Press, Cambridge, UK (2018)
44. Keagy, J., Savard, J.F., Borgia, G.: *Complex relationship between multiple measures of cognitive ability and male mating success in satin bowerbirds, ptilonorhynchus violaceus*. Anim. Behav. **81**, 1063–1070 (2011)
45. Kolata, S., Light, K., Grossman, H.C., Hale, G., Matzel, L.D.: *Selective attention is a primary determinant of the relationship between working memory and general learning ability in outbred mice*. Learn. Mem. **14**, 22–28 (2007)
46. Kotrschal, A., Taborsky, B.: *Environmental change enhances cognitive abilities in fish*. PLoS Biol. **8**, e1000,351 (2010)
47. Laubach, Z.M., Perng, W., Dolinoy, D.C., Faulk, C.D., Holekamp, K.E., Getty, T.: *Epigenetics and the maintenance of developmental plasticity: Extending the signaling theory framework*. Biological Reviews **93**, 1323–1338 (2018)
48. Lefebvre, L., Reader, S.M., Sol, D.: *Innovating innovation rate and its relationship with brains, ecology and general intelligence*. Brain, Behavior and Evolution **81**, 143–145 (2013)
49. Lehmann, K.D.S., Montgomery, T.M., MacLachlan, S.M., Parker, J.M., Spagnuolo, O.S., VandeWetering, K., Bills, P.S., Holekamp, K.E.: *Lions, hyenas and mobs (oh my!)*. Current Zoology **63**(3), 313–322 (2017)

50. Lewin, N., Swanson, E.M., Williams, B.L., Holekamp, K.E.: *IGF-1 concentrations during early life predict life-history trade-offs in a wild mammal.* Functional Ecology **31**(4), 894–902 (2017)
51. Lowry, H., Lill, A., Wong, B.: *Behavioural responses of wildlife to urban environments.* Biol. Rev. **88**, 537–549 (2013)
52. MacLean, E.L., et al.: *The evolution of self-control.* Proceedings of the National Academy of Sciences of the United States of America **111**, E2140–E2148 (2014)
53. Macphail, E.M., Bolhuis, J.J.: *The evolution of intelligence: Adaptive specializations versus general process.* Biol. Rev. **76**, 341–364 (2001)
54. Mather, J.A.: *Home choice and modification by juvenile octopus vulgaris: Specialized intelligence and tool use?* Journal of Zoology, London **233**, 359–368 (1994)
55. Mather, J.A., Anderson, R.C.: *Exploration, play and habituation in octopus dofleini.* Journal of Comparative Psychology **113**, 333–338 (1999)
56. Mather, J.A., Leite, T., Battista, A.T.: *Individual prey choices of octopus: Are they generalists or specialists?* Current Zoology **58**, 597–603 (2012)
57. Matzel, L.D., Townsend, D.A., Grossman, H., Han, Y.R., Hale, G., Zappulla, M., Light, K., Kolata, S.: *Exploration in outbred mice covaries with general learning abilities irrespective of stress reactivity, emotionality, and physical attributes.* Neurobiol. Learn. Mem. **86**, 228–240 (2006)
58. Matzel, L.D., Wass, C., Kolata, S.: *Individual differences in animal intelligence: Learning, reasoning, selective attention and inter-species conservation of a cognitive trait.* Int. J. Comp. Psychol. **24**, 36–59 (2011)
59. McGrew, K.S.: *CHC theory and the human cognitive abilities project: Standing on the shoulders of the giants of psychometric intelligence research.* Intelligence **37**, 1–10 (2009)
60. Milton, K.: *Diversity of plant foods in tropical forests as a stimulus to mental development in primates.* Amer. Anthropol. **83**, 534–548 (1981)
61. Miranda, A.C., Schielzeth, H., Sonntag, T., Partecke, J.: *Urbanization and its effects on personality traits: A result of microevolution or phenotypic plasticity?* Global Change Biol. **19**, 2634–2644 (2013)
62. Montgomery, S.: *The Soul of an Octopus.* Simon and Schuster, New York (2015)
63. Nippak, P.M., Milgram, N.W.: *An investigation of the relationship between response latency across several cognitive tasks in the beagle dog.* Prog. Neuropsychopharmacol. Biol. Psychiatry **29**, 371–377 (2005)
64. Oudeyer, P.Y., Kaplan, F., Hafner, V.: *Intrinsic motivation systems for autonomous mental development.* IEEE Transactions on Evolutionary Computation **11**, 265–286 (2007)
65. Papp, S., Vincze, E., Preiszner, B., Liker, A., Bokony, V.: *A comparison of problem-solving success between urban and rural house sparrows.* Behav. Ecol. Sociobiol. **69**, 471–480 (2015)
66. Prinz, J.J.: Is the mind really modular? In: R.J. Stainton (ed.) Contemporary Debates in Cognitive Science, pp. 22–36. Blackwell Publishing, Malden (2006)
67. Quartz, S.: Toward a developmental evolutionary psychology: Genes, development and the evolution of cognitive architecture. In: S. Scher, F. Rauscher (eds.) Evolutionary Psychology: Alternative Approaches, pp. 185–210. Kluwer, Boston (2003)
68. Rajagopalan, P., Holekamp, K.E., Miikkulainen, R.: *The evolution of general intelligence.* In: Proceedings of the Fourteenth International Conference on the Synthesis and Simulation of Living Systems (ALIFE'14, NY). MIT Press (2014)
69. Rajagopalan, P., Rawal, A., Holekamp, K.E., Miikkulainen, R.: *General intelligence through prolonged evolution of densely connected neural networks.* Proceedings of the Genetic and Evolutionary Computation Conference (GECCO-2014), Vancouver, Canada (2014)
70. Rajagopalan, P., Rawal, A., Miikkulainen, R., Wiseman, M.A., Holekamp, K.E.: *The role of reward structure coordination mechanism and net return in the evolution of cooperation.* In: IEEE Conference on Computational Intelligence and Games, 2011, pp. 258–265. IEEE Press (2011)
71. Rawal, A., Rajagopalan, P., Miikkulainen, R., Holekamp, K.E.: *Evolution of a communication code in cooperative tasks.* Artificial Life **13**, 243–250 (2012)

72. Reader, S.M., Hager, Y., Laland, K.N.: *The evolution of primate general and cultural intelligence*. Phil. Trans. Roy Soc. B **366**, 1017–1027 (2011)
73. Rosenzweig, M.R., Bennett, E.L.: *Psychobiology of plasticity: Effects of training and experience on brain and behavior*. Behav. Brain Res. **78**, 57–65 (1996)
74. van Schaik, C.P., Isler, K., Burkart, J.M.: *Explaining brain size variation: From social to cultural brain*. Trends Cog. Sci. **16**, 277–284 (2012)
75. Schmidhuber, J.A.: *Possibility for implementing curiosity and boredom in model-building neural controllers*. In: Proceedings of the International Conference on Simulation of Adaptive Behavior: From Animals to Animats, pp. 222–227. MIT Press (1991)
76. Sharkey, N.E., Ziemke, T.: *A consideration of the biological and psychological foundations of autonomous robotics*. Connection Science **10**, 361–391 (1998)
77. Shaw, R.C., Boogert, N.J., Clayton, N.S., Burns, K.C.: *Wild psychometrics: Evidence for "general" cognitive performance in wild New Zealand robins, petroica longipes*. Anim. Behav. **109**, 101–111 (2015)
78. Sherry, D.F.: *Neuroecology*. Ann. Rev. Psychol. **57**, 167–197 (2006)
79. Shettleworth, S.J.: *Cognition, Evolution, and Behavior*. Oxford University Press (2010)
80. Shumway, C.A.: *Habitat complexity, brain, and behavior*. Brain, Behav. Evol. **72**, 123–134 (2008)
81. Sinn, D.L., Perrin, N.A., Mather, J.A., Anderson, R.C.: *Early temperamental traits in an octopus (octopus bimaculoides)*. Journal of Comparative Psychology **115**, 351–364 (2001)
82. Smith, J.E., Swanson, E.M., Reed, D., Holekamp, K.E.: *Evolution of cooperation among mammalian carnivores and its relevance to hominins*. Current Anthropology **53, Supplement 6**, S436–S452 (2012)
83. Snell-Rood, E.C., Wick, N.: *Anthropogenic environments exert variable selection on cranial capacity*. Proc. Roy Soc. B. **280**, 20131,384 (2013)
84. Sol, D.: The cognitive-buffer hypothesis for the evolution of large brains. In: R. Dukas, J. Ratcliffe (eds.) Cognitive Ecology II, pp. 111–134. University of Chicago Press (2009)
85. Sol, D.: *Revisiting the cognitive buffer hypothesis for the evolution of large brains*. Biol. Lett. **5**, 130–133 (2009)
86. Sol, D., Bacher, S., Reader, S.M., Lefebvre, L.: *Brain size predicts the success of mammal species introduced into novel environments*. Amer. Natur. **172**, S63–S71 (2008)
87. Sol, D., Duncan, R.P., Blackburn, T.M., Cassey, P., Lefebvre, L.: *Big brains, enhanced cognition, and response of birds to novel environments*. Proceedings of the National Academy of Sciences of the United States of America **102**, 5460–5465 (2005)
88. Sol, D., Laperdra, O., Gonzalez-Lagos, C.: *Behavioural adjustments for a life in the city*. Anim. Behav. **85**, 1101–1112 (2013)
89. Stanton, C., Clune, J.: *Curiosity search: Producing generalists by encouraging individuals to continually explore and acquire skills throughout their lifetime*. PLoS ONE **11**(9), e0162,235 (2016)
90. Sternberg, R.J., Grigorenko, E.L.: *The general factor of intelligence: How general is it?* Psychology Press, United Kingdom (2002)
91. Strauss, E.D., Holekamp, K.E.: *Inferring longitudinal hierarchies: Framework and methods for studying the dynamics of dominance*. Journal of Animal Ecology **88**(4), 521–536 (2019)
92. Strauss, E.D., Holekamp, K.E.: *Social alliances improve rank and fitness in convention-based societies*. Proceedings of the National Academy of Sciences of the United States of America **116**, 8919–8924 (2019)
93. Swanson, E.M., Dworkin, I., Holekamp, K.E.: *Lifetime selection on a hypoallometric size trait in the spotted hyena*. Proceedings of the Royal Society of London B **278**, 3277–3285 (2011)
94. Swanson, E.M., McElhinny, T.L., Dworkin, I., Weldele, M.L., Glickman, S.E., Holekamp, K.E.: *Ontogeny of sexual size dimorphism in the spotted hyena (crocuta crocuta)*. Journal of Mammalogy **94**, 1298–1310 (2013)
95. Theis, K.R., Venkataraman, A., Dycus, J.A., Koonter, K.D.S., Schmitt-Matzen, E.N., Wagner, A., Holekamp, K., Schmidt, T.: *Symbiotic bacteria appear to mediate hyena social odors*. Proceedings of the National Academy of Sciences of the United States of America **110**, 19,832–19,837 (2013)

96. Theis, K.R., Venkataraman, V., Wagner, A., Holekamp, K.E., Schmidt, T.: Age-related variation in the scent pouch bacterial communities of striped hyenas (hyaena hyaena). In: B. Schulte, T. Goodwin, M. Ferkin (eds.) Chemical Signals in Vertebrates 13, pp. 87–103 (2015)

97. Turner, J.W., Bills, P.S., Holekamp, K.E.: *Ontogenetic change in determinants of social network position in the spotted hyena.* Behavioral Ecology and Sociobiology **72**(10), 1–15 (2018)

98. Vonk, J., Jett, S.E., Mosteller, K.W.: *Concept formation in American black bears, ursus americanus.* Anim. Behav. **84**, 953–964 (2012)

99. Whiten, A.: Social complexity and social intelligence. In: G.R. Bock, J.A. Goode, K. Webb (eds.) The nature of intelligence: Novartis Foundation Symposium, Vol 233, pp. 185–201. J. Wiley, New York, NY (2008)

100. Zuberbuehler, K., Janmaat, K.: Foraging cognition in non-human primates. In: M. Platt, A. Ghazanfar (eds.) Primate Neuroethology, pp. 64–83. Oxford University Press (2010)

101. Zullo, L., Hochner, B.: *A new perspective on the organization of an invertebrate brain.* Communicative & Integrative Biology **4**, 26–29 (2011)

SUPPLEMENTARY MATERIALS

Post-docs

The post-doctoral fellows in my lab who have benefitted directly from the presence of BEACON at MSU include:

Dr. Agathe M. Laurence, PhD University of Rennes.

Dr. Amiyaal Ilany, PhD Tel Aviv University. Currently an Assistant Professor of Biology at Bar-Ilan University in Israel.

Dr. Ariana Strandburg-Peshkin, PhD Princeton University. Currently an assistant professor at the University of Konstanz, Germany.

Dr. Frants H. Jensen, PhD Aarhus University. Currently a lecturer at the University of St Andrews, Scotland, UK.

Dr. Andrew S. Gersick, PhD University of Pennsylvania. Currently a post-doc at Princeton and MSU.

Graduate Students

The PhD students in my lab who have benefitted directly from the presence of BEACON at MSU, listed in chronological order, include:

Dr. Kevin R. Theis, PhD 2008. Currently an Assistant Professor in the Department of Immunology at the Wayne State University School of Medicine, Detroit, MI.

Dr. Jennifer E. Smith, PhD 2010. Currently an Associate Professor of Biology at Mills College, Oakland, CA.

Dr. Sarah R. Benson-Amram, PhD 2011. Currently an assistant professor of Zoology and Physiology at the University of Wyoming, Laramie, WY.

Dr. Ben Dantzer, PhD 2012. Currently an assistant professor of Psychology & EEB at the University of Michigan, Ann Arbor.

Dr. Kate C. Shaw Yoshida, PhD 2012. Currently Curator of Biology at the Las Vegas Natural History Museum in Las Vegas, Nevada.

Dr. Andrew S. Flies, PhD 2012. Currently a post-doctoral research fellow at the Menzies Institute for Medical Research, University of Tasmania, Hobart, Tasmania.

Dr. Eli M. Swanson, PhD 2013. Currently a post-doc in Ecology & Evolution at the University of Minnesota.

Dr. Katy J. Califf, PhD 2013. Currently teaching high school biology in Indianapolis, IN.

Dr. Nora S. Lewin, PhD 2017. Currently a Data Architect at Northwestern University, Chicago, IL.

Dr. Julia R. Greenberg, PhD 2017. Currently an Assessment Biologist in the Natural Resources Department for the Sault Tribe of Chippewa Indians in Sault Sainte Marie, MI.

Dr. Julie W. Turner, PhD 2018. Currently a post-doctoral fellow at Memorial University, Newfoundland, Canada.

Eli D. Strauss, PhD 2019. Currently a post-doctoral fellow at University of Nebraska.

Kenna D. S. Lehmann, PhD expected 2019

Zachary M. Laubach, PhD expected 2019

Lily Johnson-Ulrich, PhD expected 2020

Tracy M. Montgomery, PhD expected 2020

S. Kevin McCormick, PhD expected 2020

Maggie A. Sawdy, PhD expected 2021

Connie A. Rojas, PhD expected 2021

Projects

Recent, current and pending grants to the Holekamp lab made possible by BEACON support of pilot projects and graduate students:

Human Frontier Science Program Grant (Pending) "Communication and the coordination of collective behavior across spatial scales in animal societies" Principal Investigator: Ariana Strandburg-Peshkin. Co-PIs: Kay E. Holekamp, Marta Manser, Benjamin Hirsch, Marie Rosch. 12/1/2019-11/30/2022. Amount: $1,350,000

NSF Grant OISE1853934 (International Research Experience for Students) (Pending) "IRES Track I: Behavioral ecology of African carnivores." Principal Investigator: Kay E. Holekamp. Amount: $305,017.

NSF grant proposal (IOS1755089) "Selection for general intelligence by novel environments" 3/1/2018 2/28/2022. Principal Investigator: Kay E. Holekamp. Co-PIs: Elise F. Zipkin & D. Zach Hambrick. Amount: $678,441.

NSF Doctoral Dissertation Improvement Grant (IOS-Animal Behavior) to Zachary Laubach: "Early social experience and epigenetic mediation of adult phenotypes." 6/1/2017 - 5/31/2019. Amount: $20,151.

NSF Grant OISE 1556407 (International Research Experience for Students) (7/1/2016-6/30/2019) "IRES: Behavioral ecology of African carnivores." Principal Investigator: Kay E. Holekamp. Amount: $249,999.

NIH Grant Number: 1R01GM105042 (2013-2017) "The Role of Emotion and Communication in Cooperative Behavior." Principal Investigator: Risto Miikku-lainen; Co-PI: Kay E. Holekamp. Amount: $1,257,029. MSU subcontract for Hole-kamp portion: $437,087.

NSF Grant OISE-1260768 (International Research Experience for Students) (2013-2016) "IRES: Behavioral ecology of African carnivores." Principal Investi-gator: Kay E. Holekamp. Amount: $250,000.

Chapter 3
BEACON: Using Diversity as an Evolutionary Tool for a High-Performing Science and Technology Center

Judi Brown Clarke and Percy Pierre

Abstract The BEACON Center for the Study of Evolution in Action is a multi-site, interdisciplinary science and technology center funded by the National Science Foundation. BEACON effectively institutionalized its two overarching diversity goals of 1) ensuring that diversity is represented as an inclusive and connecting thread throughout the consortium, and 2) exceeded national norms for diversity at all levels across the consortium. As BEACON approaches its *sunsetting* 10th year, there is great pride in our sustained diversity outcomes and accomplishments. It was very important for us to capture our shared core values, implementation strategies, administrative infrastructure, and programmatic outcomes so that subsequent multi-site partnership can use our strategies and tools as a blueprint.

3.1 Introduction

The BEACON Center for the Study of Evolution in Action is a National Science Foundation (NSF) funded, Science and Technology Center (STC) founded in 2010 with the mission of illuminating and harnessing the power of *evolution in action* to advance science and technology and benefit society. It is a consortium of universities led by Michigan State University, with member institutions including North Carolina Agriculture & Technology State University, the University of Idaho, the University of Texas at Austin, and the University of Washington. BEACON promotes the transfer of discoveries from biology into computer science and engineering design, while using novel computational methods and artificial evolutionary systems

Judi Brown Clarke
BEACON Center for Evolution in Action, Michigan State University, East Lansing, MI, USA e-mail: jbc@egr.msu.edu

Percy Pierre
Department of Electrical and Computer Engineering, Michigan State University, East Lansing, MI, USA e-mail: pierre@msu.edu

© Springer Nature Switzerland AG 2020 33
W. Banzhaf et al. (eds.), *Evolution in Action: Past, Present and Future*,
Genetic and Evolutionary Computation, https://doi.org/10.1007/978-3-030-39831-6_3

to address complex biological questions that are difficult or impossible to study with natural organisms.

BEACON's initial *Strategic Implementation Plan* sets goals in six key areas: *education, human resources & diversity (EHRD); leadership and management; knowledge transfer; integrative research; ethical research; and research output*. These overall goals and optimal outcomes have not changed since its inception. Specifically, BEACON's EHRD goal was to integrate cutting-edge, multidisciplinary research, education, and outreach efforts across the Center that will advance innovative training, the diversity of the Center and scientific workforce, and public education to promote greater understanding of evolution and the nature of science.

The BEACON's *leadership and management goals* were to envision and enable the Center's mission through inclusive and transparent decision-making, as well as effective and responsible implementation; to inspire all Center participants; and to facilitate collaborative efforts within and beyond the Center. The *integrative research goal* was to produce transformative, synergistic research through an inclusive collaborative culture that crosses disciplinary and institutional boundaries and was embedded throughout the Center's activities.

BEACON effectively institutionalized its two overarching diversity goals of: 1) ensuring that diversity is represented as an inclusive and connecting thread through all aspects of BEACON, and 2) exceeding national norms for diversity at all levels across the consortium. These outcomes were embedded in a clear and consistent message, and demonstrated within research opportunities, grant submissions, broader impacts efforts, educational and outreach activities, formal mentoring training/support, fellowships, and direct student support funding.

3.2 BEACON's Diversity Profile

At the time of writing (January 2019), BEACON's overall membership is 53% White, 24% Black, 10% Asian, 5% Hispanic/Latino, 1% Native American/Alaskan Native, and 7% two or more races. See Figure 3.1 for a categorical breakdown of BEACON's diverse participants. Initially BEACON wanted to establish an annual goal of surpassing all the other STC's diversity outcomes. Unfortunately there were no collective reporting mechanisms available for comparative data. In response, BEACON established a National Norms baseline for diversity measures and ensured its accuracy by collecting applicable data on the numbers and percentages of undergraduate students, graduate students, post-docs, and faculty.

Figure 1 *BEACON's Diversity Profile*		
Diversity Categories	**2018 BEACON Participants**	**NSF Weighted National Norms** (BEACON-Specific Disciplines)
Females Overall	43.0%	32.8%
Female Faculty	39.0%	31.0%
Female Postdocs	38.0%	35.5%
Female Graduate Stds	37.0%	32.9%
Female Undergraduates	43.0%	36.6%
URMs Overall	37.0%	9.2%
URM Faculty	13.0%	7.7%
URM Postdocs	11.0%	6.7%
URM Graduate Stds	17.0%	8.1%
URM Undergraduates	68.0%	14.2%
Disability Overall	6.4%	3.3%
Disability Faculty	2.7%	2.7%
Disability Postdocs	2.0%	0.1%
Disability Graduate Stds	8.2%	5.7%
Disability Undergraduates	7.1%	4.7%

Fig. 3.1: BEACON's Diversity Profile

BEACON was confident that exceeding national norms would surpass the diversity results at most other research-intensive universities. A strong indication of BEACON's diversity success, compared to other STCs, is the fact that it was selected to present its activities and results to other STC's on various occasions.

BEACON targeted the NSF's mandated reporting groups of females, underrepresented minorities (URMs), and individuals with disabilities. In an effort to ensure our comparisons were replicable, BEACON captured and weighted national norm data for BEACON-specific disciplines using the NSF 2011, and then 2017 data tables found at www.nsf.gov/statistics/2017/nsf17310/.

The BEACON-specific disciplines included:

- Biological/Natural Science: includes anatomy, biochemistry, biology, epidemiology, botany, cell biology, ecology, entomology, genetics, mathematics/applied math, microbiology/immunology/virology, neuroscience, pathology, physiology, and zoology
- Computer Science
- Engineering: includes biomedical, electrical and mechanical

Figure 3.2 represents BEACON's nine-year evolution of participants compared to the National Norms. This process included data collected on the numbers and percentages of undergraduates, graduate students, post-docs, and faculty participants from NSF's recognized underrepresented groups.

Figure 2 *BEACON's 9-Year Evolution of Participants*				
Annual Reporting Data	**URM Membership**	**Multiracial Participants**	**Female Participants**	**Ind. w/ Disabilities**
Weighted National Norms	30.8%	N/A	32.8%	1.1%
2018 BEACON	34.1%	6.7%	41.9%	5.6%
2017 BEACON	34.1%	6.7%	41.8%	5.6%
2016 BEACON	33.9%	7.0%	42.0%	5.1%
2015 BEACON	33.2%	8.0%	39.0%	5.2%
2014 BEACON	33.0%	7.0%	35.0%	3.0%
2013 BEACON	33.2%	7.1%	35.2%	2.7%
2012 BEACON	32.5%	7.0%	32.6%	2.2%
2011 BEACON	27.8%	N/A	32.9%	1.4%
2010 BEACON	20.5%	N/A	20.9%	0.0%

Fig. 3.2: BEACON's 9-year Evolution of Participants

It is important to acknowledge that these accomplishments did not happen by chance, there were very deliberate and strategic steps taken. These positive results required well-designed strategies, not just programming, which made all the difference. These strategies were driven by organizational goals and structures in the focused priority areas, and then subsequently integrated into sound business strategies and human resource policies with effective planning and implementation.

While we are very proud of achieving our quantitative goals, we even more proud of the BEACON participants embracing a welcoming diversity-centric environment as a core value. The proceeding programs described in this article will illustrate the strategic efforts made for our diverse body of participants. To measure our success in these non-quantitative areas, we had a process of continuous evaluation. These annual evaluations captured and guided the ongoing evolution of the diversity efforts over time. Our goal was not only to recruit numbers of students that exceeded national norms, but also to make their experiences in BEACON productive and fulfilling across all aspects of the multisite consortium.

3.3 People First, Then Programs

It is important to take a step back and recognize some key precursors to BEACON's success. It was the right people, with the right skills, in the right place, at the right time! In 2009, when crafting the original proposal, the following critical building blocks exemplified the right *people first* model. The initial proposal's diversity section was written by two diversity champions: Center Director Dr. Erik Goodman, who has an extensive history of mentoring, recruiting and partnering with diverse students, post-docs and faculty; and Executive Committee Member and Chair of the Diversity Committee of BEACON Dr. Percy Pierre, the founder and PI of MSU's Alfred P. Sloan Foundation's Program of Exemplary Mentoring Grant, where he has mentored fifty-five engineering doctoral graduates, of which forty-six were URMs.

NSF's former Acting Deputy Office Head, Office of Integrative Activities, Dr. Joan Frye advocated that Diversity Director position for the 2010 STCs be a full-time, dedicated position with no other job responsibilities. As a result, NSF's request for proposal (RFP) boilerplate language mandated this model. This allowed BEA-CON to establish the diversity director position reporting directly to the Center Director and focusing solely on diversity, equity and inclusion across the consortium. This was a unique situation since the prior, and now subsequent STC RFP's have combined the diversity and education director positions into one blended position.

When selecting a HBCU/HSI partner school, great effort was taken to select a school that had a rigorous curriculum and established research agenda around computation, bioinformatics and evolutionary biology and could participate fully across all aspects of BEACON. North Carolina A&T State University (NCA&T) met these criteria. Faculty relationships between MSU and NCA&T researchers already existed. NCA&T was brought into the consortium as a partner at the decision making table, with full privileges, allocated funding, and participation at all levels.

The above-stated strategy can be described as a People-First Model, where the foundational architects of BEACON included the progressive insight of the NSF Program Officer, having diversity champions in executive leadership positions, and selecting a HBCU/HIS partner that aligned with the consortium's research agenda and were at full partner status. Once the "right people" provided the baseline leadership, then sustainable strategies were put in place to institutionalize the overarching diversity goals:

1. Ensure diversity is represented as an inclusive and connecting thread through all aspects of BEACON; and,
2. Exceed national norms for diversity at all levels in the Center.

3.4 BEACON's Evolution Across Time

Figure 3.3 shows an evolutionary timeline of BEACON's strategies and accomplishments in diversity efforts and activities across nine years.

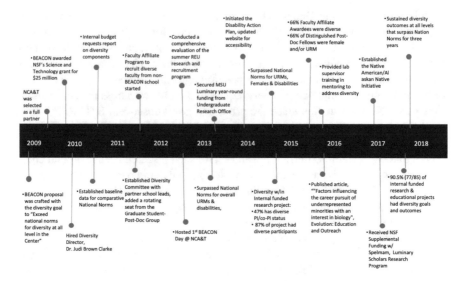

Fig. 3.3: BEACON's Evolutionary Timeline

It should be noted that BEACON's diversity and inclusion accomplishments tran-
scend the perfunctory goals of just increasing the number of diverse participants.
The outcomes demonstrated a deliberate culture that embodied efforts within our
research labs, classrooms, grant submissions, outreach efforts, formal mentoring,
fellowships, and direct student support funds. Diversity was truly a core value within
the consortium and discussed at all monthly Executive Committee meetings.

Like corporate and industrial settings, BEACON was cognizant that *state of the
art* diversity trainings were critical to demonstrated competency and sustained success. Participants were formally trained in various issues such as unconscious bias,
cultural awareness, improving engagement, sexual harassment, and fostering an inclusive climate, to name a few.

Annually, BEACON self-assessed and incorporated data-driven results and *lessons
learned* into its strategic efforts. This resulted in BEACON's positive diversity outcomes through its inclusive efforts of continuous conversations, inclusive recruiting,
supportive and adaptive environments, and the leveraging of strong partnerships and
programming to ensure consistency and sustainability.

3.5 BEACON's Diversity Programs in Action

One of the challenges of a Science and Technology Center is not having the privilege
of hiring tenue track or university funded post-docs. Majority of the time BEACON
collaborated with their respective colleges/departments on hires, however there were
times when the skills or recruitment windows weren't aligned with those partner-

ships. As a result, BEACON created the following two programs to hire faculty and postdocs.

3.5.1 Faculty Affiliates Program

BEACON created its Faculty Affiliates Program as a way to involve more diverse faculty who could also increase the diversity of trainees. BEACON recruited faculty from outside of its five institutions who were themselves from an under-represented group or who worked extensively with under-represented trainees. The awardees received a mini-grant of up to $100,000 over a two-year period to conduct research and explore the possibility of becoming a permanent member of BEACON. Two of the three awardees of this program were women, two were URMs, and one was a faculty member at a minority-serving institution. In 2016, BEACON concluded this successful program, as the two-year funding guarantee would otherwise extend into the period in which NSF funding began to be phased down.

3.5.2 Postdoctoral Fellowship Programs

BEACON offered the Distinguished Postdoctoral Fellowship, which funded exceptional postdoctoral scholars to pursue collaborative interdisciplinary research with multiple BEACON faculty members. The second was BEACON's Professional Development Postdoctoral Fellowship, which supported postdoctoral scholars with both their research agenda and the professional development of administrative and leadership skills around science education, diversity, broader impacts, science communication and diversity of careers. Collectively, these two programs supported six postdoctoral fellows, including four women and two URMs.

Additionally, BEACON thoughtfully created two initiatives to strategically recruit and support two extremely vulnerable populations. These efforts were key to BEACON's success in creating and expanding our safe and welcoming environment.

3.5.3 Disability Action Plan "ThisAbility"

The creators of this plan included undergraduate and graduate students, postdocs, leadership, faculty, and staff. They worked hard to establish best-practice activities for increasing adaptive learning and research/lab strategies; and as a result, the learning environments (e.g., classrooms, fieldwork and labs) were enriched and affirming to all participants. This effort rebranded BEACON's faculty, labs, and classrooms as inclusive destination for participants with disabilities.

An informational page was added to BEACON's internal website with opportunities to blog about the issues around disabilities. Participants across the consortium could submit content or ask for strategic assistance. The public website created an accessible repository of best practices and adaptive software in an effort to provide scalable models and tools.

3.5.4 Native American (NA)/Alaskan Native (AN) Initiative called (NAANI)

This initiative had the following goals – create collaborations in support of STEM education for NA/AN students, establish a research agenda aimed at closing knowledge gaps on barriers and best practices related to NA/AN participation in STEM, increase participation by NA/AN in setting the national research agenda on issues that impact Native lands, provide a forum to communicate educational opportunities for NA/AM American students, and to understand and respect indigenous Traditional Knowledge. One of BEACON's former Professional Development Fellows is an Alaskan Native and now a prestigious American Association for the Advancement of Science (AAAS) Science and Technology Policy Fellow. She continues to work collaboratively with BEACON's Diversity Director to drive these efforts through her established relationships and programming efforts throughout the United States.

These efforts have resulted in a national survey of NA/AN students and members from Indian Country, which has informed two publications addressing key cultural challenges in STEM education and traditional education and knowledge (TEK). This includes examining the lack of inclusion and understanding of NA/ANs when creating strategies to educate a new generation of Native scientists, and the resulting negative impacts and ethical issues raised when working in Indian country. BEACON also created a NAANI book using educational standards to teach science to NA/AN elementary and secondary students that cross walks TEK with western science (STEM) using traditional stories in the language of the respective NA/AN community. This book has been presented and well-received at conferences nationally.

3.6 Leveraged Diversity Efforts and Programming

BEACON continually looked for opportunities to leverage funding and secure grants to support diversity efforts and research opportunities for URMs. These funding opportunities included travel awards for students to attend professional conferences and present their research. Many of BEACON's URM students took full advantage of this funding opportunity and received valuable exposure to professional networks, discipline peers, and content experts. Additionally, BEACON provides year-round formal mentoring training for all graduate students and post-docs in an effort to develop personalized formal mentoring plan and/or diversity philosophy.

3.6.1 Year-Round and Summer Research Opportunities

BEACON blended, braided, and leveraged funding from multiple sources to create research opportunities at MSU, Idaho, UW, NCA&T, UTA, Kellogg Biological Station, and Friday Harbor Laboratories. We found that many URM students were underprepared for research experiences, and therefore created a two-tiered program by funding:

- Research Experiences for Undergraduates (REUs) for students with strong lab/research backgrounds that were able to conduct their own research project, and
- Undergraduate Research Apprenticeships (URAs) for students with little to no lab/research background that conduct a limited project on an existing research project.

Collectively, BEACON has served approximately 500 undergraduate students (freshman to 5th-year seniors) and spent over $2 million by leveraging funding from several sources. For every $1 dollar spent, BEACON has consistently obtained approximately $3 from external sources.

3.6.2 Luminary Scholars Program

This intensive undergraduate research program supports URM students each year to come to MSU for summer research opportunities, and then return to their minority-serving institutions (MSI) and continue their undergraduate research for up to 10-hours per week during the academic year. BEACON worked with Spelman College and awarded a subcontract to a Spelman faculty mentor who directly supervised the students working in their research lab. These funds were used to financially support the Spelman student research throughout the academic year.

Five undergraduates participated in research at Spelman during the school year and then attended MSU's 2017 & 2018 summer research program on the East Lansing, Michigan campus. BEACON provided support for the Spelman faculty to attend and present at BEACON's annual Congress. All five of the Spelman students have presented their summer research posters at different professional conferences throughout the subsequent fall; and three have submitted a paper for publication with their Spelman faculty mentor on their Luminary research projects.

3.7 Structural Reflections

BEACON was extremely fortunate to have nine years of stability within its executive management staff. This was accomplished by open communication, competitive pay, and continued support for professional development; e.g., teaching, grant solicitation, field/lab research, publications, and attendance at professional conferences.

The Diversity Committee consisted of lead representative from each partner school that monitored efforts at each site. There were points of feedback at every level for these "gatekeepers" including undergraduates, graduates, post-docs, faculty, and staff. This flow of information created an ability to identify unmet needs and gaps quickly in the classrooms, labs and field experiences. They put respective systems in place for active feedback and to create solutions for continuous improvement.

The Diversity Director worked "top down" with the Diversity Committee, faculty, and staff, and "bottom up" with undergraduates, graduate students and post-docs using active involvement and collective "decision-maker". Getting people "leaning in" versus "leaning back" on diversity establishes collective buy-in and ownership across the participants. Diversity and inclusion is more than just "the right thing to do", it is the foundation of high performance. Encompassing diverse perspectives enhances the robustness of discussions, depth of research, span of options, and effectiveness in problem solving. This is key to facilitating innovation and creativity in scientific discovery.

3.8 Lessons Learned

Our definition and efforts around diversity goes beyond NSF's protected populations. We see the intersectionality of all the social identities and experiences in our membership and provide supportive and a nurturing environment to thrive. Here are the key strategies to BEACON's Successes.

- Administrative People, Process & Policies

    - Attract and retain top talent
    - Leverage diverse ideas to develop innovative policies and inclusive ideas
    - Increase efficacy of the existing initiatives, programs, and communications
    - Use ongoing organizational evaluations/assessments as continuous improvement processes/tools

- Recruiting/Retention Networks, Mentoring, REUs

    - Use undergraduate research opportunities as recruiting tools/pathways for URM graduate students
    - Provide leadership training and professional development opportunities

    – Leverage relationships with other graduate/postdoc programs targeting diverse students; e.g. AGEP, SLOAN, AAAS, etc.

- Funding Grants and Support

    – Address costs associated with recruitment, training, and retention
    – Use annual internal budget request process to facilitate inter-institutional and interdisciplinary research and educational projects with diversity components
    – Leverage collaborative recognition (knowledge transfer relationships) to attract new participants and partners

- Outcomes Reporting and Innovation

    – Conduct benchmarking, research and analysis to drive excellence, efficiency and sustainability of efforts
    – Enhance scientific communication skills (e.g., blogs, white papers, policy briefs, publications) and tools to "tell our stories" effectively

Figure 3.4 outlines the administrative structure for BEACON's strategic core values.

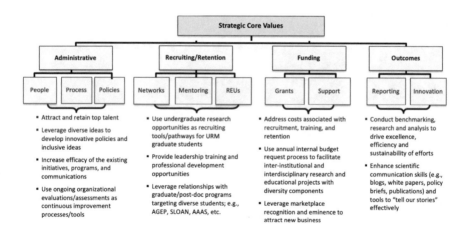

Fig. 3.4: Foundation of "High Performance"

3.9 Conclusion

As BEACON approaches its *sunsetting* 10th year, there is great pride in our sustained diversity outcomes and accomplishments. It was very important for us to

capture and share our core values, implementation strategies, administrative infrastructure, and programmatic outcomes so that subsequent multi-site partnership can
use our strategies and tools as a blueprint. We wish you the very best on your journey
to becoming a "high performance" research center, with full and inclusive access to
empowered diverse participants!

Chapter 4
Threading Together a Successful NSF-Funded Science and Technology Center: The Impact of Dr. Erik Goodman

Patricia L. Farrell-Cole and Marilyn J. Amey

Abstract Funders want people to work together, yet successful collaborations take more than wanting it to work out. Our ongoing organizational evaluation of the Bio/-computational Evolution in Action CONsortium (BEACON – a multi-institutional NSF-funded science and technology center) found that creating a successful multi-institutional research collaborative takes forethought and ongoing effort in order to thrive and achieve its mission and goals. A strategic and servant leader is necessary for a successful collaboration. The leader needs to be respected in the field; be able to work effectively with and motivate key stakeholders, faculty, and students; work collaboratively and with a coalition; recognize the importance of organizational evaluation; and be respected and trusted for his leadership abilities. This article articulates the key attributes of Dr. Erik Goodman's leadership of the BEACON Center.

> Life's natural tendency is to organize. Life organizes into greater levels of complexity to support more diversity and greater sustainability [12, p.3]

4.1 Introduction

The National Science Foundation (NSF) funnels millions of dollars to universities for empowering current and future researchers to advance innovation, discovery, and education beyond the boundaries of what is currently known [11]. Specifically, the NSF-funded Science and Technology Centers (STCs) provide a means to undertake innovative and potentially transformative research through significant investigations

Patricia L. Farrell-Cole, Organization Innovation and Learning Manager
Van Andel Institute, 333 Bostwick Ave. SE, Grand Rapids, Michigan 49503, USA e-mail: plfarrellmsu@gmail.com

Marilyn J. Amey, Chair, Department of Educational Administration
College of Education, Michigan State University, East Lansing, Michigan 48824, USA e-mail: amey@msu.edu

© Springer Nature Switzerland AG 2020   45
W. Banzhaf et al. (eds.), *Evolution in Action: Past, Present and Future*,
Genetic and Evolutionary Computation, https://doi.org/10.1007/978-3-030-39831-6_4

at the interfaces of disciplines, fresh methods within disciplines, and across institutions [9].

Bringing together researchers from diverse disciplines and multiple institutions with significant resources to answer the difficult questions is not simple; it is a "puzzle of complexity" [5]. Social scientists who study interdisciplinary research collaborations have found that even with significant resources, no guarantee exists that funding or agency accountability measures are sufficient to create, sustain, and realize success [2, 6, 7, 10]; this includes concerns for the success of STCs. Therefore, leadership is critical especially when there is a finite amount of time and resources. The organizational leadership must have a "strong awareness, strong consciousness of self. Who are we? What are we trying to do?" [12, p.357-358]. In this article, we offer our knowledge and insights on why the Bio/computational Evolution in Action CONsortium (BEACON) Center has been successful with Dr. Erik Goodman at the helm. Our insights are based on the data we have collected from our organizational effectiveness and impact evaluations over the past nine years, the literature on leadership, and our interactions with Erik Goodman and BEACON staff.

4.2 BEACON Center—The Beginning

The idea for the Bio/computational Evolution in Action CONsortium (BEACON) Center was originally conceived by Drs. Richard Lenski, Robert Pennock, and Charles Ofria from Michigan State University (MSU). They were known for their research in biological science, computer science, and philosophy. How they came together could be called serendipitous, but they also had mutual respect for one another and learned that they could accomplish more with their research if they formally organized and worked together towards common goals. As Margaret Wheatley stated,

> People organize to do more. And then life surprises us with new capacities. Until we organize, we can't know what we can accomplish together. We can't plan who we will be. Any time we join with others, newness and creativity pop up to astonish us. "The surprise within the surprise of every new discovery is that there is ever more to be discovered [12, p.69].

The three men realized that they needed a leader for the undertaking of a complex organization. They told us that Erik Goodman was the natural fit for the director of such an enterprise, not only because of his research stature, but for his leadership attributes. In talking with Robert Pennock on the philosophy behind the Center and Erik Goodman as leader, he stated that they had to make a case for the STC and Erik fit naturally because he was a pioneer of the work the STC was setting out to do.

> He was a student of John Holland, and Holland was one of the three independent originators of the idea of genetic algorithms or evolutionary computation. Really, Erik is in the first generation of this kind of work. He is someone who has a kind of natural position of recognition for being able to say, 'I've been in this from the beginning.' That was the first thing. He was in a position where he would be able to do it. He was the logical person to do it. *We just thought he would be really good at it.*

The original members worked through their networks and recruited significant researchers at other institutions who they believed would bring colleagues, resources, and similar vested interests to the Center. These included partners from the University of Idaho, North Carolina A&T State University, University of Texas at Austin and University of Washington. The group failed at their first NSF proposal, but the team including the core partners believed that their Center concept could have a "powerful legacy" upon science, education, and outreach. Their hopes were granted by NSF in 2010 [1, p.4].

> When we say yes to an organization, we awaken strong responses. [Catholic theologian] Steindl-Rast observes that if we agree to belong, we will feel called to new ways of living. We will notice what is required of us now that we are a community. We will act differently [12, p.69].

Forming a community was apparent at the start-up of the Center. BEACON leadership involved all key players, including staff, to participate in a 2-day planning meeting. Erik Goodman, working with a consulting company (hired by NSF), asked approximately 30 individuals to create the mission and vision statements, ground rules for which to work and be part of the community, and strategic goals. As an organizational evaluator, I had never witnessed an organization come together so easily, which illustrated that the founding researchers were tactical in forming the Center and they had the right individuals at the table. Everyone exhibited their belief and commitment to the Center.

4.3 BEACON Center Evaluation

NSF requires evaluation as a facet of the STC funding proposal, with a heavy emphasis on research, outreach, education, and diversity accountability measures. Fortunately, Erik Goodman realized that on-going evaluations of the organizational effectiveness and impact were important, too, due to the Center's complex, multifaceted purpose. Since its beginning, we have been engaged in the organizational effectiveness and impact evaluation of the BEACON Center. Our previous work studying leadership and organizational issues across institutional sectors allowed us this opportunity.

Why do we stay involved? The answer is easy. Erik Goodman and the team. To explain further, we have an exceptional opportunity to study leadership and collaborate with a leader who questions, guides, and utilizes our findings and recommendations to improve the organization, the Center. Again, he recognizes what we bring to the table for the Center and makes space for our work. He has also allowed us to gather data and reflect on foundational organizational elements necessary for a successful multi-institutional, multidisciplinary research collaboration.

In 2010, we conducted a baseline organizational evaluation. Through an online survey, we asked all BEACONites (approximately 120 participating faculty, postdocs, graduate students, staff) who the leader is of BEACON and greater than 95%

named Erik Goodman. Then, every other year we conducted a follow-up organizational effectiveness and impact evaluation asking BEACONites to provide insight into the Center's leadership. In 2017, 99% of the members rated leadership as satisfied/very satisfied on a five-point scale.

Based on our quantitative and qualitative studies, we attribute the success of the BEACON Center to the leadership of Erik Goodman and his colleagues, including the original members, who comprise the Executive Committee. Through our evaluation surveys and interviews, we have experienced Erik as a visionary with a deep personal conviction for academic collaboration; who has integrity and energy; who is a motivator and mentor; and who can work effectively with and relate to key stakeholders, faculty, and students. The next section will reflect more on Erik Gooman's leadership styles.

4.4 Strategic and Servant Leadership

Multi-institutional research partnerships are frequently formed by researchers who have similar interests and proven track records in obtaining grant funding, and who are respected in their research disciplines. Through our work, we learned that being excellent investigators does not ensure being successful leaders or good managers. A productive and respected multi-institutional collaboration requires a senior researcher as its leader, with specific attributes that cross several leadership attributes, characteristics, and models.

Based on our work, we categorize Erik as a strategic and servant leader. Greenleaf states that a servant-leader "is servant first. It begins with the natural feeling that one wants to serve, to serve first. Then conscious choice brings one to aspire to lead" [4, p.23]. Erik Goodman accepted the Director position not out of his own need to be at the pinnacle of hierarchy, but out of the needs for uniting the work of "biologists, computer scientists and engineers in joint study of natural and artificial evolutionary processes and in harnessing them to solve real-world problems" [1, p.4]. He is a servant-leader because he insures the welfare and needs of the Center and its members are taken care of.

In addition to seeing the needs of this nascent organization and taking on that shepherding role, Erik Gooman is simultaneously a strategic leader, insuring that senior administrators at Michigan State University understood the reasoning of the NSF Science and Technology Center proposal and its importance to the university as a role model for other multi-institutional collaborations and for national recognition.

As the leader of a five-university research collaboration, Erik Goodman had to instill a strong sense of purpose and energy into the organization. The US military requires its high-ranking officers to participate in professional development activities to learn about their responsibilities as strategic leaders including, "the skillful formulation, coordination, and application of ends (objectives), ways (courses of action), and means (supporting resources) to promote and defend the national interests across a joint, interagency, intergovernmental, and multinational environment"

[3, p.2]. In higher education, we may not think that research collaborations need the same kind of organizational leadership as the military and universities do; yet, we argue that they really DO need strategic leadership. As a strategic leader of a multi-institutional research partnership, Erik's responsibilities included spreading individualized expertise, knowledge, strategic goals, and values throughout the organization. This process involved clearly communicating a vision, influencing culture, shaping climate, and exemplifying appropriate behaviors in the areas of mentoring, teaching, and collaborating for research.

4.5 Organizational Development

The leader needs a coalition [8], or team of principal investigators and other administrators with integrity and leadership qualities to work together with the leader to ensure that the partnership succeeds. The processes, procedures, and actions developed and implemented by the leaders and team are critical to the success of the partnership. These individuals need to understand that they are building an organization. The leadership team must have trust within it; be able to think together; develop, monitor and be critical of processes; feel ownership of the initiative and responsibility for the outcomes. The members of the team need to have the different skills and abilities to move the project along.

In BEACON, the team set up an executive committee to include the original members and the four university partner representatives, along with other key faculty members who are thought leaders in specific areas (e.g., diversity, education, and industry partnerships). The executive committee then hired three administrative directors to oversee key aspects of the Center — management, education and outreach, and diversity — each person with content and process expertise. The committee also named one of the MSU co-PI researchers as Deputy Director. It is important to note that there has been no turnover in any of the administrative ranks despite being on a grant-funded position.

As a servant leader of a research collaboration, Erik Goodman has had his thumb on the pulse of the organization while knowing that an effective leadership team of competent people has been built to fill specific roles and to work together on the decisions and direction of the collective. He understood how to step away from control, to delegate and share leadership among others. For example, the Managing Director, who oversees the day-to-day activities of the organization, is an up-and-coming biologist conducting her own research, which she was afforded the time to do so that her academic identity remained intact while she provided administrative leadership to the STC. The roles and responsibilities of these two individuals have undertaken illustrate an example of succession planning devised by Erik Goodman and his team.

> Care taken by the servant first to make sure that other people's highest priority needs are being served. The best test, and most difficult to administer, is this: Do those served grow

as persons? Do they, while being served, become healthier, wiser, freer, more autonomous, more like themselves to become servants? [4, 22]

We would answer "yes" to Greenleaf's questions.

Moreover, by instilling a sense of purpose throughout the organization, the strategic leader of an interdisciplinary, multi-institutional research collaboration, along with the leadership coalition, must think ahead. The leader must not only maintain funding agency requirements, but also respond to the developmental needs of the organization, such as adapting quickly to establish different and better communication strategies, to providing and shifting support where needed, and to showcasing the contributions of others. We have learned that Erik Goodman is that leader. In addition, he and his leadership team are open to learning: seeking information and criticism, actively integrating feedback, effectively navigating organizational politics, and demonstrating the ability to deal with dissonance around ideas and processes. They have been strategic, but also stewards, in building a learning organization from its inception, while anticipating future needs and preparing partnership members for opportunities and challenges ahead without drawing undue attention to the leadership team.

Having world-renowned faculty, ideal procedures and policies, or the latest technology will not ensure a successful collaboration. A number of foundational organization development elements are necessary to result in a successful collaboration, including strategic planning, clear communication, organizational learning, and developing a sense of ownership for all participants. Having a strategic plan in place—and allowing it to continually evolve—provides a means to measure the partnership and its effectiveness. After developing its leadership and management goal statement for the strategic plan, Erik Goodman and his team created outcomes that demonstrated benchmarks for and achievement of these goals, being mindful that for many members, the Center was not their top responsibility. They kept in mind that they wanted to be an exemplary Center in NSF's eyes, as well as a beneficial intellectual partnership for its members. And BEACON has exceeded its goals, including being an exemplary Center.

4.6 Closing

We found that creating a successful multi-institutional research collaborative takes forethought and on-going effort to thrive and achieve its mission and goals. Unlike university leaders, those involved in research collaborations often focus only on the work of the project and assume the context of the work derives organically or without much thought. We can attest that bringing Erik Goodman on as the director was more than "he would be really good at it." He is a true strategic and servant leader. Greenleaf posits such leaders, "will freely respond only to individuals who are chosen as leaders because they are proven and trusted as servants" [4, p.12]. Erik Goodman has been successful because he knows how to bring people to the table who are committed to the cause, enthusiastic about the work, and willing to approach tasks with respect and authenticity. He provides the space for ideas and challenges to process and adapts in productive ways. As a servant leader, he pays attention to all aspects of the Center and is willing to go to bat for an individual or group before they need it; he is someone who shares the accolades, but will take the hits; he role models the core values and expects others to do the same; and he listens and takes feedback. The Center is making its mark on science because its members have been encouraged to see their value in all aspects of the Center's work and have been acknowledged for the leadership roles — large and small — they have taken in helping BEACON emerge a leading STC. Each has done their part; Erik Goodman has threaded them together into whole cloth.

References

1. BEACON: Beacon Center for the Study of Evolution in Action strategic and implementation plan. http://beacon.msu.edu. Accessed 15 July 2018 (2017)
2. Eddy, P.: *Partnerships and collaborations in higher education. ASHE Higher Education Report*. Jossey-Bass (2010)
3. Gerras, S.: *Strategic leadership primer*, 3rd edn. U.S. Army War College, Lexington, KY (2010)
4. Greenleaf, R.: Essentials of servant-leadership. In: L. Spears, M. Lawrence (eds.) Focus on Leadership Servant-Leadership for the Twenty-First Century, pp. 19–25. Wiley (2002)
5. Horton, D.: http://www.brainyquote.com/quotes/quotes/d/douglashor152740.html, accessed 15 July 2018 (late 1800)
6. Kezar, A.: *Redesigning for collaboration within higher education institutions: An exploration into the developmental process.* Research in Higher Education **46**, 831–860 (2005)
7. Klein, J.: *Creating interdisciplinary campus cultures: A model for strength and sustainability.* Jossey-Bass, San Francisco (2010)
8. Kotter, J.: *Leading change.* Harvard Business School Press, Boston (1996)
9. NSF: Science and Technology Centers (STCs): Integrative Partnerships. http://www.nsf.gov/od/oia/programs/stc/, accessed 20 April 2011
10. Rhoten, D.: *Interdisciplinary research: Trend or transition.* The Social Science Research Council Items and Issues **5**, 6–11 (2004)
11. Salter, A., Martin, B.: *The economic benefits of publicly funded research: A critical review.* Research Policy **30**, 509–532 (2001)
12. Wheatley, M., Kellner-Rogers, M.: *A Simpler Way.* Berrett-Koehler Publishers, San Francisco (1998)

Chapter 5
How BEACON Shaped my Research and Career Trajectory

Connie A. Rojas

Abstract This chapter reports on my graduate study experience, from having little background in computational biology and bioinformatics to a PhD studying host-microbe interactions and the microbiome. As a first-generation Latina college student, the financial, academic, and professional support provided by the BEACON Center for the Study of Evolution in Action was instrumental in allowing my journey and growth in the PhD program. BEACON is an excellent example of a Center that has programming initiatives that truly support students and do increase diversity and inclusion.

Attending Michigan State University for my PhD studies, working with Dr. Kay Holekamp on her long-term study on wild spotted hyenas, and joining the BEACON community have been the three most transformative decisions of my professional career. BEACON has not only provided me with financial, academic, and professional development resources but it has also strongly influenced the focus of my doctoral research, and shaped me into the scientist that I am today. I came into MSU uncertain about the research I wanted to pursue, and with little background in computational biology, but with the help of my adviser Dr. Holekamp, my guidance committee, and BEACON, I am now conducting exciting research on the hyena gut microbiome using next-generation sequencing technologies and robust bioinformatics pipelines.

Upon learning that I wanted to study host-associated microbial communities, Dr. Holekamp encouraged me to join BEACON. During my first semester at MSU, I enrolled in BEACONs Computational Science for Evolutionary Scientists course and learned to program in Python. During my second semester, I took the follow-up course on Multidisciplinary Methods for the Study of Evolution, and collaborated

Connie A. Rojas
BEACON Center for the Study of Evolution in Action and Department of Integrative Biology, Michigan State University, East Lansing, MI, USA e-mail: rojascon@msu.edu

© Springer Nature Switzerland AG 2020

W. Banzhaf et al. (eds.), *Evolution in Action: Past, Present and Future*,
Genetic and Evolutionary Computation, https://doi.org/10.1007/978-3-030-39831-6_5

with computer scientists on a project utilizing Markov brains to study animal foraging in relation to vegetation patch size; we are currently preparing that project for
journal submission. The two BEACON courses were instrumental to my pursuit of a
bioinformatics-heavy PhD: I realized that computational biology and programming
were fun and interesting, and that the BEACON center would serve as a very valuable resource throughout my tenure as a PhD student. That same summer, I worked
in the laboratory of fellow BEACON member Dr. Theis in the Department of Immunology and Microbiology at Wayne State University, where I extracted DNA
from hyena bacterial swabs, learned the basics of microbial sequence processing
using the popular software program mothur, and started analysis of data from my
very first dissertation research project on the microbial communities inhabiting various anatomical body sites in wild spotted hyenas. Thanks to BEACON and the
Institute for Cyber-Enabled Research (iCER), I was able to run my analyses rapidly,
using their high-power computing, handy tutorials and job submission scripts, and I
was also able to store my large raw and processed data files on their servers.

During my second year, when I was confronting the problem of how to fund
the expensive lab components of my Ph.D. research, Dr. Holekamp advised me to
apply for BEACON research funding. Along with Dr. Theis, Dr. Holekamp and
I drafted a proposal investigating the socio-ecological drivers of gut microbiome
structure and function in spotted hyenas, as well as its stability, its transmission
across generations, and its potential as a reservoir for antibiotic resistance. The proposal was funded, and those funds are now being used to cover my major lab expenses, including DNA extractions, 16S rRNA gene sequencing, shotgun metagenomics sequencing, and data analysis software and storage. I have analyzed the
data and written the manuscript for one of the two projects associated with this
grant and will soon submit the manuscript to a journal. My preliminary findings
were shared with the BEACON community during NSFs Annual Site Visit in De-
cember, 2017 and a BEACON seminar during the Spring semester of 2018. My most
recent findings will be communicated at the 2018 MSU Alliances for Graduate Education and the Professoriate (AGEP) Fall Conference, and at the annual meeting
of the Society of the Advancement of Hispanics/Chicanos and Native Americans
in Science (SACNAS) in October, 2018. To allow me to accomplish all this, BEA-
CON has awarded me fellowship funds, research funds, travel funds for national
academic conferences, and funds for academic and professional development workshops. During the Fall semester of 2018, I have been working on the bioinformatics
analysis portion of my second project, which assays gut microbiome function using
shotgun metagenomics. To accomplish this work, I am currently residing in Mexico
City, MX and working with Dr. Valeria Souza, Dr. Luis Eguiarte, and their graduate
students from the Institute of Ecology at the Universidad Nacional Autonoma de
Mexico (UNAM). The outcome of my visit will lead to significant progress on two
dissertation chapters, one submitted manuscript, a research seminar to the UNAM
community, substantial development of my bioinformatics skills and programming,
and many additional opportunities for professional development.

Apart from financial assistance, with which BEACON has been very generous,
BEACON has also provided me with a support system of brilliant, ambitious, and

like-minded scientists working in the areas of evolution and computational biology. BEACON has made me feel welcomed, supported, and most importantly, valued, which has motivated me to continue pursuit of my research, and not shy away from this work. When I gave my 2018 BEACON seminar, BEACONites asked questions and were genuinely interested in my work. When the NSF directors came for their annual visit and evaluation in 2017 BEACON invited me to a luncheon to speak about my experience with the center. When I have requested funds to cover an array of research-related expenses, BEACON has supported me without hesitation. Most recently, BEACON sponsored my trip to the 2018 STC Directors Meeting, entitled "Engaging Diverse Audiences: Broadening Participation through Science Communication," at the University of California, Berkeley. In addition, I was invited by fellow BEACONites Alexa Warwick, Maurine Neiman, and Eve Humphrey to join their symposium on host-microbe interactions at the 2018 SACNAS Conference in October, which I look forward to doing! These actions have demonstrated that BEACON values me and my work, is invested in my development as a scientist, and will support me until the very end of this arduous journey toward a PhD. As a member of an underrepresented group, that has made all of the difference.

For the first time in my graduate career, I am contemplating a bioinformatics-heavy post-doc, continuing my research in host-microbe interactions, and an academic position at an R1 institution, which was unimaginable to me as a first-year Ph D student. I sincerely believe I would not be pursuing this research, or have advanced to this stage in my graduate career, if I had attended an institution other than MSU for my PhD and thus if I did not have constant support and mentorship from Dr. Holekamp, Dr. Theis, and BEACON! I thank them all for shaping me into the scientist that I am today.

Chapter 6
The Man Behind the Leader

Constance James

Abstract This is an edited transcript of my speech on the occasion of Erik Goodman's 75th birthday celebration at BEACON Congress 2018.

Here is the story of Erik and how the man became the leader he is today. Erik was born in Palo Alto, CA at Stanford University Hospital where his dad was serving in the Army studying German to work as a translator. Interestingly enough, Erik's Dad studied genetics and received his Ph.D from the Dight Institute of Human Genetics at the University of Minnesota.

We can see where Erik got his love for science: His dad loved talking about his work at the dinner table and Erik loved to listen. Erik's family moved to East Lansing when he was a young boy for his father to become a faculty member at MSU in the Department of Zoology. During his time in East Lansing Erik attended East Lansing Jr High (this was formerly East Lansing High, then later to become Hannah Middle School, and now the Hannah Community Center).

The family lived in East Lansing for four years before moving to Winston-Salem, NC for his dad to become a faculty member at Wake Forest University where Erik attended R.J. Reynolds High School. Upon completion of high school Erik chose to come to MSU when he was offered an Alumni Distinguished Scholarship even though Duke had offered him a full ride. His years of living on the MSU campus and wanting to be back North won out. He then went on to finish his Master's Degree at MSU and his Ph.D at the University of Michigan in the Computers and Communications Sciences Program. In 1978 Erik married his wife, Cheryl, and they became parents to their son, David, in 1990.

Constance James
BEACON Center for the Study of Evolution in Action, Michigan State University, East Lansing, MI, USA e-mail: jamesc@egr.msu.edu

© Springer Nature Switzerland AG 2020

W. Banzhaf et al. (eds.), *Evolution in Action: Past, Present and Future*,
Genetic and Evolutionary Computation, https://doi.org/10.1007/978-3-030-39831-6_6

I met Erik in 2010 when he hired me as BEACON's first official hire. I was impressed with him immediately. I could see he was an intelligent and impressive gentleman. I respected him right away because he included me in the NSF required Strategic Planning Session. I realized he didn't have to include me, but he thought enough of my position to allow me to sit in to hear theses discussions, which were very valuable for me. It spoke of his respect for people, not just their status or knowledge. I've never asked him, but I believe he took a chance on hiring me. I don't think I met all of the qualifications he was looking for but I remember letting him know I was willing to work hard to learn whatever was needed to be successful in my position. Working for Erik has been a great learning experience ever since.

There are many attributes I could talk about but I will just touch on a few. Erik is kind and supportive. He cares about the success of BEACON and realizes it is directly related to the success of the students. He cares about people and tries to respect them. He goes over and beyond to help people. He continually goes out of his way to find funding for grad students in the BEACON Program. Whenever we have a PhD student in BEACON close to completing his or her program but for various reasons may be struggling to finish within time constraints, it is inevitable that Erik will call me into his office to brainstorm how to find funding.

I also remember an occasion when BEACON was in the midst of a very busy season preparing for a new funding cycle. Erik was depending on me to be there, but a dear friend of our family had been in a car accident and Erik so graciously allowed me to go be with the mother of our friend, even though it was a very bad time to leave. He has supported me many times in this way because he leads with his heart.

His caring was not just with me. When I spoke with someone else who worked with Erik — Joyce Foley — she shared a couple of memories she had of Erik that speaks to the kind of man he is. Joyce remembered the time when she was home because her son had passed on his chicken-pox to her, Erik sent her a get-well bouquet with a polka dotted ribbon (he's also funny). And then there was the time he was willing to work with an office temp longer than the standard 6 weeks, so Joyce could have an extended maternity leave with her daughter.

Erik is an intelligent and brilliant man. While I acknowledge that BEACON is full of 'super-minds', I'd like to acknowledge how I get to witness Erik's brilliance in everyday situations. I often marvel and joke about how he works through financial figures so easily. I can walk into one of our budget meetings with a calculator in hand, but before I can enter the numbers Erik has already figured it out in his head. Danielle, BEACON's Managing Director, and I will sometimes look at each other with a smile of wonder. Needless to say, I no longer bring my calculator to our financial meetings. Then there are his students whom he gives such attention. I know this might not seem much, but it amazes me how Erik can meet with the multitude of students he advises with varied research focuses. I've had the joy of sitting outside of his office to witness him sitting at the table in his office to walk these students through the specific steps of their research, with everything from breast cancer to spatial optimization and Greenhouse effects.

Erik is dedicated and he perseveres. This was very evident as he continued to work through a cancer diagnosis. Erik's battle with cancer was inspiring, the leader 'with heart' surprised us time after time as he would persevere through his sickness. He would sometimes come in looking ghostly. He would get so exhausted that he'd have to sometimes shut his door to lie down on his couch and take a quick power nap to make it through the day. He was determined, to not allow the cancer to get him down. Everyone in the office watched him win his battle with cancer, one day at a time, and what an inspiration he was. Remarkably, he remained kind and concerned for everyone else during this time. Not to mention that he was also helping his wife Cheryl win her fight against cancer as well. There would be weeks where he'd have treatment for himself and then also manage to get Cheryl to her treatments. It was an incredible thing to watch. They persevered through that difficult time together. I sent up many prayers for them during this time.

And there are many other things I could go on and on about Erik. As you can see, I have a pretty high affinity for my boss. I do often refer to Erik as my boss but recently I read an article that made a distinction between boss and leader. A boss is someone who is in charge of a workplace, but a leader is someone who inspires, guides, and influences his followers by setting an example. While they both can be the same, I know from experience not all bosses can be leaders, but I have been blessed to have a true leader with a 'heart of gold'.

Part II
Evolution of Genomes and Evolvability

Chapter 7
Limits to Predicting Evolution: Insights from a Long-Term Experiment with *Escherichia coli*

Jeffrey E. Barrick

Abstract Our inability to predict how populations of cells will evolve is a fundamental challenge to human health and biological engineering. In medicine, one would like to predict and thwart, or at least have time to adequately prepare for potentially catastrophic events such as the emergence of new pathogens, the spread of drug resistance, and the progression of chronic infections and cancers. In bioengineering, one would like to stop, or at least delay, evolution that inactivates a designed function, in order to make genetic engineering and synthetic biology more reliable and efficient. On a larger scale, one would also like to predict when the presence of recombinant DNA or a certain species might pose a threat to nature or civilization if it has the potential to evolve to become harmful.

Key words: limits of prediction, adaptive laboratory evolution, fitness trajectory, genetic basis of adaptation, chance and necessity, replaying evolution, nutrient cross-feeding, evolution of ecology, arresting evolution

7.1 Bohr's Hydrogen Atom for Evolution?

Many of these examples of biological systems in which we would like to predict evolution are complex: they involve interactions between heterogeneous populations of cells and our immune system or between cells and entire ecosystems. To make headway on this difficult problem, let's first examine what we can predict in a stripped-down evolving system that includes just a single type of relatively simple cell. Perhaps a good working analogy from chemistry is that we'd like to come up with a system and theory on the order of Bohr's model of the hydrogen atom [61].

Jeffrey E. Barrick
Department of Molecular Biosciences, The University of Texas at Austin, Austin, TX USA; and BEACON Center for the Study of Evolution in Action, Michigan State University, East Lansing, MI USA; e-mail: jbarrick@cm.utexas.edu

© Springer Nature Switzerland AG 2020   63
W. Banzhaf et al. (eds.), *Evolution in Action: Past, Present and Future*,
Genetic and Evolutionary Computation, https://doi.org/10.1007/978-3-030-39831-6_7

This approach is meant as a first step. We will know from the outset that the study system itself lacks some details that are relevant in real world situations (atoms with more electrons in our analogy). We will also only be able to predict some aspects of evolutionary dynamics and not others (the Rydberg formula but not the Zeeman effect in our analogy). Further development on both fronts (systems and models) will ultimately be needed to achieve completeness and accuracy, but this model is still an instructive waypoint on the path to more complex systems.

A population of *Escherichia coli* bacteria in an Erlenmeyer flask is our evolutionary hydrogen atom. There is little doubt that *E. coli* is the best-characterized free-living organism due to its long history as a model system for molecular biology [24]. In 1988, Richard Lenski and colleagues began propagating twelve *E. coli* populations in the laboratory under carefully controlled conditions to study evolution [30]. Every day, 1/100th of each culture is transferred to a new flask with fresh nutrients and the *E. coli* repopulate this flask through ∼6.6 generations of binary cell division. These *E. coli* reproduce asexually, with no means for genetic recombination between cells. Evolutionary dynamics in this environment are dominated by competition for a limiting supply of the sugar glucose. These twelve microcosms, each its own (simplified) world in a flask, began from an identical starting point and has now evolved in isolation for more than 60,000 cell generations.

For our discussion of predictability here, we will focus almost entirely on examining the Lenski long-term evolution experiment (LTEE) with *E. coli* and a handful of very similar setups. Before proceeding, it is important to acknowledge that there is a vibrant field of experimental evolution that has developed over the past few decades. Many similar, and equally iconic, experiments have been carried out with viruses, bacteria, yeasts, fruit flies, and mice [16]. Most of these other experiments have additional layers of complexity. They purposefully include sex, development, parasites, ecology, social behavior, and more. As a result, they have far richer dynamics than are possible in the "hydrogen atom" of the LTEE.

7.2 Levels of Prediction in Biology

What does it mean to predict evolution? There are different levels of detail at which this question can be approached. In the LTEE, the evolutionary process can be described numerically in terms of how well-adapted *E. coli* cells have become to their environment over time. Indeed, fitness is the only quantity that is directly visible to natural selection; its relevance to evolution is fundamental. One can measure fitness in the LTEE as the relative number of offspring that two different cells contribute to the final population when they compete against one another in the same flask [30]. As each population evolves over many growth cycles, more-fit *E. coli* carrying beneficial mutations arise and displace their ancestors and competitors. Thus, the fitness of cells in the population increases over time. Can we predict the future course of this upward fitness trajectory, given historical measurements covering previous generations?

Changes in fitness may reflect a wide array of possibilities in how an *E. coli* cell functions. All of these qualities are summarized as its phenotype. Phenotype encompasses the whole range of observable properties of a cell. Some changes in the LTEE are readily visible (under a microscope), such as a cell's size and shape [31, 45]. Some reflect a cell's simple behaviors: how quickly it starts growing when nutrients first become available each day, how rapidly it replicates while nutrients remain abundant, and to what extent it is able to survive once nutrients become scarce [42, 50]. Finally, a cell's properties at a molecular level are also part of its phenotype: for example, how many copies of an enzyme are in a cell or how one of its proteins responds to an environmental signal [10]. These various types of phenotypic characteristics are often interdependent. A change in the activity of a protein in a key metabolic or regulatory network may lead to more rapid growth of cells, which in turn may lead to a correlated increase in cell size. Can we predict which growth strategies will dominate and how the physiology of cells will evolve?

At the most basic level, phenotypic evolution is determined by changes in an *E. coli* cell's genome, i.e., its genotype. Many mutations that alter this DNA sequence will change the activity of a gene, leading to differences in cellular physiology, behavior, and ultimately fitness. In some cases, it might take multiple mutational steps to rewire cellular networks to achieve a new phenotype. In the LTEE, all cells started with the same genotype and evolution is driven by natural selection on de novo mutations that arise due to errors in copying and repairing DNA as cells replicate. Genetic variation often transiently builds up in the population due to competition between genotypes that are descended from the same cell but have since acquired different mutations. Then, genetic diversity typically declines when one genotype has a fitness that is so superior to others for long enough that it drives them extinct [2, 35]. One outcome is certain: over time, mutations will accumulate in the genomes of the successful lineages of cells. Can we predict which genes will mutate? Can we predict how rapidly mutations will accumulate over time in the *E. coli* genome?

7.3 Mutational Stochasticity Limits Predictability

One major challenge in predicting evolution, at any level, is that the appearance of new genotypes due to mutations is random. It turns out that in the LTEE we can mostly ignore this stochasticity when making certain types of predictions (much like Bohr could ignore the probabilistic parts of quantum theory in his model of the hydrogen atom). But, it's important to understand why this is the case for thinking more broadly about limits to predicting evolution.

A mutation anywhere in the entire *E. coli* genome occurs just once in every $\sim$1000 cell divisions [29]. Among these rare mutations, those that happen to be beneficial in a given environment are even rarer. On the order of 1% may give a fitness benefit in a laboratory flask [43]. Those mutations that are at the leading edge of being the most beneficial of these, the ones that will drive adaptation and have a reasonable chance of fixing in a population, are much rarer still. Fewer than one in a million mutations ($< 0.0001\%$) may really matter as far as determining the ultimate winners [17, 19, 64]. Thus, one expects variation in evolutionary outcomes due simply to the uncertainty as to whether a particular new mutation creating a novel genotype will appear in a population.

Counterbalancing the astronomical odds against an important beneficial mutation appearing in any given cell is the fact that each bacterial culture as a whole has many cells. Having a large population size makes evolution more predictable [55]. For example, an LTEE population grows up to approximately half a billion (5×10^8) cells each day, and about five million (5×10^6) of these will be transferred to the next flask [30]. According to the estimates here (multiplying cell number times mutation rate), there will be on the order of 5×10^5 mutations generated in each LTEE flask every day. The genome size of E. coli is only ~5 million (5×10^6) base pairs and it has ~5000 genes. Many mutations will have similar effects on a gene and a cell's phenotype, so most of the next moves in the evolutionary game will be sampled each day!

Of course, many of these mutations, even highly beneficial ones, are lost each day due to the 1/100 dilution bottleneck. Less obviously, competition between diverged lineages of E. coli that are accumulating different sets of beneficial mutations further limits the chances that any one mutation will matter in determining how the population as a whole evolves [15]. For example, even a very "good" mutation that is unlucky enough to arise in a "bad" genome—i.e., in the company of a cohort of prior mutations that is already lagging in fitness in the evolutionary race—is unlikely to win [28]. Still, in the aggregate, many different and similar mutations will appear and have a chance to win in every LTEE population. The influence of rare events on the overall outcome, in terms of fitness evolution, is thus relatively weak compared to what it would be in a smaller population.

Evolutionary unpredictability from mutational stochasticity is not, in and of itself, insurmountable or even unusual for a complex system. The LTEE and similar microbial evolution experiments offer two main ways of dealing with the resulting uncertainties. First, even very large bacterial populations require minimal feeding and upkeep, so many replicate populations can be started from precisely the same initial conditions to survey the array of possible outcomes. Thus, one very important aspect of the LTEE was that it consists of not one, but twelve separate populations that have all evolved in precisely the same environment. We can attribute variation between these cultures in how evolution progresses to chance sampling of initial mutations that may cascade into larger differences over time. Other evolution experiments have used even more populations to define the degree to which mutational stochasticity leads to different evolutionary solutions dominating in different populations [59].

As is also common in complex systems, the mutations and phenotypes that are successful in the longer term in a population also sometimes critically depend on the initial conditions (and subsequent events that, in effect, become new initial conditions for yet later dynamics). Here, mutations that appear and dominate at early generations set up a genetic background in each population in which further mutational steps can only appear in genomes with these initial mutations. Interactions between the fitness effects of mutations are common in the LTEE. For example, combining the first few mutational steps taken in one winning lineage in all possible orders showed that one mutation that was highly beneficial when it occurred would have been neutral if it had happened before a certain earlier mutation was already present [26]. The second way that microbial evolution experiments can deal with mutational stochasticity is that frozen samples of entire *E. coli* populations can be revived to "replay" the dynamics, with additional replication, at various critical points in this process or from genetically defined starting points to reveal these types of contingency in the evolutionary process [5, 64]. Imagine the implications for weather and earthquake prediction if we could watch for patterns in these phenomena time and time again on different earths that were nearly identical before they were set into motion!

7.4 Rates of Fitness and Genome Evolution Are Predictable

Once we recognize the inherent stochasticity of evolution and quantify the uncertainties in the exact outcomes by studying replicate and replay populations, we can now put our ability to predict the future trajectory of evolution to the test. We will begin at the two levels of prediction that we discussed: changes in the competitive fitness of *E. coli* cells and how quickly mutations accumulate in surviving genomes over time. The in-between predictions of phenotypic characteristics are harder. We'll revisit them in a later section.

Remarkably, most of the replicate populations of *E. coli* in the LTEE display very similar fitness trajectories over the course of the entire $> 60,000$ generation experiment [32, 63]. With a couple of exceptions (described in the next section), fitness measurements are surprisingly robust to possible artifacts that could complicate their interpretation, such as non-transitivity and frequency-dependence [13, 63]. Precise fitness trajectories were measured at various points in the history of the LTEE as it was in progress. At each point, modeling of the trajectory was done in an attempt to predict how fitness would continue to increase in the future. Originally, it was noted that a rectangular hyperbolic curve fit the data well at both 2,000 generations [30] and through 10,000 generations [31]. However, a hyperbolic curve assumes an asymptote, a maximal fitness ceiling that can never be broken. The asymptote calculated for the data through 2,000 generations was broken by 10,000 generations. The asymptote predicted with the data through 10,000 generations was also later surpassed, so it became clear that a hyperbolic model has a fatal shortcoming in its functional form and leads to poor long-term predictions.

More recently, the rate of fitness increase through 50,000 generations has been fit to an improved "diminishing returns" power law curve [63]. This model reflects an intuitive aspect of evolution toward a fitness optimum: it typically becomes harder and harder to improve fitness over time with each new beneficial mutation. Even though it neglects the details of interactions between the fitness effects of individual mutations that are known to be more complex [8, 26] and the detailed dynamics of competition between mutations in a population [2, 35], this model makes remarkably accurate predictions. Fitting the model to the LTEE fitness data from 0 to 5,000 generations can predict quite accurately the fitness trajectory out to at least 50,000 generations. Furthermore, extrapolating the model's predictions, even to exceptionally long time horizons (2.5 billion bacterial generations), still makes physiologically reasonable predictions (an evolved *E. coli* doubling time of ∼23 minutes) [63]. Thus, the average trajectory of fitness evolution into the future can be predicted surprisingly well for a typical population in the LTEE.

With the revolution in next-generation DNA sequencing [9, 3], it became possible to comprehensively reconstruct the dynamics with which new mutations accumulate over time in the genomes of *E. coli* sampled at different generations from the LTEE. In an initial study, the rate at which mutations accumulated was found to be indistinguishable from a linear model based on data from one population [2]. However, it was linear in a discontinuous fashion, with two different rates early and late in the experiment, and the linearity was for two different reasons within each time period. Before 20,000 generations, the near linear rate of increase appears to be due to the fitness advantage of the best new genotypes in the population over second-best "also-ran" genotypes remaining near constant, leading to their takeover and fixation in the population happening more or less regularly, except possibly for an initial burst when the first mutations are mainly competing versus the ancestral genotype.

After 20,000 generations, the linear rate at which mutations accumulated in genomes in this LTEE population steeply increased by a factor of more than 20-fold. This acceleration was due to *E. coli* with a much higher mutation rate—due to a defect in a gene that normally prevents the incorporation of damaged nucleotides into DNA—evolving and taking over this population. Similarly high mutation rates have evolved at some point in five of the other eleven LTEE populations [53, 57]. Mutations like these, which lead to hypermutation, can be successful in asexual microbial populations because genomes that contain them have a larger per-capita chance of sampling other beneficial mutations that enable them to be successful [58, 62]. The mutational trajectories in each of these hypermutator LTEE populations become constant in way that is typical of the clock-like genetic evolution of neutral models [27, 41]. That is, the accumulation of neutral or nearly neutral mutations in the hypermutators now so greatly outpaces the dynamics with which beneficial mutations appear and sweep through these populations, that it defines the overall rate.

Here, too, analyzing more data has led to a more refined model of the mutational dynamics in the LTEE [57]. Specifically, after sequencing a total of 264 genomes from all twelve populations, a model that mixes in a diminishing rate of beneficial mutations over time with the normally low clock-like rate of neutral mutations was found to fit the curve for the non-mutator populations better than the original linear model. The overall effect is a slight decrease in the rate of mutations that accumulate over time, though not as strong a deceleration as was found for the fitness trajectories. The form of this model, which combines two evolutionary processes, is supported by various genetic signatures of neutral versus adaptive evolution and by an ancillary evolution experiment that observed genome dynamics under conditions of relaxed selection to estimate mutation rates [3, 57]. In conclusion, the rates at which new mutations accumulate in genomes over time in the LTEE can also be predicted into the future, except for when hypermutators evolve. Even in these cases, after the switch to a different mutation rate, the trajectories settle on new, at least transiently predictable, rates of genome evolution. In time, however, these new rates may further change as a result of compensatory changes or reversions that readjust the mutation rate to lower values [57, 62].

7.5 ... Except When Ecology and Innovations Appear

The environment that the *E. coli* cells experience has been kept constant for the duration of the LTEE, although not in the sense that it remains entirely unchanging. Rather, it has seasonal regularity. Every 24 hours the same amount of glucose appears, cells "wake up" and grow until this nutrient is depleted over the course of several hours, and then they become quiescent and "sleep" until the next day. The low concentration of glucose, which limits the cell density to about 1/100th of what it would be in a typical microbial culture, leaves few opportunities for complex interactions between cells. In dense populations such interactions are often mediated by excreted metabolic byproducts released by some cells becoming a food source or toxin to other cells, but the LTEE cultures are so sparsely populated that the opportunities for these indirect effects are limited. Each cell is essentially competing for glucose on its own without any other influence from the rest of the cells in the mixture. This is one very important reason that the LTEE behaves so well as a "hydrogen atom" for evolution. Nonetheless, in two of the twelve LTEE populations more complicated ecology has crept back into the experiment. The populations with these deviations were ignored in considering the fitness trajectories in the previous section. These departures from the standard model of evolution in this environment would have been difficult to predict a priori.

In the first case, one population diversified into two types of cells by 6,000 generations. Each type accumulated a different set of mutations, and these types continued to co-exist for tens of thousands of generations [46, 51]. Apparently, one type was a superior competitor during growth on glucose while the other type was better at surviving and scavenging byproducts, and perhaps nutrients from dead cells, after most or all of the glucose had been exhausted [50]. These different behaviors led to a situation in which each type had an advantage over the other when it became rare, thus, stabilizing their long-term co-existence. Populations of *E. coli* in another evolution experiment, in an environment that includes a higher concentration of glucose mixed with acetate (which transiently accumulates as a byproduct of glucose metabolism in the LTEE), nearly always diversify into two specialist types: one that grows fastest on glucose and one that switches more rapidly to utilizing acetate in a second growth phase [20, 54]. Thus, unpredictability in this case in the LTEE likely stems from conditions (low nutrient concentrations) that are on the cusp of a domain in which a more complex ecology is a likely evolutionary outcome. The upshot is that stable coexistence of diverged *E. coli* types with different growth strategies requires a rare sequence of mutations and/or interactions within a population to develop in the LTEE.

In a second population, an even bigger deviation from the typical evolutionary dynamics occurred. This population evolved to utilize citrate, a second potential nutrient that has been present in every flask on every day of the LTEE at a much higher concentration than glucose. *E. coli* cannot normally metabolize citrate under these oxygen-rich conditions, and the citrate innovation was rare—it appeared only after $\sim$30,000 generations of evolution ($\sim$15 years) and has remained unique to this one population so far of the twelve [4, 5]. Citrate utilization enabled these newly evolved bacteria to colonize a vacant nutrient niche, essentially giving them a private and highly abundant resource. So, it was "big league" beneficial. The citrate innovation is rare in the LTEE because, in part, it is contingent upon a certain set of earlier mutations that alter *E. coli* metabolism in this particular population in a way that pre-adapts them, such that a subsequent mutation that turns on a pump that can exchange citrate into these cells is beneficial, rather than neutral or even deleterious [33, 47, 48].

After efficient citrate utilization arose in this population, a complex ecology also evolved, one related to how the citrate users export other carbon compounds into their environment in exchange for this nutrient. These efflux byproducts accumulate and can be utilized in turn by other genotypes that evolved to specialize on them [60]. The citrate-eating subpopulation also evolved a high mutation rate shortly after it arose [4]. The dynamics of adaptation had effectively been reset, such that it was back at the beginning of an increasing fitness trajectory for optimizing growth on citrate instead of deep into the diminishing tail of adaptation to glucose.

7.6 Predicting the Genetic Basis of Adaptation Is Difficult

Now, let's consider where our predictions start to fail. We can also look not just at how many mutations there are in an *E. coli* genome, but at what genes they affect. In general, there is a lot of parallelism (i.e., convergent molecular evolution) in what genes acquire beneficial mutations among the twelve populations of the LTEE [2, 18, 57]. Though the exact changes to the DNA sequences of those genes are rarely the same, it is likely that the mutations in the same gene from different populations have the same, or at least very similar, effects on molecular and cellular phenotypes. There are even predictable dependencies within some of these genes, such that an earlier mutation in a certain gene can change the probabilities of further mutations accumulating in other genes [18, 64]. So, we might be able to build up a model of the expected probabilities of different mutational paths impacting certain genes and cellular pathways in this system, but currently we can only do this on a post hoc basis by looking at enough replicate and replay experiments.

What about predicting what mutations in which genes will evolve beforehand? For this aim, we would need a mechanistic model that connects mutations to fitness through our previously neglected level of cellular phenotypes. Advanced models of metabolic and gene regulatory networks exist for bacterial cells [25, 40]. In certain cases, one can indeed identify "traffic jams" in metabolic "highways" that are alleviated by "road-widening" mutations in specific enzymes during the course of adaptive laboratory evolution experiments [22, 39]. One can also construct cells with specific changes in gene functions (knockouts, especially) that are predicted by these whole-cell models and show that they are often beneficial to fitness. However, even the complexity of "just" an *E. coli* cell makes it rare that we can predict a priori which genes will harbor the best beneficial mutations that will drive adaptation during an evolution experiment.

Why? Often, it is mutations in global regulatory processes instead of single enzymes that are the most impactful [9, 36, 44]. The effects of these mutations are difficult to predict because they change many of the links in a cellular network at once. Even though our systems biology knowledge of the *E. coli* strain used in the LTEE continues to improve [6, 7, 21], we don't have anywhere near a complete accounting of these subtle effects. Examples of global regulators include RNA polymerase, nucleoid-like DNA-binding proteins, and enzymes that wind and unwind DNA. A change in any one of these proteins may up- or down-regulate the expression of hundreds or thousands of other genes. Some of the changes in the levels of these affected proteins may be beneficial, neutral, or even deleterious on their own. Since the net effect is the sum of many of these weak interactions, regulatory genes may be particularly likely targets for adaptive mutations. It is currently difficult to predict when a specific global regulator will be an effective target for selection. One remaining question is: will mutations in global regulators remain just as common or diminish in importance as LTEE populations reach higher fitness? On the one hand, one might expect evolution to give way to mutations that more precisely adjust the activities of individual genes. On the other hand, mutations in global regulators may still be just as beneficial to fitness overall as those one-gene mutations because they can fine-tune many targets simultaneously.

There are further complications in predicting what mutations will occur in an evolution experiment. There are different mutational target sizes for different types of changes in gene function, and mutations leading to different types of genome variation may arise at vastly different rates [52]. For example, many single-base edits to a gene will inactivate it or reduce its activity, whereas very few may result in greater activity or novel functions. Thus, mutations that inactivate genes will appear more often than other types of mutations, and they may have a short-term advantage for this reason in the evolutionary race, even though they may not be optimal in the long run. There are also certain genomic regions and DNA sequences that are especially prone to mutations compared to others. Since they mutate more rapidly, they may reliably contribute to evolution above their "weight class" (i.e., even when they are not highly beneficial). For example, deletions of the ribose-utilization operon due to a mutational hotspot occur at a high rate in the LTEE and rapidly fix in successful genomes, even though this mutation is only slightly beneficial [11]. Though some types of mutational hotspots can be identified computationally [23], we are far from being able to comprehensively predict these types of sites accurately enough to take them into account and predict which mutations are likely to contribute to evolution within a gene or genome.

7.7 Predicting, Preparing for, and Preventing Evolution

In the end, what can we say about Lenski's flasks as Bohr's hydrogen atom for evolution? First, it truly is possible to predict (or forecast) certain aspects of the evolutionary trajectory fairly well at a non-mechanistic, non-molecular level. Changes in fitness and in the numbers of mutations that accumulate over time follow reliable trajectories. If we have data for a short initial period and fit a few parameters to get the shape of the curve, these predictions hold well into the future. This approach works even though our models of the underlying evolutionary dynamics are relatively coarse-grained (e.g., we have one generic term for the diminishing fitness returns of each new mutation and don't need to know any details of genetic diversity in the population).

These types of predictions for the LTEE fail, however, when evolution finds a way to break the rules of the game. In these instances, some of our first-order assumptions are violated, e.g., by the appearance of hyper-mutators, ecological interactions, or metabolic innovations. The outcome is outside of the realm in which evolution is gradually optimizing fitness on glucose as a limiting nutrient with the ancestral mutation rate. Unfortunately, it is just these types of unanticipated events that we are most concerned about when predicting evolution in the real world. It's not that we are worried that evolution will proceed a bit faster or slower than we planned, it's that an out-of-scope danger will arise: a species will mutate to become invasive in a new environment or an especially virulent pathogen will emerge. It's the risk of these rare events and chance encounters that we struggle to define and mitigate; they are the "hopeful monsters" that keep us up at night.

In closing, let's consider these rare and potentially destructive evolutionary events more directly. In terms of prediction, they are more akin to earthquakes than to electrons, and a shift in analogy at this point is helpful for changing our perspective. It may be near impossible to forecast the exact moment at which an earthquake will occur, but this limitation does not mean that we are helpless against them. Not all of the danger from an earthquake is immediate. In particular, earthquakes may trigger tsunamis that travel thousands of miles across the ocean before wreaking havoc on distant communities. While the propagation of these massive tidal waves also cannot be precisely predicted, providing early warning to at-risk locations by triggering an alarm immediately after an earthquake will save lives.

So, what if we are asking the wrong (or at least a harder than necessary) question of biology in trying to predict outcomes in a complex system of evolving cells? In many cases, such as the evolution of drug resistance, it may be almost as useful as prediction if we could just receive an early warning about what resistance mutations have appeared in a population while they are still very rare. Then we could prepare for these contingencies, by switching the drugs we use before the troublesome variants ever begin to matter. The tsunami warning that we may be after could be deeply sequencing an evolving bacterial or tumor population to profile its rare genetic diversity [37]. Early warning may be enough in these cases.

Another strategy for mitigating the destruction of earthquakes is disaster preparedness. Although we cannot predict exactly when and where an earthquake will happen, we know that certain areas of the world are far more prone to seismic activity than others. We have mapped out the fault systems of the world based on a long history of recording earthquakes and found they delineate the Ring of Fire. As more and more genomic information becomes available from sequencing tumors and microbial populations involved in chronic infections, such as those in cystic fibrosis [34, 38], we are building up similar genetic maps of how problem cells are likely to evolve over time. The resolution of these maps could potentially be improved by augmenting the outcomes of "natural experiments" or unplanned infections with laboratory experiments like the LTEE, for example with the flu or bacterial pathogens. By repeating and recording many outcomes of evolution in the lab, we would theoretically know what problems to expect and could tailor treatments to undercut those evolutionary paths. This strategy is not unlike enforcing robust building codes in earthquake-prone areas in order to mitigate dangerous forces that are unpredictable and exceptionally rare on a daily basis but essentially certain to occur in the long run.

Finally, what if we invert the prediction problem and seek to purposefully re-engineer an organism's genome to make its evolution more predictable? One could perhaps unravel the tangled network of weak links in cellular networks that has been the product of mindless evolution in a much more complex environment and replace it with a simpler gene expression scheme [56]. Given that entire microbial genomes are now being constructed or mutated on a large-scale [14], this level of re-design is becoming a possibility. Just as mutation rates can evolve to be higher than normal, it is also possible to engineer and evolve "antimutator" organisms that have lower than natural mutation rates [12, 49]. The limits of this approach have not yet been fully explored, and it might also be possible to block some cellular process that lead to stress-induced mutagenesis with drugs to lower cellular mutation rates [1]. With these final interventions, evolution of any kind would be expected to be less of a danger, simply because harmful genetic variants would be less likely to ever appear in the first place.

Acknowledgements The inspiration for this chapter came from attending a Santa Fe Institute workshop on "Limits to Prediction". I thank David Krakauer and Joshua Grochow for organizing this event and my fellow participants for stimulating talks about topics ranging from weather and earthquakes to viruses and the evolution of complexity. I would also like to acknowledge many colleagues and collaborators from the National Science Foundation BEACON Center for the Study of Evolution in Action, particularly Erik Goodman for his leadership in creating this one-of-a-kind environment and Richard Lenski for helpful comments on a draft of this chapter. Research in my lab related to the predictability of evolution has been sponsored by BEACON (DBI-0939454), an NSF CAREER Award (CBET-1554179), the National Institutes of Health (GM087550), the

Army Research Office (W911NF-12-1-0390), the Defense Advanced Research Projects Agency (HR0011-15-C0095), and the Welch Foundation (F-1780).

References

1. Al Mamun, A., et al.: *Identity and function of a large gene network underlying mutagenic repair of DNA breaks.* Science **338**, 1344–1348 (2012)
2. Barrick, J.E., Lenski, R.E.: *Genome-wide mutational diversity in an evolving population of Escherichia coli.* Cold Spring Harb Symp Quant Biol **74**, 119–129 (2009)
3. Barrick, J.E., Lenski, R.E.: *Genome dynamics during experimental evolution.* Nature Rev Gen **14**, 827–839 (2013)
4. Blount, Z.D., Barrick, J.E., Davidson, C.J., Lenski, R.E.: *Genomic analysis of a key innovation in an experimental Escherichia coli population.* Nature **489**, 513–518 (2012)
5. Blount, Z.D., Borland, C.Z., Lenski, R.E.: *Historical contingency and the evolution of a key innovation in an experimental population of Escherichia coli.* Proc Natl Acad Sci USA **105**(23), 7899–7906 (2008)
6. Brown, C.W., Sridhara, V., Boutz, D.R., Person, M.D., Marcotte, E.M., Barrick, J.E., Wilke, C.O.: *Large-scale analysis of post-translational modifications in E. coli under glucose-limiting conditions.* BMC Genomics **18**, 301 (2017)
7. Caglar, M.U., Houser, J.R., Barnhart, C.S., Boutz, D.R., Carroll, S.M., Dasgupta, A., Lenoir, W.F., Smith, B.L., Sridhara, V., Sydykova, D.K., Vander Wood, D., Marx, C.J., Marcotte, E.M., Barrick, J.E., Wilke, C.O.: *The E. coli molecular phenotype under different growth conditions.* Sci. Rep. **7**, 45303 (2017)
8. Chou, H.H., Chiu, H.C., Delaney, N.F., Segrè, D., Marx, C.J.: *Diminishing returns epistasis among beneficial mutations decelerates adaptation.* Science **332**, 1190–1192 (2011)
9. Conrad, T.M., Lewis, N.E., Palsson, B.Ø.: *Microbial laboratory evolution in the era of genome-scale science.* Mol Syst Biol **7**(1), 509 (2011)
10. Cooper, T.F., Rozen, D.E., Lenski, R.E.: *Parallel changes in gene expression after 20,000 generations of evolution in Escherichia coli.* Proc Natl Acad Sci USA **100**, 1072–1077 (2003)
11. Cooper, V.S., Schneider, D., Blot, M., Lenski, R.E.: *Mechanisms causing rapid and parallel losses of ribose catabolism in evolving populations of Escherichia coli B.* J Bacteriol. **183**, 2834–2841 (2001)
12. Deatherage, D.E., Leon, D., Rodriguez, Á., et al.: *Directed evolution of Escherichia coli with lower-than-natural plasmid mutation rates.* Nucleic Acids Res **46**, 9236–9250 (2018)
13. Elena, S.F., Lenski, R.E.: *Long-term experimental evolution in Escherichia coli. VII. Mechanisms maintaining genetic variability within populations.* Evolution **51**, 1058–1067 (1997)
14. Esvelt, K.M., Wang, H.H.: *Genome-scale engineering for systems and synthetic biology.* Mol Syst Biol **9**, 641 (2013)
15. Fogle, C.A., Nagle, J.L., Desai, M.M.: *Clonal interference, multiple mutations and adaptation in large asexual populations.* Genetics **180**, 2163–2173 (2008)
16. Garland, T., Rose, M. (eds.): *Experimental Evolution: Concepts, Methods, and Applications of Selection Experiments.* University of California Press (2009)
17. Gerrish, P.J., Lenski, R.E.: *The fate of competing beneficial mutations in an asexual population.* Genetica **102–103**, 127–144 (1998)
18. Good, B.H., McDonald, M.J., Barrick, J.E., Lenski, R.E., Desai, M.M.: *The dynamics of molecular evolution over 60,000 generations.* Nature **551**, 45–50 (2017)
19. Hegreness, M., Shoresh, N., Hartl, D., Kishony, R.: *An equivalence principle for the incorporation of favorable mutations in asexual populations.* Science **311**, 1615–1617 (2006)
20. Herron, M.D., Doebeli, M.: *Parallel evolutionary dynamics of adaptive diversification in Escherichia coli.* PLoS Biol **11**, e1001490 (2013)

21. Houser, J.R., et al.: *Controlled measurement and comparative analysis of cellular components in E. coli reveals broad regulatory changes in response to glucose starvation.* PLoS Comput Biol **11**, e1004400 (2015)
22. Ibarra, R.U., Edwards, J.S., Palsson, B.O.: *Escherichia coli K-12 undergoes adaptive evolution to achieve in silico predicted optimal growth.* Nature **420**, 186–189 (2002)
23. Jack, B.R., et al.: *Predicting the genetic stability of engineered DNA sequences with the EFM calculator.* ACS Synth Biol **4**, 939–943 (2015)
24. Judson, H.: *The Eighth Day of Creation.* Cold Spring Harbor Press (1996)
25. Karr, J.R., et al.: *A whole-cell computational model predicts phenotype from genotype.* Cell **150**, 389–401 (2012)
26. Khan, A.I., Dinh, D.M., Schneider, D., Lenski, R.E., Cooper, T.F.: *Negative epistasis between beneficial mutations in an evolving bacterial population.* Science **332**, 1193–1196 (2011)
27. Kimura, M.: *The Neutral Theory of Molecular Evolution.* Cambridge University Press (1985)
28. Lang, G.I., et al.: *Pervasive genetic hitchhiking and clonal interference in forty evolving yeast populations.* Nature **500**, 571–574 (2013)
29. Lee, H., Popodi, E., Tang, H., Foster, P.L.: *Rate and molecular spectrum of spontaneous mutations in the bacterium Escherichia coli as determined by whole-genome sequencing.* Proc Natl Acad Sci USA **109**, E2774–E2783 (2012)
30. Lenski, R.E., Rose, M.R., Simpson, S.C., Tadler, S.C.: *Long-term experimental evolution in Escherichia coli. I. adaptation and divergence during 2,000 generations.* Am Nat **138**(6), 1315–1341 (1991)
31. Lenski, R.E., Travisano, M.: *Dynamics of adaptation and diversification: a 10,000-generation experiment with bacterial populations.* Proc Natl Acad Sci USA **91**(15), 6808–6814 (1994)
32. Lenski, R.E., Wiser, M.J., Ribeck, N., Blount, Z.D., Nahum, J.R., Morris, J.J., Zaman, L., Turner, C.B., Wade, B.D., Maddamsetti, R., Burmeister, A.R., Baird, E.J., Bundy, J., Grant, N.A., Card, K.J., Rowles, M., Weatherspoon, K., Papoulis, S.E., Sullivan, R., Clark, C., Mulka, J.S., Hajela, N.: *Sustained fitness gains and variability in fitness trajectories in the long-term evolution experiment with Escherichia coli.* Proc R Soc B **282**(1821), 20152292 (2015)
33. Leon, D., D'Alton, S., Quandt, E.M., Barrick, J.E.: *Innovation in an E. coli evolution experiment is contingent on maintaining adaptive potential until competition subsides.* PLoS Genet **14**(4), e1007348 (2018)
34. Lieberman, T.D., et al.: *Parallel bacterial evolution within multiple patients identifies candidate pathogenicity genes.* Nat Genet **43**, 1275–1280 (2011)
35. Maddamsetti, R., Lenski, R.E., Barrick, J.E.: *Adaptation, clonal interference, and frequency-dependent interactions in a long-term evolution experiment with Escherichia coli.* Genetics **200**, 619–631 (2015)
36. Maharjan, R., Seeto, S., Notley-McRobb, L., Ferenci, T.: *Clonal adaptive radiation in a constant environment.* Science **313**, 514–517 (2006)
37. Maley, C.C., et al.: *Genetic clonal diversity predicts progression to esophageal adenocarcinoma.* Nat Genet **38**, 468–473 (2006)
38. Marvig, R.L., Sommer, L.M., Molin, S., Johansen, H.K.: *Convergent evolution and adaptation of pseudomonas aeruginosa within patients with cystic fibrosis.* Nat Genet **47**, 57–64 (2015)
39. McCloskey, D., Xu, S., Sandberg, T.E., et al.: *Evolution of gene knockout strains of E. coli reveal regulatory architectures governed by metabolism.* Nat Commun **9**, 3796 (2018)
40. Monk, J.M., et al.: *Genome-scale metabolic reconstructions of multiple Escherichia coli strains highlight strain-specific adaptations to nutritional environments.* Proc Natl Acad Sci USA **110**, 20338–20343 (2013)
41. Ohta, T.: *The nearly neutral theory of molecular evolution.* Annu Rev Ecol Syst **23**, 263–286 (1992)
42. Oxman, E., Alon, U., Dekel, E.: *Defined order of evolutionary adaptations: experimental evidence.* Evolution **62**, 1547–1554 (2008)
43. Perfeito, L., Fernandes, L., Mota, C., Gordo, I.: *Adaptive mutations in bacteria: high rate and small effects.* Science **317**, 813–815 (2007)
44. Philippe, N., Crozat, E., Lenski, R.E., Schneider, D.: *Evolution of global regulatory networks during a long-term experiment with Escherichia coli.* BioEssays **29**, 846–860 (2007)

45. Philippe, N., Pelosi, L., Lenski, R.E., Schneider, D.: *Evolution of penicillin-binding protein 2 concentration and cell shape during a long-term experiment with Escherichia coli.* J Bacteriol **191**, 909–921 (2009)

46. Plucain, J., Hindré, T., Le Gac, M., Tenaillon, O., Cruveiller, S., Médigue, C., Leiby, N., Harcombe, W.R., Marx, C.J., Lenski, R.E., Schneider, D.: *Epistasis and allele specificity in the emergence of a stable polymorphism in Escherichia coli.* Science **343**(6177), 1366–1369 (2014)

47. Quandt, E.M., Deatherage, D.E., Ellington, A.D., Georgiou, G., Barrick, J.E.: *Recursive genomewide recombination and sequencing reveals a key refinement step in the evolution of a metabolic innovation in Escherichia coli.* Proc Natl Acad Sci USA **111**, 2217–2222 (2014)

48. Quandt, E.M., Gollihar, J., Blount, Z.D., Ellington, A.D., Georgiou, G., Barrick, J.E.: *Fine-tuning citrate synthase flux potentiates and refines metabolic innovation in the Lenski evolution experiment.* eLife **4**, e09696 (2015)

49. Renda, B.A., Hammerling, M.J., Barrick, J.E.: *Engineering reduced evolutionary potential for synthetic biology.* Mol Biosyst **10**, 1668–1678 (2014)

50. Rozen, D.E., Philippe, N., Arjan De Visser, J., Lenski, R., Schneider, D.: *Death and cannibalism in a seasonal environment facilitate bacterial coexistence.* Ecol Lett **12**, 34–44 (2009)

51. Rozen, D.E., Schneider, D., Lenski, R.E.: *Long-term experimental evolution in Escherichia coli. XIII. phylogenetic history of a balanced polymorphism.* J Mol Evol **61**(2), 171–180 (2005)

52. Ryall, B., Eydallin, G., Ferenci, T.: *Culture history and population heterogeneity as determinants of bacterial adaptation: the adaptomics of a single environmental transition.* Microbiol Mol Biol Rev **76**, 597–625 (2012)

53. Sniegowski, P.D., Gerrish, P.J., Lenski, R.E.: *Evolution of high mutation rates in experimental populations of E. coli.* Nature **387**, 703–705 (1997)

54. Spencer, C.C., Tyerman, J., Bertrand, M., Doebeli, M.: *Adaptation increases the likelihood of diversification in an experimental bacterial lineage.* Proc Natl Acad Sci USA **105**, 1585–1589 (2008)

55. Szendro, I.G., Franke, J., de Visser, J., Krug, J.: *Predictability of evolution depends nonmonotonically on population size.* Proc Natl Acad Sci USA **110**, 571–576 (2012)

56. Temme, K., Zhao, D., Voigt, C.A.: *Refactoring the nitrogen fixation gene cluster from Klebsiella oxytoca.* Proc Natl Acad Sci USA **109**, 7085–7090 (2012)

57. Tenaillon, O., Barrick, J.E., Ribeck, N., Deatherage, D.E., Blanchard, J.L., Dasgupta, A., Wu, G.C., Wielgoss, S., Cruveiller, S., Médigue, C., Schneider, D., Lenski, R.E.: *Tempo and mode of genome evolution in a 50,000-generation experiment.* Nature **536**, 165–170 (2016)

58. Tenaillon, O., Taddei, F., Radman, M., Matic, I.: *Second-order selection in bacterial evolution: selection acting on mutation and recombination rates in the course of adaptation.* Res Microbiol **152**, 11–16 (2001)

59. Tenaillon, O., et al.: *The molecular diversity of adaptive convergence.* Science **335**, 457–461 (2012)

60. Turner, C.B., Blount, Z.D., Lenski, R.E.: *Replaying evolution to test the cause of extinction of one ecotype in an experimentally evolved population.* PLoS ONE **10**(11), e0142050 (2015)

61. Turner, J.E.: *Chapter 2: Atomic structure and atomic radiation.* In: Atoms, Radiation, and Radiation Protection, pp. 15–53. Wiley-VCH (2007)

62. Wielgoss, S., et al.: *Mutation rate dynamics in a bacterial population reflect tension between adaptation and genetic load.* Proc Natl Acad Sci USA **110**, 222–227 (2013)

63. Wiser, M., Ribeck, N., Lenski, R.: *Long-term dynamics of adaptation in asexual populations.* Science **342**, 1364–1367 (2013)

64. Woods, R.J., et al.: *Second-order selection for evolvability in a large Escherichia coli population.* Science **331**, 1433–1436 (2011)

Chapter 8

A Test of the Repeatability of Measurements of Relative Fitness in the Long-Term Evolution Experiment with *Escherichia coli*

Jeffrey E. Barrick, Daniel E. Deatherage and Richard E. Lenski

Abstract Experimental studies of evolution using microbes have a long tradition, and these studies have increased greatly in number and scope in recent decades. Most such experiments have been short in duration, typically running for weeks or months. A venerable exception, the long-term evolution experiment (LTEE) with *Escherichia coli* has continued for 30 years and 70,000 bacterial generations. The LTEE has become one of the cornerstones of the field of experimental evolution, in general, and the BEACON Center for the Study of Evolution in Action, in particular. Science laboratories and experiments usually have finite lifespans, but we hope that the LTEE can continue far into the future. There are practical issues associated with maintaining such a long-term experiment. One issue, which we address here, is whether key measurements made at one time and place are reproducible, within reasonable limits, at other times and places. This issue comes to the forefront when one considers moving an experiment like the LTEE from one lab to another. To that end, the Barrick lab at The University of Texas at Austin, measured the fitness values of samples from the 12 LTEE populations at 2,000, 10,000, and 50,000 generations and compared the new data to data previously obtained at Michigan State University. On balance, the datasets agree very well. More generally, this finding shows the value of simplicity in experimental design, such as using a chemically defined growth medium and appropriately storing samples from microbiological experiments. Even

Jeffrey E. Barrick
Department of Molecular Biosciences, The University of Texas at Austin, Austin, TX USA; and
BEACON Center for the Study of Evolution in Action, Michigan State University, East Lansing,
MI USA; e-mail: jbarrick@cm.utexas.edu

Daniel E. Deatherage
Department of Molecular Biosciences, The University of Texas at Austin, Austin, TX USA; and
BEACON Center for the Study of Evolution in Action, Michigan State University, East Lansing,
MI USA

Richard E. Lenski
Department of Microbiology and Molecular Genetics, and BEACON Center for the Study of Evolution in Action, Michigan State University, East Lansing, MI USA; e-mail: lenski@msu.edu

© Springer Nature Switzerland AG 2020
W. Banzhaf et al. (eds.), *Evolution in Action: Past, Present and Future*,
Genetic and Evolutionary Computation, https://doi.org/10.1007/978-3-030-39831-6_8

so, one must be vigilant in checking assumptions and procedures given the potential for uncontrolled factors (e.g., water quality) to affect outcomes. This vigilance is perhaps especially important for a trait like fitness, which integrates all aspects of organismal performance and may therefore be sensitive to any number of subtle environmental influences.

Key words: Bacteria, experimental evolution, fitness, long-term studies, reproducible research

8.1 Introduction

Microorganisms have been used to study evolution in action for well over a century, dating back to work by William Dallinger, who corresponded with Charles Darwin about his research in the 1880s [18, 26], and continuing with pioneering modern experiments by Novick and Szilard [33] and Atwood et al. [2]. The field of microbial experimental evolution has expanded greatly in recent years with many laboratories using diverse viruses, bacteria, and fungi to address a wide range of questions [5, 14, 25]. The ability to sequence and compare the complete genomes of ancestral and experimentally derived samples has led to an even faster expansion of this field in the last decade [3, 5, 9, 12].

The speed with which many microbes can reproduce is one of the main reasons they have become experimental models for studying evolution. It is often possible to observe evolutionary changes within days, weeks, or months, depending on the organisms and environments used in an experiment. Even so, there are advantages to running such experiments for much longer periods. Some evolutionary phenomena, like speciation, may take thousands or even millions of generations to play out. Also, resolving subtly different models of evolutionary dynamics and how adaptive fitness landscape are structured may require very long time-series of data [50, 52].

In 1988, Richard Lenski began the long-term evolution experiment—the LTEE—with 12 replicate populations of *Escherichia coli*, all started from the same ancestor, except for a genetic marker embedded in the experimental design. The LTEE was intended to address three overarching questions [15, 28]. First, for how long can the bacteria continue to improve before they reach some limit to their fitness? Second, are adaptive changes repeatable across the replicate populations, or do the populations adapt in different ways to the same environment? Third, how tightly coupled are the dynamics of phenotypic adaptation and genetic change?

Lenski called this experiment 'long-term' from the outset, including in the title of the first paper on the LTEE [15, 28]. He did so because he previously performed similar experiments that lasted for several hundred generations, and he realized from the fitness trajectories and resulting gains that the evolving populations in those experiments had probably experienced only one or two selective sweeps [8, 22]. So few fixations seemed unsatisfactory for addressing the questions that motivated the LTEE.

The LTEE started at the University of California, Irvine, in February of 1988. After Lenski accepted a faculty position at Michigan State University, the 12 populations were frozen in April of 1992, after reaching 10,000 generations. The LTEE was then restarted from the frozen samples at Michigan State University in October of 1992, where it has continued ever since. Competition experiments were performed in the experiment's old and new homes to confirm that the improvements in fitness seen in the original laboratory would be reproducible in its new environment. That experiment used samples from only a single time point, namely 10,000 generations, and it was not formally analyzed or published. However, the correspondence in fitness values was judged as being satisfactory.

Over its long history, samples from the LTEE populations have been shared with dozens of laboratories around the world. In general, observations made in the original lab have been highly reproducible in other labs. For example, extensive genome sequencing by other labs shows that the populations have been successfully maintained without cross-contamination [17, 45]. Major phenotypic changes, including the evolution of hypermutability in some populations [4, 17, 43, 45] and the surprising appearance of citrate utilization (Cit$^+$ phenotype) in one population [6, 7, 31, 38, 39] have also been confirmed in other labs. Even more subtle phenomena, such as crossfeeding between two lineages that evolved and coexisted for tens of thousands of generations in one population, have been confirmed in studies by other labs [21, 37, 41, 42].

Nonetheless, there have been occasional unexpected results in the LTEE. Notably, it has not been possible, even in the lab at Michigan State University, to repeat or explain the extinction of the Cit$^-$ lineage in the LTEE population that evolved the Cit$^+$ phenotype [46]. When frozen populations from before this extinction occurred in the LTEE were revived and evolution was "replayed" many times from them, this extinction did not re-occur. The best guess in this case is that some unintended, transient perturbation in the conditions of the experiment caused the Cit$^-$ ecotype to go extinct. It may have been especially susceptible to these types of fluctuations because the Cit$^-$ ecotype had a small population size relative to the dominant Cit$^+$ ecotype.

Measurements of how *E. coli* fitness evolves on long timescales are arguably some of the most important and unique data from the LTEE. These fitness trajectories and the related mutational dynamics have been used to examine different models of evolution [4, 16, 20, 36, 48, 49, 50, 52]. Yet, it is rare in microbial evolution experiments to repeat fitness measurements in different labs or at intervals separated by decades in the same lab to examine the repeatability of these values and their evolutionary trajectories. Testing repeatability is especially important for a trait like fitness, which integrates across all aspects of organismal performance and might therefore be sensitive to any number of environmental influences, such as water quality or trace impurities in chemical components [13, 34], in an otherwise defined culture medium and environment. In light of the fact that all research groups have finite lives and the LTEE will move, sooner or later to another laboratory and likely another location, it seems important to examine the repeatability of these measurements. To that end, we measured the fitness values of samples from the 12 LTEE populations at 2,000, 10,000, and 50,000 generations at The University of Texas at Austin and compared these new data to data previously obtained at Michigan State University [51, 52]. As we show below, the two datasets agree quite well, given the inherent measurement noise associated with these assays.

8.2 Methods

8.2.1 Long-Term Evolution Experiment

The LTEE consists of 12 populations of E. coli, all founded from essentially the same ancestor and propagated under identical conditions [28]. Six populations began with strain REL606 [19], and the other six started with REL607. These two strains differ by a mutation that allows the former, but not the latter, to grow on arabinose [28, 44]. This marker serves to distinguish competitors during assays of relative fitness (see Section 8.2.2). The marker also helps guard against undetected cross-contamination of the populations. The six REL606-derived Ara$^-$ and six REL607-derived Ara$^+$ populations are strictly alternated during the serial-transfer procedure. As a consequence, cross-contamination during successive transfers would introduce cells with the wrong arabinose-marker state, which could be detected during quality-control checks performed when samples are frozen.

Every day, 1% of each population is removed and transferred to fresh medium. The 100-fold dilution and subsequent re-growth allow $\log_2 (100)$ generations (i.e., doublings), which is rounded to 6 2/3 generations per daily transfer. Every 75 days (500 generations), after the transfers are performed and the cells are plated on various test media, glycerol (a cryoprotectant) is added to each culture from the previous day. These whole-population samples are then split and stored in duplicate vials at $-80\,^\circ$C, providing a "frozen fossil record" of cells that remain viable and can be revived for later analyses. After incubation, the test plates are used to inspect the bacteria for growth and colony appearance, with particular attention given to whether there is any evidence of cells with the wrong arabinose-marker state. In recent years, genome analysis methods discovered unique mutations that have arisen in each of the LTEE populations [35, 45, 53], and those loci can now also be checked for suspicious colonies. In the event of suspected or confirmed contamination, the population in question is re-started from an earlier frozen sample; as a result, some LTEE populations are 500 generations or more behind the leaders, which recently passed 70,000 generations. Various other disruptions have occurred over the years, including moving the experiment from California to Michigan, so that the LTEE as a whole runs several thousand generations behind the maximum number that could have been achieved after some 30 years.

The LTEE populations live in 50 mL glass Erlenmeyer flasks containing 10 mL of culture medium, with small glass beakers placed over the flask openings. The flasks are incubated at 37 $^\circ$C with orbital shaking at 120 r.p.m. The culture medium is Davis Minimal medium [10] supplemented with thiamine at 2 µg per mL and glucose at 25 µg per mL (DM25), where glucose is the limiting resource. (Note: The first paper on the LTEE [28] misstated the concentration of thiamine, but cited an earlier paper [23] with the correct concentration; the recipe itself has never varied.) The ancestral strains reach a stationary-phase density of $\sim 5 \times 10^7$ cells per mL in DM25 [28]. The evolved bacteria tend to produce larger cells, and they reach somewhat lower numerical densities at stationary phase [27, 29, 47], with one conspicuous exception. That exception is the population, called Ara−3, that evolved the capacity to use the citrate in the medium as an additional carbon source, which allows it to reach a substantially higher cell density than the ancestors or other evolved populations [7].

8.2.2 Assays of Relative Fitness

To assess relative fitness, we conducted head-to-head competition assays in the same medium and other conditions as used for the LTEE. Each pairwise assay competed a whole-population sample taken from one of the 12 populations at 2,000, 10,000, or 50,000 generations against the reciprocally marked ancestral strain, except as noted below. Thus, the Ara$^+$ ancestor REL607 competed against the six populations founded by the Ara$^-$ REL606 ancestor, while REL606 competed against the six populations founded by REL607. The arabinose-utilization marker is neutral in the LTEE environment, and the ancestral strains have indistinguishable relative fitness [28]. When grown on tetrazolium arabinose (TA) indicator agar plates, Ara$^-$ and Ara$^+$ cells form red and white colonies, respectively, thereby allowing the number of cells of each type in a mixture to be estimated [28]. Over time, however, some LTEE populations evolved such that they no longer produce colonies on TA agar, or they produce many colonies that are difficult to accurately score as Ara$^-$ versus Ara$^+$. For this reason, Wiser et al. [52] excluded the 10,000-generation sample for population Ara+6 and the 50,000-generation samples for populations Ara+6 and Ara−2. Additionally, this previous study omitted the Ara−3 population at 50,000 generations because the utilization of a new nutrient pool by Cit$^+$ cells that is inaccessible to the Cit$^-$ ancestor complicates both the measurement and interpretation of relative fitness values. In this study, we excluded those same four samples as well as the 50,000-generation samples for populations Ara+1 and Ara+4 because we could not reliably estimate the evolved cell numbers from their colony counts. Thus, our dataset has relative fitness estimates for a total of 30 evolved population samples, with two-fold replication of each estimate.

Prior to the competitions, 0.12 mL aliquots of frozen stocks of each whole-population sample and each of the two ancestral strains were transferred to 50 mL Erlenmeyer flasks containing 10 mL Luria Broth and grown overnight. Each revived culture was then diluted 100-fold in 10 mL of sterile saline solution, before using 0.1 mL to inoculate 9.9 mL cultures of DM25 in 50 mL Erlenmeyer flasks. Two independent cultures were inoculated for each evolved strain, 28 were inoculated for

the Ara⁻ REL606 ancestor, and 32 were inoculated for the Ara⁺ REL607 ancestor.
These cultures were grown for 24 h to acclimate them to the LTEE conditions. Each
competition assay was then initiated by combining 50 μL of an evolved population
with 50 μL of the ancestral strain with the opposite arabinose-marker state in 9.9
mL of DM25 in a 50 mL Erlenmeyer flask. Each acclimated culture was used to
inoculate only a single competition culture, providing technical independence. After vortexing the competition culture, 0.1 mL was removed, diluted into 9.9 mL of
saline solution, and 40 μL was spread on a TA plate. After 24 h of growth under
LTEE conditions, 0.1 mL of culture was serially diluted twice in 9.9 mL of saline
(10,000-fold total dilution), and 40 μL was spread on TA plates for the competitions involving 2,000- and 10,000-generation populations; 80 μL was used for the
assays involving the 50,000-generation populations due to their lower cell density.
For both the initial and final TA plates, red and white colonies were counted after
incubation for 16 to 24 h at 37 °C to estimate the abundances of the ancestral and
evolved competitors.

We calculated a relative fitness value from each competition assay as the ratio of
the realized growth rates for each competitor in that assay [28], as follows:

$$w = \frac{\ln(E_f/E_i)}{\ln(A_f/A_i)},$$

where E and A are the evolved population and ancestor, respectively, and the f and
i subscripts indicate final and initial densities, respectively, as estimated from the
plate counts. These fitness values integrate any and all differences in growth of the
two competitors across all of the physiological states experienced by cells during
the serial-transfer cycle used in the LTEE [28, 47].

8.2.3 Statistical Analysis

The relative fitness values and power-law model calculations from the paper by
Wiser et al. [52] were previously deposited in the Dryad Digital Repository [51]. We
deposited the new fitness data collected for this study in the Dryad Digital Repository [1]. For convenience, we copied the relevant information from that earlier paper
into the same data file.

We obtained two estimates of the relative fitness for each of 30 evolved population samples (see Section 8.2.2). We chose to perform two replicates for each
sample because that was the same level of technical replication as in the earlier
dataset [51, 52]. However, our interest in this study is not in the statistical noise
among technical replicates, but rather in the correspondence between the estimates
of relative fitness obtained in the two different labs. Therefore, all statistical analyses were performed on the geometric means of the relative fitness values obtained
from each set of two technical replicates. Calculations were performed in R version
3.5.0 [40].

8.3 Results

In this study we measured the fitness of each of the 12 LTEE populations at 2,000, 10,000, and 50,000 generations relative to the ancestral *E. coli* strains at The University of Texas at Austin (UTA). A total of 30 fitness values were obtained after some evolved population samples were omitted for technical reasons (see Section 8.2.2). We compared these fitness values to a matched set of 30 fitness values obtained previously at Michigan State University (MSU) [51, 52] (Fig. 8.1). Overall, there was a strong correlation between the UTA and MSU values (Pearson's $r = 0.830$, $p < 0.001$). The fact that this correlation is imperfect could reflect measurement error alone, or it might reflect some consistent difference in the experimental conditions between the two locales. Therefore, we next considered whether we could detect any systematic bias between the two datasets.

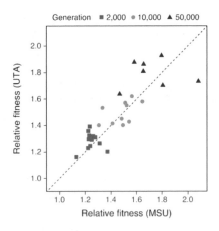

Fig. 8.1: Comparison of fitness values for LTEE population samples from 2,000, 10,000, and 50,000 generations, measured relative to the ancestral *E. coli* strain at The University of Texas at Austin (UTA, this study) and previously at Michigan State University (MSU, [51])

First, there might be some difference between the laboratories that affected all measurements of fitness in the same direction. In this case, one set of values would consistently overestimate fitness relative to the other. We found no evidence for this type of bias. The UTA fitness values were slightly higher, on average, than the corresponding MSU values, but this difference was not significant ($p = 0.169$, two-tailed paired t-test). The 95% confidence interval for this difference ranged from -0.015 to 0.080.

Second, there might be a bias that somehow depends on the generational time point tested. There has been widespread parallel evolution in the LTEE [24, 45, 53], and it is possible that the populations accumulated mutations over time in a way that led to generation-specific changes in the sensitivity of fitness measurements to the two different lab environments. However, we saw no systematic difference in fitness between the UTA and MSU measurements at any of the generations assayed ($p = 0.246$, $p = 0.645$, and $p = 0.446$ for the samples from 2,000, 10,000, and 50,000 generations, respectively, by two-tailed paired t-tests).

We then considered an alternative way of evaluating the UTA data. The true fitness values of the LTEE populations are not known, of course, but models of fitness trajectories that integrate information from fitness measurements across many generations should provide a more accurate estimate of the actual fitness at any generation than the values used above. In particular, a power-law model (PLM) has been shown to describe and even predict the fitness trajectories for each LTEE population over time much better than an alternative hyperbolic model. The PLM was first evaluated by Wiser et al. [52] using a dataset of fitness measurements for each population at 41 time points through 50,000 generations of the LTEE. This model was subsequently extended and supported through 60,000 generations [30].

If we assume the fitness value predicted by the PLM for each population at each time point represents the true value (or at least is closer to the true value than the measurements from that generation alone), then we would expect there to be a better correlation between the UTA data and the PLM predictions than there is between the UTA and MSU data. Indeed, this is the case. The correlation between the UTA data and PLM predictions ($r = 0.882$) is somewhat stronger than the correlation between the UTA and MSU data ($r = 0.830$). However, the improvement in the strength of this correlation is not significant ($p = 0.125$, one-tailed paired t-test comparing the squared residuals).

Under the assumption that the PLM predicts the true fitness values with reasonable precision, we can also perform a regression to test whether we can discriminate between the measured data and the predicted values. A linear regression of the MSU data against the PLM predictions that is forced to pass through a relative fitness of one on both axes gives a slope that is not significantly different from one (Fig. 8.2A, $p = 0.769$). Regressing the UTA data against the PLM predictions in this way also gives a slope that is not significantly different from one (Fig. 8.2B, $p = 0.113$).

8.4 Discussion

The LTEE has been running for more than 30 years and 70,000 generations, and we hope it will continue far into the future. More than 80 papers have been published using the bacteria and data derived from this one experiment; several other evolution experiments have been spun off from it; and countless other experiments have been influenced by it. Arguably, no other evolution experiment has been studied in as many ways and with as much quantitative rigor as the LTEE. Much of the value of any model system, including the LTEE, rests on its reproducibility. The LTEE was designed to be simple in terms of the culture medium and other environmental conditions used, and such simplicity undoubtedly helps promote reproducibility. The fact that the bacteria can be stored frozen and later revived is also critical,

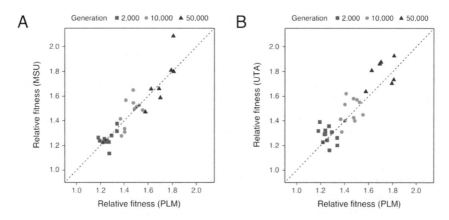

Fig. 8.2: Comparisons of (A) MSU and (B) UTA relative fitness measurements to predictions of a power-law model for the fitness trajectories in the LTEE (PLM). PLM parameters fit separately to each population were used to predict fitness values at 2,000, 10,000, and 50,000 generations

because it allows samples from one time and location to be analyzed in new ways or re-analyzed at different times and places.

In evolutionary biology, the relative fitness of different genotypes, including the ancestral and derived bacteria in the LTEE, is a quantity of central interest and importance. It effectively integrates everything about the organisms' genomes and their phenotypes (at least those that are relevant for performance in a given environment) into a single measure of reproductive success. However, that integrative aspect also raises the possibility that measurements of relative fitness might be especially sensitive to subtle variations in the test conditions. The simplicity of the conditions used in the LTEE means that many possible sources of variation are well controlled. Nonetheless, there is the potential for small fluctuations and unintended perturbations to have an outsized impact, especially on an integrative and quantitative trait like fitness.

To address this issue, we performed competition assays at The University of Texas at Austin to measure the fitness of population samples from generations 2,000, 10,000, and 50,000 of the LTEE relative to their common ancestor. We compared these new estimates to values obtained for the same populations several years earlier at Michigan State University using the same methods. In short, the new and old data agreed well, with deviations fully consistent with ordinary sampling error. It is highly encouraging for the future of the LTEE that such a potentially sensitive metric as relative fitness can be reliably estimated even after population samples have been stored frozen for years, "copied" by re-culturing the samples, and then shipped to and analyzed in another laboratory. This consistency is important because it means that the experiment can be passed to future generations of researchers, who can continue to monitor and study the evolution of the populations to analyze both the long-term trends that are in common to all of them and any surprises that may be in store for particular populations.

It is also interesting that, while the methods and resulting estimates are reproducible, the bacteria themselves have sometimes evolved in ways that make obtaining those estimates more challenging. In particular, several populations have evolved such that cells no longer produce colonies that can be reliably counted on the indicator agar used to distinguish competitors in the assays of relative fitness. This colony feature is not a phenotype that was selected or contributes to fitness in the LTEE; on the contrary, this trait can decay precisely because it has no bearing on fitness in that environment. Other challenges that have evolved include cross-feeding, in which some cells secrete metabolites that others can use as resources, and the ability of one lineage to use citrate. As a consequence of these changes, the fitness of some evolved lineages relative to their ancestor or to one another depends (albeit only subtly in most cases) on the frequency of the competitors at the start of a competition assay. It is now becoming possible to decouple making fitness measurements from the requirement that cells from the LTEE form colonies on indicator agar by using DNA sequencing to read out the frequencies of different alleles [11] and to multiplex fitness measurements using DNA barcoding techniques [32]. These and other new technologies will continue to enhance our understanding of evolutionary dynamics in the LTEE. The phenotypic changes in the LTEE may complicate some experimental procedures, but they also enrich the LTEE by revealing the interesting ways that the bacteria can evolve and adapt, and thus the experiment's lasting potential to generate new insights and discoveries.

Acknowledgements We thank Erik Goodman for his inspirational and effective leadership of the BEACON Center, Mike Wiser for sharing data, and Caroline Turner for catching the error in the published recipe for the DM25 medium. This work has been supported by an NSF grant (DEB-1451740), the BEACON Center for the Study of Evolution in Action (NSF Cooperative Agreement DBI-0939454), and the John Hannah endowment at Michigan State University. Any opinions, findings, and conclusions or recommendations in this paper are those of the authors and do not necessarily reflect the views of the National Science Foundation.

References

1. Barrick, J.E., Deatherage, D.E., Lenski, R.E.: Data from: A test of the repeatability of measurements of relative fitness in the long-term evolution experiment with *Escherichia coli*. Dryad Digital Repository. https://doi.org/10.5061/dryad.cz8w9ghzm (2019)
2. Atwood, K.C., Schneider, L.K., Ryan, F.J.: *Periodic selection in Escherichia coli*. Proc Natl Acad Sci USA **37**, 146–155 (1951)
3. Barrick, J.E., Lenski, R.E.: *Genome dynamics during experimental evolution*. Nature Rev Gen **14**, 827–839 (2013)
4. Barrick, J.E., Yu, D.S., Yoon, S.H., Jeong, H., Oh, T.K., Schneider, D., Lenski, R.E., Kim, J.F.: *Genome evolution and adaptation in a long-term experiment with Escherichia coli*. Nature **461**, 1243–1247 (2009)

5. Van den Bergh, B., Swings, T., Fauvart, M., Michiels, J.: *Experimental design, population dynamics, and diversity in microbial experimental evolution.* Microbiol Mol Biol Rev **82**, e00008–18 (2018)

6. Blount, Z.D., Barrick, J.E., Davidson, C.J., Lenski, R.E.: *Genomic analysis of a key innovation in an experimental Escherichia coli population.* Nature **489**, 513–518 (2012)

7. Blount, Z.D., Borland, C.Z., Lenski, R.E.: *Historical contingency and the evolution of a key innovation in an experimental population of Escherichia coli.* Proc Natl Acad Sci USA **105**, 7899–7906 (2008)

8. Bouma, J.E., Lenski, R.E.: *Evolution of a bacteria/plasmid association.* Nature **335**, 351–352 (1988)

9. Bruger, E.L., Marx, C.J.: *A decade of genome sequencing has revolutionized studies of experimental evolution.* Curr Opin Microbiol **45**, 149–155 (2018)

10. Carlton, B.C., Brown, B.J.: Gene mutation. In: P. Gerhardt (ed.) Manual of Methods for General Bacteriology, pp. 222–242. American Society for Microbiology, Washington (1981)

11. Chubiz, L.M., Lee, M.C., Delaney, N.F., Marx, C.J.: *FREQ-Seq: A rapid, cost-effective, sequencing-based method to determine allele frequencies directly from mixed populations.* PLoS ONE **7**, 1–9 (2012)

12. Conrad, T.M., Lewis, N.E., Palsson, B.Ø.: *Microbial laboratory evolution in the era of genome-scale science.* Mol Syst Biol **7**, 509 (2011)

13. Egli, T.: *Microbial growth and physiology: A call for better craftsmanship.* Frontiers Microbiol **6**, 287 (2015)

14. Elena, S.F., Lenski, R.E.: *Evolution experiments with microorganisms: The dynamics and genetic bases of adaptation.* Nature Rev Genet **4**, 457–469 (2003)

15. Fox, J.W., Lenski, R.E.: *From here to eternity - the theory and practice of a really long experiment.* PLoS Biol **13**, e1002815 (2015)

16. Good, B.H., Desai, M.M.: *The impact of macroscopic epistasis on long-term evolutionary dynamics.* Genetics **199**, 177–190 (2015)

17. Good, B.H., McDonald, M.J., Barrick, J.E., Lenski, R.E., Desai, M.M.: *The dynamics of molecular evolution over 60,000 generations.* Nature **551**, 45–50 (2017)

18. Haas, J.W.: *The Reverend Dr William Henry Dallinger, F.R.S. (1839-1909).* Notes Rec R Soc Lond **54**, 53–65 (2000)

19. Jeong, H., Barbe, V., Lee, C.H., Vallenet, D., Yu, D.S., Choi, S.H., Couloux, A., Lee, S.W., Yoon, S.H., Cattolico, L., Hur, C.G., Park, H.S., Sgurens, B., Kim, S.C., Oh, T.K., Lenski, R.E., Studier, F.W., Daegelen, P., Kim, J.F.: *Genome sequences of Escherichia coli B strains REL606 and BL21(DE3).* J Mol Biol **394**, 644–652 (2009)

20. Kryazhimskiy, S., Tkačik, G., Plotkin, J.B.: *The dynamics of adaptation on correlated fitness landscapes.* Proc Natl Acad Sci USA **106**, 18638–18643 (2009)

21. Le Gac, M., Plucain, J., Hindré, T., Lenski, R.E., Schneider, D.: *Ecological and evolutionary dynamics of coexisting lineages during a long-term experiment with Escherichia coli.* Proc Natl Acad Sci USA **109**, 9487–9492 (2012)

22. Lenski, R.E.: *Experimental studies of pleiotropy and epistasis in Escherichia coli. II. Compensation for maladaptive effects associated with resistance to virus T4.* Evolution **42**, 433–440

23. Lenski, R.E.: *Experimental studies of pleiotropy and epistasis in Escherichia coli. I. Variation in competitive fitness among mutants resistant to virus T4.* Evolution **42**, 425–432 (1988)

24. Lenski, R.E.: *Convergence and divergence in a long-term experiment with bacteria.* Am Nat **190**, S57–S68 (2017)

25. Lenski, R.E.: *Experimental evolution and the dynamics of adaptation and genome evolution in microbial populations.* ISME Journal **11**, 2181–2194 (2017)

26. Lenski, R.E.: *What is adaptation by natural selection? Perspectives of an experimental microbiologist.* PLoS Genet **13**, e1006668 (2017)

27. Lenski, R.E., Mongold, J.A.: *Cell size, shape and fitness in evolving populations of bacteria.* In: J.H. Brown, G.B. West (eds.) Scaling in Biology, pp. 221–235. Oxford University Press, Oxford, UK (2000)

28. Lenski, R.E., Rose, M.R., Simpson, S.C., Tadler, S.C.: *Long-term experimental evolution in Escherichia coli. I. Adaptation and divergence during 2,000 generations.* Am Nat **138**, 1315–1341 (1991)

29. Lenski, R.E., Travisano, M.: *Dynamics of adaptation and diversification: A 10,000-generation experiment with bacterial populations.* Proc Natl Acad Sci USA **91**, 6808–6814 (1994)

30. Lenski, R.E., Wiser, M.J., Ribeck, N., Blount, Z.D., Nahum, J.R., Morris, J.J., Zaman, L., Turner, C.B., Wade, B.D., Maddamsetti, R., Burmeister, A.R., Baird, E.J., Bundy, J., Grant, N.A., Card, K.J., Rowles, M., Weatherspoon, K., Papoulis, S.E., Sullivan, R., Clark, C., Mulka, J.S., Hajela, N.: *Sustained fitness gains and variability in fitness trajectories in the long-term evolution experiment with Escherichia coli.* Proc R Soc B **282**, 20152292 (2015)

31. Leon, D., D'Alton, S., Quandt, E.M., Barrick, J.E.: *Innovation in an E. coli evolution experiment is contingent on maintaining adaptive potential until competition subsides.* PLoS Genet **14**, e1007348 (2018)

32. Levy, S.F., Blundell, J.R., Venkataram, S., Petrov, D.A., Fisher, D.S., Sherlock, G.: *Quantitative evolutionary dynamics using high-resolution lineage tracking.* Nature **519**, 181–186 (2015)

33. Novick, A., Szilard, L.: *Experiments with the chemostat on spontaneous mutations of bacteria.* Proc Natl Acad Sci USA **36**, 708–719 (1950)

34. O'Keefe, K.J., Morales, N.M., Ernstberger, H., Benoit, G., Turner, P.E.: *Laboratory-dependent bacterial ecology: A cautionary tale.* Appl Environ Microbiol **72**, 3032–3035 (2006)

35. Papadopoulos, D., Schneider, D., Meier-Eiss, J., Arber, W., Lenski, R.E., Blot, M.: *Genomic evolution during a 10,000-generation experiment with bacteria.* Proc Natl Acad Sci USA **96**, 3807–3812 (1999)

36. Passagem-Santos, D., Zacarias, S., Perfeito, L.: *Power law fitness landscapes and their ability to predict fitness.* Heredity **121**, 482–498 (2018)

37. Plucain, J., Hindré, T., Le Gac, M., Tenaillon, O., Cruveiller, S., Médigue, C., Leiby, N., Harcombe, W.R., Marx, C.J., Lenski, R.E., Schneider, D.: *Epistasis and allele specificity in the emergence of a stable polymorphism in Escherichia coli.* Science **343**, 1366–1369 (2014)

38. Quandt, E.M., Deatherage, D.E., Ellington, A.D., Georgiou, G., Barrick, J.E.: *Recursive genomewide recombination and sequencing reveals a key refinement step in the evolution of a metabolic innovation in Escherichia coli.* Proc Natl Acad Sci USA **111**, 2217–2222 (2014)

39. Quandt, E.M., Gollihar, J., Blount, Z.D., Ellington, A.D., Georgiou, G., Barrick, J.E.: *Fine-tuning citrate synthase flux potentiates and refines metabolic innovation in the Lenski evolution experiment.* eLife **4**, e09696 (2015)

40. R Core Team: R: A Language and Environment for Statistical Computing. R Foundation for Statistical Computing, Vienna, Austria (2018)

41. Rozen, D.E., Lenski, R.E.: *Long-term experimental evolution in Escherichia coli. VIII. Dynamics of a balanced polymorphism.* Am Nat **155**, 24–35 (2000)

42. Rozen, D.E., Schneider, D., Lenski, R.E.: *Long-term experimental evolution in Escherichia coli. XIII. Phylogenetic history of a balanced polymorphism.* J Mol Evol **61**, 171–180 (2005)

43. Sniegowski, P.D., Gerrish, P.J., Lenski, R.E.: *Evolution of high mutation rates in experimental populations of E. coli.* Nature **387**, 703–705 (1997)

44. Studier, F.W., Daegelen, P., Lenski, R.E., Maslov, S., Kim, J.F.: *Understanding the differences between genome sequences of Escherichia coli B strains REL606 and BL21(DE3) and comparison of the E. coli B and K-12 genomes.* J Mol Biol **394**, 653–680 (2009)

45. Tenaillon, O., Barrick, J.E., Ribeck, N., Deatherage, D.E., Blanchard, J.L., Dasgupta, A., Wu, G.C., Wielgoss, S., Cruveiller, S., Médigue, C., Schneider, D., Lenski, R.E.: *Tempo and mode of genome evolution in a 50,000-generation experiment.* Nature **536**, 165–170 (2016)

46. Turner, C.B., Blount, Z.D., Lenski, R.E.: *Replaying evolution to test the cause of extinction of one ecotype in an experimentally evolved population.* PLoS ONE **10**, e0142050 (2015)

47. Vasi, F., Travisano, M., Lenski, R.E.: *Long-term experimental evolution in Escherichia coli. II. Changes in life-history traits during adaptation to a seasonal environment.* Am Nat **144**, 432–456 (1994)

48. de Visser, J.A.G.M., Elena, S.F., Fragata, I., Matuszewski, S.: *The utility of fitness landscapes and big data for predicting evolution.* Heredity **121**, 401–405 (2018)

49. de Visser, J.A.G.M., Krug, J.: *Empirical fitness landscapes and the predictability of evolution.* Nature Rev Gen **15**, 480–490 (2014)
50. Wiser, M.J., Dolson, E.L., Vostinar, A., Lenski, R.E., Ofria, C.: *The boundedness illusion: Asymptotic projections from early evolution underestimate evolutionary potential.* PeerJ Preprints **6**, e27246v2 (2018)
51. Wiser, M.J., Ribeck, N., Lenski, R.E. Data from: Long-term dynamics of adaptation in asexual populations. Dryad Digital Repository. https://doi.org/10.5061/dryad.0hc2m (2014)
52. Wiser, M.J., Ribeck, N., Lenski, R.E.: *Long-term dynamics of adaptation in asexual populations.* Science **342**, 1364–1367 (2013)
53. Woods, R., Schneider, D., Winkworth, C.L., Riley, M.A., Lenski, R.E.: *Tests of parallel molecular evolution in a long-term experiment with Escherichia coli.* Proc Natl Acad Sci USA **103**, 9107–9112 (2006)

Chapter 9
Experimental Evolution of Metal Resistance in Bacteria

Joseph L. Graves Jr, Akamu J. Ewunkem, Misty D. Thomas, Jian Han, Kristen L. Rhinehardt, Sada Boyd, Rasheena Edmundson, Liesl Jeffers-Francis and Scott H. Harrison

Abstract There has been an increased usage of metallic antimicrobial materials to control pathogenic and multi-drug resistant bacteria, yet there is a corresponding need to know if this usage may lead to genetic adaptations that produce even more dangerous bacterial strains. In this paper we examine important recent results from the literature as well as report results from a series of our own studies. In that work, we utilized experimental evolution to produce strains of *Escherichia coli* K-12 MG1655 resistant to silver (Ag^+), excess copper (Cu^{2+}), excess iron (II, III), and the iron analog gallium (Ga^{3+}). Silver and gallium are toxic to bacteria, whereas iron and copper are essential micronutrients that can be toxic in excess amounts. In all cases, the evolution of metal resistance was rapid and resulted in pleiotropic effects that included resistance to other metals as well as traditional antibiotics. Genomic analysis identified mutations in several genes associated with metal resistance, falling in several broad classes: genes that prevent entry of metal into the cell, genes involved in the energy-dependent efflux of metal out of the cell, genes associated with ROS-induced membrane damage, and genes associated with transcription.

Joseph L. Graves Jr · Akamu J. Ewunkem · Kristen L. Rhinehardt
Department of Nanoengineering, Joint School of Nanoscience & Nanoengineering North Carolina A&T State University & UNC Greensboro, Greensboro, NC 27401 and BEACON Center for the Study of Evolution in Action, Michigan State University, East Lansing, MI 48824 e-mail: gravesjl@ncat.edu

Misty D. Thomas · Jian Han · Liesl Jeffers-Francis · Scott H. Harrison
Department of Biology, North Carolina A&T State University and BEACON Center for the Study of Evolution in Action, Michigan State University, East Lansing, MI 48824

Sada Boyd
Graduate Research Student, Energy and Environment Program, North Carolina A&T State University and BEACON Center for the Study of Evolution in Action, Michigan State University, East Lansing, MI 48824

Rasheena Edmundson
Department of Biology, Bennett College, Greensboro, NC 27401

All authors contributed equally to this work.

© Springer Nature Switzerland AG 2020
W. Banzhaf et al. (eds.), *Evolution in Action: Past, Present and Future*,
Genetic and Evolutionary Computation, https://doi.org/10.1007/978-3-030-39831-6_9

Several of the genes found to be associated with resistance have known relationships to metal metabolism, although some have not previously been identified in this role. Our studies demonstrate the power of evolution in action and are a powerful warning to those who assume that the use of metals represents such a steep hill for bacteria to climb that it will prevent the evolution of resistance.

9.1 Introduction

Metallic and metal oxide ions and nanoparticles have recently been called the "powerful nanoweapon" against multidrug resistant bacteria [50]. However, these claims were made primarily by chemists and materials scientists who had little background or understanding of microbes and their evolution [23]. Indeed, a warning concerning the potential spread of silver resistance throughout the microbial world was sounded as early as 1998:

> "Silver resistance is important to monitor because modern technology has developed a wide range of products that depend upon silver as a key microbial component." [26]

Commercial applications of antimicrobial silver include baby-changing stations, cleaning supplies, high chairs, dispensers, keyboards, elevator surfaces, food preparation, food processing, storage, surface coatings, and hand rails (see [3]). Curad makes bandages with antimicrobial silver [2]. Copper is also sold commercially to control bacteria [1]. However, what the manufacturers of these commercial products may not recognize is the widespread occurrence of metal resistance in microbes.

Toxic metals are abundant on our planet, and microbes have been exposed to them for $\sim$4 billion years [59]. Metal resistance genes have been found in bacteria dating from long before the worshippers of Yahweh began their craft. For example, such genes have been found in 1.6 – 1.8 million-year old permafrost [44, 59]. Genes conferring resistance to toxic metal ions (Ag^+, Cd^{2+}, Hg^{2+}, Ni^{2+}) are widespread in bacteria allowing them to maintain appropriate homeostasis in changing environments [59]. Mechanisms of resistance include energy-dependent efflux (as found with Ag^+ resistance in *E. coli*), enzymatic transformations (oxidation, reduction, methylation, and demethylation) and upregulation of metal-binding proteins (metallothionein, SmtA; chaperone, CopZ; or SilE; [24, 51, 59, 62]). Additional strategies include changes in overall growth rate and formation of biofilms allowing populations to tolerate elevated levels of the toxic metal [28, 39].

In particular, arsenic resistance shows strong evidence of ancient origin [38]. These investigators examined sequence homology of arsenite oxidase from two β-proteobacteria and compared these to Archaeans. They obtained phylogenetic trees that indicated an early origin of arsenite oxidase, pre-dating the divergence of Archaea and Bacteria. In addition to arsenic, Silver and Phuong [59] reviewed and showed widespread existence of resistance to mercury, cadmium, lead, silver, tellurium, as well as to excess amounts of zinc, copper, and nickel. Resistance mechanisms included oxidation and reduction, ATPase and chemiosmotic membrane pumps, and binding of toxic ions. Numerous genes have been implicated in these metal resistance mechanisms (e.g., *mer, ars, aso, arr, cad, czc, cnr, ncc, nre, cop, pco, chr, pbr, sil, tel, teh,* and *kil*). The genes are often pleiotropic (conferring resistance to different metals), and they can be located on the bacterial chromosome or plasmid borne [43, 59].

On the other hand, several metals (iron, copper, zinc, and manganese) are required for crucial biological processes and the physiological function of all organisms. They are found exclusively as components of proteins including enzymes, transcription factors, and storage proteins [32, 49]. The unique redox potential of some of the transition metals allows them to serve important roles as cofactors in enzymes, and it is estimated that 30 – 45% of enzymes are metalloproteins [35]. However, the same transition metals can be toxic in high intracellular concentrations, perturbing cellular redox potentials as well as producing highly reactive hydroxyl radicals and reactive oxygen species. Thus, all organisms require mechanisms for sensing small variations in metal levels to maintain a controlled balance of uptake, efflux, and sequestration to ensure that metal availability is in accordance with proper physiological function [49].

Our research group has utilized the methods of experimental evolution to better understand how bacteria may evolve resistance to toxic metals (silver, gallium; [24, 21, 62]), and to excesses of essential micronutrients (iron, copper; [4, 13, 64]). These studies have found that both Gram-negative (*E. coli*) and Gram-positive (*Staphylococcus aureus*) are capable to rapid evolution of silver resistance and that this resistance is accomplished with small changes in the genome. The studies on genomic analysis of gallium is not complete at the time of this writing, but the seems to be no reason to believe that the genomic architecture of its resistance if fundamentally different.

With regards to excess micronutrients, we were first concerned with iron, as it is one of the most important metal micronutrients and fulfills many biological roles [12]. Iron-containing proteins (heme-proteins, iron-sulfur cluster proteins, and di-iron and mononuclear enzymes) play roles in nitrogen fixation and metabolism and serve as electron carriers for respiration [41]. Despite its critical role in metabolism, acquiring iron is one of the greatest challenges for bacterial growth [15, 20], and iron deficiency is one of the most common nutritional stresses [8]. However, excess iron can also be extremely toxic under aerobic conditions [57].

In aerobic environments, iron occurs mainly as ferric iron (Fe^{3+}), although $Fe(OH)_3$ is poorly soluble in aqueous solution (as low as 10^{-18} M at pH 7.0). Under anaerobic conditions, the equilibrium shifts to ferrous iron (Fe^{2+}), which is more readily bioavailable to microorganisms. That availability is a key to pathogenesis for a variety of microbes, and thus many innate immunity mechanisms utilize iron sequestration (e.g., serum albumin, calprotectin; [36, 45]).

Under anaerobic and microaerobic conditions, members of the Enterobacteriaceae [14] primarily use the FeoB pathway [5] to transport free Fe^{2+} across the inner membrane into the cytoplasm in a GTP-dependent manner. Due to the difficulty of obtaining iron from the environment, many microbes have evolved means to sequester iron from their extracellular surroundings. This sequestration is accomplished by means of proteins called siderophores. Enterobactin is a siderophore synthesized in *E. coli*, via the *entC* and *fep* genes, which are tightly controlled by the global iron homeostasis regulator Fur [15]. The export of enterobactin into the external environment often relies on the outer membrane channel protein TolC [18], and the subsequent import of the enterobactin-bound iron uses chaperone proteins and an ABC-transporters [37].

Under aerobic conditions, excess iron has been shown to produce oxidative damage. Both Fe^{2+} and Fe^{3+} can interact with hydrogen peroxide and super oxides respectively and thereby generating highly reactive hydroxyl radicals leading to cell damage and eventual cell death [57]. This damage is in addition to the other effects summarized in Table 9.1. When oxygen free radicals form in the cell, bacteria have genes that encode products to defend themselves from these stressors. This defense is aided by many proteins including: OxyR, which responds to the presence of hydrogen peroxide; SoxS and SoxR, which respond to redox active compounds; and RpoS, which responds to general oxidative stress [46, 69, 70]. Frawley et al. [17] demonstrated that iron and citrate export by the efflux pump MdtD enable Salmonella to survive excess iron stress, showing the importance of regulating intracellular iron to prevent toxicity [9, 58].

With all this evidence pertaining to bacterial responses to metals, we performed a series of studies to examine whether and how *Escherichia coli* evolves resistance to excess metals. In addition, we tested whether evolved resistance to one particular metal conferred resistance to other metals with similar physiological effects. These studies included comparisons between silver and iron (Table 9.1) as well as with copper and gallium. In all of these studies, we utilized experimental evolution to examine the genomic changes that allowed *E. coli* K-12 MG1655 to survive in environments containing these excess metals. Most importantly, our work showed that bacteria, such as *E. coli*, can rapidly evolve resistance to silver (ionic and spherical nanoparticles; [24, 62]); copper (ionic, [4]) and iron (II, III, and magnetite nanoparticles, [13, 21, 64]).

Table 9.1: **Mechanisms of excess iron and silver toxicity**. Mechanisms of cellular damage resulting from excess iron (Fe) and silver (Ag) toxicity are listed below. Because these systems are common to almost all bacteria, it is likely that resistance mechanisms are also conserved across many taxa.

Mechanism	Fe	Ag
Reactive oxygen species	+	+
Disruption of transcription/translation	+	+
Damage to cell wall/membrane	+	+
Interfering with respiration	+	+
Release of cellular components	+	+
Binding to thiol groups	?	+

+: mechanism established; ?: mechanism unknown

9.2 Resistance to Metals

The physiological effects of toxic metals on bacteria has been studied for decades [6, 68]. Ionic silver has been shown to be effective in low concentrations against a variety of Gram-negative, Gram-Positive bacteria, and viruses. However these studies were generally conducted over short time periods with no attention to the potential of evolution [22, 43]. This was in part due to the belief that since silver impacted so many arenas of bacterial physiology that resistance was unlikely (see Table 9.1). Similarly, the use of silver nanoparticles as potential antimicrobial agents gained traction in the latter portion of the 1st decade of the 21st century with similar claims of the innate sustainability of silver nanoparticles as antimicrobial agents [50]. Yet both beliefs were inconsistent with observations of silver resistant microbes isolated from burn wards, silver mines, and coastal water from the early 1960 – 1990s [7, 27, 34]. Li, Nikaido, and Williams [40] was one of the first experiments that isolated silver resistant mutants in *E. coli*. They found that these strains were deficient in porins, but without whole genome sequencing they were not capable of demonstrating what genomic features accounts for this. In 1999 Gupta et al. [25] showed discovered the silver resistance region of the pMG101 plasmid (this plasmid was responsible for the outbreak of silver resistant *Salmonella* that killed several patients in a Massachusetts hospital in 1975). The region sil-CFBA(ORF105)PRSE comprises 9 genes that were characterized based on homologies to other known silver resistance determinants [59]. SilP functions as a P-type ATPase pump, transports silver from the cytoplasm to the periplasm. SilF, is a periplasmic protein that functions as a chaperone, transporting Ag^+ from SilP to the SilCBA complex [43]. These plasmid genes are homologous to chromosomal elements that have been found in all *E.coli* strains [16]. The chromosomal elements homologous to *sil* are the *cus* genes. For example, Lok [42] utilized step-wise selection in increasing concentrations of silver to isolate a mutant strain that constitutively expressed *cusB* and *cusF* as their primary silver resistance mechanism.

9.2.1 Experimental Evolution of Metal Resistance

Experimental evolution is the study of populations, in the laboratory or in nature, eover multiple generations, under defined and reproducible conditions [19]. This method has been used to study adaptation in a variety of organisms and conditions, including the study of antimicrobial resistance to antibiotics [10, 65, 66] and metals [24, 54, 62]. Other methods have been used, such as step-wise selection [40] to isolate metal-resistant mutants in *E. coli*, but the underlying principle of artificial selection is the same as that employed in experimental evolution. The advantage to experimental evolution is that it allows for the study of adaptation to toxic metal concentrations over time, which does not occur using the step-wise selection protocol.

In our studies, the Gram-negative bacterium *E. coli* K-12 MG1655 (ATCC #4707 6) was used, due to the paucity of known genes conferring metal resistance in this strain (silver, [24, 62]; copper, [4]; iron II, III, and magnetite nanoparticles, [13, 21, 64]; gallium, [21]). Its chromosome has 4,641,652 base pairs (GenBank: NC_000913.3), and this strain contains no plasmids. In our experiments *E. coli* was cultured in Davis Minimal Broth (DMB; Difco, Sparks, MD) with 1 gram of dextrose per liter (Fisher Scientific, Fair Lawn, NJ) carbon source. The medium is enriched with thiamine hydrochloride with 10 μL in the final volume of 10 mL. This is kept in 50-mL Erlenmeyer flasks and placed in a shaking incubator at 115 rpm at 37°C. Each replicate population was founded from a unique colony on an agar plate isolated via serial dilution from the ancestral stock culture. A minimum inhibitory concentration (MIC) assay was performed to determine the initial sub-lethal concentration of each metal species to be utilized for selection purposes. Populations were placed in separate flasks and allowed to grow for 24 h before being transferred to fresh media. In general, five replicate populations of the controls (C1 – C5), which grew and evolved in standard DMB medium without the addition of metal, were compared to five replicates of metal-selected populations. These populations were founded and propagated by diluting 0.1 mL into 9.9 mL of fresh sterile DMB daily. The population density increased from ~107 cells per mL to ~109 cells per mL over the course of each daily transfer cycle. Then, every seven days, after the corresponding transfers, the remainder of each population was frozen at −80°C for future analysis. Replication of selection treatments is important as it allows for the observation of parallelism; that is, the adaptations that are consistent responses to the selection regime.

9.2.2 Phenotypic Assays: 24-Hour Growth

In our studies, measurements of final density after 24 h in excess metal (Ag^+, Fe^{2+}, Fe^{3+}, and Ga^{3+}) for the metal-selected and control populations were conducted, with measurements taken at multiple time points during the evolution experiments. These time points ranged from 15 to 200 days, depending on what was being measured. The range used for each compound was determined by the original MIC for each selection treatment after 24 h growth. Test concentrations varied between 0 – 2,500 mg/L for the metal assays (Ag^+, Fe^{2+}, Fe^{3+}, and Ga^{3+}) with silver being the most toxic 0 – 250 mg/L and iron species being least toxic 0-1750 mg/L. Of special note for the iron studies, the DMB medium does contain the minimal amount of iron required for bacterial growth (0.1 M or about 5.5×10^{-3} mg/L; [47]), thus the values listed in our studies of resistance to this metal are in addition to that minimum requirement present in the standard medium.

Table 9.2 summarizes the overall results of our evolution experiments with *E. coli* K-12 MG1655 for silver, copper, iron, and gallium. The first column gives the number of days required for the first observation of a significant change in population density after 24 h in increasing concentrations of the metal; the second column

Table 9.2: Phenotypic changes observed in metal-selected *E. coli* populations

Treatment	Populations	Days	MIC (increase)	Correlated responses to other metals	Correlated responses to antibiotics
AgNP	5	15	750 μg/L	Modest increase $Fe^{2/3}$	B/R reduced
Ag^+	18	15	40 – 400 mg/L	Modest increase $Fe^{2/3}$	B/R reduced
Fe^{2+}	5	51	50 – 1750 mg/L[*]	Fe^{3+}; FeNPs, Ga^{3+}, Ag^+	A, C, P, R, S, T increased
Fe^{3+}	5	51	50 – 1750 mg/L[*]	Fe^{2+}; FeNPs, Ga^{3+}, Ag^+	C, P, R, S, T increased
FeNP	5	14	250 – 1750 mg/L[*]	Fe^{2+}; magnetite NPs, Ga^{3+}, Ag^+	C, P, R, S, T increased
Cu^{2+}	5	117	60 – 500 mg/L[*]	Reduced Ag^+ and Ga^{3+}	None increased
Ga^{3+}	5	10	250 – 1000 mg/L[*]	Fe^{2+}, Fe^{3+}, Ag^+	R reduced

Treatments: Populations selected for resistance to AgNP (spherical silver nanoparticles: [24]); Ag^+ (ionic silver: [62]); Fe^{2+} = (iron II: [13, 64]); FeNP (magnetite nanoparticles: [13]); Ga^{3+} (gallium III: [21]; also, elongated cells were observed under electron microscope compared to controls); Cu^{2+} (copper II; [4]), also showing the duration of each selection experiment in days.

[*] Range over which metal-selected populations display statistically significantly higher 24-h growth compared to controls.

A = ampicillin, B = bacitracin, C = chloramphenicol, P = polymyxin B, R = rifampicin, S = sulfanilamide, T = tetracycline.

shows the increase in resistance relative to the controls; the third column shows whether resistance to other metals evolved as a correlated response; and the fourth shows whether there was correlated resistance to traditional antibiotics.

Table 9.3 shows the genomic changes following selection for resistance for each metal. Column one shows the metal selection treatment, the second column shows genes that were associated with metal resistance, the third column shows whether any mutation in that gene under went a hard selective sweep in at least one replicate population, the fourth column indicates whether a soft selective sweep was observed in at least one of the replicate populations, and the fifth column indicates whether any mutation in that gene was observed in more than one replicate population. Table 9.4 provides the functional description for each gene identified as associated with metal resistance. Across all experiments we were able to show that *E. coli* could evolve resistance to the specific metal and that resistance was often associated with a correlated change in resistance to traditional antibiotics. In the case of silver selection resistant populations lost resistance to traditional antibiotics, whereas with iron-selection populations gained resistance to one or more traditional antibiotics.

Table 9.3: Genomic changes observed in metal-selected *E. coli* populations

Treatments	Populations	Genes	H Sweep	S Sweep	Multiple
AgNP	5	*cusS*	Yes	No	Yes
		purL	Yes	No	Yes
		rpoB	Yes	No	Yes
Ag$^+$	18	*cusS*	Yes	Yes	Yes
		ompR	Yes	Yes	Yes
		yfhM	Yes	No	No
		rpoA	Yes	Yes	Yes
		rpoB	Yes	No	No
		rpoC	Yes	Yes	Yes
Fe^{2+}	5	*murC*	Yes	No	No
		cueR	Yes	No	No
		yeaG	Yes	No	No
		fliP	Yes	No	No
		ptsP	Yes	Yes	Yes
		ilvG	Yes	Yes	Yes
		fecA	Yes	No	Yes
		ilvL/ilvX	Yes	No	No
Fe^{3+}	5	*dnaK*	Yes	No	Yes
		tolC	No	Yes	No
		nusA	Yes	No	Yes
		crp	Yes	Yes	Yes
		ompC	Yes	No	No
		yjiH	Yes	No	No
		nlpl/pnp	No	Yes	No
		ompF/asnS	Yes	Yes	Yes
		nudS/tolC	No	Yes	No
		tdcR/rhaB	Yes	Yes	Yes
FeNP	5	Data not yet collected			
Cu^{2+}	5	*cpxA*	Yes	Yes	Yes
		rhsD	Yes	No	No
		rspQ	Yes	No	No
Ga^{3+}	5	Data not yet collected			

Treatments: Same as shown in Table 9.2.

H Sweep: Hard selective sweep observed in at least one replicate population. A hard sweep is defined as a mutation not present in the ancestral strain, observed at a frequency of 1.00 in the metal-selected population.

S Sweep: Soft selective sweep observed in at least one replicate population. A soft sweep is defined as a mutation not present in the ancestral strain, observed at a frequency of > 0.25 and < 1.00 in the metal-selected population.

Multiple: This means that selective sweeps (hard, soft, or both) were observed in more than one replicate population in at least one of the genes listed.

There is no evidence yet whether gallium resistant populations change their correlated resistance to traditional antibiotics. In addition, in all cases the acquisition of metal resistance occurred through hard selective sweeps in a few genes (silver, 3 – 6 genes; 2 – 3 for copper; 15 for iron; 3 – 5 for gallium.) These numbers represent those found in the entire selection treatments. In the case of silver nanoparticle resistance, all populations displayed the 3 resistance mutations, but for ionic silver resistance each population had one or two of the resistance mutations. In iron resistance (II, III, magnetite) each population had a subset of the resistance mutations, but never all of them. We have also shown that in the case of iron resistance that there is an alteration in gene expression patterns in resistant compared to controls populations in the presence of excess iron. While this was not examined in our silver research, previous studies [40, 42] support that this is true.

9.3 Conclusion

Our studies have clearly shown that *E. coli* K-12 MG1655 can rapidly evolve resistance to toxic metals including silver (both ionic and nanoparticles), ionic copper (Cu^{2+}), gallium, and iron as iron II in the form of sulfate heptahydrate ($FeSO_4$ $7H_2O$), iron III in the form of iron sulfate ($Fe_2(SO_4)_3$), and magnetite nanoparticles (which are a combination of iron II and III). The similarity of the results for ionic and nanoparticle exposures of metal indicates that the primary toxicity of the nanoparticles results from the release of ions. Indeed, studies have shown that in the case of silver, in which bacteria exposed to silver nanoparticles, under reducing conditions, show no adverse impact of nanoparticles [67]. However, this equivalence may not be true with regards to iron nanoparticles, as it is possible that small (< 10 nm) iron particles may cross the cell membrane into bacteria [61]. Also, in the iron studies, the populations were grown under aerobic conditions, which indicates that it is likely that most of the ionic iron experienced in the environment of the Fe^{2+}-selected populations was in the form of Fe^{3+}. This expectation is supported by gene expression data: once iron is added, control strains reduce *fecA* expression $\sim$50-fold (data not shown, [64]). This change is important because the *FecA* protein is responsible for Fe^{3+} transport in aerobic conditions [49], thereby limiting excess Fe^{3+} import in these strains. In addition, resistant strains express $\sim$65-fold less *fecA* than the controls in the absence of iron, suggesting a preemptive mechanism of defense [64].

Our studies found that silver and iron resistance were correlated traits in *E. coli* K-12 MG1655. This finding is not overly surprising given the similarity of the toxicity of silver and excess iron in bacteria. However, given the difference in the mechanisms of silver compared to iron resistance, the former only slightly increased the latter; but the opposite effect was seen for iron-selection's effect on silver. In the case of silver resistance, our data suggest that it was driven mainly by mutations in the *cusS* gene. Part of a two-component response system, this gene encodes a sensory histidine kinase that works in conjunction with *CusR* to express the *cusCFBA* ef-

Table 9.4: Description of genes in Table 9.3

Gene	Description
cpxA	Histidine kinase member of the two-component regulatory system CpxA/CpxR which responds to envelope stress response
crp	A global transcription regulator. Complexes with cyclic AMP (cAMP) which allosterically activates DNA binding
cueR →	Copper-responsive regulon transcriptional regulator
cusS	Sensory histidine kinase, senses Ag^+ and Cu^+ ions
dnaK →	Chaperone Hsp70, with co^- chaperone DnaJ
fecA ←	Ferric citrate outer membrane transporter
fliP →	Flagellar biosynthesis protein
ilvL → / → *ilvX*	ilvG operon leader peptide/uncharacterized protein
ilvG →	Pseudogene, acetolactate synthase 2 large subunit
murC →	UDP-N-acetylmuramate: L-alanine ligase
nudF ← / → *tolC*	ADP-ribose pyrophosphatase/transport channel
nusA ←	Transcription termination/antitermination L factor
ompC ←	Outer membrane porin protein C
ompF ← / ← *asnS*	Outer membrane porin 1a (Ia;b;F)/asparaginyl tRNA synthetase
ompR	Outer membrane protein R, regulates ompC, ompF, and porins.
ptsP ←	Fused PTS enzyme: PEP-prot. phosphotransferase (enz. I)/GAF domain containing prot
purL	Phosphoribosylformylglycinamidine synthase involved in the purines biosynthetic pathway
rpo	RNA polymerase, sub units A, B, C
tolC →	Transport channel
tdcR → / →yhaB	L-threonine dehydratase operon activator protein/uncharacterized protein
yeaG →	Protein kinase, endogenous substrate unidentified; autokinase
yfhm	Protects the bacterial cell from host peptidases
yjiH ←	Nucleoside recognition pore and gate family putative inner membrane transporter
hspQ	Heat shock protein, degradation of mutant DnaA; hemimethylated oriC DNA-binding protein
rhsD	Rhs family putative polymorphic toxin

flux pump. Similarly, in the copper resistant populations, resistance seems strongly tied to mutations in *cpxA*, which also encodes a sensory histidine kinase in a two-component response system. On the other hand, both Fe^{2+} and Fe^{3+} selection produced a statistically significant increase in their correlated resistance to ionic silver. However, the correlated gain in silver resistance is small compared to this same strain when directly selected for resistance to nanoparticle and ionic silver [24, 62]. In the latter study, *E. coli* K-12 MG1655 increased its resistance to ionic silver by ∼10-fold in 30 days of selection. Silver resistant strains also showed genomic changes in *ompR*, as seen in other studies of silver resistance in *E. coli* [51].

We also found that iron-resistance is correlated to resistance to gallium (an iron analog) as well as to five traditional antibiotics with varying mechanisms of action (ampicillin, polymyxin B, sulfanilamide, rifampicin, and tetracycline; [64]). This broad set of correlated traits might in part reflect the number of mutations associated with different physiological systems in iron selection (Table 9.3). Finally, we succeeded in evolving gallium resistant *E.coli*; by contrast, a similar experiment using *Pseudomonas aeruginosa* failed to achieve this resistance [54]. The plan of the Ross-Gillepsie experiment was to quench bacterial siderophores with gallium, which does not function in required iron-dependent pathways. However, our Ga^3-resistant populations were also resistant to excess iron (II, III) and, silver, but they became more sensitive to rifampicin relative to controls.

Co-selection of metal and antibiotic resistance could, in principle, result from linkage as well as pleiotropy. In the case of metals and antibiotics, however, pleiotropy is well known and appears also to be the more relevant process in our experiments, where the initial genetic uniformity and the small number of fixed mutations observed in most experiments greatly limits the potential for hitchhiking and linkage. With respect to pleiotropy, resistance mechanisms to metals such as reduction in cell-wall permeability, chemical modification, efflux, alteration of cellular targets, and sequestration are widespread. These same traits also often produce resistance to traditional antibiotics [11, 56].

In several cases, our selection studies found genomic changes consistent with what is known about general metal metabolism in *E. coli*. For example, the relationship of *fecA* and *cueR* to iron metabolism has been established by other studies [48, 52], and based on new gene-expression data we think changes in the expression of these genes may be one of the main mechanisms of preventing the entry of iron into the cell [64]. The genes yeaG and ilvX are known to be up-regulated in response to a variety of stresses, including nitrogen starvation, and we suspect that the metal resistant strains may be better equipped to deal with stress [60]. Finally, the *murC* gene is indicated to play a role in cell wall biosynthesis as well as in repairing oxidative damage [33].

The five Fe^{3+}-resistant populations showed three selective sweeps (i.e., fixations) of mutations in the gene *dnaK*, two complete sweeps of mutations in *nusA*, two mutations with one sweep being complete in *crp*, one sweep each in *ompC* and *yjiH*, and a near sweep in *tolC*. The gene *dnaK* encodes a chaperone protein (UniProtKB-P0A6Y8) that is involved in a variety of processes including DNA replication and responses to osmotic shock, both of which may have some role in iron resistance.

OmpC, an outer membrane protein (UniProtKb-P06996) encoded by *ompC*, enables the passive diffusion of small molecules across the outer membrane. Sandrini et al. [55] have shown that both *OmpC* and *OmpF* play a role in the diffusion of transferrin (an iron siderophore) across the membrane. This function might also account for the selective sweep of a mutation seen in the intergenic region between *ompF* and *asnS* in one of the Fe3+-selected replicates. The TolC protein (UniProtKB-P02930), encoded by *tolC*, forms an outer membrane channel required for the function of several efflux systems involved in the export of antibiotics and other toxic compounds from the cell. The mutation in this gene is thus a likely candidate for being involved in excess iron resistance, and it might also contribute to the increased resistance to traditional antibiotics observed in the Fe^{3+}-resistant lines [64]. The transcription termination/anti-termination protein NusA (UniProtKB-P0Aff6), encoded by *nusA*, participates in both transcription termination and antitermination. It is involved in a variety of crucial cellular processes including playing a role in rho-dependent transcriptional termination and coordinating cellular responses to DNA damage by coupling nucleotide excision repair and translation synthesis to transcription. The fact that mutations in nusA are associated with iron resistance is not surprising in that they have already been shown to contribute to general metal resistance [30, 31]. The cAMP-activated global transcriptional regulator CRP (UniProtKB-P0ACJ8), encoded by the crp gene, is involved in the regulation of over 300 genes in E. coli. This result is also consistent with the profound changes in patterns of gene expression we have observed [64]. Protein YjiH (UniProtKB-P39379, encoded by *yjiH*, remains uncharacterized, but it is thought to be involved in arginine transport. Whether the mutation in this gene plays any role in iron resistance is unclear; it might well be a case where an unrelated mutation hitchhiked to high frequency by virtue of linkage.

9.4 Summary

We have found that *E. coli* K-12 MG1655 can rapidly evolve resistance to toxic metals (Ag^+, Ga^{3+}), to excess amounts of metals that are required micronutrients (Fe^{2+}, Fe^{3+}, Cu^{2+}), and to metallic nanoparticles (FeNPs). In each case, only one or a few mutations were required to produce the resistant phenotypes. Several of the genes associated with metal resistance underwent selective sweeps in multiple populations, indicative of parallel evolution and consistent with those genes being targets of selection [63]. Although we have so far only measured gene expression changes in the Fe^{2+}-resistant populations, we suspect that the genomic changes in the other metal resistant lines will also have led to important changes in gene expression, which might be associated with the specific metal, with general metal metabolism or both. This expectation is consistent with the notion that bacteria, when first exposed to stressful environments, respond with physiological acclimation, and this state then becomes the target for subsequent genetic evolution [53].

Finally, our studies have important applications from the view of informing attempts to utilize excess ionic and nanoparticle metals to control pathogenic and multi-drug resistant bacteria (e.g. [29, 61]). Specifically, we have observed that resistance to metals evolves readily, contrary to some expectations. At the same time, we have uncovered possible constraints in the form of trade-offs between specific metal resistance traits across certain genetic backgrounds, which might impede the evolution of resistance [21]. For example, having an iron-resistant background seems to impede the evolution of silver resistance, whereas having a silver-resistant genetic background did not impede the evolution of iron resistance [21]. These and other trade-offs, if generalizable, might more effectively be of service in developing sustainable antimicrobials [23].

Acknowledgements We thank the many undergraduate students who gave dedicated attention to these experiments over the last few years: Adero Campbell, Jaminah Norman, Anna Tapia, Bobi Yang, Fidaa Almuyhaysh, Sarah Hamood, Morgan Thorton, Mariama Ibrahim, Constance Staley, and Telah Wingate. Mr. Jason Ward participated in this work via the BEACON Research Experiences for Teachers Program. This work was funded via support from the Joint School of Nanoscience & Nanoengineering, North Carolina A&T State University and UNC Greensboro, by the BEACON: An NSF Center for the Study of Evolution in Action (National Science Foundation Cooperative Agreement No. DBI-0939454), Characterizing the Evolutionary Behavior of Bacteria in the Presence of Iron Nanoparticles, NSF No. CBET-1602593, and further by an undergraduate training grant, ACE Implementation Project: Data Science and Analytics Advancing STEM Education at North Carolina A&T State University, NSF Grant No. HRD-1719498. Any opinions, findings, and conclusions or recommendations expressed in this material are those of the authors and do not necessarily reflect the views of the National Science Foundation.

References

1. https://cuverro.com/how-cuverro-works?gclid=CjwKCAjwx7DeBRBJEiwA9MeX_MwKzBp QM4_p41ku4P_ltaL9FbtFHzP8hHfW5HYrl9z6xaT0xMYikxoC2XYQAvD_BwE
2. https://www.amazon.com/curad-silver-natural-antibacterial-bandages/product-reviews/b0002 7cuzy
3. https://www.microban.com/micro-prevention/applications/commercial
4. Boyd, S., Rhinehardt, K., Thomas, M., Ewunkem, J., Graves, J.: *Evolution of excess copper resistance in Escherichia coli.* in prep (2019)
5. Cao, J., Woodhall, M.R., Alvarez, J., Cartron, M.L., Andrews, S.C.: *EfeUOB (YcdNOB) is a tripartite, acid-induced and CpxAR-regulated, low-pH Fe2+ transporter that is cryptic in Escherichia coli K-12 but functional in E. coli O157: H7.* Molecular Microbiology **65**(4), 857–875 (2007)
6. Chambers, C.W., Proctor, C.M., Kabler, P.W.: *Bactericidal effect of low concentrations of silver.* Journal-American Water Works Association **54**(2), 208–216 (1962)
7. Choudhury, P., Kumar, R.: *Multidrug-and metal-resistant strains of Klebsiella pneumoniae isolated from Penaeus monodon of the coastal waters of deltaic Sundarban.* Canadian Journal of Microbiology **44**(2), 186–189 (1998)
8. Coale, K.H., Johnson, K.S., Fitzwater, S.E., Gordon, R.M., Tanner, S., Chavez, F.P., Ferioli, L., Sakamoto, C., Rogers, P., Millero, F., et al.: *A massive phytoplankton bloom induced by an ecosystem-scale iron fertilization experiment in the equatorial Pacific Ocean.* Nature **383**(6600), 495 (1996)

9. Delany, I., Rappuoli, R., Scarlato, V.: *Fur functions as an activator and as a repressor of putative virulence genes in Neisseria meningitidis.* Molecular Microbiology **52**(4), 1081–1090 (2004)

10. Dettman, J.R., Rodrigue, N., Aaron, S.D., Kassen, R.: *Evolutionary genomics of epidemic and nonepidemic strains of Pseudomonas aeruginosa.* Proceedings of the National Academy of Sciences **110**(52), 21,065–21,070 (2013)

11. Di Cesare, A., Eckert, E., Corno, G.: *Co-selection of antibiotic and heavy metal resistance in freshwater bacteria.* Journal of Limnology **75**(s2) (2016)

12. Escolar, L., Pérez-Martín, J., De Lorenzo, V.: *Opening the iron box: Transcriptional metal-loregulation by the Fur protein.* Journal of Bacteriology **181**(20), 6223–6229 (1999)

13. Ewunkem, J., Thomas, M., Rhinehardt, K., Van Beveren, E., Yang, B., Boyd, S., Han, J., Graves, J.: *Evolution of excess iron resistance in Escherichia coli.* in prep (2018)

14. Fetherston, J.D., Kirillina, O., Bobrov, A.G., Paulley, J.T., Perry, R.D.: *The yersiniabactin transport system is critical for the pathogenesis of bubonic and pneumonic plague.* Infection and Immunity **78**(5), 2045–2052 (2010)

15. Fillat, M.F.: *The FUR (ferric uptake regulator) superfamily: Diversity and versatility of key transcriptional regulators.* Archives of Biochemistry and Biophysics **546**, 41–52 (2014)

16. Franke, S., Grass, G., Rensing, C., Nies, D.H.: *Molecular analysis of the copper-transporting efflux system CusCFBA of Escherichia coli.* Journal of Bacteriology **185**(13), 3804–3812 (2003)

17. Frawley, E.R., Crouch, M.L.V., Bingham-Ramos, L.K., Robbins, H.F., Wang, W., Wright, G.D., Fang, F.C.: *Iron and citrate export by a major facilitator superfamily pump regulates metabolism and stress resistance in Salmonella Typhimurium.* Proceedings of the National Academy of Sciences **110**(29), 12,054–12,059 (2013)

18. Garénaux, A., Caza, M., Dozois, C.M.: *The Ins and Outs of siderophore mediated iron uptake by extra-intestinal pathogenic Escherichia coli.* Veterinary Microbiology **153**(1-2), 89–98 (2011)

19. Garland, T., Rose, M.R.: *Experimental evolution: Concepts, methods, and applications of selection experiments.* University of California Press Berkeley (2009)

20. Grass, G.: *Iron transport in Escherichia coli: All has not been said and done.* Biometals **19**(2), 159–172 (2006)

21. Graves, J., Ewunkem, J., Thomas, M., K., R., Van Beveren, E., Yang, B., Boyd, S., Tapia, A., Han, J., Harrison, S.: *Silver-resistance epistasis influences iron resistance in Escherichia coli.* in prep (2019)

22. Graves, J.L.: *A grain of salt: Metallic and metallic oxide nanoparticles as the new antimicrobials.* JSM Nanotechnol Nanomed **2**(2), 1026–30 (2014)

23. Graves, J.L., Thomas, M., Ewunkem, J.A.: *Antimicrobial nanomaterials: Why evolution matters.* Nanomaterials **7**(10), 283 (2017)

24. Graves Jr, J.L., Tajkarimi, M., Cunningham, Q., Campbell, A., Nonga, H., Harrison, S.H., Barrick, J.E.: *Rapid evolution of silver nanoparticle resistance in Escherichia coli.* Frontiers in Genetics **6**, 42 (2015)

25. Gupta, A., Matsui, K., Lo, J.F., Silver, S.: *Molecular basis for resistance to silver cations in Salmonella.* Nature Medicine **5**(2), 183 (1999)

26. Gupta, A., Silver, S.: *Molecular genetics: Silver as a biocide: Will resistance become a problem?* Nature Biotechnology **16**(10), 888 (1998)

27. Haefeli, C., Franklin, C., Hardy, K.: *Plasmid-determined silver resistance in Pseudomonas stutzeri isolated from a silver mine.* Journal of Bacteriology **158**(1), 389–392 (1984)

28. Harrison, J.J., Ceri, H., Roper, N.J., Badry, E.A., Sproule, K.M., Turner, R.J.: *Persister cells mediate tolerance to metal oxyanions in Escherichia coli.* Microbiology **151**(10), 3181–3195 (2005)

29. He, S., Feng, Y., Gu, N., Zhang, Y., Lin, X.: *The effect of γ-Fe_2O_3 nanoparticles on Escherichia coli genome.* Environmental Pollution **159**(12), 3468–3473 (2011)

30. Hobman, J.L., Crossman, L.C.: *Bacterial antimicrobial metal ion resistance.* Journal of Medical Microbiology **64**(5), 471–497 (2015)

31. Hoegler, K.J., Hecht, M.H.: *Artificial gene amplification in Escherichia coli reveals numerous determinants for resistance to metal toxicity.* Journal of Molecular Evolution **86**(2), 103–110 (2018)
32. Hood, M.I., Skaar, E.P.: *Nutritional immunity: Transition metals at the pathogen–host interface.* Nature Reviews Microbiology **10**(8), 525 (2012)
33. Humnabadkar, V., Prabhakar, K., Narayan, A., Sharma, S., Guptha, S., Manjrekar, P., Chinnapattu, M., Ramachandran, V., Hameed, S., Ravishankar, S., et al.: *UDP-N-Acetylmuramic Acid L-Alanine ligase (MurC) inhibition in E. coli tolC-leads to cell death.* Antimicrobial agents and chemotherapy pp. AAC–02,890 (2014)
34. Jelenko 3rd, C.: *Silver nitrate resistant e. coli: Report of case.* Annals of Surgery **170**(2), 296 (1969)
35. Klein, J.S., Lewinson, O.: *Bacterial ATP-driven transporters of transition metals: Physiological roles, mechanisms of action, and roles in bacterial virulence.* Metallomics **3**(11), 1098–1108 (2011)
36. Konopka, K., Neilands, J.: *Effect of serum albumin on siderophore-mediated utilization of transferrin iron.* Biochemistry **23**(10), 2122–2127 (1984)
37. Köster, W., Braun, V.: *Iron (III) hydroxamate transport into Escherichia coli. Substrate binding to the periplasmic FhuD protein.* Journal of Biological Chemistry **265**(35), 21,407–21,410 (1990)
38. Lebrun, E., Brugna, M., Baymann, F., Muller, D., Lièvremont, D., Lett, M.C., Nitschke, W.: *Arsenite oxidase, an ancient bioenergetic enzyme.* Molecular Biology and Evolution **20**(5), 686–693 (2003)
39. Lewis, K.: *Persister cells.* Annual Review of Microbiology **64**, 357–372 (2010)
40. Li, X.Z., Nikaido, H., Williams, K.E.: *Silver-resistant mutants of Escherichia coli display active efflux of Ag+ and are deficient in porins.* Journal of Bacteriology **179**(19), 6127–6132 (1997)
41. Lill, R.: *Function and biogenesis of iron–sulphur proteins.* Nature **460**(7257), 831 (2009)
42. Lok, C.N., Ho, C.M., Chen, R., Tam, P.K.H., Chiu, J.F., Che, C.M.: *Proteomic identification of the Cus system as a major determinant of constitutive Escherichia coli silver resistance of chromosomal origin.* Journal of Proteome Research **7**(6), 2351–2356 (2008)
43. Mijnendonckx, K., Leys, N., Mahillon, J., Silver, S., Van Houdt, R.: *Antimicrobial silver: Uses, toxicity and potential for resistance.* Biometals **26**(4), 609–621 (2013)
44. Mindlin, S., Petrenko, A., Kurakov, A., Beletsky, A., Mardanov, A., Petrova, M.: *Resistance of permafrost and modern Acinetobacter lwoffii strains to heavy metals and arsenic revealed by genome analysis.* BioMed Research International **2016** (2016)
45. Nakashige, T.G., Zhang, B., Krebs, C., Nolan, E.M.: *Human calprotectin is an iron-sequestering host-defense protein.* Nature Chemical Biology **11**(10), 765 (2015)
46. Nunoshiba, T., Hidalgo, E., Cuevas, C.A., Demple, B.: *Two-stage control of an oxidative stress regulon: The Escherichia coli SoxR protein triggers redox-inducible expression of the SoxS regulatory gene.* Journal of Bacteriology **174**(19), 6054–6060 (1992)
47. Outten, C.E., O'halloran, T.V.: *Femtomolar sensitivity of metalloregulatory proteins controlling zinc homeostasis.* Science **292**(5526), 2488–2492 (2001)
48. Piggot, T.J., Holdbrook, D.A., Khalid, S.: *Conformational dynamics and membrane interactions of the E. coli outer membrane protein FecA: A molecular dynamics simulation study.* Biochimica et Biophysica Acta (BBA)-Biomembranes **1828**(2), 284–293 (2013)
49. Porcheron, G., Garénaux, A., Proulx, J., Sabri, M., Dozois, C.M.: *Iron, copper, zinc, and manganese transport and regulation in pathogenic enterobacteria: Correlations between strains, site of infection and the relative importance of the different metal transport systems for virulence.* Frontiers in Cellular and Infection Microbiology **3**, 90 (2013)
50. Rai, M., Deshmukh, S., Ingle, A., Gade, A.: *Silver nanoparticles: the powerful nanoweapon against multidrug-resistant bacteria.* Journal of Applied Microbiology **112**(5), 841–852 (2012)
51. Randall, C.P., Gupta, A., Jackson, N., Busse, D., O'neill, A.J.: *Silver resistance in Gram-negative bacteria: A dissection of endogenous and exogenous mechanisms.* Journal of Antimicrobial Chemotherapy **70**(4), 1037–1046 (2015)

52. Rensing, C., Grass, G.: *Escherichia coli mechanisms of copper homeostasis in a changing environment.* FEMS Microbiology Reviews **27**(2-3), 197–213 (2003)

53. Rodríguez-Verdugo, A., Tenaillon, O., Gaut, B.S.: *First-step mutations during adaptation restore the expression of hundreds of genes.* Molecular Biology and Evolution **33**(1), 25–39 (2015)

54. Ross-Gillespie, A., Weigert, M., Brown, S.P., Kümmerli, R.: *Gallium-mediated siderophore quenching as an evolutionarily robust antibacterial treatment.* Evolution, Medicine and Public Health **2014**(1), 18–29 (2014)

55. Sandrini, S., Masania, R., Zia, F., Haigh, R., Freestone, P.: *Role of porin proteins in acquisition of transferrin iron by enteropathogens.* Microbiology **159**(12), 2639–2650 (2013)

56. Seiler, C., Berendonk, T.U.: *Heavy metal driven co-selection of antibiotic resistance in soil and water bodies impacted by agriculture and aquaculture.* Frontiers in Microbiology **3**, 399 (2012)

57. Seo, S.W., Kim, D., Latif, H., OBrien, E.J., Szubin, R., Palsson, B.O.: *Deciphering Fur transcriptional regulatory network highlights its complex role beyond iron metabolism in Escherichia coli.* Nature Communications **5**, 4910 (2014)

58. Shea, C., McIntosh, M.: *Nucleotide sequence and genetic organization of the ferric enterobactin transport system: Homology to other peripiasmic binding protein-dependent systems in Escherichia coli.* Molecular Microbiology **5**(6), 1415–1428 (1991)

59. Silver, S., Phung, l.T.: *A bacterial view of the periodic table: Genes and proteins for toxic inorganic ions.* Journal of Industrial Microbiology and Biotechnology **32**(11-12), 587–605 (2005)

60. Singh, V., Chandra, D., Srivastava, B.S., Srivastava, R.: *Biochemical and transcription analysis of acetohydroxyacid synthase isoforms in Mycobacterium tuberculosis identifies these enzymes as potential targets for drug development.* Microbiology **157**(1), 29–37 (2011)

61. Sun, H., Lu, X., Gao, P.: *The exploration of the antibacterial mechanism of Fe3+ against bacteria.* Brazilian Journal of Microbiology **42**(1), 410–414 (2011)

62. Tajkarimi, M., Rhinehardt, K., Thomas, M., Ewunkem, J., Campbell, A., Boyd, S., Turner, D., Harrison, S., Graves, J.: *Selection for ionic-confers silver nanoparticle resistance in Escherichia coli.* JSM Nanotechnol Nanomed **5**, 1047 (2017)

63. Tenaillon, O., Barrick, J.E., Ribeck, N., Deatherage, D.E., Blanchard, J.L., Dasgupta, A., Wu, G.C., Wielgoss, S., Cruveiller, S., Médigue, C., et al.: *Tempo and mode of genome evolution in a 50,000-generation experiment.* Nature **536**(7615), 165 (2016)

64. Thomas, M., K., R., Ewunkem, J., Van Beveren, E., Yang, B., Boyd, S., Tapia, A., Han, J., Harrison, S., Jr., J.L.G.: *Gene expression associated with excess iron (II) resistance in Escherichia coli.* in prep (2019)

65. Vogwill, T., Kojadinovic, M., Furió, V., MacLean, R.C.: *Testing the role of genetic background in parallel evolution using the comparative experimental evolution of antibiotic resistance.* Molecular Biology and Evolution **31**(12), 3314–3323 (2014)

66. Wong, A., Kassen, R.: *Parallel evolution and local differentiation in quinolone resistance in Pseudomonas aeruginosa.* Microbiology **157**(4), 937–944 (2011)

67. Xiu, Z., Zhang, Q., Puppala, H.L., Colvin, V.L., Alvarez, P.J.: *Negligible particle-specific antibacterial activity of silver nanoparticles.* Nano Letters **12**(8), 4271–4275 (2012)

68. Yudkin, J.: *The effect of silver ions on some enzymes of bacterium coli 2.* Enzymologia **2**, 161–170 (1937)

69. Zheng, M., Åslund, F., Storz, G.: *Activation of the OxyR transcription factor by reversible disulfide bond formation.* Science **279**(5357), 1718–1722 (1998)

70. Zheng, M., Doan, B., Schneider, T.D., Storz, G.: *OxyR and SoxRS regulation of Fur.* Journal of Bacteriology **181**(15), 4639–4643 (1999)

Chapter 10
Probing the Deep Genetic Basis of a Novel Trait in *Escherichia coli**

Tanush Jagdish, J. Jeffrey Morris, Brian D. Wade and Zachary D. Blount

Abstract Evolution innovates by repurposing existing genetic elements to produce new functions. However, the range of new functions and traits this evolutionary tinkering can produce is limited to those that are supported and enabled by the rest of the genome. The full complement of genes in a genome required for a novel trait to manifest constitutes the trait's "deep" genetic basis. The deep genetic basis of novel traits can be very difficult to determine under most circumstances, leaving it understudied despite its critical importance. Novel traits that arise during highly tractable microbial evolution experiments present opportunities to correct this deficit. One such novel trait is aerobic growth on citrate (Cit^+), which evolved in one of twelve populations in the Long-Term Evolution Experiment with *Escherichia coli* (LTEE). We sought to uncover the deep genetic basis of this trait by transforming 3,985 single gene knockout mutants from the Keio collection with a plasmid that can confer

Tanush Jagdish
Department of Microbiology and Molecular Genetics, Michigan State University; BEACON Center for the Study of Evolution in Action, East Lansing, MI 48824, USA; and Department of Physics and of Biology, Kalamazoo College, Kalamazoo, MI 49006, USA. Currently at: Program for Systems Biology, Harvard University, Cambridge, MA 02138, USA e-mail: `tanush@g.harvard.edu`

J. Jeffrey Morris
Department of Microbiology and Molecular Genetics, Michigan State University; BEACON Center for the Study of Evolution in Action, East Lansing, MI 48824, USA; and Department of Biology, University of Alabama at Birmingham, Birmingham, AL 35924, USA

Brian D. Wade
Department of Plant, Soil and Microbial Sciences, Michigan State University, East Lansing, MI 48824, USA

Zachary D. Blount
Department of Microbiology and Molecular Genetics, Michigan State University; BEACON Center for the Study of Evolution in Action, East Lansing, MI 48823, USA; and Department of Biology, Gambier, OH 43022, USA e-mail: `zachary.david.blount@gmail.com`

* This paper was externally peer-reviewed.

© Springer Nature Switzerland AG 2020   107
W. Banzhaf et al. (eds.), *Evolution in Action: Past, Present and Future*,
Genetic and Evolutionary Computation, https://doi.org/10.1007/978-3-030-39831-6_10

aerobic growth on citrate. In our preliminary screen, we identified 111 genes putatively necessary for expression of the Cit$^+$ trait. Of these, $\sim 32\%$ are involved in core metabolic pathways, including the TCA and glycolysis pathways. Another $\sim 22\%$ encode a variety of transporter proteins. The remaining genes are either of unknown function or uncertain involvement with citrate metabolism. Our work demonstrates how novel traits that are built upon pre-existing functions can depend on the activity of a large number of genes, hinting at an unappreciated level of complexity in the evolution of relatively simple new functions.

Key words: Experimental Evolution, Microbial Evolution, Evolutionary Innovation, Evolution Experiments, Cit$^+$

10.1 Introduction

The living world is astonishingly diverse. Ecologically significant, qualitatively new traits have played an important role in the origin of this diversity [1]. Evolutionary innovations allow lineages to escape competitive pressures in their ancestral niches by invading new niches. Adaptation to the new niches can then drive divergence, speciation, and increased diversity [50]. Diversification driven by novel trait evolution has likely been particularly consequential in microbes, where the origin of novel traits and speciation are thought to be synonymous [12] and thus responsible for the estimated billions to trillions of extant microbial clades [37, 49].

Microbial lineages can evolve novel traits in two distinct ways: acquisition of genes from other lineages through horizontal gene transfer (HGT) or the origination of new genes through modification of existing sequences. HGT can rapidly disseminate new traits among a diverse and distantly related community of microbes and has been implicated in the spread of antibiotic resistance [32, 55], virulence factors [19, 21], and even entire metabolically related gene clusters [2]. However, truly novel traits ultimately arise by the modification of genetic information that did not originally encode them [23].

Such genetic modification, or 'evolutionary tinkering', occurs through four distinct mechanisms [1, 16]. First, new genes can arise *de novo* from mutations that induce expression of previously non-coding DNA that fortuitously yield functional polypeptides [6, 20]. Second, duplications can lead to neofunctionalization, in which mutations of redundant gene copies confer novel functions [13, 41, 43]. Third, preexisting gene components recombine to yield new functions in a process called 'domain shuffling' [8, 23, 26]. And finally, via a mechanism directly relevant to our work reported here, mutation or recombination can place existing genes under new regulatory control, co-opting functionality by changing the physiological or developmental context within which the genes are expressed [8].

The mutational event that immediately causes the manifestation — or "actualization" — of a novel trait has been the principal concern of most research into the mechanistic bases of evolutionary innovations [1, 16], but it is only one part of a larger evolutionary process [7, 8]. A new trait can only be actualized if prior evolution has "potentiated" its emergence, either by mechanistically increasing the rate of actualizing mutations, or by epistatically enabling those mutations to produce the trait upon their occurrence [8, 9]. Moreover, new traits almost always first appear in a weak form. The effectiveness of a new trait therefore requires a potentially open-ended, selection-mediated accumulation of mutations that "refines" its functionality and improves its fitness contribution [8, 44, 45]. In this process, potentiation makes the innovation possible, actualization brings it into being, and refinement makes it functionally effective [8].

Prior evolution crucially determines the potential for evolving a novel trait [10]. A new trait can only be actualized if it is mutationally reachable from an organism's existing genetic state [7, 8]. This is to say, it must be possible for the genetic information needed for the trait to arise from sequences that already exist in the genome. One facet of this principle is the necessary interrelationships between genes within a genome. All genes exist in a genome that includes both the set of regulatory elements that govern their expression and that of the other genes with which they interact. Genes also function in a broader organismal context in which their expressed products interact with those encoded by other genes. Typically, gene products only produce a given function when working in tandem with many other gene products. One consequence of this interdependence is that a new trait can only evolve in a lineage in which it will be supported by the existing genomic and organismal context. This broader context that is a necessary part of the potentiation of a novel trait's evolution may be called its "deep" genetic basis.

Consider, for example, an organism that grows on substrate A. However, the organism's environment also contains substrate B, which the organism cannot metabolize. Suppose a mutation in a gene produces an enzyme for converting substrate B to substrate A once it is in the cell, thus allowing the organism to survive on substrate B. The capacity to grow on substrate B will only manifest if the organism has the means to transport substrate B into the cell. Evolution of growth on substrate B is hence contingent on and potentiated by not only the presence of the gene that can be mutated to produce the new enzyme, but also the broader genetic context that includes the requisite transporter gene.

In real organisms, the broader context that determines evolutionary potential is exponentially more complicated [51, 56]. Organisms possess a wide array of integrated traits ranging from survival and stress tolerance to mating and reproduction [52, 54], and the genes and pathways coding for any given trait often also affect other functions at different levels of causality [25, 33, 34]. These interactions between traits and their underlying genetic bases are highly complex and difficult to understand [24, 29, 30]. Adding to this challenge is the fact that novel traits emerge over evolutionary time, which can range from hundreds to thousands of generations [10, 39, 40]. This system-wide, multi-level complexity coupled with the infeasible timescales of novel trait evolution makes it inherently challenging to identify the full set of interacting genetic elements underlying a new trait.

Novel traits sometimes arise during long-term evolution experiments with highly tractable microbial model organisms, providing opportunities to examine their deep genetic bases [17]. One such instance arose during the Long-Term Evolution Experiment with *Escherichia coli* (LTEE). The LTEE was begun in 1988 with the founding of twelve populations of *E. coli* from a single clone. These populations have since been evolved for more than 70,000 generations of daily 1:100 serial transfer in Davis and Mingioli minimal medium supplemented with glucose (DM25) [15, 34, 35]. Throughout the experiment, population samples have been frozen every 500 generations, providing a complete, viable fossil record of the evolution in each population [35]. DM25 also contains a high concentration of citrate (500 mg/L), which is added as an iron chelating agent and constitutes a potential second carbon and energy source. However, despite possessing a complete TCA cycle, *E. coli* is partly defined as a species by its inability to grow aerobically on citrate (Cit⁻) [22, 47]. This Cit⁻ phenotype is due to an inability to transport citrate into the cell when oxygen is present, but most *E. coli* strains can grow fermentatively on citrate using a transporter expressed only under anoxia [42].

Despite having the cellular machinery to potentially evolve aerobic growth on citrate (Cit⁺), spontaneous Cit⁺ mutants of *E. coli* are extremely rare under most conditions [22, 53]. Nonetheless, a weak Cit⁺ variant appeared in one LTEE population after 31,000 generations [9]. Later genomic analysis showed that the trait was actualized by a duplication that placed the previously silent citrate transporter gene, *citT*, under the control of a promoter that directs expression under aerobic metabolism [8]. The Cit⁺ subpopulation remained a minority until shortly after 33,000 generations, when refined variants better able to exploit the citrate resource evolved and rose to high frequency in the population, concurrently leading to a several-fold increase in the size of the population. This process of refinement is ongoing and has involved further adaptation to the idiosyncratic physiology of the Cit⁺ trait [5].

The long-delayed and singular evolution of the Cit⁺ trait in the LTEE was contingent upon the particular history of the population in which it arose. The ongoing research into this history has revealed the complexity of the interactions underlying evolutionary potential. A series of "replay" experiments with clones isolated from the population's fossil record showed that later clones had a significantly higher rate of mutation to Cit⁺, and thus a greater potential to evolve the trait than did earlier ones [9]. This potentiation arose in part from adaptation to acetate-based ecological interactions that evolved in the population. Quandt et al. [45] identified a series of mutations that occurred in the citrate synthase gene, *gltA*, of the lineage in which Cit⁺ eventually arose. These mutations altered carbon flow into the TCA cycle, improving growth on acetate. As citrate is also metabolized via the TCA cycle, they also pre-adapted the lineage to growth on citrate, and rendered the Cit⁺ actualizing mutation beneficial when it eventually arose. These mutations hence allowed the weak initial Cit⁺ variant lineage to remain in the population long enough to accumulate refining mutations that improved growth on citrate [45]. Ironically, the potentiating mutations in *gltA* compensated for the anti-potentiating effects of earlier mutations that were beneficial to growth on glucose [36]. Moreover, even after potentiation, the Cit⁺ actualizing mutation's fitness benefit prior to 30,000 generations was too low to permit it to outcompete other beneficial mutations available to the population [36].

Much has therefore been gleaned about the evolutionary potentiation of the Cit⁺ trait during the LTEE. However, the trait's evolution also depended on the full complement of genes necessary for citrate metabolism that existed in the ancestral genome prior to the start of the LTEE. These genes that interact and support aerobic growth on citrate, the trait's deep genetic basis, remain a mystery. Here we describe an initial exploration of the deep genetic basis of the novel Cit⁺ trait, in which we sought to identify the non-essential genes required for aerobic growth on citrate in the presence of the actualizing mutation.

10.2 Methods

10.2.1 Long-Term Evolution Experiment with E. coli

The LTEE has been described in greater detail elsewhere [35]. Briefly, twelve populations of *E. coli* B were founded in 1988 and have since been evolved for more than 70,000 generations of serial batch culture in Davis and Mingioli minimal medium supplemented with 25 mg/L glucose ("DM25"; [15, 34]. The populations are diluted 100-fold every 24 hours into fresh medium to a final volume of 10 mL and maintained at 37 °C with 120 rpm orbital aeration.

10.2.2 Screening for Genes Necessary for Aerobic Growth on Citrate

The pZB*rnk-citT* plasmid is a pUC19-based recombinant plasmid that contains a copy of the novel *rnk-citT* module that actualized the Cit⁺ trait in Ara-3 [8]. This high copy number plasmid confers a Cit⁺ phenotype in both *E. coli* B and *E. coli* K12. The Keio collection is made up of 3985 strains of *E. coli* K12 in which a single, non-essential gene has been deleted and replaced with a kanamycin resistance cassette [4]. Genes required for aerobic growth on citrate were screened by chemically transforming each Keio strain with pZB*rnk-citT* and testing each transformant for the capacity to grow aerobically on citrate using the procedure described below.

Each knock-out strain to be transformed was grown on LB plates supplemented with kanamycin at a final working concentration of 50 μg/mL. Three to five colonies of each strain were then resuspended in 300 μL of 50 mM CaCl₂. The suspensions were incubated on ice for 15 minutes, after which ~100 ng of plasmid DNA was added to the suspensions. Following 45 minutes of further incubation on ice, the cells were heat-shocked at 42 °C for 1 minute before being returned to ice for 5 minutes. A 500 μL volume of LB was added to the cells, shaken to mix, and the full volume directly spread on Christensen's citrate indicator plates supplemented with 50 μg/mL kanamycin and 100 μg/mL ampicillin [3, 11]. The plates were then incubated for 10 days at 37 °C. Plates were assessed after 5 and 10 days of incubation for development of a hot pink coloration of the medium indicative of growth on citrate. Those Keio strain transformants able to grow on citrate as indicated by a color change were presumed to have deletions in genes that were not required for aerobic growth on citrate. Those that did not produce a color change were considered to have deletions of candidate genes required for aerobic growth on citrate. Thirty six mutants from the Keio collection were unused as viable transformants for those mutants could not be obtained. This screen was carried out twice to rule out false positives.

10.2.3 Curation and Analysis of Gene Functions

Functions of candidate genes were first manually curated using the EcoCyc database [28]. All 111 functions were then grouped into five major categories: Metabolism, Membrane-Related Proteins, Stress Response, Transcription, and Motility. Genes that could not be classified into groups of three or more were labelled as "Other", and those with undefined functions were labelled "Unknown".

10.2.4 Analysis of the Mutational History in Candidate Genes

Good et al. [18] conducted whole genome, whole population sequencing of all frozen populations between 0 and 60,000 generations across all 12 LTEE lines. We downloaded the annotated sequence data made publicly available by Good et al. [18] and looked for mutations in any of the 111 candidate genes in all 12 LTEE lines over the course of 60,000 generations. In order to ensure we only looked at mutations that were nearing or had already reached fixation, we only considered mutations that had reached a frequency of 0.95. All mutational analysis was carried out using R (version 3.5.1). Analysis scripts and raw data are deposited at https://github.com/tjagdish/DeepGeneticBasis

10.3 Results

We screened the entire Keio *E. coli* K-12 gene knockout collection for non-essential genes required for aerobic growth on citrate by transforming each constituent knockout strain with a plasmid, pZB*rnk-citT*, which can confer a Cit$^+$ phenotype in a wild type genetic background. We identified 111 candidate genes that are putatively necessary for aerobic growth on citrate (Appendix). We manually categorized each candidate gene by the primary function identified for it by the EcoCyc *E. coli* database ([27], Fig. 10.1). Thirty-six percent of the candidates are involved in core

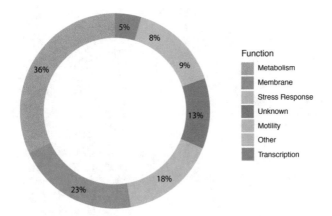

Fig. 10.1: Functional Categories of Genes Putatively Necessary for Aerobic Growth on Citrate. Genes grouped by functions assigned by EcoCyc [27]. Genes assigned to the 'other' category include those associated with fimbriae (2%), metalloproteases (2%), lipoproteins (2%), cell division (1%), cell shape (1%), and curli formation (1%)

metabolic pathways, including *sucA, sucB* and *sdhB*, which encode TCA cycle enzymes, and others that encode glycolysis pathway enzymes (Fig. 10.1). Another ~23% encode for membrane-associated proteins, such as *tatB, macB,* and *garP*, which are involved in protein translocation, antibiotic export, and galactarate transport, respectively (Fig. 10.1).

The remaining candidates are either of unknown function or uncertain involvement with citrate metabolism. These genes include a substantial number that are likely involved in bacterial stress response, such as multi-drug efflux pumps (*emrD* and *ybhF*), and regulators of acid resistance and biofilm formation (*ymgB*). Other candidates include transcription regulators (5), metalloproteases (3), lipoproteins (3), and genes involved in regulating fimbriae (3), cell division (1), cell shape (1), and curli formation (2).

Genes necessary for growth on citrate would be logical targets of selection in the Cit$^+$ population. To determine if this has been the case for the 111 candidate mutations we identified, we searched the whole-metagenome sequence data for the citrate-using population from Good et al. [18]. We found that mutations had fixed or nearly fixed in 8 of the candidate genes by 60,000 generations (Table 10.2). However, all mutations occurred in Ara-3 after ~35,000 generations, at which point Ara-3 had evolved a mutator phenotype [8].

All 12 LTEE populations have been evolving for over 70,000 generations. Ara-3 remains the only population in which aerobic growth on citrate has evolved. An ongoing question is that of whether it might evolve in any of the other 11 populations. Mutations that impair or eliminate the function of genes required for the Cit$^+$ trait would presumably reduce the likelihood of evolving it. We therefore examined

Ara-1	Ara-2	Ara-3	Ara-4	Ara-5	Ara+2	Ara+3		Ara+4	Ara+6	
bisC	ilvG	aroG	emrD	iscR	sucB	phoR	aroG	aroG	aroG	ybaY
rarD	nadB	atpl	flgl			rarD	atpl	phoR	csgG	ycbS
sucB	phoR	csgG	gatY			rimL	dinJ		dedA	ydiJ
yahK	treA	emrD	mltD			sdhB	flgE		erfK	yegV
ycbS	ybhF	flgF	sucA			sfmA	flgF		exbD	yejB
yegV	ydhP	ybhF	yahB			sucA	flgl		flgE	yfiB
	ydiZ	ycbS	ydiS			sucB	garL		hcaE	ygbl
	yhjK	ydiJ	ydiZ			tauC	garP		ilvG	yadC
	yneJ		yegV			thiD	hcaE		pfkA	wcaK
			yihF			yadC	ilvD		phoR	treA
			yigZ			yahB	mltD		rimL	sucB
						ybaY	ygbl			
						ydhP	yggN			
						yejB	yibQ			
						ypdE	yijB			

Fig. 10.2: Genes putatively identified as necessary for aerobic growth on citrate in which mutations have fixed in LTEE populations by 60,000 generations. Ara-1, Ara-2, Ara-3, Ara-4, Ara+3 and Ara+6 have all evolved heightened mutation rates over the course of evolution. No fixed mutations in candidate genes were identified in populations Ara-6, Ara+1, and Ara+5

the metagenome sequences for the other LTEE populations for mutations in the 111 genes our screen identified.

We found an array of mutations in genes putatively essential for aerobic growth on citrate across the other LTEE populations. The vast majority of these mutations occurred in the five populations, Ara-1, Ara-2, Ara-3, Ara-4, Ara+3, and Ara+6, in which elevated mutation rates have also evolved over the course of the experiment (Fig. 10.3; Fig. 10.2). However, we did identify mutations in three non-mutator populations: a single mutation in a noncoding region of the transcription regulator gene, *iscR*, fixed in Ara-5 by 45,000 generations; a missense mutation in a key oxoglutarate dehydrogenase component gene, *sucB*, fixed in Ara+2 by 47,000 generations; and in Ara+4 a missense mutation in the glycolysis regulating phosphofructokinase gene, *phoR*, fixed by 18,000 generations, as well as the insertion of a mobile genetic element in the aromatic amino acid biosynthesis gene *aroG*. We identified no mutations in any candidate genes in Ara-6, Ara+1, or Ara+5.

10.4 Discussion

The Cit⁺ trait that arose in the LTEE would seem to be a simple innovation, given that the trait can be conferred in the ancestral genetic background by activating expression of the *citT* gene that encodes a citrate-C4-dicarboxylate antiporter [8]. Despite this apparent simplicity, the trait's manifestation required the activity of other gene products, and thus the presence of other preexisting genes. These genes

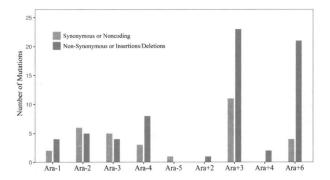

Fig. 10.3: Fixed mutations in candidate genes across 12 LTEE populations by 60,000 generations.

constitute the deep genetic basis of the Cit$^+$ trait, and we sought to identify them. In total, our preliminary screen showed that 3%, or 111 out of 4288 non-essential genes in *E. coli* were required for aerobic growth on citrate. Moreover, considering the possible role of epistatic interactions and the conservative nature of our screen, this number is likely to be an underestimate [14, 46]. Our findings therefore suggest that the deep genetic basis of the novel Cit$^+$ trait is quite broad, highlighting the integrated nature of the organism.

The largest group of genes that we identified as putatively necessary for aerobic growth on citrate belong to core metabolism. *E. coli* metabolizes exogenous citrate via the TCA cycle, making the genes that encode the steps in the cycle necessary [31, 38, 47]. Aerobic growth on citrate as a sole carbon source also creates problems for biosynthesis as glycolysis is bypassed. Growth on citrate as a sole carbon source thus requires the capacity to feed carbon into the gluconeogenesis pathways, as well as to produce the TCA intermediates and key amino acid precursors 2-oxyglutarate and succinyl-CoA. A further complication is caused by the physiology of the citT transporter. Every molecule of citrate imported requires the simultaneous export of a C4-dicarboxylate, and specifically the TCA intermediates of succinate, fumarate, or malate [42]. Consequently, as few as 2 carbons per citrate molecule are available for both catabolism and anabolism (Fig. 10.4). Given these considerations, several genes might be predicted as necessary for aerobic growth on citrate.

Predicted genes include isocitrate dehydrogenase (*icd*), which is necessary to yield 2-oxoglutarate because the 2-oxoglutarate decarboxylase (SucAB) reaction is irreversible; isocitrate lyase (*aceA*), which is necessary to bypass the CO2-producing SucAB and SucCD reactions via the glyoxylate shunt; and citrate synthase (*gltA*), which is necessary to pass carbons harvested via glyoxylate back into the TCA cycle to reach 2-oxoglutarate and succinyl-CoA. Interestingly, our screen did not yield any of these genes. This odd result might be an artifact of the screen's design. These genes would be essential in minimal medium with citrate as a sole carbon source. However, the Christensen's citrate agar on which we conducted our phenotypic screen likely contains sufficient amino acids and other metabolites to compensate for anabolic deficiencies caused by loss of any one of these key en-

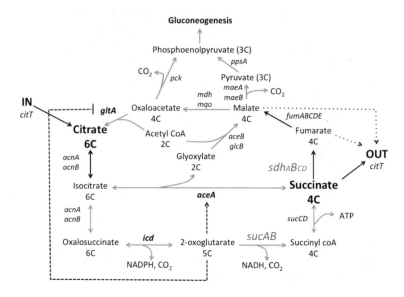

Fig. 10.4: Citrate metabolic pathways in Ara-3 Cit^+. Extracellular citrate is exchanged for the intracellular C4-dicarboxylate TCA intermediates, succinate, fumarate, and malate by the citT antiporter. This physiology means perhaps only 2 carbon atoms are available per citrate molecule for both catabolism and anabolism. The citrate can be metabolized in two ways. In the catabolic pathway (orange arrows), both carbons are lost as CO_2 (leaving none for biosynthesis), but substantial energy is conserved in the form of NAD(P)H. In the anabolic pathway (green arrows), no energy is gained, but the carbons are harvested as glyoxylate, and can either be passed back into the TCA cycle or into gluconeogenesis for biosynthesis of amino acids and other necessary metabolites. Where two gene names are given (e.g. *acnA, acnB*) for a reaction, the genes are redundant; where genes are listed as a complex (e.g. *sucAB*), all gene products are needed to catalyze the reaction. Because of redundant genes and/or pathways, only 3 genes (*gltA, icd*, and *aceA*, shown in bold) may be predicted as essential for aerobic growth on citrate as a sole carbon source. None of these genes were identified in our screen, but others, shown in blue, were. Regulatory effects of 2-oxoglutarate are indicated with dashed lines.

zymes. We plan to follow up on these preliminary findings with further screens using Christensen's citrate broth from which we will exclude yeast extract, which should allow us to evaluate this hypothesis.

Some of the TCA genes we identified in our screen are not strictly essential, but their absence could lead to the over-accumulation of intermediates, negatively affecting regulation, and causing unbalanced growth. For instance, deletion of either of the subunits of 2-oxoglutarate decarboxylase (*sucA* or *sucB*) eliminated the capacity to grow aerobically on citrate. This effect could be due to an accumulation of 2-oxoglutarate levels that would inhibit citrate synthase, preventing the use of citrate-derived carbon for amino acid biosynthesis. Similarly, loss of one of the four subunits of succinate dehydrogenase, SdhB, eliminated the Cit^+ trait. SdhB is the cytoplasmic subunit responsible for passing electrons from SdhA, which binds both succinate and FAD, to the electron transport chain [48]. It is possible that loss of SdhB sequesters succinate in a form that cannot be accessed by citT, stopping the flow of citrate into the cell.

The second largest subset of essential gene candidates were membrane proteins and cell wall synthesis enzymes. The 22 membrane proteins include ABC transporter families such as *macB* and *yhhJ*, and a diversity of importers, exporters, and symporters. Several genes involved in peptidoglycan synthesis and remodeling (*ddlA, erfK, dacA*) were also identified. This subset is more puzzling, as there is no obvious direct relationship between these genes and citrate metabolism. While Cit$^+$ requires an aerobically functioning citT membrane transporter, an *rnk-citT* module (where rnk is an aerobic promoter) was available to the tested cells via a high-copy-number plasmid [8]. Thus, the removal of the citT gene from the genome, or any other transporter gene, should in principle not affect Cit$^+$.

The role of most of the genes we have identified in aerobic growth on citrate is unclear. This lack of obvious connection is perhaps most clearly seen in the cell appendage biosynthesis genes our screen showed as putatively necessary to the Cit$^+$ trait. These genes included those involved in flagellar biosynthesis (*flgE, flgF, flgI,* and *flgJ*), fimbrial assembly (*ycbS*), and curli synthesis (*csgABFG*). Similarly, 18 candidate genes are involved in stress response functions in *E. coli*, including multidrug efflux pumps (*emrD* and *ybhF*), reactive oxygen defenses (*sodB*), acid resistance regulators (*ymgB*), and response regulators for phosphate starvation (*phoR*). At least some of these genes may be experimental artifacts of a screening procedure that exposed the cells to multiple stressors, including exposure to two antibiotics, treatment in high concentrations of calcium chloride, and cold.

Might the candidate genes we identified be targets of refining mutations that improved aerobic growth on citrate after the evolution of the Cit$^+$ trait? To answer this question, we examined the metagenome sequence data generated by Good et al. [18] for the Ara-3 population through 60,000 generations. We found mutations in 8 candidate genes, though none reached high frequency until well after the Cit$^+$ clade evolved an elevated mutation rate around 35,000 generations [8]. The earliest mutation to rise to high frequency occurred at around 37,000 generations in the gene *csgG*, which encodes an outer membrane lipoprotein involved in curli biosynthesis. Due to the elevated mutation rate, however, it is unclear if these mutations are beneficial or reached high frequency by hitchhiking with some other beneficial allele.

Only 3 of the 8 mutated candidate genes in Ara-3, *aroG, atpI*, and *ydiJ*, are related to central metabolism. These genes have a total of six mutations, all of which are either synonymous or occur in noncoding regions. While it is possible that these mutations might have beneficial fitness effects, it seems more likely that they are hitchhikers. The five other candidate genes, *csgG, flgF, ycbS, emrD*, and *ybhF*, which are involved in biofilm, flagellar, and fibrial biosynthesis, and encode multidrug efflux pumps, respectively, have a total of 12 mutations. Five of these mutations are synonymous or noncoding, but the remaining seven are missense or indel mutations. The latter seven mutations would presumably affect gene function, and potentially conflict with our findings that they are necessary for the Cit$^+$ trait. Later work will examine these mutations, their effects, and will determine if their identification in the screen was perhaps an experimental artifact of some sort. Broadly, however, the rarity of mutations in candidate genes in Ara-3 after the evolution of Cit$^+$ is consistent with their being necessary for the trait and suggests that mutations in them are generally detrimental.

If the 111 genes we identified are actually necessary for aerobic growth on citrate, then mutations that impair or eliminate their function would seemingly reduce the likelihood of evolving the Cit$^+$ trait. We identified nonsynonymous mutations and indels in multiple candidate genes in 7 of the other 11 LTEE populations. We do not yet know enough about the effects of these mutations on the function of the genes in question. However, any that impair or eliminate function would likely reduce or foreclose the possibility of the evolution of Cit$^+$ in the respective populations. It will be interesting to examine how the evolvability of Cit$^+$ varies between the populations in which mutations have occurred in candidate genes, and Ara-6, Ara+1, and Ara+5, in which they have not.

Novel traits are not the result solely of the genetic changes that immediately underlie them. Those genetic changes that actualize a trait must always occur in a genetic background containing an integrated set of genes and gene products that allow them to produce that new trait. Though the importance of this deep genetic basis of novel traits is in a sense obvious, it has rarely been examined. The preliminary work we have described gives a glimpse into the deep genetic basis of the Cit$^+$ trait that arose in the LTEE. The trait was actualized by a mutation that activated the expression of a citrate transporter gene, *citT*, when oxygen was present via the cooption of an alternate promoter, that of the aerobically expressed gene *rnk*. Despite this apparent simplicity, our results show that the manifestation of the Cit$^+$ trait depends on the activity of more than 100 other genes. This finding shows how the additive nature of novel traits in evolution means that even relatively minor, seemingly simple novel traits nonetheless depend on a foundation of many pre-existing elements that interact in complex and highly integrated ways.

Indeed, our findings suggest that what genes might be involved in the manifestation of a trait may be anything but obvious due to this complexity and integration. Prior work has shown that the evolution of the Cit$^+$ trait was historically contingent upon several potentiating mutations that arose over the course of the Ara-3 population's history during the experiment, and the twisted paths of evolution during that history. Our findings here show that the trait was contingent upon not simply this history during the experiment. Indeed, it was contingent upon the much longer history that preceded the experiment, over which evolution constructed an organism with the full complement of genes that interacted in such a way as to support the manifestation of the trait once the actualizing mutation took place. Such contingent histories are necessary to provide the proper deep genetic basis that underlies the evolution of all novel traits. Given the role of novel traits in evolution, our work argues that it is time to take this deeper historical contingency seriously.

Acknowledgements We thank Neerja Hajela for her outstanding technical expertise, logistical aid, and general moral support. This project would not have been possible without her. We also thank the undergraduate technicians who provided logistical support for our work, including Camorrie Bradley, Michelle Mize, Maia Rowles, Kiyana Weatherspoon, Rafael Martinez, Jamie Johnson, Devin Lake, and Jessica Baxter. We are grateful to Richard Lenski for helpful discussions, advice, funding, lab space, and his continued mentorship. We are especially grateful to Erik Goodman for his extensive and path-defining leadership of the BEACON Center for the Study of Evolution in Action, and in whose honor this work is submitted. Mark Kauth, Kyle Card, and Nkrumah Grant made critical contributions to the design of this project. This research was supported in part by grants from the National Science Foundation (currently DEB-1451740) and the BEACON Center for the Study of Evolution in Action (DBI-0939454). T.J. acknowledges support from the Internship Stipend Program at the Center for Career and Professional Development at Kalamazoo College.

References

1. Andersson, D.I., Jerlstroem-Hultqvist, J., Naesvall, J.: *Evolution of new functions de novo and from preexisting genes*. Cold Spring Harb Perspect Biol **7**(6), a017996 (2015)
2. Araviad, L., Tatusov, R.L., Wolf, Y.I., Roland Walker, D., Koonin, E.: *Evidence for massive gene exchange between archaeal and bacterial hyperthermophiles*. Trends in Genetics **14**, 442–444 (1999)
3. Atlas, R.: *Handbook of Microbiological Media for the Examination of Food*. CRC, Boca Raton, FL (2006)
4. Baba, T., Ara, T., Hasegawa, M., Takai, Y., Okumura, Y., Baba, M., Datsenko, K.A., Tomita, M., Wanner, B.L., Mori, H.: *Construction of Escherichia coli K-12 in-frame, single-gene knockout mutants: The Keio collection*. Molecular Systems Biology **2**, 2006.0008 (2006)
5. Bajic, D., Vila, J.C.C., Blount, Z.D., Sanchez, A.: *On the deformability of an empirical fitness landscape by microbial evolution*. Proceedings of the National Academy of Sciences of the United States of America **115**, 11,286–11,291 (2018)
6. Begun, D.J., Lindfors, H.A., Kern, A.D., Jones, C.D.: *Evidence for de novo evolution of testis-expressed genes in the Drosophila yakuba/Drosophila erecta clade*. Genetics **176**(2), 1131–1137 (2007)
7. Blount, Z.D.: *A case study in evolutionary contingency*. Stud Hist Philos Sci Part C: Stud Hist Philos Biol Biomed Sci **58**, 82 – 92 (2016)
8. Blount, Z.D., Barrick, J.E., Davidson, C.J., Lenski, R.E.: *Genomic analysis of a key innovation in an experimental Escherichia coli population*. Nature **489**, 513–518 (2012)
9. Blount, Z.D., Borland, C.Z., Lenski, R.E.: *Historical contingency and the evolution of a key innovation in an experimental population of Escherichia coli*. Proceedings of the National Academy of Sciences of the United States of America **105**(23), 7899–7906 (2008)
10. Blount, Z.D., Lenski, R.E., Losos, J.B.: *Contingency and determinism in evolution: Replaying life's tape*. Science **362**, eaam5979 (2018)
11. Christensen, W.: *Hydrogen sulfide production and citrate utilization in the differentiating of enteric pathogens and coliform bacteria*. Res Bull Weld County Health Dept Greeley Colo **1**, 3–16 (1949)
12. Cohan, F.M.: *Bacterial species and speciation*. System Biology **50**(4), 513–524 (2001)
13. Conant, G.C., Wolfe, K.H.: *Turning a hobby into a job: How duplicated genes find new functions*. Nature Reviews Genetics **9**, 938–950 (2008)
14. Dasmeh, P., Girard, E., Serohijos, A.W.R.: *Highly expressed genes evolve under strong epistasis from a proteome-wide scan in E. coli*. Scientific Reports **7**, 15844 (2017)
15. Davis, B.D., Mingioli, E.S.: *Mutants of Escherichia coli requiring methionine or vitamin B12*. Journal of Bacteriology **60**, 17–28 (1950)

16. Ding, Y., Zhou, Q., Wang, W.: *Origins of new genes and evolution of their novel functions.* Annual Review of Ecology, Evolution, and Systematics **43**, 345–363 (2012)
17. Elena, S.F., Lenski, R.E.: *Evolution experiments with microorganisms: The dynamics and genetic bases of adaptation.* Nature Reviews Genetics **4**, 457–469 (2003)
18. Good, B.H., McDonald, M.J., Barrick, J.E., Lenski, R.E., Desai, M.M.: *The dynamics of molecular evolution over 60,000 generations.* Nature **551**, 45–50 (2017)
19. Groisman, E.A., Ochman, H.: *Pathogenicity islands: Bacterial evolution in quantum leaps.* Cell **87**, 791–794 (1996)
20. Guerzoni, D., McLysaght, A.: *De novo origins of human genes.* PLoS Genetics **7**(11), e1002381 (2011)
21. Hacker, J., Blum-Oehler, G., Muehldorfer, I., Tschaepe, H.: *Pathogenicity islands of virulent bacteria: Structure, function and impact on microbial evolution.* Molecular Microbiology **23**, 1089–1097 (1997)
22. Hall, B.G.: *Chromosomal mutation for citrate utilization by Escherichia coli k-12.* Journal of Bacteriology **151**, 269–273 (1982)
23. Jacob, F.: *Evolution and tinkering.* Science **196**(4295), 1161–1166 (1977)
24. Jerison, E.R., Desai, M.M.: *Genomic investigations of evolutionary dynamics and epistasis in microbial evolution experiments.* Current Opinion in Genetics & Development **35**, 33–39 (2015)
25. Jerison, E.R., Kryazhimskiy, S., Mitchell, J.K., Bloom, J.S., Kruglyak, L., Desai, M.M.: *Genetic variation in adaptability and pleiotropy in budding yeast.* eLife **6**, e27167 (2017)
26. Kawashima, T., Kawashima, S., Tanaka, C., Murai, M., Yoneda, M., Putnam, N.H., Rokhsar, D.S., Kanehisa, M., Satoh, N., Wada, H.: *Domain shuffling and the evolution of vertebrates.* Genome Research **19**(8), 1393–1403 (2009)
27. Keseler, I.M., Collado-Vides, J., Gama-Castro, S., Ingraham, J., Paley, S., Paulsen, I.T., Peralta-Gil, M., Karp, P.D.: *EcoCyc: A comprehensive database resource for Escherichia coli.* Nucleic Acids Research **33**(suppl-1), D334–D337 (2005)
28. Keseler, I.M., Mackie, A., Santos-Zavaleta, A., Billington, R., Bonavides-Martinez, C., Caspi, R., et al.: *The EcoCyc database: Reflecting new knowledge about Escherichia coli K-12.* Nucleic Acids Research **45**, D543–D550 (2016)
29. Khan, A.I., Dinh, D.M., Schneider, D., Lenski, R.E., Cooper, T.F.: *Negative epistasis between beneficial mutations in an evolving bacterial population.* Science **332**, 1193–1196 (2011)
30. Kryazhimskiy, S., Rice, D.P., Jerison, E.R., Desai, M.M.: *Global epistasis makes adaptation predictable despite sequence-level stochasticity.* Science **344**, 1519–1522 (2014)
31. Lara, F.J.S., Stokes, J.L.: *Oxidation of citrate by Escherichia coli.* Journal of Bacteriology **63**(3), 415–420 (1952)
32. Lawrence, J.G., Ochman, H.: *Reconciling the many faces of lateral gene transfer.* Trends in Microbiology **10**, 1–4 (2002)
33. Lee, S.H., Yang, J., Goddard, M.E., Visscher, P.M., Wray, N.R.: *Estimation of pleiotropy between complex diseases using single-nucleotide polymorphism-derived genomic relationships and restricted maximum likelihood.* Bioinformatics **28**, 2540–2542 (2012)
34. Lenski, R.E.: *Experimental studies of pleiotropy and epistasis in Escherichia coli. i. variation in competitive fitness among mutants resistant to virus T4.* Evolution **42**, 425–432 (1988)
35. Lenski, R.E., Rose, M.R., Simpson, S.C., Tadler, S.C.: *Long-term experimental evolution in Escherichia coli. I. adaptation and divergence during 2,000 generations.* American Naturalist **138**(6), 1315–1341 (1991)
36. Leon, D., D'Alton, S., Quandt, E.M., Barrick, J.E.: *Innovation in an E. coli evolution experiment is contingent on maintaining adaptive potential until competition subsides.* PLoS Genetics **14**, 1–22 (2018)
37. Locey, K.J., Lennon, J.T.: *Scaling laws predict global microbial diversity.* Proceedings of the National Academy of Sciences of the United States of America **113**(21), 5970–5975 (2016)
38. Luetgens, M., Gottschalk, G.: *Why a co-substrate is required for anaerobic growth of Escherichia coli on citrate.* Journal of General Microbiology **119**, 63–70 (1980)
39. Meyer, J.R., Dobias, D.T., Medina, S.J., Servilio, L., Gupta, A., Lenski, R.E.: *Ecological speciation of bacteriophage lambda in allopatry and sympatry.* Science **354**, 1301–1304 (2016)

40. Meyer, J.R., Dobias, D.T., Weitz, J.S., Barrick, J.E., Quick, R.T., Lenski, R.E.: *Repeatability and contingency in the evolution of a key innovation in phage lambda.* Science **335**, 428–432 (2012)
41. Ohno, S.: *Evolution by gene duplication.* Springer-Verlag (1970)
42. Pos, K.M., Dimroth, P., Bott, M.: *The Escherichia coli citrate carrier CitT: A member of a novel eubacterial transporter family related to the 2-oxoglutarate/malate translocator from spinach chloroplasts.* Journal of Bacteriology **180**, 4160–4165 (1998)
43. Qian, W., Zhang, J.: *Genomic evidence for adaptation by gene duplication.* Genome Research **24**, 1356–1362 (2014)
44. Quandt, E.M., Deatherage, D.E., Ellington, A.D., Georgiou, G., Barrick, J.E.: *Recursive genomewide recombination and sequencing reveals a key refinement step in the evolution of a metabolic innovation in Escherichia coli.* Proceedings of the National Academy of Sciences of the United States of America **111**, 2217–2222 (2014)
45. Quandt, E.M., Gollihar, J., Blount, Z.D., Ellington, A.D., Georgiou, G., Barrick, J.E.: *Fine-tuning citrate synthase flux potentiates and refines metabolic innovation in the Lenski evolution experiment.* eLife **4**, e09696 (2015)
46. Remold, S.K., Lenski, R.E.: *Pervasive joint influence of epistasis and plasticity on mutational effects in Escherichia coli.* Nature Genetics **36**, 423–426 (2004)
47. Reynolds, C.H., Silver, S.: *Citrate utilization by Escherichia coli: Plasmid- and chromosome-encoded systems.* Journal of Bacteriology **156**, 1019–1024 (1983)
48. Ruprecht, J., Yankovskaya, V., Maklashina, E., Iwata, S., Cecchini, G.: *Structure of Escherichia coli succinate: Quinone oxidoreductase with an occupied and empty quinone-binding site.* Journal of Biological Chemistry **284**, 29836–29846 (2009)
49. Schloss, P.D., Handelsman, J.: *Status of the microbial census.* Microbiol Mol Biol Rev **68**(4), 686–691 (2004)
50. Schluter, D.: *Evidence for ecological speciation and its alternative.* Science **323**(5915), 737–741 (2009)
51. Segrè, D., DeLuna, A., Church, G.M., Kishony, R.: *Modular epistasis in yeast metabolism.* Nature Genetics **37**, 77–83 (2005)
52. Suding, K.N., Goldberg, D.E., Hartman, K.M.: *Relationships among species traits: Separating levels of response and identifying linkages to abundance.* Ecology **84**, 1–16 (2003)
53. Van Hofwegen, D.J., Hovde, C.J., Minnich, S.A.: *Rapid evolution of citrate utilization by Escherichia coli by direct selection requires citT and dctA.* Journal of Bacteriology **198**, 1022–1034 (2016)
54. Vasi, F., Travisano, M., Lenski, R.E.: *Long-term experimental evolution in Escherichia coli. ii. changes in life-history traits during adaptation to a seasonal environment.* The American Naturalist **144**, 432–456 (1994)
55. von Wintersdorff, C.J.H., Penders, J., van Niekerk, J.M., Mills, N.D., Majumder, S., van Alphen, L.B., Savelkoul, P.H.M., Wolffs, P.F.G.: *Dissemination of antimicrobial resistance in microbial ecosystems through horizontal gene transfer.* Front Microbiol **7**, 173 (2016)
56. Wang, Z., Liao, B.Y., Zhang, J.: *Genomic patterns of pleiotropy and the evolution of complexity.* Proceedings of the National Academy of Sciences of the United States of America **107**, 18034–18039 (2010)

Appendix

Genes identified as putatively necessary for aerobic growth on citrate and their annotations as reported by EcoCyc.

Genes	Group	Function
ada	stress response	Adaptive response transcriptional regulator
aroG	metabolism	Amino acid biosynthesis
-	-	-
atpI	metabolism	ATP synthase accessory factor
bisC	stress response	Biotin scavenging and protection from oxygen damage
chaC	metabolism	γ-glutamylcyclotransferase with relatively low catalytic efficiency
chpB	stress response	ChpB is the toxin component of the ChpB-ChpS toxin-antitoxin system
csgA	motility	Curlin subunit
csgB	motility	Curlin subunit
csgF	membrane	Curlin subunit
csgG	membrane	Membrane lipoprotein
dacA	membrane	Penicillin-binding protein, antibiotic resistance
ddlA	metabolism	Peptidoglycan biosynthesis
dedA	membrane	Inner membrane protein with trans-membrane domains
dinJ	stress response	Antitoxin to YafQ toxin
dksA	transcription	Initiates transcription by binding to RNA polymerase
ecnA	stress response	Antidote/toxin system which may function as a starvation adaptation
emrD	stress response	Multidrug efflux pump
eptB	metabolism	Encodes the Ca^{2+}-induced phosphoethanolamine transferase
erfK	cell shape	Covalent attachment of peptidoglycan to the outer membrane
exbD	metabolism	Active transport of iron-siderophore complexes and vitamin B12 across the outer membrane
FabR	metabolism	Fatty acid biosynthesis regulator
flgE	motility	Flagellar hook protein
flgF	motility	Flagellar basal-body rod protein
flgI	motility	Flagellar P-ring protein
fliJ	motility	Flagellar biosynthesis protein
fur	transcription	DNA-binding transcriptional dual regulator
galP	membrane	Galactose symporter
garL	metabolism	α-dehydro-β-deoxy-D-glucarate aldolase
garP	membrane	Galactarate/glucarate/glycerate transporter
gatY	stress response	Catalytic subunit of the GatYZ tagatose-1,6-bisphosphate aldolase complex. Responds to low pH, acetate stress, and glucose limitation
gmX	metabolism	Catabolism of DNA as carbon source
hcaE	unknown	Unknown
hslO	stress response	Stress regulated chaperone
htrL	unknown	Unknown
hyfC	metabolism	Hydrogenase-4 component C
ilvD	metabolism	Isoleucine and valine biosynthesis
ilvG	metabolism	Isoleucine biosynthesis
iscR	transcription	Iron-sulfur cluster regulator
ligB	mismatch repair	NAD—dependent DNA ligase
letC	membrane	ABC transporter membrane subunit
mac-B	membrane	ABC exporter
mltD	membrane	Membrane-bound transglycosylase
murP	membrane	N-acetylmuramic acid PTS transport system
nadB	metabolism	NAD biosynthesis I pathway
nohA	transcription	Transcription regulator of nohA
ompW	membrane	Receptor for Colicin S4
oppC	membrane	ABC transporter
paaK	metabolism	Phenylacetate degradation
pfkA	metabolism	6-phosphofructokinase I (PFK I)
phoR	stress response	Responds to variations in the level of extracellular inorganic phosphate
ppdB	membrane	Conformational isomerization of prolyl residues in peptides
rarD	membrane	Putative transporter
rimL	metabolism	Acetyltransferase
sdhB	metabolism	Succinate:quinone oxidoreductase
sfmA	motility	Putative fimbrial protein
yngB	unknown	Unknown
sfmA	motility	Putative fimbrial protein
sodB	stress response	Superoxide dismutases, implicated in the response to a large number of environmental changes
sucA	metabolism	2-oxoglutarate decarboxylase, thiamine-requiring
sucB	metabolism	Conversion of 2-oxoglutarate (2-ketoglutarate) to succinyl-CoA and carbon dioxide
sufC	metabolism	Assembly of iron-sulfur clusters
tatB	membrane	Inner membrane component of the twin arginine translocation (Tat) complex
tauC	membrane	ABC transporter
thiD	metabolism	Biosynthesis of thiamine
treA	metabolism	Hydrolysis of trehalose into two molecules of D-glucose
trkD	membrane	Potassium ion uptake under hyper-osmotic stress at a low pH
ulaE	metabolism	Isomerization of L-ribulose 5-phosphate to D-xylulose 5-phosphate
ulaF	metabolism	Isomerization of L-ribulose 5-phosphate to D-xylulose 5-phosphate
weaK	metabolism	Colanic acid biosynthesis
yadC	motility	Fimbrial operon
yahB	unknown	Unknown
yahK	metabolism	NADPH-dependent aldehyde reductase activity
ybaO	stress response	Cysteine detoxification
ybaY	stress response	ybaY has supercoiling-dependent transcription, which is associated with the osmotic stress response
ybhA	metabolism	Pyridoxal phosphate phosphatase
ybhF	stress response	Multidrug ABC exporter
ychS	motility	Fimbrial operon
yceO	stress response	Acid stress response
ycgB	unknown	Unknown
ydcS	stress response	Toxic protein; expression of YdcS leads to loss of cell viability
ydbP	membrane	Putative transporter
ydfJ	metabolism	FAD-linked oxidoreductase
ydiS	metabolism	Putative oxidoreductase with FAD/NAD(P)-binding domain
ydjJ	metabolism	Putative zinc-binding dehydrogenase
ydjZ	membrane	Putative transmembrane protein
yeaM	stress response	DNA-binding transcriptional repressor
yeaW	metabolism	2Fe-2S cluster-containing protein
yegV	metabolism	Putative sugar kinase
ygjB	membrane	Inner membrane subunit of a putative ATP-dependent oligopeptide uptake system
yfiB	biofilm	Biofilm formation
ygbI	unknown	Unknown
ygeN	biofilm	Biofilm formation
yhaC	unknown	Unknown
yhbJ	membrane	ABC transporter
yhjK	membrane	Inner membrane protein
yiaY	metabolism	NAD-dependent L-threonine dehydrogenase activity
yibQ	metabolism	Nucleoside diphosphatase
yidQ	unknown	Putative lipoprotein
yihF	unknown	Unknown
yjgZ	unknown	Unknown
yjbE	membrane	Putative membrane protein
yjjB	metabolism	Putative succinate exporter
ymfD	metabolism	e14 prophage; Putative SAM-dependent methyltransferase
ymfJ	unknown	e14 prophage; protein YmfJ
ymfI	transcription	Putative LysR-type DNA-binding transcriptional regulator; protein YmfI
yobD	membrane	Inner membrane protein
ypdE	metalloprotease	Metalloendopeptidase with broad specificity
ypdB	stress response	Acid stress response
zapA	cell division	Cell division protein
taeE	unknown	Unknown
yfcG	unknown	Unknown
yfeG	transcription	Transcriptional regulator
yrhC	unknown	Unknown

Chapter 11
Fitness Costs and Benefits of Resistance to Phage Lambda in Experimentally Evolved *Escherichia coli**

Alita R. Burmeister, Rachel M. Sullivan and Richard E. Lenski

Abstract Fitness tradeoffs play important roles in the evolution of organisms and communities. One such tradeoff often occurs when bacteria become resistant to phage at the cost of reduced competitiveness for resources. Quantifying the cost of phage resistance has frequently relied on measuring specific traits of interest to industrial applications. In an evolutionary context, however, fitness encompasses the effects of all traits relevant to an organism's survival and reproductive success in a particular environment. Therefore, measurements of the net reproduction and survival of alternative genotypes offer greater power for predicting the fate of different genotypes in complex and dynamic communities. In this study, we measured the fitness of experimentally evolved, λ-resistant *Escherichia coli* isolates relative to their sensitive progenitor in both the absence and presence of phage. We also characterized certain phage-related phenotypes and obtained complete genome sequences of the bacteria. All of the evolved bacteria exhibited tradeoffs, such that they were more fit than the ancestor in the presence of phage, but less fit than the ancestor in the absence of phage. The fitness benefit of evolved resistance in the presence of abundant phage was generally much larger than the cost, and these effects appear to be driven by only a few resistance mutations. This asymmetrical benefit-to-cost relation is consistent with the observation that sensitive cells did not persist in the experimental communities. Quantifying fitness effects in both the presence and

Alita Burmeister
Department of Ecology and Evolutionary Biology, Yale University, New Haven, CT; and BEA-CON Center for the Study of Evolution in Action, Michigan State University, East Lansing, MI, USA; e-mail: alita.burmeister@yale.edu

Rachel Sullivan
Department of Microbiology and Molecular Genetics; Michigan State University, East Lansing, MI; present address: Waisman Center, University of Wisconsin-Madison, Madison, WI, USA

Richard Lenski
Department of Microbiology and Molecular Genetics, and BEACON Center for the Study of Evolution in Action, Michigan State University, East Lansing, MI, USA

* This paper was externally peer-reviewed.

© Springer Nature Switzerland AG 2020
W. Banzhaf et al. (eds.), *Evolution in Action: Past, Present and Future*,
Genetic and Evolutionary Computation, https://doi.org/10.1007/978-3-030-39831-6_11

absence of phage may thus provide a useful approach for predicting evolutionary outcomes in both natural and engineered microbial communities.

Key words: Coevolution, phage, fitness, tradeoffs, experimental evolution

11.1 Introduction

The grim reaper looms large for microbes, which face frequent and intense selection by viruses and other antagonists. Bacterial cells often pay the ultimate price – death – for sensitivity to phage, but nevertheless resistant and sensitive bacteria typically coexist in natural ecosystems, including the ocean [53] and phyllosphere [25], as well as in engineered systems, including wastewater treatment facilities [16] and laboratory microcosms [15, 29]. One hypothesis to explain this coexistence is that bacteria pay a cost for resistance, one that is incurred whether or not phages are present. Such costs are typically manifest as a loss of competitiveness for resources, for example caused by the modification or loss of sugar-transport proteins [12]. As consequences of the tradeoff, resistant and sensitive host populations may coexist in the presence of phage (provided that the dynamics equilibrate such that neither host type has an overall fitness advantage [29]), and independently coevolving populations may diverge from one another [38, 46].

Resistance costs have been reported in many phage-bacteria systems [2, 3, 4, 8, 20, 21, 23, 24, 26, 27, 28, 29, 32, 41, 52]. However, not all evolved resistance traits incur costs [26, 28, 52], costs may depend on the environmental context and mechanism of resistance [2, 5, 32], and in some cases phage may even increase bacterial competitiveness for resources [34, 43]. Gaining more knowledge of whether, when, and why resistance is costly will improve understanding of the role of phage-bacteria interactions in maintaining diversity in microbial ecosystems. Such knowledge may also eventually aid in the preservation of natural communities, the maintenance of industrial fermentations, and the engineering of synthetic ecosystems.

In addition to its interest to the fields of ecology and evolutionary biology, the study of resistance tradeoffs has a rich history rooted in applied microbiology, where the cost of phage resistance often has implications for strain performance. Early studies sought to characterize trait-specific costs, such as the rate of acid production and cell doubling time. This focus on functional traits influenced later studies, including those in ecology and evolutionary biology. Therefore, we will first briefly review the history of tradeoffs in applied microbiology and its implications. We will then discuss evolutionary-based approaches that aim to encompass all aspects of bacterial fitness, which may offer greater potential for understanding and managing natural and applied microbial communities.

Attempts to obtain high-performing but phage-resistant bacterial strains date to the 1930s, when Whitehead and Cox allowed *Lactococcus lactis* ("*Str. cremoris*" at the time) to grow for two days in the presence of phage and then screened for isolates that rapidly produced lactic acid [54]. This approach was refined throughout the 20th century, including strain modifications generated by conjugation and transformation of resistance plasmids [37] in attempts to obtain low-cost resistance. The emergence of theoretically motivated studies of microbial ecology and experimental evolution in the 1960s and 1970s led to experimental analyses of resistance tradeoffs for their own sake. Continuous culture systems allowed for the study of host-phage community dynamics over extended periods, including the coexistence of sensitive bacteria, resistant bacteria, and phage [19, 39, 49]. Levin *et al.* added dynamical population modeling to these studies, demonstrating that the stable coexistence of phage with both sensitive and resistant bacteria depended on resistance tradeoffs [29]. Other recent studies have examined phage resistance and its costs to bacterial hosts in natural systems of applied importance including the plant pathogen *Pseudomonas syringae* [32] and the human pathogen *Pseudomonas aeruginosa* [20].

These studies often focused on various fitness components, such as the bacteria's growth rate and the rate of phage attack. For example, Levin *et al.* measured the maximum growth rate of *E. coli* in the absence of phage T2 along with the adsorption rate of T2 to resistant and sensitive hosts [5]. Such fitness-component (i.e., trait-specific) approaches have proven to be useful for parameterizing mathematical models of phage-host interactions, similar in that respect to models of parasites and long-lived animal and plant hosts [31]. Measuring specific fitness components is also important when the costs of phage resistance affect specific bacterial traits of interest, such as virulence determinants [20, 21, 48] and the ability to use various carbon sources [36]. Although these approaches have clearly demonstrated tradeoffs associated with phage resistance, much remains to be learned about the net effect of resistance mutations on host fitness.

Most bacteria and phage have short generation times, and therefore the relative fitness of mutants and their progenitors can be readily determined by measuring their differential survival and reproduction across multiple generations. This approach captures the costs and benefits associated with all life stages including lag, exponential, stationary, and death phases for bacteria, as well as adsorption, intracellular replication, lysis, and extracellular survival for phage. It also accounts for the different selection pressures that may exist in a given environment over time, such as those caused by nutrient depletion, the accumulation of cell debris, pH changes, and the like. Therefore, multi-generation assays offer a powerful and general method for quantifying evolutionary fitness. Such assays been used extensively in the field of experimental evolution for bacteria [4, 27, 55] and phage [7, 47], and we employ them here to quantify the impact of phage resistance on bacterial fitness.

We isolated bacterial clones from seven *E. coli* populations that were serially passaged and coevolved with phage λ, including two clones from one population and one from each of six others. For these isolates and their ancestor, we obtained certain qualitative phage-related phenotypes, complete genome sequences, and quantitative fitness measurements in both the presence and absence of phage. These integrated data provide new insights into the benefits, costs, genetics, and mechanisms of bacterial resistance to phage.

11.2 Materials and Methods

11.2.1 Experimental Evolution

All of the resistant strains used in this study are experimentally evolved descendants of *E. coli* B strain REL606. Replicate populations founded with REL606 previously coevolved with phage λ strain cI26 for 20 rounds of serial transfers, with each community archived by freezing samples on a daily basis [35]. The cI26 strain is obligately lytic and uses the host's outer membrane protein LamB (maltose transport porin) as a receptor. In this study, we isolated several individual bacterial clones from the day 19 frozen samples of seven of the experimentally evolved communities. We chose communities based on two criteria: (i) the final communities had phage present; and (ii) at least some of the phage present had evolved the ability to infect the bacteria through a second outer membrane protein, OmpF, as previously reported [35]. We also isolated and used the experimentally evolved phage λ G9-15b, which can use OmpF, from a frozen sample of one of the evolved communities.

11.2.2 Growth Conditions and Phenotypic Assays

All bacterial fitness assays were conducted in the same medium and other conditions as the original evolution experiment, except that phage were present or absent as described below. Specifically, bacteria and phage (when added) were propagated in modified M9 (mM9) medium (M9 salts with 1 g/L magnesium sulfate, 1 g/L glucose, and 0.02% LB), as used previously [35]. We assessed phage sensitivity by plating a lawn of the bacterial strain in a soft-agar overlay on an LB base agar and spotting 10 μL of a concentrated phage stock on the lawn. If a plaque formed, then we scored the bacterial strain as sensitive to the phage. Resistance to phage λ is associated with the partial or complete loss of the ability to grow on maltose, mannose, or both, because wild-type λ requires the LamB maltose transporter for initial attachment [17, 51] and the ManXYZ mannose permease for its DNA to enter the host cell [12, 13, 45]. Therefore, we used colony phenotypes on tetrazolium-maltose and tetrazolium-mannose indicator media to characterize resistance-associated diversity among evolved isolates. Bacteria that are deficient in growth on a given sugar will typically produce red colonies on the corresponding indicator agar, although we saw some exceptions that will be discussed later. In cases where we saw the same phenotypic pattern for multiple isolates from the same community, we selected a representative clone for additional phenotypic characterization, genome sequencing, and reporting.

11.2.3 Genome Sequencing

We isolated genomic DNA from ten evolved bacterial clones (four from population A8 and one from each of the other populations) using the Qiagen Genomic-tip 100/G kit. Short-read sequences were generated by the Michigan State University Research Technology Support Facility on an Illumina MiSeq platform. We aligned the sequences and compared them to the annotated sequence of the ancestral strain REL606 using the *breseq* pipeline [1].

11.2.4 Fitness Assays

We estimated the fitness of each bacterial clone by performing competition assays similar to those described previously [55]. During the competitions, the evolved phage-resistant descendants of REL606 (Ara$^-$) compete for resources with REL607, which is a genetically marked Ara$^+$ derivative of REL606. REL606 and its descendants form red colonies on tetrazolium-arabinose agar plates, whereas REL607 forms pink colonies, allowing the abundance of each competitor to be quantified. We confirmed that REL606 and REL607 have equal fitness in mM9 medium in both the presence and absence of phage λ (Appendix, "Marker neutrality experiments" and Fig. 11.5). To initiate the fitness assays, we revived the bacterial strains from the stocks (frozen at $-80°$C) by streaking for isolated colonies on LB agar plates. To precondition the cultures, we picked individual colonies into 10 mL of mM9 medium and incubated them overnight at $37°$C with orbital shaking at 120 rpm. We began each replicate fitness assay with a different individual colony of the evolved competitor and another of the reference competitor REL607; this approach ensures independence with respect to any mutations that might arise and affect fitness, such as ones that confer phage resistance. We sampled the bacteria by plating on tetrazolium-arabinose plates from all of the fitness assays at 0, 4, 8, and 24 h. We also sampled the assays performed without phage after three and six days, with 1:100 dilutions into fresh mM9 medium each day. The final fitness measurements reported in the main text were determined using colony counts obtained at 24 h with phage present and after six days without phage (except for strain AB113, where we used the three-day counts). These times were constrained by the speed with which one competitor excluded the other, which limits enumeration of the rarer competitor on the indicator plates. For the fitness assays conducted in the presence of phage, we added 10^5 pfu (plaque-forming units) per mL of the ancestral phage strain cI26 immediately after sampling the initial bacterial composition; the initial phage density corresponds to a multiplicity of infection of ~ 0.01. We added an equivalent volume (100 μL) of mM9 for assays performed without phage.

11.2.5 Statistical Analysis

We tested the following predictions about the fitness of the bacteria: (i) the evolved isolates are more fit than their ancestor in the presence of phage, and (ii) the evolved isolates are less fit than the ancestor in the absence of phage. To test these directional predictions, we used one-tailed t-tests. To account for multiple tests, we used sequential Bonferroni corrections [18] for each environment. We tested for differences in the variance of fitness measurements between the two environments for each bacterial strain using two-tailed F-tests based on the competition data obtained at 24 h. All data were analyzed using custom R scripts [42].

11.3 Results

11.3.1 Phenotypic Properties of the Evolved Bacteria

The bacterial isolates varied phenotypically in terms of their phage resistance and sugar use (Table 11.1, Fig. 11.1). All of the evolved isolates were resistant to the ancestral λ phage, and clones from population A8 were also resistant to the evolved λ strain that can infect through the outer membrane protein OmpF. Resistance to λ is typically caused by the reduced expression or loss of the LamB porin and the ManXYZ permease that allow the cell to import maltose and mannose across the outer and inner membranes, respectively [12, 17, 35, 45]. Therefore, we examined the ability of the bacteria to use these sugars when growing on tetrazolium indicator agar (Fig. 11.1). In our experience, mutations that eliminate the ability to use a sugar result in red colonies on tetrazolium agar, whereas strains that can grow on that sugar form white or pink colonies. The ancestral host, REL606, is both Mal$^+$ and Man$^+$, forming pink colonies on the tetrazolium maltose and tetrazolium mannose agar, respectively (Fig. 11.1). All of the evolved isolates formed darker colonies on tetrazolium maltose agar, indicating either complete or partial loss of growth on maltose. Some of the isolates (AB114, AB125) also formed darker colonies on tetrazolium mannose agar, but most (AB113, AB117, AB124, AB128, and AB131) retained the ancestral coloration. Isolate AB121 formed colonies with a magenta hue on both indicator media. AB113 colonies were typically smaller than the ancestor, while AB121 colonies were typically as large as or larger than the ancestor.

11.3.2 Mutational Signatures of Phage Resistance

We sequenced the genomes of the ten evolved bacterial clones and compared them to the ancestral strain REL606 (Table 11.2, Appendix Table 11.4). We observed several instances of parallel evolution at the gene level. All sequenced isolates contained

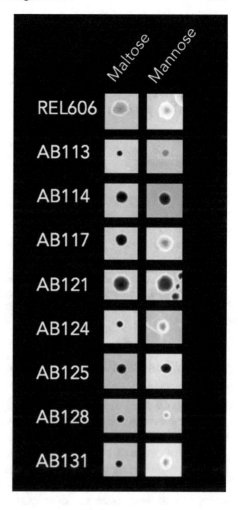

Fig. 11.1: Bacterial phenotypes associated with sugar use on two types of tetrazolium indicator agar. The ancestral strain, REL606, is Mal$^+$ and Man$^+$ and forms pink colonies on both types. The evolved isolates produce more deeply pigmented colonies on one or both agar types, which indicates that their ability to use maltose, mannose, or both was diminished or lost during evolution.

a *malT* mutation, and those from populations A8 (AB114), C4 (AB121), and D7 (AB125) had a mutation in or deletion of *manY*, *manZ*, or both. Isolates from two populations (AB113 and AB114 from A8 and AB117 from C3) had mutations of either *ompF* or *ompR*, which is a positive regulator of *ompF* [22]. Population A8 harbors two genotypes (AB113 and AB114) that share no mutations, indicating that they represent lineages that arose independently and early during the coevolution experiment. Two additional isolates from A8 had the same genome sequences as AB113 and AB114 (see footnotes to Table 11.2).

Table 11.1: Phenotypic properties of evolved, phage-resistant bacterial isolates. Isolates that metabolize a specific sugar form pink colonies on tetrazolium agar containing the sugar; those that lack the ability form red colonies. Fig. 11.1 shows representative colonies.

Isolate	Population	λ_{LamB}	λ_{OmpF}	Tetrazolium maltose colony morphology	Tetrazolium mannose colony morphology
AB113	A8	R	R	Small red	Small Pink
AB114	A8	R	R	Red	Red
AB117	C3	R	S	Red	Pink
AB121	C4	R	S	Large magenta	Large magenta
AB124	D4	R	S	Red	Pink
AB125	D7	R	S	Red	Red
AB128	F6	R	S	Red	Pink
AB131	H2	R	S	Red	Pink

Population: Population from which the isolate was obtained; λ_{LamB}: Plaque spot test for ancestral phage that uses the LamB porin as its receptor; λ_{OmpF}: Plaque spot test for an evolved phage that can also use the OmpF porin as its receptor. S: Sensitive; R: Resistant.

11.3.3 Bacterial Fitness in the Presence and Absence of Phage

We ran multi-generation competition assays to estimate the fitness of the evolved bacteria relative to their ancestor in both the presence and absence of phage. In all cases, the mean fitness of the evolved clone was higher than the ancestor in the presence of phage and lower in the absence of phage (Table 11.3, Fig. 11.2), indicating an evolutionary tradeoff between phage resistance and competitiveness. After performing sequential Bonferroni corrections to account for the multiple statistical comparisons, this pattern was significant in most cases (Table 11.3). The fitness estimates were much more variable in the presence of phages than in their absence (Table 11.3); we consider the explanation for this difference in Section 11.4.

An important feature of the competition assays run in the presence of phage is that the phage replicate during the experiment, thus subjecting their hosts to dynamic changes in the strength of selection. We therefore expected that selection for the evolved resistant bacteria would increase over the course of the assays. To test this prediction, we counted the ancestral and evolved competitors after 4, 8, and 24 h during the first day of the assays. All of the genotypes, including the sensitive ancestor, had net positive growth over the full 24 h period, even in the presence of λ (Fig. 11.3A-B). However, between 8 and 24 h, the resistant bacteria increased (Fig. 11.3A) while their sensitive ancestor declined in density (Fig. 11.3B), consistent with the expected increase in the phage density over time. In the absence of phage, both the ancestor and evolved bacteria populations increased throughout the growth cycle (Fig. 11.4A-B).

Table 11.2: Mutations in sequenced genomes of evolved λ-resistant isolates of *E. coli*. Resistance Associated Mutations: Mutations in genes known from previous studies to confer resistance to phage λ. Other Mutations: Mutations at other loci not known to confer resistance to phage λ. Asterisks (*) indicate stop codons, plus signs (+) indicate insertions, and triangles (Δ) indicate deletions. Appendix Table 11.4 provides further details on all of the mutations.

Isolate (Population)	Resistance Associated Mutations			Other Mutations
	malT	*manY, manZ*	*ompF, ompR*	
AB113 (A8)[a]	T47P (A<u>C</u>C→<u>C</u>CC)	none	Δ1 bp (805/1089 nt) in *ompF*	*ampH*
AB114 (A8)[b]	25-bp duplication	Δ*manXYZ*	Y26* (TA<u>T</u>→TA<u>G</u>) in *ompF*	*ybaK-ybaP* intergenic, Δ*rbs* operon
AB117 (C3)	+A (435/2706 nt)	none	+A (237/720 nt) in *ompR*	Δ*ecb_00726 -00739*, Δ*rbs* operon
AB121 (C4)	T154P (A<u>C</u>C→<u>C</u>CC)	G21* (GGA→<u>T</u>GA) *manY*	none	Δ*rbs* operon
AB124 (D4)	Δ89 bp	none	none	none
AB125 (D7)	W334* (T<u>G</u>G→T<u>A</u>G)	IS3 (rev) in *manZ*	none	*ecb_01992*, Δ*rbs* operon
AB128 (F6)	25 bp duplication	none	none	*ecb_01992*, *ecb_02825*, Δ*rbs* operon
AB131 (H2)	Δ4 bp	none	none	*ecb_02825*, Δ*rbs* operon

[a] Isolate AB116 had the identical genome sequence to AB113.
[b] Isolate AB115 had the identical genome sequence to AB114.

To calculate the changes in relative fitness over the course of the competition assays, we subtracted the net growth of the ancestor from the net growth of the evolved bacteria at each time point, yielding the difference shown in Fig. 11.3C and Fig. 11.4C. As in Fig. 11.2, when the resulting difference is *greater* than zero, the evolved competitor has the fitness advantage, whereas a value *below* zero means the ancestor has the advantage. In all cases, the evolved resistant strain had the advantage after 24 h in the presence of phages (Fig. 11.3C), and in all cases the ancestral sensitive strain had the 24-h advantage in the absence of phages (Fig. 11.4C). To express these data in terms of the selection rate shown in Fig. 11.2, the difference in net growth was divided by the duration of the assay in days. Thus, the 24-h data in Fig. 11.3C divided by time (here, 1 day) yields precisely the selection rates shown by the filled circles in Fig. 11.2. However, the selection-rate values in the absence of phage, shown as open circles in Fig. 11.2, were computed from the differences

Table 11.3: Fitness costs and benefits of the evolved resistant isolates. In most cases, the evolved isolates were more fit than the progenitor (selection rate > 0) in the presence of phage and less fit (selection rate < 0) in the absence of phage. Most values were significantly different from the null hypothesis (selection rate = 0) even after performing sequential Bonferroni corrections for multiple tests of the same hypothesis; however, one test was non-significant even before the corrections (indicated in bold), and three others were non-significant after the corrections (indicated in italics). In all cases, the variance associated with measuring the selection rate after one day was significantly greater in the presence than in the absence of phage.

| | Fitness expressed as selection rate (day^{-1}) | | | | | | Measurement Variance | |
| | Without Phage | | | With Phage | | | | |
Isolate	N	Mean	p-value	N	Mean	p-value	F	p-value
AB113	7	-1.018	<0.0001	8	2.139	*0.0210*	18.4	0.0010
AB114	8	-0.550	<0.0001	7	2.452	*0.0280*	52.0	<0.0001
AB117	8	-0.404	<0.0001	7	1.487	0.0100	11.2	0.0054
AB121	8	-0.038	**0.1860**	8	1.785	0.0002	7.1	0.0191
AB124	7	-0.196	0.0001	8	2.366	0.0007	29.8	0.0002
AB125	8	-0.143	0.0027	7	1.959	0.0010	21.4	0.0017
AB128	8	-0.155	<0.0001	8	1.709	0.0053	38.9	<0.0001
AB131	8	-0.151	*0.0150*	8	2.278	0.0008	6.5	0.0249

between the competitors' growth rates over longer periods than shown in Fig. 11.3 and Fig. 11.4, in order to increase the resolution of the much smaller differences in net growth observed in that environment (see Section 11.2). As expected, in the presence of phage, the growth-rate differential between the evolved and ancestral bacteria increased as the phage replicated over the 24-h period (Fig. 11.3C).

11.4 Discussion

Tradeoffs play a major role in the evolution and maintenance of diversity, allowing the coexistence of distinct community members with diverse traits by constraining how well any given member can perform in the face of multiple selective forces. To examine the prevalence and importance of tradeoffs in a microbial community, we characterized the genotypes, phenotypes, and fitness tradeoffs across two environments for a set of experimentally evolved phage-resistant *E. coli* isolates and their ancestor.

We observed a variety of phenotypes and genotypes among the evolved isolates (Tables 11.1 and 11.2). All of the evolved isolates had mutations in the *malT* gene, and some of them also had mutations in one or more of the *manY*, *manZ*, *ompF*, and *ompR* genes. These mutations almost certainly underlie the evolved phage re-

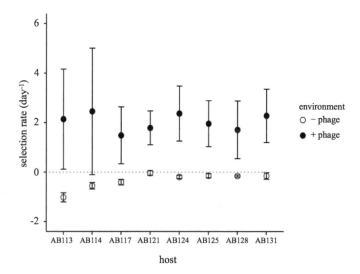

Fig. 11.2: Tradeoff in fitness of evolved resistant bacteria in the presence and absence of phage λ. Selection rates greater or less than 0 indicate that the evolved bacteria are more or less fit, respectively, relative to their sensitive ancestor. Error bars are 95% CIs, based on the number of replicates shown in Table 11.3.

sistance, as these genes affect the expression or structure of known phage receptors involved in λ's adsorption to the cell's outer membrane or the entry of λ's DNA across the inner membrane [7, 12, 13, 17, 22, 35, 45, 49, 51]. Other mutations in the evolved isolates (Table 11.1, Appendix Table 11.4) have no obvious relation to phage resistance; they might be neutral mutations, or they may confer benefits unrelated to the phage. For example, deletions of ribose genes (*rbs*) also arose, albeit more slowly, in an evolution experiment without phage [10], and *malT* mutations (which reduce *lamB* expression) also conferred a benefit in that experiment [40].

Within population A8, the two sequenced genotypes AB113 and AB114 share no mutations, indicating that they arose independently from the ancestral strain and that this population had not undergone any hard selective sweeps that would result in mutations achieving fixation [33]. Two processes could, in principle, explain this coexistence. First, clonal interference might have prevented either lineage from fixing [14]. Alternatively, the two lineages might have coexisted by negative frequency-dependent selection caused by phage specialization or other ecological feedbacks. We saw no evidence for phage specialization between these isolates (Table 11.1), although we did not characterize enough phage or analyze other phenotypic traits in enough detail to exclude this hypothesis.

More generally, our finding of a fitness tradeoff when the ancestral and evolved bacteria compete in the presence and absence of phages (Table 11.3, Fig. 11.2) is consistent with their resistance phenotypes (Table 11.1), as well as with many previous studies using other pairs of bacteria and phages [2, 3, 4, 8, 20, 21, 23, 24,

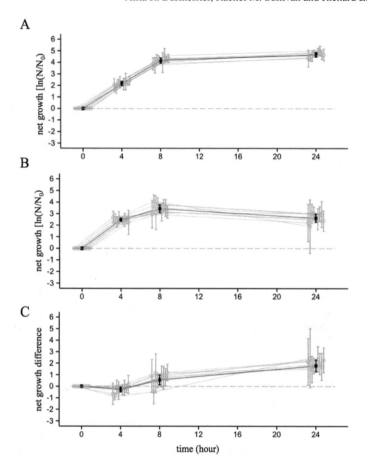

Fig. 11.3: Growth dynamics of evolved resistant and ancestral sensitive bacteria during
competition over a 24-h cycle in the presence of phage λ. A) Net growth of each of the
eight evolved isolates (gray) and their grand mean (black). B) Net growth of the
ancestral strain during the eight sets of competitions with the evolved isolates (gray)
shown in A, and the ancestor's grand mean (black). C) The difference in net growth
between the evolved and ancestral competitors shown in panels A and B. All time
points have been systematically shifted and arranged in order of the isolate numbers,
with AB113 furthest to the left and AB131 furthest to the right; the actual time points at
which all data were collected are 0, 4, 8, and 24 h. Error bars are 95% CIs, based on the
number of replicates shown in Table 11.3.

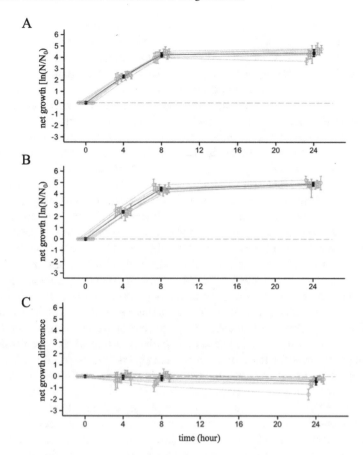

Fig. 11.4: Growth dynamics of evolved resistant and ancestral sensitive bacteria during competition over a 24-h cycle in the absence of phage. See Fig. 11.3 for panel descriptions and details. Note that AB113 is the evolved isolate that showed the decline in density between 8 and 24 h (panel A), as discussed in the text.

26, 27, 28, 29, 32, 41, 48, 52]. Bacterial fitness has multiple components, including competitiveness for limiting resources as well as survival. Many phages exploit bacterial transporters for sugars and other resources in order to initiate infections, and resistance to λ can occur through the loss, reduced expression, or modification of such structures [12, 17, 35, 45] . The effects of the resistance-conferring mutations on such transporters (Table 11.1, Fig. 11.1) evidently reduced the bacteria's competitiveness for glucose, which is the most important fitness component in the absence of phage.

Some of the evolved bacteria may also face a tradeoff between resistance and osmotic regulation. In *E. coli* K12, OmpF and OmpC, regulated by the EnvZ/OmpR two-component system [44], allow glucose and electrolyte to flow across the cell envelope. However, the ancestral strain, REL606, is a derivative of *E. coli* B and has a truncated *ompC* gene (coding 935/1101 base pairs), leaving the task of osmoregulation to OmpF [11, 50]. Changes to OmpF or its expression may therefore adversely affect the ability to regulate osmotic stress, thereby reducing growth rate and perhaps even increasing rates of cell death. Indeed, evolved isolates AB113, AB114, and AB117 have mutations in either *ompF* or *ompR*, and they have the highest fitness costs in the absence of phage (Table 11.2, Fig. 11.2). The most severe of these, AB113, has a 1-bp deletion in *ompF* that causes a premature stop codon and truncated protein; this isolate has the lowest fitness of all those tested (Fig. 11.2), it produces small colonies (Fig. 11.2), and it shows signs of an early onset of death phase between 8 and 24 h (Fig. 11.4A).

The fitness benefits of the evolved resistant bacteria in the presence of phage λ were many times the magnitude of their fitness costs in the absence of phage (Fig. 11.2). In another study, experimentally evolved *Pseudomonas aeruginosa* isolates with resistance to phage PP7 showed a similar fitness asymmetry [4]. It is not surprising that resistant bacteria should at least initially outperform their sensitive ancestors when phage are present [28, 29], although the subsequent dynamics may be complex. In our experiments, the phage population began at 10^5 pfu/mL, but its density would have increased owing to infections of the sensitive bacteria. In chemostat culture, the resistant bacteria, sensitive bacteria, and phage often coexist in a dynamic equilibrium if the bacteria experience a tradeoff between resistance and competitiveness for the limiting resources [3, 28, 29]. In serial batch culture, such as used in our experiments, the outcome is more uncertain and complex. For example, if the phage density rises too high during its initial growth, then it might drive the sensitive bacteria extinct; and if the phage cannot evolve to overcome the resistant population's defenses, then the phage might also go extinct. In any case, the management of bacterial strains for industrial applications often requires finding or designing resistant strains that are also strong performers in terms of their growth and production of desired products [37]. The existence of tradeoffs between resistance and other desired traits makes strain development both more important and more challenging.

Conversely, some applications may benefit when the costs of resistance outweigh the benefits. For example, the use of antibiotics and phage to treat pathogens selects for drug- and phage-resistant bacteria [6, 9]. If the resistant bacteria pay a sufficiently high fitness cost, then they might be unable to outcompete their sensitive ancestor, thereby slowing or halting the evolution of resistance [8, 48]. Therefore, therapies for which resistance tradeoffs are more severe could provide greater benefits than therapies for which those tradeoffs are weak. As in our experiment, the magnitude of tradeoffs in applications will depend on the bacterial genotype as well as the phage density (or drug concentration in the case of an antibiotic), so using relevant phage densities (or drug concentrations) to quantify tradeoffs should yield more accurate predictions.

We also observed that replicate estimates of fitness were much more variable in the presence than in the absence of phage (Table 11.3). One possible explanation for the unequal variance is the spontaneous occurrence of resistance mutations in the sensitive ancestral population during the preconditioning step of the fitness assays. The presence, timing, number, and genotypes of such resistance mutations will randomly vary from replicate to replicate [30]. When selection is strong, as it is in the presence of phage, such mutations can meaningfully affect the subsequent dynamics [7]. Conversely, in the absence of phage, selection is weaker and new mutations would have less impact on any given fitness assay and hence on the variation across the replicate assays.

In summary, all of the *E. coli* isolates that coevolved with phage λ had improved fitness in the presence of phage and reduced fitness in the absence of phage relative to their sensitive ancestor, providing clear evidence of a tradeoff. In the glucose-limited minimal medium used in this study, and at the initial phage density employed, the fitness gain in the presence of phage was much greater than the cost of resistance in the absence of phage. Whole-genome sequencing of the bacteria also revealed parallel evolution across the resistant isolates affecting several loci including *malT*, *manYZ*, *ompF*, and *ompR*. These mutations caused additional tradeoffs including reduced abilities to grow on maltose and mannose. Future studies might investigate how these tradeoffs in bacterial performance are affected by changes in the density and genetic composition of the coevolving phage population over time.

Acknowledgements We thank Justin Meyer for sharing strains and for helpful discussions, and Neerja Hajela for assistance in the laboratory. We thank Caroline Turner, Mike Wiser, Rohan Maddamsetti, and Kayla Miller for feedback that improved this manuscript. Last but not least, we thank Erik Goodman for his outstanding leadership of the BEACON Center. This paper is based upon work supported by the National Science Foundation Graduate Research Fellowship (DGE-1424871) to A.R.B., the BEACON Center for the Study of Evolution in Action (NSF Cooperative Agreement DBI-0939454), and the John Hannah endowment from Michigan State University. Any opinions, findings, and conclusions or recommendations expressed in this paper are those of the authors and do not necessarily reflect the views of the National Science Foundation.

References

1. Barrick, J.E., Colburn, G., Deatherage, D.E., Traverse, C.C., Strand, M.D., Borges, J.J., Knoester, D.B., Reba, A., Meyer, A.G.: *Identifying structural variation in haploid microbial genomes from short-read resequencing data using breseq.* BMC Genomics **15**, 1039 (2014)
2. Bohannan, B.J.M., Kerr, B., Jessup, C.M., Hughes, J.B., Sandvik, G.: *Trade-offs and coexistence in microbial microcosms.* Antonie van Leeuwenhoek **81**, 107–115 (2002)
3. Bohannan, B.J.M., Lenski, R.E.: *Effect of prey heterogeneity on the response of a model food chain to resource enrichment.* Am Nat **153**, 73–82 (1999)
4. Brockhurst, M.A., Buckling, A., Rainey, P.B.: *The effect of a bacteriophage on diversification of the opportunistic bacterial pathogen, Pseudomonas aeruginosa.* Proc R Soc Lond B Biol Sci **272**, 1385–1391 (2005)
5. Buckling, A., Wei, Y., Massey, R.C., Brockhurst, M.A., Hochberg, M.E.: *Antagonistic coevolution with parasites increases the cost of host deleterious mutations.* Proc R Soc Lond B Biol Sci **273**, 45–49 (2006)
6. Burmeister, A.R., Bender, R.G., Fortier, A., Lessing, A.J., Chan, B.K., Turner, P.E.: *Two lytic bacteriophages that depend on the Escherichia coli multi-drug efflux gene tolC and differentially affect bacterial growth and selection.* bioRxiv **397695** (2018)

7. Burmeister, A.R., Lenski, R.E., Meyer, J.R.: *Host coevolution alters the adaptive landscape of a virus.* Proc R Soc Lond B Biol Sci **283**, 20161528 (2016)

8. Chan, B.K., Sistrom, M., Wertz, J.E., Kortright, K.E., Narayan, D., Turner, P.E.: *Phage selection restores antibiotic sensitivity in MDR Pseudomonas aeruginosa.* Sci Rep **6**, 26717 (2016)

9. Chan, B.K., Turner, P.E., Kim, S., Mojibian, H.R., Elefteriades, J.A., Narayan, D.: *Phage treatment of an aortic graft infected with Pseudomonas aeruginosa.* Evol Med Public Health **2018**, 60–66 (2018)

10. Cooper, V.S., Schneider, D., Blot, M., Lenski, R.E.: *Mechanisms causing rapid and parallel losses of ribose catabolism in evolving populations of Escherichia coli B.* J Bacteriol **183**, 2834–2841 (2001)

11. Crozat, E., Hindré, T., Kühn, L., Garin, J., Lenski, R.E., Schneider, D.: *Altered regulation of the OmpF porin by Fis in Escherichia coli during an evolution experiment and between B and K-12 strains.* J Bacteriol **193**, 429–440 (2011)

12. Elliott, J., Arber, W.: *E. coli K-12 pel mutants, which block phage λ DNA injection, coincide with ptsM, which determines a component of a sugar transport system.* Mol Gen Gen **161**, 1–8 (1978)

13. Erni, B., Zanolari, B., Kocher, H.P.: *The mannose permease of Escherichia coli consists of three different proteins: Amino acid sequence and function in sugar transport, sugar phosphorylation, and penetration of phage λ DNA.* J Biol Chem **262**, 5238–5247 (1987)

14. Gerrish, P.J., Lenski, R.E.: *The fate of competing beneficial mutations in an asexual population.* Genetica **102**, 127 (1998)

15. Gratia, A.: *Studies on the d'Herelle phenomenon.* J Exp Med **34**, 115–126 (1921)

16. Hantula, J., Kurki, A., Vuoriranta, P., Bamford, D.H.: *Ecology of bacteriophages infecting activated sludge bacteria.* Appl Environ Microbiol **57**, 2147–2151 (1991)

17. Hofnung, M., Jezierska, A., Braun-Breton, C.: *lamB mutations in E. coli K12: Growth of λ host range mutants and effect of nonsense suppressors.* Mol Gen Gen **145**, 207–213 (1976)

18. Holm, S.: *A simple sequentially rejective multiple test procedure.* Scan J Stat **6**, 65–70 (1979)

19. Horne, M.T.: *Coevolution of Escherichia coli and bacteriophages in chemostat culture.* Science **168**, 992–993 (1970)

20. Hosseinidoust, Z., Tufenkji, N., van de Ven, T.G.M.: *Predation in homogeneous and heterogeneous phage environments affects virulence determinants of Pseudomonas aeruginosa.* Appl Environ Microbiol **79**, 2862–2871 (2013)

21. Hosseinidoust, Z., van de Ven, T.G.M., Tufenkji, N.: *Evolution of Pseudomonas aeruginosa virulence as a result of phage predation.* Appl Environ Microbiol **79**, 6110–6116 (2013)

22. Kato, M., Aiba, H., Tate, S.i., Nishimura, Y., Mizuno, T.: *Location of phosphorylation site and DNA-binding site of a positive regulator, OmpR, involved in activation of the osmoregulatory genes of Escherichia coli.* FEBS Lett **249**, 168–172 (1989)

23. Koskella, B., Brockhurst, M.A.: *Bacteria-phage coevolution as a driver of ecological and evolutionary processes in microbial communities.* FEMS Microbiol Rev **38**, 916–931 (2014)

24. Koskella, B., Lin, D.M., Buckling, A., Thompson, J.N.: *The costs of evolving resistance in heterogeneous parasite environments.* Proc R Soc Lond B Biol Sci **279**, 1896–1903 (2012)

25. Koskella, B., Lin, D.M., Buckling, A., Thompson, J.N.: *The evolution of bacterial resistance against bacteriophages in the horse chestnut phyllosphere is general across both space and time.* Phil Trans R Soc B Biol Sci **370**, 1896–1903 (2015)

26. Lennon, J.T., Khatana, S.A.M., Marston, M.F., Martiny, J.B.H.: *Is there a cost of virus resistance in marine cyanobacteria?* ISME J **1**, 300–312 (2007)

27. Lenski, R.E.: *Experimental studies of pleiotropy and epistasis in Escherichia coli. I. Variation in competitive fitness among mutants resistant to virus T4.* Evolution **42**, 425–432 (1988)

28. Lenski, R.E., Levin, B.R.: *Constraints on the coevolution of bacteria and virulent phage: A model, some experiments, and predictions for natural communities.* Am Nat **125**, 585–602 (1985)

29. Levin, B.R., Stewart, F.M., Chao, L.: *Resource-limited growth, competition, and predation: A model and experimental studies with bacteria and bacteriophage.* Am Nat **111**, 3–24 (1977)

30. Luria, S.E., Delbrück, M.: *Mutations of bacteria from virus sensitivity to virus resistance.* Genetics **28**, 491–511 (1943)
31. May, R., Anderson, R.: *Epidemiology and genetics in the coevolution of parasites and hosts.* Proc R Soc Lond B Biol Sci **219**, 281–313 (1983)
32. Meaden, S., Paszkiewicz, K., Koskella, B.: *The cost of phage resistance in a plant pathogenic bacterium is context-dependent.* Evolution **69**, 1321–1328 (2015)
33. Messer, P., Petrov, D.: *Population genomics of rapid adaptation by soft selective sweeps.* Trends Ecol Evol **28**, 659 – 669 (2013)
34. Meyer, J.R., Agrawal, A.A., Quick, R.T., Dobias, D.T., Schneider, D., Lenski, R.E.: *Parallel changes in host resistance to viral infection during 45,000 generations of relaxed selection.* Evolution **64**, 3024–3034 (2010)
35. Meyer, J.R., Dobias, D.T., Weitz, J.S., Barrick, J.E., Quick, R.T., Lenski, R.E.: *Repeatability and contingency in the evolution of a key innovation in phage lambda.* Science **335**, 428–432 (2012)
36. Middelboe, M., Holmfeldt, K., Riemann, L., Nybroe, O., Haaber, J.: *Bacteriophages drive strain diversification in a marine Flavobacterium: implications for phage resistance and physiological properties.* Environ Microbiol **11**, 1971–1982 (2009)
37. Moineau, S.: *Applications of phage resistance in lactic acid bacteria.* Antonie van Leeuwenhoek **76**, 377–382 (1999)
38. Paterson, S., Vogwill, T., Buckling, A., Benmayor, R., Spiers, A.J., Thomson, N.R., Quail, M., Smith, F., Walker, D., Libberton, B., Fenton, A., Hall, N., Brockhurst, M.A.: *Antagonistic coevolution accelerates molecular evolution.* Nature **464**, 275 – 278 (2010)
39. Paynter, M., Bungay, H.: Dynamics of coliphage infections. In: Fermentation Advances, pp. 323–336. Academic Press, New York, NY (1969)
40. Pelosi, L., Kühn, L., Guetta, D., Garin, J., Geiselmann, J., Lenski, R.E., Schneider, D.: *Parallel changes in global protein profiles during long-term experimental evolution in Escherichia coli.* Genetics **173**, 1851–1869 (2006)
41. Quance, M.A., Travisano, M.: *Effects of temperature on the fitness cost of resistance to bacteriophage T4 in Escherichia coli.* Evolution **63**, 1406–1416 (2009)
42. R Core Team: R: A Language and Environment for Statistical Computing. R Foundation for Statistical Computing, Vienna, Austria (2014)
43. Rice, S.A., Tan, C.H., Mikkelsen, P.J., Kung, V., Woo, J., Tay, M., Hauser, A., McDougald, D., Webb, J.S., Kjelleberg, S.: *The biofilm life cycle and virulence of Pseudomonas aeruginosa are dependent on a filamentous prophage.* ISME J **3**, 271–282 (2009)
44. Russo, F.D., Silhavy, T.J.: *EnvZ controls the concentration of phosphorylated OmpR to mediate osmoregulation of the porin genes.* J Mol Biol **222**, 567 – 580 (1991)
45. Scandella, D., Arber, W.: *An Escherichia coli mutant which inhibits the injection of phage λ DNA.* Virology **58**, 504 – 513 (1974)
46. Scanlan, P.D., Hall, A.R., Blackshields, G., Friman, V.P., Davis Jr, M.R., Goldberg, J.B., Buckling, A.: *Coevolution with bacteriophages drives genome-wide host evolution and constrains the acquisition of abiotic-beneficial mutations.* Mol Biol Evol **32**, 1425–1435 (2015)
47. Shao, Y., Wang, I.N.: *Bacteriophage adsorption rate and optimal lysis time.* Genetics **180**, 471–482 (2008)
48. Smith, H.W., Huggins, M.B.: *Successful treatment of experimental Escherichia coli infections in mice using phage: its general superiority over antibiotics.* J Gen Microbiol **128**, 307–18 (1982)
49. Spanakis, E., Horne, M.T.: *Co-adaptation of Escherichia coli and coliphage λvir in continuous culture.* J Gen Microbiol **133**, 353–360 (1987)
50. Studier, F.W., Daegelen, P., Lenski, R.E., Maslov, S., Kim, J.F.: *Understanding the differences between genome sequences of Escherichia coli B strains REL606 and BL21(DE3) and comparison of the E. coli B and K-12 genomes.* J Mol Biol **394**, 653 – 680 (2009)
51. Thirion, J.P., Hofnung, M.: *On some genetic aspects of phage λ resistance in E. coli K12.* Genetics **71**, 207–216 (1972)

52. Vale, P.F., Lafforgue, G., Gatchitch, F., Gardan, R., Moineau, S., Gandon, S.: *Costs of CRISPR-Cas-mediated resistance in Streptococcus thermophilus*. Proc R Soc Lond B Biol Sci **282**, 2015.1270 (2015)
53. Waterbury, J.B., Valois, F.W.: *Resistance to co-occurring phages enables marine synechococcus communities to coexist with cyanophages abundant in seawater*. Appl Environ Microbiol **59**, 3393–3399 (1993)
54. Whitehead, H.R., Cox, G.A.: *Bacteriophage phenomena in cultures of lactic streptococci*. J Dairy Res **7**, 55–62 (1936)
55. Wiser, M.J., Ribeck, N., Lenski, R.E.: *Long-term dynamics of adaptation in asexual populations*. Science **342**, 1364–1367 (2013)

Appendix

Marker neutrality experiments. Before conducting the fitness assays on the evolved bacterial isolates, we ran three experiments to test whether the arabinose-utilization marker used to distinguish competitors is neutral in the presence and absence of phage. In Experiment 1, we competed the Ara^- ancestral strain, REL606, against its Ara^+ derivative, REL607, measuring relative fitness after one and six days in the absence of phage (Fig. 11.5, Exp. 1). In Experiment 2, we competed the same Ara^- and Ara^+ strains in the presence of 10^5 pfu/mL of phage cI26 using a single "preconditioning culture" for each strain to begin all of the replicate competitions. Experiment 2 suggested that the Ara marker might not be neutral, and so we performed Experiment 3 using separately preconditioned cultures for each replicate competition. By separately growing each of the cultures used to start the competitions, we limited the influence of any resistance mutations that arose at random before the start of the competition. Indeed, by doing so the apparent fitness effect of the marker disappeared (Fig. 11.5, Exp. 3). To further ensure neutrality, we conducted tests of the marker's fitness effect alongside the competitions (reported in the main text) that assessed the fitness of the evolved resistant bacteria relative to their sensitive progenitor (Fig. 11.5, Exp. 4). For each experiment, we tested for neutrality using a two-tailed, one-sample t-test against the null hypothesis (selection rate = 0).

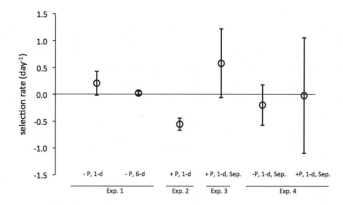

Fig. 11.5: Tests of Ara marker neutrality (Ara$^-$ relative to Ara$^+$) in mM9 medium with and
without phage λ. +P, with phage; –P, without phage; 1-d, one-day competition; 6-d,
six-day competition; Sep, separate preconditioning step. Error bars depict 95%
confidence intervals. Experiment 1: REL606 (Ara$^-$) vs. REL607 (Ara$^+$) competed in
the absence of phage for one ($p = 0.0646$, $N = 7$) and six days ($p = 0.1543$, $N = 7$).
Experiment 2: REL606 vs. REL607 competed in the presence of phage using common
bacterial preconditioning cultures ($p < 0.0001$, $N = 8$).
Experiment 3: REL606 vs. REL607 competed in the presence of phage using separate
bacterial preconditioning cultures for each replicate ($p = 0.0690$, $N = 8$).
Experiment 4: REL606 vs. REL607 competed in the absence ($p = 0.2537, N = 8$) and
presence of phage ($p = 0.9646$, $N = 7$), using separate bacterial preconditioning
cultures and performed alongside the competitions reported in Fig. 11.2 of the main
text.

Table 11.4: Detailed description of mutations in sequenced genomes of evolved λ-resistant isolates of *E. coli*. Asterisks (*) indicate stop codons, plus signs (+) indicate insertions, and triangles (Δ) indicate deletions. IS*150*-mediated indicates that an IS*150* insertion-sequence element caused the mutation. The last column lists the gene product's annotation or, in the case of multi-gene deletions, shows the set of deleted genes. Brackets denote partial deletions of the named gene.

Isolate	Genome Position	Mutation	Protein Change	Gene and Gene Description
AB113[a]	363,781	C→T	M51I (ATG→ATA)	*ampH* beta-lactamase/D-alanine carboxypeptidase
	1,003,270	Δ1 bp	coding (805/1089 nt)	*ompF* outer membrane porin 1a (Ia;b;F)
	3,481,823	A→C	T47P (ACC→CCC)	*malT* transcriptional regulator MalT
AB114[b]	479,323	IS1 (rev) +9 bp duplication[c]	intergenic (-174/+22)	*ybaK* ← / ← *yba* Hypothetical proteins
	1,003,997	A→C	Y26* (TAT→TAG)	*ompF* outer membrane porin 1a (Ia;b;F)
	1,879,350	Δ5,451 bp		[*yoaE*]–[*yebN*] [*yoaE*], *manX, manY, manZ, yobD,* [*yebN*]
	3,482,676	25 bp duplication		*malT* transcriptional regulator MalT
	3,894,997	Δ3,773 bp, IS*150*-mediated		*rbsD*–[*rbsB*] *rbsD, rbsA, rbsC,* [*rbsB*]
AB117	787,879	Δ12,090 bp		*ECB_00726–ECB_00739* 14 unannotated hypothetical genes
	3,465,032	+A	coding (237/720 nt)	*ompR* osmolarity response regulator
	3,482,119	+A	coding (435/2706 nt)	*malT* transcriptional regulator MalT
	3,894,997	Δ653 bp	IS*150*-mediated	*rbsD*–[*rbsA*] *rbsD,* [*rbsA*]
AB121	1,881,721	G→T	G21* (GGA→TGA)	*manY* mannose-specific enzyme IIC component of PTS
	3,482,144	A→C	T154P (ACC→CCC)	*malT* transcriptional regulator MalT
	3,894,997	Δ3,773 bp, IS*150*-mediated		*rbsD*–[*rbsB*] *rbsD, rbsA, rbsC,* [*rbsB*]
AB124	3,481,852	Δ89 bp	coding (168-256/2706 nt)	*malT* transcriptional regulator MalT
AB125	1,883,007	IS*3* (rev) +4 bp duplication::+TC[d]	coding (543-546/861 nt)	*manZ* mannose-specific enzyme IID component of PTS
	2,103,887	+CAGCCAGC	coding (154/216 nt)	*ECB_01992* hypothetical protein, in possible regulatory region of *ECB_01993* hypothetical gene
	3,482,685	G→A	W334* (TGG→TAG)	*malT* transcriptional regulator MalT
	3,894,997	Δ3,773 bp, IS*150*-mediated		*rbsD*–[*rbsB*] *rbsD, rbsA, rbsC,* [*rbsB*]

contd.

Isolate	Genome Position	Mutation	Protein Change	Gene and Gene Description
AB128	2,103,888	Δ4 bp	coding (155-158/216 nt)	*ECB_01992* hypothetical protein, in possible regulatory region of *ECB_01993* hypothetical gene
	3,023,945	Δ777 bp		*insB-22–[ECB_02825]* *insB-22, insA-22, [ECB_02825]*
	3,482,676	25 bp x 2	duplication	*malT* transcriptional regulator MalT
	3,894,997	Δ4,631 bp, IS*150*-mediated		*rbsD–[rbsK]* *rbsD, rbsA, rbsC, rbsB, [rbsK]*
AB131	3,023,945	Δ777 bp		*insB-22–[ECB_02825]* *insB-22, insA-22, [ECB_02825]*
	3,481,938	Δ4 bp	coding (254-257/2706 nt)	*malT* transcriptional regulator MalT
	3,894,997	Δ6,796 bp, IS*150*-mediated		*rbsD–[yieO]* *rbsD, rbsA, rbsC, rbsB, rbsK, rbsR, [yieO]*

[a] Isolate AB116 had the identical genome sequence to AB113.
[b] solate AB115 had the identical genome sequence to AB114.
[c] IS*1* insertion in the reverse orientation, in an intergenic region 22 bp downstream of *ybaP* and 174 bp downstream of *ybaK*. Nine bases (479,323 through 479,332) were duplicated and now occur on each side of the new IS*1* location.
[d] IS*3*-mediated disruption of *manZ* in the reverse orientation. Four bases (bases 543-546 of the *manZ* coding sequence) are duplicated and now occur on each side of the IS*3* location. Two additional bases (TC) were inserted at the right side of the new IS*3* location.

Chapter 12
Experimental Evolution of Human Rhinovirus Strains Adapting to Mouse Cells

Bethany R. Wasik, Brian R. Wasik, Ellen F. Foxman, Akiko Iwasaki and Paul E. Turner

Abstract Experimental evolution studies offer the possibility to examine whether closely related populations converge versus diverge in phenotype and genotype, when challenged to adapt under identical selection pressures. Human rhinoviruses (RV) are largely responsible for common cold illnesses, but little is known on whether different strains of these RNA viruses tend to evolve similarly when cultured on cells in laboratory tissue culture. Here we compared and contrasted evolution of two RV-A serotype strains (RV-1A, RV-1B) grown for ~25 passages on mouse-derived LA-4 cells, expanding on a previous study concerning temperature-dependent innate immunity in mouse-adapted rhinovirus strains. Results showed faster adaptation of the population founded by RV-1B, consistent with classic theory predicting more rapid improvement in populations poorly adapted to their environment. Moreover, we observed increased molecular divergence between the two lineages as they adapted to mouse cells, demonstrated by selection targeting different viral capsid genes and different substitutions in affected viral replication genes,

Bethany R. Wasik
Department of Ecology and Evolutionary Biology, Yale University, New Haven, CT 06511, USA, and Department of Ecology and Evolutionary Biology, Cornell University, Ithaca, NY 14853, USA

Brian R. Wasik
Department of Ecology and Evolutionary Biology, Yale University, New Haven, CT 06511, USA, and Baker Institute for Animal Health, Department of Microbiology and Immunology, College of Veterinary Medicine, Cornell University, Ithaca, NY 14853, USA

Ellen F. Foxman
Department of Laboratory Medicine and Department of Immunobiology, Yale School of Medicine, New Haven, CT, 06520, USA

Akiko Iwasaki
Department of Immunobiology, Yale School of Medicine, New Haven, CT, 06520, USA and Department of Molecular, Cellular and Developmental Biology, Yale University, New Haven, CT, 06511, USA and Howard Hughes Medical Institute, Chevy Chase, MD, 20815, USA

Paul E. Turner
Department of Ecology and Evolutionary Biology, Yale University, New Haven, CT 06511, USA

© Springer Nature Switzerland AG 2020
W. Banzhaf et al. (eds.), *Evolution in Action: Past, Present and Future*,
Genetic and Evolutionary Computation, https://doi.org/10.1007/978-3-030-39831-6_12

across the two populations. Our findings identify the genetic changes associated with human rhinovirus adaptation to a novel mouse host, which furthers the understanding of loci contributing to RV host jumps in mammals and efforts to develop mice as useful animal models for studying RV infection and pathogenicity.

Key words: Adaptation, Divergence, RNA virus, Experimental Evolution, Mutation

12.1 Introduction

Evolutionary biologists have long relied on the comparative method to elucidate similarities and differences among closely related taxa in their evolutionary responses to common selection pressures. Experiments in virology sometimes compare and contrast the phenotypic performance (fitness) of virus strains across environments, to gauge whether genotype-by-environment effects cause viruses to respond similarly versus differently to growth challenges (e.g., [13, 26, 35]). A frequent focus of experimental evolution studies in viruses and other microbes is to test whether replicate lineages of the same strain (or population) evolutionarily converge versus diverge, when selected in an identical environment (e.g., [4, 37, 38], see reviews by Lenski [20] and Blount [2]). However, fewer studies have examined a related topic in comparative virology: how do different virus strains change phenotypically and molecularly when challenged to adapt to the same laboratory environment (e.g., [3, 24]). These studies can represent "historical difference experiments" [2] that examine the effect of divergent evolutionary histories on subsequent evolution; lineages that previously diverged (e.g., in the wild, or in the laboratory) are challenged to adapt to the same experimental environment. Here we capitalized on a prior virology study that allowed two independent lineages of different rhinoviruses to evolve on the same host in tissue culture [12], to examine whether homologous loci are responsible for adaptation of distinct human rhinoviruses on a common novel host.

Recent comparative virology studies have focused on human rhinovirus (RV), which is largely responsible for the familiar upper respiratory illness, known as the common cold [32, 33]. RV (genus *Enterovirus*, family *Picornaviridae*) is a single-stranded, plus-sense RNA virus with a genome size ranging between ~7 and 8.5 kb; at least three RV serotypes (A, B, C) circulate in humans [22, 34]. The RV genome contains a single open reading frame encoding a large polyprotein that contains four structural capsid proteins (1A-1D or alternatively, Vp4-Vp1) and seven replication proteins (2A-2C,3A-3D). The RV genome also contains functional and structurally significant 5' and 3' UTR regions (see [22]). Experimental virology studies with the RV-1A serotype identified mutations associated with RV adaptation to mouse cells in laboratory tissue culture [25], and additional studies uncovered differences in selective pressures and mutation rates among different polyprotein regions at the nucleotide level [6, 18, 22, 29]. However, previous studies have not directly compared the evolution of different RV-A strains, to determine whether they evolutionarily converge versus diverge in phenotype and genotype, when challenged to adapt under controlled laboratory conditions to the same new environment.

Here we provide an analysis that capitalizes on experimental virus lineages from a recent study on temperature-dependent innate immunity of cells in response to RV infection [12]. The prior study used each of two different RV-A strains to found a lineage that was experimentally evolved on mouse-derived cells in the laboratory, and the current study examines adaptive convergence versus divergence between these two evolved lineages. In particular, we provide detailed analyses of phenotypic (reproductive growth) and genomic changes during adaptive trajectories of the two different RV-A strains when serially passaged ∼25 times on mouse-derived LA-4 cells in tissue culture. Consistent with earlier studies highlighting the role of changes in RV replication genes when adapting to mouse cells [12, 25], we show that these genes are targeted by selection in independently evolving RV-A strains cultured on LA-4 mouse cells. However, we find that the overarching pattern is molecular divergence between different RV-A populations as they adapt to this same host challenge. We show that selection targets different capsid genes and causes different substitutions in affected replication genes, as the RV-1A and RV-1B populations phenotypically improve on mouse cells.

12.2 Materials and Methods

12.2.1 Viruses and Cells

Virus strains were obtained from American Type Culture Collection: RV-1A (ATCC #VR-1559; GenBank #FJ445111); RV-1B (ATCC #VR-1645; GenBank #D00239). Viruses were evolved on the same stock of mouse lung-cell derived LA-4 cells (ATCC #CCL-196), grown at 33°C in T-75 tissue culture flasks containing 10-12 mL of F12K medium supplemented with 15% fetal bovine serum (FBS), 1% penicillin/streptomycin solution (mixture containing 5000 units/mL penicillin and 5000 μg/mL streptomycin), and 30 mM $MgCl_2$. H1-HeLa cells (provided by W.M. Lee, University of Wisconsin) were grown similarly in T-75 flasks or 6-well plates using Eagle's minimum essential medium (MEM) plus 10% FBS, supplemented with 1% penicillin/streptomycin solution and twofold concentration of non-essential amino acids (relative to standard MEM media recipe) and used for plaque assays that estimated virus titers (plaque-forming units [pfu] per mL) (see [10, 12]).

12.2.2 Serial Passage on Mouse Cells

Strains RV-1A and RV-1B were cultured independently on LA-4 mouse cells in parallel experiments for 25 and 27 passages, respectively. Briefly, viruses were introduced to freshly grown cell monolayers, and allowed to incubate for 3 days at 33°C, to ensure sufficient population growth of these viruses, which were not optimally adapted to LA-4 cells. In early passages, supernatant containing virus progeny was harvested, and 0.5 mL of this lysate was used to re-infect a new cell monolayer. Once cytopathic effect (CPE) was observed (RV-1A at passage 7; RV-1B at passage 9), lysates were then titered to initiate passages at a specific multiplicity of infection (MOI; ratio of viruses to cells) of 0.01; such adjustments are consistent with other virus evolution studies (e.g. [31]). Cell lysates were harvested once CPE was observed, at 2-3 days post infection. At the end of the study, the evolved lineages were designated RV-1AM and RV-1BM, where the 'M' indicated that these endpoint lineages were selected on mouse cells (see [12]).

12.2.3 Growth Assays and Statistical Analyses

Per the method of Foxman et al. [12], virus growth was measured by infecting freshly grown (24-h old) LA-4 cells in 6-well plates, at an MOI of 0.001. Samples from each test population were isolated at four time points (1, 18, 41, and 72 h post infection) and frozen at -80°C for later analysis. The growth assay was repeated two or three times for each test population. Data were analyzed by ANOVA, performed using JMP statistical software package (v.10, SAS Institute, Cary, NC) where *time point* and *population* were treated as model effects. Additional ANOVAs were performed at each time point with *population* as a model effect (Table 12.1), and t-tests were used to compare titers at 72 h for each ancestor and its endpoint population. Means and standard errors of the mean (SEM) were calculated and graphed using GraphPad Prism (v. 6.00, GraphPad Software, La Jolla, CA).

12.2.4 Sequencing

Viruses were sequenced as previously described [12], using consensus PCR primers (available upon request) constructed from wildtype RV-1A (GenBank #ACK37367, [17]). Ancestral and evolved genomes were amplified via touchdown PCR [9], using primers to generate seven overlapping fragments, each ~2kb in length. All sequences were submitted to GenBank (#KC894166-KC894169) and were aligned using CLC DNA Workbench (v 6.0, CLC Bio). Genetic data were acquired via Sanger sequencing at the DNA Analysis Facility on Science Hill at Yale University. Observed differences in the consensus sequence of an evolved population compared to the ancestral sequence were defined as fixed mutational substitutions. Manual

Table 12.1: Results from Analysis of Variance (ANOVA) at different time points of mouse LA-4 cell infection with population and time point as model effects.

Time point (h)	RV-1A Populations				RV-1B Populations			
	df	SS	F	P	df	SS	F	P
1: Model	1	0.050	3.48	0.0990	1	0.012	1.74	0.2350
Error	8	0.115			6	0.041		
18: Model	1	0.139	15.72	0.0033*	1	2.440	12.79	0.0050*
Error	9	0.080			10	1.910		
41: Model	1	0.779	53.43	0.0003*	1	2.770	17.02	0.0021
Error	6	0.087			10	1.630		
72: Model	1	0.422	42.77	$< 0.0001**$	1	1.790	10.58	0.0174
Error	10	0.099			6	1.020		
Population Effect	1	0.57	4.46	0.0416*	1	5.832	20.79	$< 0.0001**$
Time Point Effect	1	40.49	316.67	$< 0.0001**$	1	28.137	100.29	$< 0.0001**$

Values followed by * indicate $P < 0.05$ and ** indicate $P < 0.0001$.

scrutiny of Sanger chromatogram spectra was used to detect the presence of reproducible secondary peaks in the sequence data, which was used to define a locus as polymorphic.

12.3 Results

12.3.1 Improved Growth of Evolved RV Populations Relative to Ancestors

We first tested whether 25 passages on LA-4 cells led to improved growth in the RV-1A lineage, similar to published results for lineage RV-1B [12]. To examine phenotypic changes in the growth capacity (population density in pfu/mL achieved in 72 h) of evolving lineages through evolutionary time, we obtained replicated ($n = 2$ or 3) growth curves over 72 h for each test population at four time points in its evolution (RV-1A: passages 0, 5, 7, and 25; RV-1B: passages 0, 5, 9, and 27); intermediate time points were chosen just before and after cytopathic effects of virus infection on host cells became visible. We observed that the RV-1A ancestor grew better on LA-4 cells by 72 h, compared to the RV-1B ancestor (RV-1A: 1.47 x 10^5 pfu/mL 0.12 x 10^5 SEM; RV-1B: 3.50 x 10^3 pfu/mL 0.00 x 10^3 SEM, Fig. 12.1). However, by the end of the experimental evolution, population RV-1BM (where 'M' indicates mouse-selected) achieved greater fitness improvement relative to its RV-1B ancestor, compared to the fitness increase shown by the mouse-adapted RV-1AM relative to its RV-1A ancestor (RV-1A: $\sim$0.5 $\log_{10}$ pfu/mL improvement measured at 72 hours; RV-1B: $\sim$1.5 $\log_{10}$ pfu/mL improvement at 72 hours, Fig. 12.1). These results show that the RV-1A ancestral strain grew better on LA-4 mouse cells compared to the ancestral RV-1B, but that experimental evolution allowed lineage RV-1BM to improve more than lineage RV-1AM in the evolutionary time allowed.

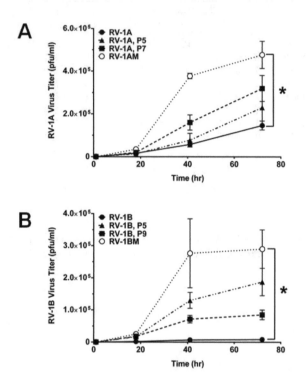

Fig. 12.1: Growth trajectories of RV-1A and RV-1B ancestors and evolved populations. Growth curves for (A) RV-1A and (B) RV-1B are shown as virus titers (pfu/mL) for the ancestor, samples from intermediate passages (labeled by passage number P), and final evolved populations (indicated by 'M', or mouse-cell evolved [12]) at 4 time points (1, 18, 41, and 72 h). Error bars represent standard errors of the mean. In both datasets, the endpoint population exhibited significantly increased growth on mouse LA-4 cells relative to its ancestor (denoted by *).

For the RV-1A lineage, ANOVA comparisons of virus titers across all generation times revealed significant differences among ancestor, intermediate, and final evolved population samples at 18 h ($P = 0.0033$), 41 h ($P = 0.0003$), and 72 h ($P < 0.0001$); however, we found no difference at 1 h ($P = 0.0990$) (Fig. 12.1A, Table 12.1 left). Comparison of titers for the RV-1A ancestor and RV-1AM evolved population showed a significant difference at 72 h (unpaired t-test with $t = 5.0988$, $df = 4$, $P = 0.007$).

Similarly, ANOVA comparisons of virus titers across generation times for the RV-1B lineage showed significant differences among ancestor, intermediate, and evolved population samples at 18 h ($P = 0.0050$), 41 h ($P = 0.0021$), and 72 h ($P = 0.0174$), but no difference at 1 h ($P = 0.2350$) (Fig. 12.1B, Table 12.1 right).

Titer comparisons for the RV-1B ancestor and RV-1BM evolved population significantly differed at 72 h (unpaired t-test with $t = 4.6917$, $df = 2$, $P = 0.0426$). Interestingly, we observed that titers for intermediate passage 5 of RV-1B exceeded titers at passage 9, except at 1 h (Fig. 12.1B). However, we concluded that the RV-1A and RV-1B lineages generally improved in growth on LA-4 cells through evolutionary time. We next examined whether or not these comparable phenotypic improvements involved molecular convergence, whereby similar molecular substitutions might have occurred at homologous loci.

12.3.2 RV-1AM and RV-1BM Populations Evolve Different Mutations in Similar Gene Regions

RNA was extracted and analyzed from the RV-1A and RV-1B ancestor viruses, the endpoint RV-1AM and RV-1BM lineages, and from population samples in the passages immediately following the time points for phenotypic assays (passages 6 and 8 for RV-1AM lineage; passages 6 and 10 for RV-1BM lineage). This analysis yielded entire genomes that code for a 2157aa polyprotein with variable UTR length (RV-1A samples: 7095-7100bp; RV-1B samples: 7116-7123bp, also see [12]). A BLAST alignment showed that the ancestral RV-1A and RV-1B strains were ∼10.1% diverged (BLAST: [1]).

As summarized in Fig. 12.2A, we observed 3 fixed non-synonymous nucleotide substitutions that separated the final evolved RV-1AM sequence from the RV-1A ancestor, occurring in a structural gene (Vp3) and in 2 non-structural genes (2B, 3A). In addition, we observed a single fixed synonymous change in non-structural gene $3C^{PRO}$. We detected 3 non-synonymous polymorphisms in non-structural gene 3A, and a single non-synonymous polymorphic site in non-structural gene $3D^{POL}$. Last, we observed 2 synonymous polymorphisms in structural gene Vp3. No mutations were observed among sequenced partial 5' and 3' UTR regions. It appeared that RV-1A genes associated with viral replication tended to evolve faster than viral capsid genes during selection on LA-4 mouse cells.

Foxman et al. [12] briefly reported non-synonymous genetic changes in the passage 10 and endpoint samples from evolved lineage RV-1B. Here, we present additional genetic data from passage 6. As summarized in Fig. 12.2B (see also Supp. Fig. 1C in [12]), we observed 13 fixed non-synonymous nucleotide substitutions that separated the final evolved RV-1BM sequence from the RV-1B ancestor. One change occurred in structural gene Vp4 and 3 changes in structural gene Vp2. The remaining 10 non-synonymous fixed substitutions were distributed among four non-structural genes, with 3 changes occurring in each of genes 2B and 3A, 2 changes in gene 2C and 1 change in gene 2A. We observed 2 fixed synonymous substitutions in structural gene Vp2, and 2 such changes in non-structural gene 2C, as well as a single change in non-structural gene $3C^{PRO}$. The sole polymorphism detected in lineage RV-1BM was a non-synonymous polymorphism in non-structural gene $3D^{POL}$. No mutations were observed among sequenced partial 5' and 3' UTR regions. In this

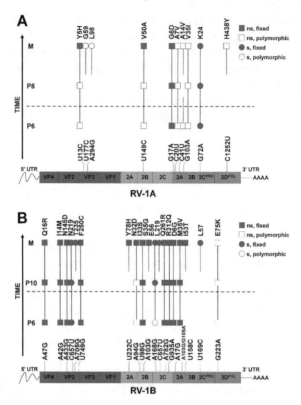

Fig. 12.2: Observed consensus mutations of evolving RV-1A and RV-1B lineages through time. Consensus derived alleles over time in evolving (A) RV-1A and (B) RV-1B lineages adapting to mouse LA-4 cells. Horizontal dashed lines show approximate time when cytopathic effects arose in each lineage. Alleles are shown as fixed non-synonymous (ns, filled square), fixed synonymous (s, filled circle), polymorphic non-synonymous (ns, open square), and polymorphic synonymous (s, open circle) changes. The organization of the RV genome is shown at the bottom of each panel comprising the open reading frames for individual proteins, including those involved in capsid (dark grey, left) and replication (light and medium grey, right) functions.

case, evolution of RV-1B on LA-4 mouse cells led to widespread changes across the RV genome, including mutations in multiple capsid and replication genes and affecting more genes compared to RV-1A evolving under identical conditions and for about the same number of passages.

Comparing the RV-1A and RV-1B genetic datasets, we see that the mutations in the capsid genes are entirely distinct. RV-1AM showed multiple changes in Vp3, whereas RV-1BM had a substitution in Vp4 and multiple changes in Vp2 (Fig. 12.2). Although RV-1AM underwent changes in most of the replication genes that also evolved in RV-1BM (2B, 3A, 3C^{PRO}, 3D^{POL}), none of the substitutions or polymorphisms were shared between the two evolved populations (Fig. 12.2). However, we note that point mutations at position 35 in the 3A gene are reciprocal changes for the evolved lineages (RV-1AM: G6D; RV-1BM: D6G) (Fig. 12.2). It appears that mutations in 4 replication proteins may have been commonly important for allowing RV-1A and RV-1B to improve in performance on LA-4 mouse cells, but that the viruses underwent different molecular changes at these loci during their evolution. As discussed below, it is currently unknown which of these mutations (alone or in combination) actually contributed to growth improvement of the evolved RV-1AM and RV-1BM lineages on mouse-derived LA-4 cells.

12.4 Discussion

Prior studies have focused on the adaptive evolution of RV on non-human host cells, using a single population or founding strain of the virus [15, 16, 25, 39, 40]. Building upon previous work examining temperature-dependent viral response to cellular innate immunity in lineage RV-1BM adapted to LA-4 mouse cells [12], we performed an experiment that compared the evolution of two different strains of RV serotype A (RV-1A, RV-1B) on this novel host. We showed that both strains could improve in growth during experimental evolution, that RV-1BM underwent greater phenotypic improvement than RV-1AM, and that the evolved lineages further diverged molecularly over the duration of the experiment.

The greater improvement in RV-1BM is consistent with evolutionary theory as well as prior experiments on the speed of adaptation in a novel environment (e.g., [7], see review by [8]). In particular, some theory suggests that a population which is reasonably well-adapted to a novel environment relies on mutations of small effect size as it evolves optimal performance (e.g., climbing a fitness peak in the adaptive landscape), whereas a poorly adapted population is likely to make larger 'fitness jumps' in the time allowed, because it can improve through mutations of large effect size without overshooting the fitness-peak target [11] (see also e.g., [5]). Also, although beneficial mutations of small effect size may be more common than those of large effect, mutations of small effect take longer to fix [7] (see also [19, 21]). Consistent with these ideas, we observed that the lineage founded by the initially more-fit RV-1A ancestor adapted more slowly (improved less in the time allowed) than the lineage founded by the less-fit RV-1B ancestor. Moreover, we observed fewer fixed mutations, but more polymorphic loci, in the RV-1A lineage; although we did not measure the fitness effects of individual or epistatic (interacting) mutations, these observations are consistent with longer fixation times of small-effect mutations in RV-1AM. Intriguing avenues for future work would be to characterize the effect sizes of spontaneous mutations in RV-1A and RV-1B and, in particular, to measure the effect sizes of the mutations that arose in our study (e.g., [27]).

Our study also suggests that the mutation supply rate was not limiting for lineages founded by RV-1A and RV-1B, regardless of their initial difference in fitness on LA-4 mouse cells. RV has a high mutation rate, typical of most viruses with ssRNA genomes [28]. Indeed, we observed early in our experiment (by passage 6) most of the mutations that were eventually fixed or polymorphic at the endpoint (passage 25 or 27). These early mutations coincided with increased cytopathic effects (i.e., visible damage to host cells, not shown by either ancestor) in the early evolution of the virus lineages on LA-4 cell monolayers. This result emphasizes the ability of RV strains to rapidly evolve when challenged to grow on a novel host. Importantly, it is currently unknown which of the observed mutations (alone or in combination) in this study are responsible for adaptive changes in virus growth on LA-4 cells; e.g., the typically high mutation rates in ssRNA viruses afford the possibility for neutral mutations to increase in frequency due to hitchhiking alongside linked beneficial mutations [23].

Several previous studies have documented mutations when a single RV strain was allowed to evolve in cell culture, although the fitness effects (neutral, beneficial or deleterious) of these observed mutations were not always discerned. For example, mutations in structural capsid proteins (Vp1, Vp3) and replication protein 3A were observed when RV-1A experienced alternating passages on human cells (or cells expressing a human receptor: ICAM-1, human intercellular adhesion molecule-1) and mouse cells (e.g. [15, 25, 39]). Rasmussen and Racaniello [25] inferred that observed changes in the Vp1 and Vp3 genes were neutral mutations that did not affect phenotype, whereas a mutation in replication protein 3A was deemed adaptive. Other results emphasized important changes in replication protein 2C [15, 16]. When Cordey et al. [6] passaged RV by in vivo immunocompetent human inoculations, observed mutations were predominantly in Vp2, Vp1, and 2C and these genes were termed "hot spots", in contrast to genes 2B and 3A that were termed "cold spots" less prone to changes. Additionally, RV-1A was previously analyzed to infer which regions of the genome were under strong selection pressure, and gene 3A was concluded to be a region of low variability between two regions of high genetic diversity, genes 2C and 3B [18]. We note that protein 3A has been additionally linked to changes in viral replication in other non-RV enterovirus studies, such as poliovirus and Aichi virus [14, 30].

Because of differences in selection conditions (i.e., human, mouse, alternating human and mouse cells), it is difficult to directly compare the results of our work to these earlier RV studies. However, we can cautiously suggest some conclusions, based on overall observed patterns. In particular, our data suggested that gene 3A was important for mouse-cell adaptation in RV-1AM and RV-1BM, regardless of the different initial fitness of the founding strains and their overall fitness gained in our study. Thus, changes in the 3A gene seem often to occur during RV selection experiments that involve mouse cells (e.g. [25], current study), whereas this gene seems to be a cold spot on human cells [18]. This difference implicates gene 3A as important for emergence of the human virus RV if it were to jump to a rodent host. While this idea is speculative and warrants further investigation, it is useful for designing future experiments using RV and host cells in tissue culture as a model for studying changes in emerging ssRNA viruses of mammals. Such efforts relate to goals of identifying loci that contribute to emergence success of RNA viruses, which seem especially capable of jumping to new hosts [36].

In general, our experiments found that evolution of RV-1A and RV-1B under the same host adaptation challenge led to divergence, rather than convergence. Although both strains fixed mutations in genes associated with capsid proteins and with replication, there was minimal similarity across the molecular datasets that measured separation from their respective ancestors. These results showed that the ~10% initial divergence between the viruses increased even during the short period in our experiments. A promising follow-up study would be a much larger experiment, simultaneously examining phenotypic and genotypic evolution of replicate populations of RV strains of differing relatedness, as they evolve in and adapt to a common environment. In addition, our study contributes to efforts for establishing a much-needed, non-human animal model system of RV infection. Understanding the mechanisms underlying RV replication and adaptation is difficult, because culturing human RV in small mammal (e.g., mouse) cells results in poor replication (Yin and Lomax 1983). Documenting the genetic basis of RV adaptation to mouse cells, as we have begun to do in this study, may help to overcome this difficulty.

Acknowledgements We thank Richard Lenski and Wolfgang Banzhaf for valuable comments on the manuscript, and Jason Shapiro for feedback on statistical analyses. This work was supported in part by the National Science Foundation award #DEB-1021243 (to P.E.T.), and by the NIH/NIAID under award #U54AI057160 to the Midwest Regional Center of Excellence for Biodefense and Emerging Infectious Diseases Research #AI054359 and #AI064705 (to A.I.), and NIH awards #AI054359S1 and #T32-HL007974-11 (to E.F.F).

References

1. Altschul, S.F., Gish, W., Miller, W., Myers, E.W., Lipman, D.J.: *Basic local alignment search tool.* Journal of Molecular Biology **215**, 403 – 410 (1990)
2. Blount, Z.D., Lenski, R.E., Losos, J.B.: *Contingency and determinism in evolution: Replaying life's tape* **362** (2018)
3. Bull, J.J., Badgett, M.R., Springman, R., Molineux, I.J.: *Genome properties and the limits of adaptation in bacteriophages.* Evolution **58**, 692–701 (2004)
4. Bull, J.J., Badgett, M.R., Wichman, H.A., Huelsenbeck, J.P., Hillis, D.M., Gulati, A., Ho, C., Molineux, I.J.: *Exceptional convergent evolution in a virus.* Genetics **147**, 1497–1507 (1997)
5. Burch, C.L., Chao, L.R.: *Evolution by small steps and rugged landscapes in the RNA virus phi6.* Genetics **151**, 921–927 (1999)
6. Cordey, S., Junier, T., Gerlach, D., Gobbini, F., Farinelli, L., Zdobnov, E.M., Winther, B., Tapparel, C., Kaiser, L.: *Rhinovirus genome evolution during experimental human infection.* PloS ONE **5**(e10588) (2010)
7. Crow, J.F., Kimura, M.: *An introduction to population genetics theory.* Harper & Row, New York (1970)
8. De Visser, J.A.G., Krug, J.: *Empirical fitness landscapes and the predictability of evolution.* Nature Reviews Genetics **15**(7), 480 (2014)
9. Don, R., Cox, P.T., Wainwright, B.J., Baker, K., Mattick, J.S.: *'Touchdown' PCR to circumvent spurious priming during gene amplification.* Nucleic acids research **19**, 4008 (1991)
10. Fiala, M., Kenny, G.E.: *Enhancement of rhinovirus plaque formation in human heteroploid cell cultures by magnesium and calcium.* Journal of bacteriology **92**, 1710–1715 (1966)
11. Fisher, R.A.: *The genetical theory of natural selection.* Clarendon Press, Oxford (1930)

12. Foxman, E.F., Storer, J.A., Fitzgerald, M.E., Wasik, B.R., Hou, L., Zhao, H., Turner, P.E., Pyle, A.M., Iwasaki, A.: *Temperature-dependent innate defense against the common cold virus limits viral replication at warm temperature in mouse airway cells.* Proceedings of the National Academy of Sciences of the United States of America **112**, 827–832 (2015)

13. Gloria-Soria, A., Armstrong, P.M., Powell, J.R., Turner, P.E.: *Infection rate of Aedes aegypti mosquitoes with dengue virus depends on the interaction between temperature and mosquito genotype.* Proceedings of the Royal Society B: Biological Sciences **284**, 20171,506 (2017)

14. Greninger, A.L., Knudsen, G.M., Betegon, M., Burlingame, A.L., DeRisi, J.L.: *The 3A protein from multiple picornaviruses utilizes the Golgi adaptor protein ACBD3 to recruit PI4KIIIβ.* Journal of virology **86**, 3605–3616 (2012)

15. Harris, J.R., Racaniello, V.R.: *Changes in rhinovirus protein 2c allow efficient replication in mouse cells.* Journal of virology **77**, 4773–80 (2003)

16. Harris, J.R., Racaniello, V.R.: *Amino acid changes in proteins 2b and 3a mediate rhinovirus type 39 growth in mouse cells.* Journal of Virology **79**, 5363–5373 (2005)

17. Hughes, P.J., North, C., Jellis, C.H., Minor, P.D., Stanway, G.: *The nucleotide sequence of human rhinovirus 1B: molecular relationships within the rhinovirus genus.* Journal of General Virology **69**, 49–58 (1988)

18. Kistler, A.L., Webster, D.R., Rouskin, S., Magrini, V., Credle, J.J., Schnurr, D.P., Boushey, H.A., Mardis, E.R., Li, H., DeRisi, J.L.: *Genome-wide diversity and selective pressure in the human rhinovirus.* Virology Journal **4**, 40 – 40 (2007)

19. Lenski, R.E.: *Quantifying fitness and gene stability in microorganisms.* Biotechnology (Reading, Mass.) **15**, 173 – 192 (1991)

20. Lenski, R.E.: *Convergence and divergence in a long-term experiment with bacteria.* The American Naturalist **190**, S57–S68 (2017)

21. Morley, V.J., Turner, P.E.: *Dynamics of molecular evolution in RNA virus populations depend on sudden versus gradual environmental change.* Evolution **71**, 872–883 (2017)

22. Palmenberg, A.C., Spiro, D., Kuzmickas, R., Wang, S., Djikeng, A., Rathe, J.A., Fraser-Liggett, C., Liggett, S.B.: *Sequencing and analyses of all known human rhinovirus genomes reveal structure and evolution.* Science **324**, 55–9 (2009)

23. Peck, K.M., Lauring, A.S.: *Complexities of viral mutation rates.* Journal of Virology **92**(e01031-17) (2018)

24. Pesko, K., Voigt, E.A., Swick, A.D., Morley, V.J., Timm, C., Yin, J., Turner, P.E.: *Genome rearrangement affects RNA virus adaptability on prostate cancer cells.* Frontiers in Genetics **6**, 121 (2015)

25. Rasmussen, A.L., Racaniello, V.R.: *Selection of rhinovirus 1a variants adapted for growth in mouse lung epithelial cells.* Virology **420**, 82–8 (2011)

26. Remold, S.K., Rambaut, A., Turner, P.E.: *Evolutionary genomics of host adaptation in vesicular stomatitis virus.* Molecular biology and evolution **25**, 1138–47 (2008)

27. Sanjuán, R., Moya, A., Elena, S.F.: *The distribution of fitness effects caused by single-nucleotide substitutions in an RNA virus.* Proceedings of the National Academy of Sciences **101**(22), 8396–8401 (2004)

28. Sanjuán, R., Nebot, M.R., Chirico, N., Mansky, L.M., Belshaw, R.: *Viral mutation rates.* Journal of Virology **84**, 9733–9748 (2010)

29. Tapparel, C., Cordey, S., Junier, T., Farinelli, L., van Belle, S., Soccal, P.M., Aubert, J., Zdobnov, E.M., Kaiser, L.: *Rhinovirus genome variation during chronic upper and lower respiratory tract infections.* PLoS ONE **6**(e21163) (2011)

30. Towner, J.S., Brown, D.M., Nguyen, J.H., Semler, B.L.: *Functional conservation of the hydrophobic domain of polypeptide 3ab between human rhinovirus and poliovirus.* Virology **314**, 432 – 442 (2003)

31. Turner, P.E., Elena, S.F.: *Cost of host radiation in an RNA virus.* Genetics **156**, 1465–70 (2000)

32. Turner, R.B., Couch, R.B.: Rhinoviruses. In: B.N. Fields, D.M. Knipe, P.M. Howley (eds.) Fields' Virology, pp. 895–909. Wolters Kluwer Health/Lippincott Williams & Wilkins, Philadelphia (2007)

33. Tyrrell, D., Parsons, R.: *Some virus isolations from common colds: III. Cytopathic effects in tissue cultures.* The Lancet **275**, 239 – 242 (1960)
34. Vlasak, M., Roivainen, M., Reithmayer, M., Goesler, I., Laine, P., Snyers, L., Hovi, T., Blaas, D.: *The minor receptor group of human rhinovirus (HRV) includes HRV23 and HRV25, but the presence of a Lysine in the VP1 HI loop is not sufficient for receptor binding.* Journal of Virology **79**, 7389–7395 (2005)
35. Wasik, B.R., Muñoz-Rojas, A.R., Okamoto, K.W., Miller-Jensen, K., Turner, P.E.: *Generalized selection to overcome innate immunity selects for host breadth in an RNA virus.* Evolution; international journal of organic evolution **70**, 270–81 (2016)
36. Wasik, B.R., Turner, P.E.: *On the biological success of viruses.* Annual review of microbiology **67**, 519–41 (2013)
37. Wichman, H.A., Badgett, M.R., Scott, L.A., Boulianne, C.M., Bull, J.J.: *Different trajectories of parallel evolution during viral adaptation.* Science **285**, 422–424 (1999)
38. Woods, R., Schneider, D., Winkworth, C.L., Riley, M.A., Lenski, R.E.: *Tests of parallel molecular evolution in a long-term experiment with Escherichia coli.* Proceedings of the National Academy of Sciences of the United States of America **103**, 9107–12 (2006)
39. Yin, F.H., Lomax, N.B.: *Host range mutants of human rhinovirus in which nonstructural proteins are altered.* Journal of Virology **48**, 410–418 (1983)
40. Yin, F.H., Lomax, N.B.: *Establishment of a mouse model for human rhinovirus infection.* Journal of General Virology **67**, 2335–2340 (1986)

Chapter 13
Normed Phase Space Model of Natural Variation in Bacterial Chromosomes

Julius H. Jackson

Abstract This study uses multidimensional, information vectors of amino acid composition and codon usage to characterize natural variation in a wide, phylogenetic spectrum of bacteria, and probes the limits of natural variation. These vectors are used to construct a phase space of natural variation that represents all possible states the evolving genes or sets of genes occupy within a viable organism.

Key words: bioinformatics, bacterial chromosomes, natural variation, chromosome evolution, gene evolution, codon usage measures, proteomics, Hilbert space, genetic information, phase space, evolutionary distance measures, inner product space

13.1 Concept and Purpose

The purpose of this study is to develop approaches to discover the limits of variation of nucleotide sequence in protein-coding genes of chromosomes in the genomes of bacterial organisms.

We first consider how to represent the functional history of a single protein-coding gene, in all organisms, through all time. Next we consider how to represent the functional history of a set of protein-coding genes over time. Using a comparative study of variation in protein-coding information, we seek to develop a mathematical model of the history of natural variation of single genes, or groups of genes in chromosomes of bacteria.

Natural variation in the nucleotide sequence of a protein-coding gene begins from the moment of birth of a new gene function and continues for the life of that gene function and beyond. The nucleotide sequence of a gene, within the structural and

Julius H. Jackson
Department of Microbiology & Molecular Genetics, and BEACON Center for the Study of Evolution in Action, Michigan State University, East Lansing, MI 48824 e-mail: jhjacksn@msu.edu

© Springer Nature Switzerland AG 2020
W. Banzhaf et al. (eds.), *Evolution in Action: Past, Present and Future*,
Genetic and Evolutionary Computation, https://doi.org/10.1007/978-3-030-39831-6_13

functional context of a genome and the cellular environment, contains all information necessary for the protein to be active in a physiological role for the organism. The specific function of such a protein depends upon the composition of amino acids comprising the protein, the arrangement of these amino acids for the structure and specific reactive properties of the protein, and the expression of the protein at a sufficient level to play its physiological role. For the purpose of this study, we develop multidimensional vectors of three types of information to follow changes in any gene or set of genes encoding an active protein: **type (*i*)**, a vector of relative amino acid coding frequency represents the amino acid composition; **type (*ii*)**, a vector of relative codon frequency for each amino acid reflects translation potential and efficiency for the coding mRNA; and **type (*iii*)**, a vector of a subset of relative amino acid coding frequencies for reactive site residues contains information for the specific functional, reaction property of the protein. Previous studies show that the average properties for **type (*i*)** and **type (*ii*)** vectors are characteristic for all genes within a chromosome [2]. Thus, a complete set or subset of heterologous genes within an organism can be represented by vector **types (*i*)** and (**ii**) of the average relative frequencies of amino acids and codon usage, and used to distinguish one organism or clade of related organisms from another in the phylogenetic spectrum.

A simple measure of nucleotide variation, such as %(G+C) composition of a single-strand of DNA in protein-coding genes, shows that nature limits bacterial and archaeal organisms to the functional range of variation of $25 \leq \%(G+C) \leq 75$ [2]. Within such a range of nucleotide composition, the coding sequence for a protein can undergo a wide range of variation and still retain protein function as long as mutations that change the amino acid composition leave the protein translatable and with the same reactive property. Thus, we infer the existence of measurable boundaries to the natural variation of protein-coding genes when such variation preserves the functional properties of the gene products.

This study uses the multidimensional vectors **type (*i*)** and (**ii**) to characterize natural variation in a wide, phylogenetic spectrum of bacteria and to visualize that multidimensional variation in three or fewer dimensions.

13.2 Relative Frequency Vectors of Sequence Information for Genes and Genomes

We previously introduced a 20-dimensional vector of relative amino acid coding frequency, $\mathbf{F}_\alpha$ to represent the amino acid composition of a gene or genome [2]. Let us define, here, the relative amino acid coding frequency as

$$f_{\alpha_j} = \frac{\alpha_j}{\sum_{i=1}^{20} \alpha_j} \tag{13.1}$$

where $\alpha =$ n amino acid, $j =$ a specific amino acid number, and $\alpha_j =$ amino acid number j, i.e. the j-th amino acid. Then,

$$\mathbf{F}_\alpha = (f_{\alpha_1}, f_{\alpha_2}, \ldots, f_{\alpha_j}) \tag{13.2}$$

is the vector of relative amino acid coding frequencies and represents information vector **type (*i*)**.

From the same prior study [2], $\mathbf{F}_\gamma$ is a vector of relative codon frequency that defines the codon usage pattern of a gene or genome. Let us define relative codon frequency as

$$f_{\gamma_j} = \frac{\gamma_j}{\sum_{i=1}^{n} \gamma_j} \tag{13.3}$$

where $1 \leq n \leq 6$ for the number of codons for an amino acid, and $\gamma =$codon, $i =$ codon number, and $j =$amino acid number. Then,

$$\mathbf{F}_\gamma = (f_{\gamma_1 \alpha_1}, f_{\gamma_2 \alpha_1}, \ldots, f_{\gamma_n \alpha_1}, \ldots, f_{\gamma_n \alpha_j}) \tag{13.4}$$

is a vector of relative codon frequencies and represents information vector **type (*ii*)**.

Assume that within the set of all fully sequenced genomes of bacteria, the variation of sequence for an orthologous set of protein-coding genes is bounded by the selective constraint to retain reaction site specificity and function. This condition limits the natural variation changes of nucleotide sequence in genes and to changes in gene organization. Information vector **type (*iii*)** can be treated as a vector $\boldsymbol{\Phi}$ of unknown composition to be addressed elsewhere without effect on the current argument.

From this point, treat the relative frequency vectors of Eqs. (13.2) and (13.4) as vectors of mean relative frequencies for some labeled subset $\{a, b, c, \ldots\}$ of protein-coding genes in any chromosome or genome in representative organisms A, B, C, and so forth. Then, an individual gene within a representative chromosomal subset is $a_i \in A$, or $b_i \in B$, and so forth. For simplicity, relative frequency vector **type (*i*)** for organism A is $\mathbf{F}_\alpha(A) = \mathbf{A}_\alpha$, and for organism B, $\mathbf{F}_\alpha(B) = \mathbf{B}_\alpha$, and so on. Likewise, relative frequency vector **type (*ii*)** for organism A is $\mathbf{F}_\gamma(A) = \mathbf{A}_\gamma$, and for organism B, $\mathbf{F}_\gamma(B) = \mathbf{B}_\gamma$, and so on.

13.3 Normed Vector Space for Relative Frequency Vectors

The geometry of 20 dimensions for $\mathbf{A}_\alpha$ and 59 dimensions for $\mathbf{A}_\gamma$ poses difficulty in visualizing the variation in 79 dimensions. Note that we have yet to consider the number of dimensions needed to represent the functional properties of the gene subsets. What may be mathematically tractable is not always satisfying or intelligible to human observers in the 3-D biological world. The use of evolutionary tree structures to represent variation provides visually intelligible relationships of organisms [1, 4], but provides no insight into the limits or boundaries that we seek to define and explore beyond the separation distances of branches. Our earlier work showed that comparison of any one genomic vector to a reference vector by a correlation measure produced a scalar value for each vector comparison, and that a 2-D map of the correlation space revealed ordered relationships among all genomes examined [2]. We seek, in this study, new ways to visualize the complete history of natural variation in representative geometries of three or fewer dimensions.

One strategy was to create a phase space on a 2-D plane to represent the evolutionary changes in relative amino acid frequencies, ΔA_α, and relative codon frequencies, ΔA_γ, associated with species divergence. In this study we took advantage of the well-characterized properties of the inner products (dot products or scalar products) of 3-D vectors of physical systems, to explore the inner product space of higher dimensional vectors of the information system under consideration here. The first step was to determine the magnitude of each A_α and A_γ, and to use these magnitudes to form a coordinate vector to represent each genome in a 2-D phase space. Since the magnitude of a vector is the vector norm, a vector norm was calculated according to Eqs. (13.5–13.8). using the inner products for each genomic vector pair, i.e. $\langle A_\alpha, A_\alpha \rangle$ and $\langle A_\gamma, A_\gamma \rangle$ for all 73 genomes in this study (Table 13.1).

$$\langle A_\alpha, A_\alpha \rangle = \sum_{\alpha=1}^{20} a_\alpha^2 \tag{13.5}$$

$$\langle A_\gamma, A_\gamma \rangle = \sum_{\gamma=1}^{59} a_\gamma^2 \tag{13.6}$$

The norm or magnitude of any vector A_α is

$$\|A_\alpha\| = \sqrt{\langle A_\alpha, A_\alpha \rangle} \tag{13.7}$$

and the norm or magnitude of any vector A_γ is

$$\|A_\gamma\| = \sqrt{\langle A_\gamma, A_\gamma \rangle} \tag{13.8}$$

Here we form a new coordinate vector, $v = u_x i + v_y j$, with magnitude u_x in direction i, and magnitude v_y in direction j, where $u_x = \|A_\gamma\|$ and $v_y = \|A_\alpha\|$. Fig. 13.1 is a plot of vectors, v, to (u_x, v_y) coordinates for each of the 73 genomes. The complete vector is drawn from the origin for only two genomes in Fig. 1(a) to unclutter the graph and illustrate the differences in vectors. Fig. 1(b) is a replot of Fig. 1(a) data using $u_x = 200$ and $u_y = 20$ as the origin in order to see more clearly the spread of genomic coordinates.

In molecular terms, this normed phase space contains the magnitude of a 59-dimensional vector in the direction of codon usage as an x component, and the magnitude of a 20-D vector in the direction of amino acid composition as a y component of a coordinate vector, v, for each genome in this study. This 2-D plane enables us to see the information state of each genome in the phase space of bacterial proteomes.

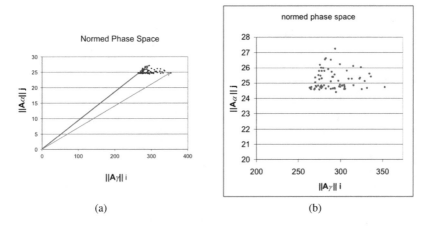

Fig. 13.1: Normed phase space. (a), complete vector plots at full scale for two different organisms; (b), replot of (a) to magnify the spread of coordinates for individual organisms.

13.4 Inner Product Space of Relative Frequency Vectors

Each of the v coordinates derives from a genomic mean and represents any specific gene within a chromosome. Thus, the record of variation in codon usage and amino acid composition for any specific gene can be represented by the difference between a query vector $\mathbf{A}_\gamma$, and a reference vector $\mathbf{B}_\gamma$; and between a query vector $\mathbf{A}_\alpha$, and a reference vector $\mathbf{B}_\alpha$.

For this study of 73 fully sequenced genomes of bacteria (Table 13.1), the relative frequency vectors of one genome were designated $\mathbf{B}_\alpha$ and $\mathbf{B}_\gamma$. These were the reference vectors. Vectors for each of the 72 other genomes were designated $\mathbf{A}_\alpha$ and $\mathbf{A}_\gamma$ for pairwise calculations of scalar products. The inner product of $\mathbf{A}_\alpha$ and $\mathbf{B}_\alpha$ is defined here as

$$\langle \mathbf{A}_\alpha, \mathbf{B}_\alpha \rangle = \sum_{\alpha=1}^{20} (a_\alpha \cdot b_\alpha) \tag{13.9}$$

where the scalar magnitudes of vector components are $a_\alpha \in \mathbf{A}_\alpha$, and $b_\alpha \in \mathbf{B}_\alpha$. Similarly, the dot product of $\mathbf{A}_\gamma$ and $\mathbf{B}_\gamma$ is

$$\langle \mathbf{A}_\gamma, \mathbf{B}_\gamma \rangle = \sum_{\gamma=1}^{59} (a_\gamma \cdot b_\gamma) \tag{13.10}$$

The inner products calculated included the Euclidean norms

$$\sqrt{\langle b_\alpha, b_\alpha \rangle} = \sqrt{b_1^2 + b_2^2 + \ldots + b_{20}^2} \tag{13.11}$$

Table 13.1: Genus and species listing of Eubacteria with complete genome sequences analyzed for this study. Databases used for these samplings were: The National Center for Biotechnology Information, National Library of Medicine, National Institutes of Health (https://www.ncbi.nlm.nih.gov/genome/); and The Institute for Genomic Research (TIGR), now part of the J. Craig Venter Institute (https://www.jcvi.org/).

1	*Aquifex aeolicus*	38	*Mycoplasma pulmonis*
2	*Agrobacterium turnefaciens* C58 Cereon	39	*Neisseria meningitidis*
3	*Agrobacterium turnefaciens* C58 Uwash	40	*Neisseria meningitidis* MC58
4	*Buchnera aphidicola* Sg	41	*Nostoc* PCC 7120
5	*Borellia burgdoferi*	42	*Oceanobacillus iheyensis*
6	*Bacillus halodurans*	43	*Pasteurella multocida*
7	*Bacillus melitensis*	44	*Pseudomonas aeruginosa*
8	*Bacillus subtilis*	45	*Pseudomonas gingivalis*
9	*Bacillus suis*	46	*Pseudomonas putida*
10	Buchnera sp. APS	47	*Rickettsia conorii*
11	*Caulobacter crescentus*	48	*Ralstonia solanacearum*
12	*Corynebacterium glutamicum*	49	*Salmonella typhimurium*
13	*Campylobacter jejuni*	50	*Samonella enterica* sv Typhi
14	*Clostridium perfingens*	51	*Streptococcus agalactiae* 2603V/R
15	*Chlamydophila pneumoniae* AR39	52	*Streptococcus agalactiae* NEM316
16	*Chlamydophila pneumoniae* CWL029	53	*Staphyloccus aureus* Mu50
17	*Chlamydophila pneumoniae* J138	54	*Staphyloccus aureus* MW2
18	*Chlorobium tepidum*	55	*Staphyloccus aureus* N315
19	*Chlamydia trachomatis*	56	*Streptomyces coelicolor*
20	*Chlamydia muridarum*	57	*Sinorhizobium meliloti*
21	*Clostridium acetobutylicum*	58	*Shewanella oneidensis*
22	*Deinococcus radiodurans*	59	*Streptococcus pnuemoniae* R6
23	*Escherichia coli*	60	*Streptococcus pnuemoniae* TIGR4
24	*Enterococcus faecalis*	61	*Streptococcus pyogenes*
25	*Fusobacterium nucleatum*	62	*Synechocystis* PCC 6803
26	*Haemophilus influenzae*	63	*Thermosynechococcus elongatus*
27	*Haemophilus influenzae*	64	*Thermotoga maritima*
28	*Helicobacter pylori* 26695	65	*Treponema pallidum*
29	*Helicobacter pylori* J99	66	*Thermoanaerobacter tengcongens*
30	*Lactococcus lactis*	67	*Ureaplasma urealyticum*
31	*Listeria innocua*	68	*Vibrio cholerae*
32	*Listeria monocytogenes*	69	*Xanthomonas axonopodis* pv. citri
33	*Mesorhizobium loti*	70	*Xanthomonas campestris* pv. campestris
34	*Mycobacterium leprae*	71	*Xylella fastidiosa*
35	*Mycobacterium tuberculosis*	72	*Yersinia pestis* CO92
36	*Mycoplasma genitalium*	73	*Yersinia pestis* KIM
37	*Mycoplasma pneumoniae*		

and

$$\sqrt{\langle b_\gamma, b_\gamma \rangle} = \sqrt{b_1^2 + b_2^2 + \ldots + b_{59}^2} \tag{13.12}$$

This inner product space is complete, and therefore is a Hilbert space and has all attributed properties [5]. This normed phase space is, then, a state space of gene and chromosome evolution and the inner product space can be used to quantify the variation and evolutionary distance between genes and chromosomes in different organisms. This makes possible the measurement of any $\Delta \mathbf{A}_\alpha$ and $\Delta \mathbf{A}_\gamma$ by comparison of a sequence vector $\mathbf{A}$ to a reference vector $\mathbf{B}$, according to relationships (13.13) and (13.14).

$$\langle \mathbf{A}_\gamma, \mathbf{B}_\gamma \rangle \mapsto \Delta \| \mathbf{A}_\gamma \| \tag{13.13}$$

$$\langle \mathbf{A}_\alpha, \mathbf{B}_\alpha \rangle \mapsto \Delta \| \mathbf{A}_\alpha \| \tag{13.14}$$

So the sequence variation that results from the evolutionary divergence of two organisms from a common ancestor, and the resulting divergence of any specific gene, is in the record of the normed state space explored by the inner product space $\langle \mathbf{A}_\alpha, \mathbf{B}_\alpha \rangle$ and $\langle \mathbf{A}_\gamma, \mathbf{B}_\gamma \rangle$.

13.5 Discussion and Conclusions

The normed phase space characterizes the totality of states that this genomic information system can occupy. As such, all sequence variation that exists in live organisms, occurs within this space. While this sampling of bacterial genomes does not define the boundaries of the variation space, it does provide a basis to explore how one genome or one gene relates to another.

In theory, the entire history and future of nucleotide variation (i.e. all states of nucleotide variation) in any protein-coding gene or set of genes is captured by changes in the relative frequency vectors, i.e. $\Delta \mathbf{A}_\alpha$ and $\Delta \mathbf{A}_\gamma$. The complete phase space of natural variation represents all existing states that evolving genes or sets of genes visit.

Measures of relative codon and amino acid frequencies are readily determined from complete, annotated genomic sequences of bacteria. Comparison of these frequencies is independent of gene types, i.e. homologs, orthologs, paralogs, and requires no sequence alignments to determine evolutionary distances between query sequences and reference sequences. Whether evolutionary processes preserve or change protein function from a gene, vector **types (i)** and **(ii)** preserve a record of the history.

Comparing this relative frequency method to a common alignment-based method to measure synonymous and non-synonymous variation presents a way to test the potential accuracy and utility of using relative frequency vectors. It is easy to see that any nucleotide change in a codon that changes an amino acid is non-synonymous (d_n) and changes the magnitude of **two** vector components, i.e. an element of both $\mathbf{A}_\alpha$ and $\mathbf{A}_\gamma$. Thus, $d_n \mapsto \Delta\mathbf{A}_\alpha \,\&\, \Delta\mathbf{A}_\gamma$. Any nucleotide change in a codon that **does not** change an amino acid is synonymous (d_s) and is captured by a change in the magnitude of **one** vector component, i.e. an element of $\mathbf{A}_\gamma$ only. Thus, $d_s \mapsto \Delta\mathbf{A}_\gamma$. Synonymous changes accumulate more frequently than non-synonymous such that the ratio $\frac{d_s}{d_n}$ is higher for closely related orthologs than for more distant relatedness. A principal question to pose, then, is whether and how $\frac{\Delta\|\mathbf{A}_\gamma\|}{\Delta\|\mathbf{A}_\alpha\|} \mapsto \frac{d_s}{d_n}$. If so, then a record of relative codon frequencies and relative amino acid coding frequencies enables calculation of $\frac{d_s}{d_n}$ without the need to align nucleotide sequences and may enable easier, perhaps better, and possibly more accurate comparisons of widely diverged sequences. This study offers the testable prospect that $\frac{\langle \mathbf{A}_\gamma, \mathbf{B}_\gamma \rangle}{\langle \mathbf{A}_\alpha, \mathbf{B}_\alpha \rangle} \mapsto \frac{\Delta\|\mathbf{A}_\gamma\|}{\Delta\|\mathbf{A}_\alpha\|} \mapsto \frac{d_s}{d_n}$.

Previous work in this lab supports a hypothesis that translation pressure is the driving force behind sequence variation and conservation reflected in the relative codon and amino acid frequency vectors predicted from sequenced bacterial genomes [3].

Acknowledgements The author presents this work as a tribute to Erik Goodman for his inspiring leadership of the BEACON Center for the Study of Evolution in Action, and his unfailing ability to inspire new thinking and to promote creative collaborations. Scott Harrison, at North Carolina A&T University, constructed the MYCROW data system that enabled much of the work presented, and his fearless ideas and extraordinary work efforts contributed greatly as a pre-doctoral student and now as a stimulating professor and colleague in the BEACON Center. Patricia Herring has presented untiring intellectual stimulation, research production and editorial support as a research and teaching colleague for all of the years preceding this report. No statement of appreciation is sufficient to acknowledge the full measure of her career sacrifices and selfless contributions to encourage exploration of unexplored horizons.

References

1. Battistuzzi, F.U., Feijao, A., Hedges, S.B.: *A genomic timescale of prokaryote evolution: insights into the origin of methanogenesis, phototrophy, and the colonization of land.* BMC Evolutionary Biology **4**(1), 44 (2004)
2. Buckley, C.O., Stephens, D., Herring, P.A., Jackson, J.H.: *%(g+ c) variation and prediction by a model of bacterial gene transfer and codon adaptation.* Omics: A Journal of Integrative Biology **6**(3), 259–272 (2002)
3. Jackson, J.H., Schmidt, T.M., Herring, P.A.: *A systems approach to model natural variation in reactive properties of bacterial ribosomes.* BMC Systems Biology **2**(1), 62 (2008)
4. Snel, B., Huynen, M.A., Dutilh, B.E.: *Genome trees and the nature of genome evolution.* Annu. Rev. Microbiol. **59**, 191–209 (2005)
5. Young, N.: *An introduction to Hilbert space.* Cambridge University Press (1988)

Chapter 14
Genome Size and the Extinction of Small Populations

Thomas LaBar and Christoph Adami

Abstract Although extinction is ubiquitous throughout the history of life, the factors that drive extinction events are often difficult to decipher. Most studies of extinction focus on inferring causal factors from past extinction events, but these studies are constrained by our inability to observe extinction events as they occur. Here, we use digital evolution to avoid these constraints and study "extinction in action". We examine the genetic mechanisms driving the relationship between genome size and population extinction. We find that genome expansions enhance extinction risk through two genetic mechanisms that increase a population's lethal mutational burden: an increased lethal mutation rate and an increased likelihood of stochastic reproduction errors. This result, contrary to the expectation that genome expansions should buffer mutational effects, suggests a role for epistasis in driving extinction. We discuss biological analogues of these digital "genetic" mechanisms and how large genome size may inform which natural populations are at an increased risk of extinction.

14.1 Introduction

The ubiquity of extinction events throughout the history of life [20] and the increasing realization that Earth's biosphere may be experiencing a sixth mass extinction [4] drive interest in determining the factors that cause certain species, but not others, to go extinct [33]. It is accepted that genetic [38, 47], demographic [32, 35], environmental [28, 50], and ecological [9, 12, 39] factors contribute to species ex-

Thomas LaBar
Department of Microbiology and Molecular Genetics, Michigan State University e-mail:
labartho@msu.edu

Christoph Adami
Department of Microbiology and Molecular Genetics, Michigan State University e-mail: adami@
msu.edu

© Springer Nature Switzerland AG 2020 167
W. Banzhaf et al. (eds.), *Evolution in Action: Past, Present and Future*,
Genetic and Evolutionary Computation, https://doi.org/10.1007/978-3-030-39831-6_14

tinctions. Beyond those deterministic factors, chance events also likely influence some extinction events [41, 49]. Here, we focus on the genetic factors influencing extinction, specifically the role of small population size and genetic drift [31].

In small populations, weak purifying selection leads to increased fixation of small-effect deleterious mutations [56]. As multiple deleterious mutations fix, the absolute fitness of the population may decrease, resulting in a decrease in population size. This decreased population size further weakens selection, leading to the fixation of additional deleterious mutations and a further decrease in population size. This process continues until population extinction occurs. This positive feedback loop between decreased population size and deleterious mutation fixation is known as a mutational meltdown [29]. Mathematical models of mutational meltdowns suggest that even intermediate-sized asexual populations (approximately 10^3 to 10^4 individuals) can quickly go extinct [15, 30]. Likewise, small sexual populations are also vulnerable to fast meltdowns [25].

The concept of a mutational meltdown provides a population-genetic mechanism for extinction. However, it is still uncertain which factors beyond population size influence the likelihood of a meltdown. If deleterious mutation accumulation drives mutational meltdowns, then species with a greater genomic mutation rate should be at a greater risk of extinction [46, 61]. Genome expansions (i.e., mutations that increase genome size) are another proposed genetic mechanism that could lead to population extinction. Indeed, there is some evidence that genome size positively correlates with extinction risk in certain clades of multicellular organisms [53, 54].

While the relationship between high mutation rates and extinction suggests that larger genome size heightens extinction risk solely by increasing mutation rates, the connection between genome size and extinction can be complicated. If genome expansions lead to increased neutrality, the overall genomic mutation rate may increase, but the deleterious mutation rate will remain constant. Species with larger genomes should only face an increased mutational burden if genome expansions lead to increased genome content under purifying selection. For example, potential detrimental molecular interactions between an original genomic region and its duplicate may result in an increased mutational burden [11]. As genome expansions are likely to lead to many alterations in the distribution of mutational effects, it is still unclear which genetic mechanisms lead genome expansions to drive population extinction.

It is difficult to test the role of genome size in extinction in both natural and laboratory model systems. Here, we use digital experimental evolution [2, 5, 18, 22, 40] to test whether genome expansions can drive population extinction. In a previous study with the digital evolution system Avida [37] on the role of population size in the evolution of complexity, we found that the smallest populations evolved the largest genomes and the most novel traits, but also had the greatest extinction rates [23]. Now, we use Avida to test explicitly the mechanisms behind the role of genome size in the extinction of small populations.

Avida differs from previous models of extinction in small populations in the mode of selection. Unlike mutational meltdown models [31], where selection is hard and the accumulation of deleterious mutations directly leads to population extinction, selection is primarily soft in Avida and deleterious mutations alter relative fitness (i.e., competitive differences between genotypes), not absolute fitness (i.e., differences in the number of viable offspring between genotypes). Extinction occurs in Avida through "lethal," or "non-viable," mutations that prevent their bearer from reproducing. These non-viable avidians occupy a portion of the limited space allocated to an avidian population, thus reducing the effective population size and potentially causing extinction over time.

We find multiple genetic mechanisms lead genome expansions to drive the extinction of small populations. Increased genome size not only leads to an increase in the genomic mutation rate, but specifically to an increase in the lethal mutation rate. Elevated lethal mutation rates in large-genome genotypes are likely due to detrimental interactions between ancestral genome regions and duplicated genome content. Additionally, we show that genotypes with large genomes have an elevated probability of stochastic replication errors during reproduction (i.e., stochastic viability), further elevating the likelihood of offspring non-viability and extinction. These results suggest that large genome size does elevate the risk of population extinction due to an increased lethal mutational burden from multiple genetic mechanisms.

14.2 Methods

14.2.1 Avida

Here we review those aspects of Avida (version 2.14; available at https:// github.com/devosoft/avida) relevant to the current study (see [37] for a complete overview). In Avida, simple computer programs ("avidians") compete for the resources required to undergo self-replication and reproduction. Each avidian consists of a genome of computer instructions drawn from a set of twenty-six available instructions in the Avida genetic code. A viable asexual avidian genome must contain the instructions to allocate a new (offspring) avidian genome, copy the instructions from the parent genome to the offspring genome, and divide the offspring genome to produce a new avidian. During this copying process, mutations may occur, introducing heritable variation into the population. This genetic variation causes phenotypic variation: avidians with different genomes may self-replicate at different speeds. As faster self-replicators outcompete slower self-replicators, this heritable variation results in differential fitness between avidians. Therefore, an Avida population undergoes Darwinian evolution [1, 40]. Avida has previously been used to test hypotheses concerning the evolution of genome size [17, 23], the role of population size in evolution [13, 23, 24, 36], and the consequences of population extinction [48, 58, 59, 60].

The Avida world consists of a grid of N cells; each cell can be occupied by at most one avidian. Thus, N is the maximum population size for the Avida environment. While avidian populations are usually at carrying capacity, the presence of lethal mutations can reduce their population size below this maximum size. Here, offspring can be placed into any cell, simulating a well-mixed environment (i.e., no spatial structure). If a cell is occupied by another avidian, the new offspring will overwrite the occupant. The random placement of offspring avidians adds genetic drift to Avida populations, as avidians are overwritten without regard to fitness.

Fitness for an avidian genotype is estimated as the ratio of the number of instructions a genotype executes per unit time to the number of instructions it needs to execute to reproduce. Therefore, there are two avenues for a population of avidians to increase fitness: 1) increase the number of instructions executed per unit time, or 2) decrease the number of instruction executions needed for self-replication. Avidian populations can increase the number of instructions executed by evolving the ability to input random numbers and perform Boolean calculations on these numbers (a "computational metabolism" [26]). They can also decrease the number of instruction executions necessary for reproduction by optimizing their replication machinery.

There are a variety of different implementations of mutations in Avida. Here, we used settings that differed from the default in order to improve our ability to analyze the causes of population extinction (see Table 14.1 for a list of changes to the default settings). Point mutations change one locus from one of the twenty-six Avida instructions to another random, uniformly chosen, instruction; these mutations occur upon division between parent and offspring. There is an equal probability that each instruction in the genome will receive a point mutation; thus, genome size determines the total genomic mutation rate. To model indels, we used so-called "slip" mutations. This mutational type will randomly select two loci in the genome and then, with equal probability, either duplicate or delete the instructions in the genome between those two loci. While the rate of indel mutations remains constant, the chance of large indel mutations increases as genome size grows. Finally, to ease our analysis, we required every offspring genotype to be identical to its parent's genotype before the above mutations were applied at division. This setting prevented the origin of deterministic "implicit" mutations that occur when certain genotypes undergo genome replication [2].

One aspect of Avida mutations that differs from traditional models of population extinction is the presence of non-viable mutations in addition to merely deleterious, but still viable, mutations. We call these mutations "lethal," but strictly speaking they do not kill their bearer. Instead, they prevent their bearer from successfully reproducing within the maximum allowed lifespan (i.e., they are non-viable). Here, we used the default maximum lifespan of $20 \times L$ instruction executions, where L is the genome size. In other words, this setting limits the number of times an avidian can cycle through their genome in an attempt to reproduce. Such a setting must exist in order to allow avidian genomes to be analyzed. Otherwise, non-reproducing avidians could be analyzed forever, as the only way to decide if an avidian can reproduce is to actually execute the code in its genome.

In Avida, it is possible to perform experiments where mutations with certain effects are prevented from appearing in a population [10]. To enable this dynamic, the Avida program analyzes the fitness of every novel genotype that enters the popula-

Table 14.1: **Notable Avida parameters changed from default value.**

Parameter	Default Value	Changed Value	Treatment
WORLD_X	60	N	All
WORLD_Y	60	1	All
BIRTH_METHOD	0	4	All
COPY_MUT_PROB	0.0075	0.0	All
DIV_MUT_PROB	0.0	μ	All
DIVIDE_INS_PROB	0.05	0.0	All
DIVIDE_DEL_PROB	0.05	0.0	All
DIVIDE_SLIP_PROB	0.0	0.01	Variable Genome Size
REQUIRE_EXACT_COPY	0	1	All
REVERT_FATAL	0.0	1.0	Lethal-reversion
REVERT_DETRIMENTAL	0.0	1.0	Deleterious-reversion

tion and, if the fitness is of the pre-set effect, the mutation is reverted. This system allows experimenters the ability to determine the relevance of certain mutational effects to evolution. However, mutations of certain effects can still enter the population if their fitness effects are stochastic. An avidian has stochastic fitness if its replication speed depends on characteristics of the random numbers it inputs in order to perform its Boolean calculations. Some stored numbers may alter the order in which certain instructions are executed or copied into an offspring's genome, thus altering fitness.

14.2.2 Experimental Design

To study the role of genome size in the extinction of small populations, we first evolved populations across a range of per-site mutation rates ($\mu = 0.01$ and $\mu = 0.1$) and population sizes ($N = \{5, 6, 7, 8, 10, 15, 20\}$ for $\mu = 0.01$ and $N = \{10, 12, 15, 16, 17, 20, 25\}$ for $\mu = 0.1$). For each combination of population size and mutation rate we evolved 100 populations for at most 10^5 generations. Each population was initialized at carrying capacity with N copies of the default Avida ancestor (which has 100 instructions) with all excess instructions removed; this resulted in an ancestor with a genome of 15 instructions (only those needed for replication). By using an ancestral genotype with an almost-minimal genome (avidian genotypes with smaller genomes do exist [3, 7]), we were better able to explore the consequences of genome expansions (i.e., the ancestor is close to the theoretical lower bound on genome size). Ancestral genotypes with per-site mutation rates of $\mu = 0.01$ and $\mu = 0.1$ thus have genomic mutation rates of $U = 0.15$ and $U = 1.5$ mutations/genome/generation, respectively. Genome size mutations (indels) occured at a fixed rate of 0.01 mutations/genome/generation for all treatments. Additionally, for each mutation rate and population size combination, an additional 100 populations were evolved in an environment where genome size was fixed. To directly test for the role of lethal and deleterious mutations in driving extinction, we evolved 100 populations at the low mutation rate population sizes under conditions where either lethal mutations or deleterious, but non-lethal, mutations were reverted (the "lethal-reversion" and "deleterious-reversion" treatments, respectively).

14.2.3 Data Analysis

For all evolution experiments, we saved data on the most abundant genotype every ten generations. The final saved genotype was used in all analyses here. All data represent either genotypes at most ten generations before extinction (in the case of extinct populations) or genotypes from the end of the experiment (in the case of surviving populations). In order to calculate the lethal mutation rate and other relevant statistics for a genotype, we generated every single point mutation for that genotype and measured these mutants' fitness using Avida's Analyze mode. The lethal mutation rate was estimated as $U_{\text{lethal}} = \mu \times L \times p_{\text{lethal}}$, where μ is the per-site mutation rate, L is the genome size, and p_{lethal} is the probability that a random mutation will be lethal.

14.2.3.1 Analysis of the relationship between genome expansions and changes in the lethal mutation rate

To test whether genome expansions themselves were directly responsible for the increase in the lethal mutation rate or whether the lethal mutation rate increased after evolution in response to a genome expansion, we first reconstructed the line-of-descents (LODs) for each of the one hundred genotypes evolved in a population of 20 individuals with a per-site mutation rate of 0.01 mutations/site/generation (the low mutation rate). An LOD contains every intermediate genotype from the ancestral genotype to an evolved genotype and allows us to trace how genome size evolved over the course of the experiment [26]. We reduced these LODs to only contain the ancestral genotype, the genotypes that changed genome size, the geno-type immediately preceding a change in genome size, and the final genotype. We measured the genome size and the lethal mutation rate for each of these remaining genotypes. Then, we measured the relationship between the change in genome size and the change in the lethal mutation rate for genome expansions, genome reductions, and the segments of evolutionary time where genome size was constant.

14.2.3.2 Analysis of stochastic viability

In order to test the possibility that some of our populations had evolved stochastic viability, we analyzed each genotype from the $N = 5$ lethal-reversion populations and each genotype from the $N = 8$, $\mu = 0.01$, original populations. These population sizes were chosen because they had the most nearly equal number of extinct and surviving populations. We performed 1000 viability trials, where a genotype was declared non-viable if it could not reproduce. A genotype was declared stochastically-viable if the number of non-viable trials was greater than 0 and less than 1000. Otherwise, it was defined as deterministically-viable.

All data analysis beyond that using Avida's Analyze Mode was performed using the Python packages NumPy version 1.12.1 [51], SciPy version 0.19.0 [21], and Pandas version 0.20.1 [34]; figures were generated using the Python package Matplotlib version 2.0.2 [19]. All Avida scripts and data analysis scripts used here are available at https://github.com/thomaslabar/LaBarAdami_GenomeSizeExtinction.

14.3 Results

14.3.1 Large genome size increases the extinction risk of small populations

To test if large genome size enhances the probability of population extinction, we evolved populations across a range of population sizes at both high (1.5 mutations/genome/generation) and low (0.15 mutations/genome/generation) mutation rates with either a fixed genome size or a variable genome size. Based on our previous work with a similar experimental setup [23], we predicted that our smallest populations would go extinct at high rates if genome size could vary (which, based on this previous study, results in genome expansion and large genome size). As expected, under the low mutation rate regime populations with variable genome sizes had greater rates of extinction than those with fixed small genomes (Fig. 14.1A). However, under the high mutation rate regime, there was no significant difference between populations with a variable genome size and populations with a constant genome size (Fig. 14.1A). Estimations of the time to extinction further support these trends: in the low mutation regime, populations where genome size could evolve went extinct in fewer generations than those where genome size was constant. There were no differences in the high mutation rate regime (Fig. 14.1B).

Next, we compared the final evolved genome size between genotypes from extinct populations and surviving populations under the Variable Genome Size treatment. Across the range of population sizes for which at least 10 populations both survived and went extinct, "extinct" genotypes evolved larger genomes than those "surviving" genotypes in the low mutation rate regime (Fig. 14.2A). In the high mutation rate regime, one population size ($N = 15$ individuals) led to surviving popu-

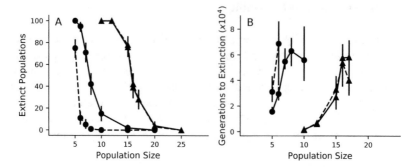

Fig. 14.1: **Possibility of genome expansions increases extinction in low mutation rate populations.** A) Number of extinct populations (out of 100) as a function of population size. Solid (dashed) lines represent variable (fixed) genome size populations. Circles (triangles) represent low (high) mutation rate populations. Error bars are bootstrapped 95% confidence intervals (10^4 samples). B) Median time to extinction for population size and mutation rate combinations. Lines and symbols same as in panel A. Error bars are bootstrapped (10^4 samples) 95% confidence intervals of the median. Data only shown for those treatments that resulted in at least ten extinct populations.

lations evolving larger genomes, while there was no statistically-significant difference for the other population sizes (Fig. 14.2B). Together, these results suggest that genome expansions and large genome size can enhance the risk of small population extinction if the initial mutation rate is too low for extinction to otherwise occur. We next focus on examining the mechanism behind the relationship between genome size and extinction in the low mutation rate populations.

14.3.2 Extinction and large genome size are associated with increases in the lethal mutational load

Avidian populations only face population-size reductions through one mechanism: parent avidians produce non-viable offspring that replace viable avidians. In other words, the lethal mutational load should drive population extinction. It is therefore possible that the increased genomic mutation rate that co-occurs with genome expansions specifically increased the genomic lethal mutation rate. The elevated lethal mutation rate then leads to an increased rate of population extinction. We first tested whether larger genomes had increased lethal mutation rates. Genome size was correlated with the lethal mutation rate across genotypes from all population sizes, supporting the hypothesis that increases in genome size result in increased lethal mutational loads and eventually population extinction (Fig. 14.3A; Spearman's $\rho \approx 0.75$, $n = 616$, $p = 1.77 \times 10^{-148}$). Next, we examined whether populations that went extinct had previously evolved greater lethal mutation rates than surviving popula-

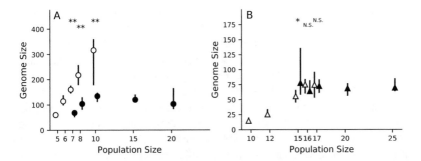

Fig. 14.2: **Extinct populations evolved larger genomes.** A) Final genome size for the low mutation rate populations from the Variable Genome Size treatment as a function of population size. Populations that survived are shown in black; populations that went extinct are shown in white. Data points are median values and error bars are bootstrapped (10^4 samples) 95% confidence intervals of the median. Data points are offset for clarity. ** indicates $p < 10^{-4}$, * indicates $p < 10^{-2}$, and N.S. indicates $p > 0.05$ for the Mann-Whitney U-test. Population sizes where fewer than ten populations went extinct (or survived) not shown. B) Final genome size for the high mutation rate populations from the Variable Genome Size treatment as a function of population size. Description same as in panel A. Population sizes where fewer than ten populations went extinct (or survived) not shown.

tions. As with the trend for genome size, extinct populations evolved greater lethal mutations rates than surviving populations (Fig. 14.3B).

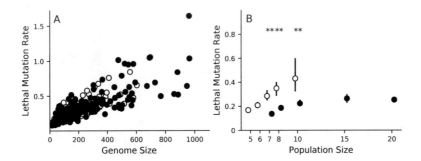

Fig. 14.3: **Lethal mutation rate correlates with genome size and population extinction.** A) The lethal mutation rate as a function of genome size for the final genotypes from each evolved low mutation rate population. B) The lethal mutation rate for extinct and surviving populations across population sizes. Error bars are bootstrapped (10^4 samples) 95% confidence intervals of the median. Population sizes where fewer than ten populations went extinct (or survived) not shown. Colors and symbols same as Fig. 14.2.

The previous data support the hypothesis that genome expansions drive population extinction by increasing the lethal mutation rate and thus the lethal mutational load. However, it is unclear whether genome expansions themselves increase the likelihood of lethal mutations (suggesting epistasis between genome expansions and the ancestral genome) or whether genome expansions merely potentiate future increases in the lethal mutation rate (due to subsequent adaptation). If genome expansions themselves increased the likelihood of lethal mutations, we expect that mutations that increase genome size should, on average, increase the lethal mutation rate. If genome expansions merely allow for the future accumulation of additional mutations that themselves increase the lethal mutation rate, there should be no relationship between mutations that increase genome size and the lethal mutation rate.

To test these two scenarios, we examined the evolutionary histories (i.e., lines-of-descent or LODs) for all $N = 20$ low mutation-rate populations. We then examined the relationship between changes in genome size and changes in the lethal mutation rate (Fig. 14.4A). When genome size was constant, the lethal mutation rate did not change on average (median change = 0.0 mutations/genome/generation). Genome size increases on average increased the lethal mutation rate (median change = 0.01 mutations/genome/generation), while genome size decreases on average decreased the lethal mutation rate (median change = -0.013 mutations/genome/generation). Additionally, the change in genome size positively correlates with the change in the lethal mutation rate (Fig. 14.4B; Spearman's $\rho = 0.67$, $n = 3600$, $p \approx 0.0$), suggesting that genome expansions directly lead to increases in the genomic lethal mutation rate.

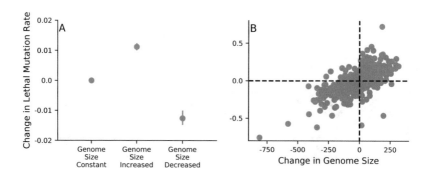

Fig. 14.4: **Insertions and deletions directly change the lethal mutation rate.** A) Change in the lethal mutation rate as a function of a mutation's effect on genome size. Each circle is the median value of all genome size alterations of a given type and error bars are bootstrapped (10^4 samples) 95% confidence intervals of the median. B) Relationship between a mutation's change in genome size and the change in the lethal mutation rate. Data same as in panel A. Dashed lines represent no change. Data points comparing genotypes with equal genome size were excluded.

14.3.3 Lethal mutation rates and stochastic viability drive population extinction

Finally, to establish the role of the lethal mutation rate in driving population extinction, we performed additional evolution experiments to test whether the absence of lethal mutations would prevent population extinction. We repeated our initial experiments (Fig. 14.1), except offspring with lethal mutations were reverted to their parental genotype (lethal-reversion treatment; see Methods for details). We also did the same experiment where deleterious, but non-lethal, mutations were reverted in order to test if deleterious mutations contributed to extinction. When populations evolved without deleterious mutations, extinction rates were similar to, if not greater than, those for populations that evolved with deleterious mutations (Fig. 14.5A). Populations that evolved with fixed-size genomes and without lethal mutations never went extinct, demonstrating how the lack of lethal mutations can prevent extinction (Fig. 14.5B). However, when these populations evolved with variable genome sizes, extinction still occurred, although at a lower rate than when lethal mutations were present (Fig. 14.5B).

While these data demonstrate that lethal mutations do primarily drive extinction risk, the fact that extinction can still occur presumably without lethal mutations is unexpected and indicates that there is a second factor that relates genome size to extinction. This is surprising, as lethal mutations are the only direct mechanism to cause extinction in Avida. One possible explanation for extinction in the lethal-reversion populations is that mutants arise in these populations that are initially viable, but later become non-viable. In other words, these populations evolve *stochastic viability*, where characteristics of the random numbers the avidians input during their life-cycle affect their ability to reproduce. These genotypes with stochastic viability would, on occasion, not be detected as lethal mutants, and thus enter the population even when lethal mutations are reverted. As they reproduce, these stochastically-viable genotypes will input other numbers and thus become, in effect, non-viable and subsequently lead to population extinction. To check if the populations that went extinct without lethal mutations did evolve stochastic viability, we tested the viability of all 100 genotypes from the lethal-reversion, variable genome size $N = 5$ populations. We also performed the same tests with the 100 genotypes from the $N = 8$ populations that evolved with lethal mutations to see if these mutants arose in our original populations.

For both sets of genotypes, we found that some genotypes were stochastically viable (Fig. 14.5C). In fact, of the 23 genotypes from populations that went extinct in the lethal-reversion treatment, 19 displayed stochastic viability. No genotypes from surviving populations were stochastically-viable. Of the 42 genotypes from populations that went extinct among our original treatment genotypes, 8 displayed stochastic viability. Two genotypes from surviving populations were stochastically-viable. Finally, we compared the genome sizes between genotypes from the lethal-reversion genotypes that were always measured as viable and those that measured as stochastic-viable. Stochastic-viable genotypes evolved larger genomes

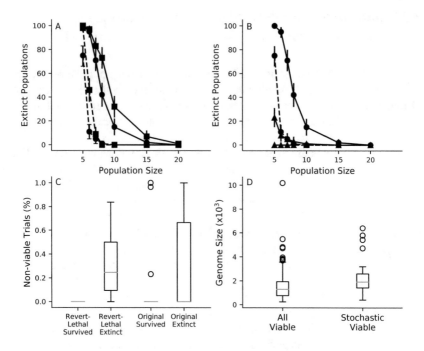

Fig. 14.5: **Evolution of stochastic viability contributes to extinction risk.** A) Number of
population extinctions (out of 100 replicates) as a function of population size for the
deleterious-reversion (squares) and no-reversion (circles) treatments. Dashed and solid
lines represent populations from fixed genome size and variable genome size
treatments, respectively. Error bars are 95% confidence intervals generated using
bootstrap sampling (10^4 samples). No-reversion treatment data same as in Fig. 14.1A.
B) Number of population extinctions (out of 100 replicates) as a function of population
size for the lethal-reversion (triangles) and no-reversion (circles) treatments. Other
symbols same as in panel A. C) Percent of viability trials (out of 1000) for which a
given genotype was not viable. Values between 0 and 1 indicate stochastic viability.
"Original" refers to the 100 genotypes from the $N = 8$ populations that evolved with
lethal mutations. "Revert-lethal" refers the 100 genotypes from the $N = 5$
lethal-reversion populations. Red lines are median values, boxes represent the first- and
third-quartile, whiskers are at-most $1.5\times$ the relevant quartile, and circles are outliers.
D) Genome size for always-viable and stochastically-viable genotypes from both the
$N = 8$ no-reversion populations and the $N = 5$ lethal reversion populations.

than deterministic-viable genotypes (median = 128 instructions versus median =
191 instructions, Mann-Whitney $U = 1968.0$, $n_1 = 161$, $n_2 = 39$, $p < 2 \times 10^{-4}$;
Fig. 14.5D), further suggesting that increased genome size can lead to the evolution of stochastic viability and eventual population extinction. We comment on the
relevance of stochastic viability in biological populations in the Discussion below.

14.4 Discussion

We explored potential genetic mechanisms behind the relationship between genome size and the extinction of small populations. Genome expansions drive extinction because they increase the lethal mutation rate of small populations. Elevated lethal mutation rates arise through two genetic mechanisms. First, genome expansions directly increase the lethal mutation rate, suggesting that epistatic interactions between ancestral genome content and novel duplicated content lead to more lethal mutations. Second, genotypes with larger genomes have a greater likelihood of evolving stochastic viability. Both mechanisms contribute to the lethal mutational burden of small populations and together heighten the risk of population extinction.

The relationship between genome expansions and increases in the lethal mutation rate is at first counterintuitive. It is classically thought that gene/genome duplications should lead to an increase in the rate of *neutral* mutations, not lethal mutations, due to increased mutational robustness [16]. Increases in the lethal mutation rate (and not the neutral mutation rate) should only occur if there are genetic interactions (i.e., epistasis) between the ancestral genome section and the duplicated genome section. Is there evidence for gene/genome duplications leading to increased mutational load, as opposed to increased robustness? Recently, it was argued that gene duplication can also result in increased mutational fragility (not just mutational robustness) if a duplicate gene evolves to interact with its ancestral version [11]. However, more empirical studies are needed to determine whether genome expansions can elevate the mutational burden of a population to such a level that population extinction becomes a possibility.

Our second proposed mechanism underlying the connection between genome size and extinction is the evolution of stochastically-viable genotypes that can only reproduce under some environmental conditions (here, particular random number inputs). The connection behind stochastic viability and extinction in small populations is intuitive. Mutations causing stochastic viability likely have a weak effect (due to their stochastic nature) and can fix in small populations due to weakened selection. After fixation, the lethality of these mutation may be stochastically revealed and then extinction occurs. However, studies on the functional consequences of mutations responsible for extinction are rare (although see [14, 44]) and it is uncertain whether these mutations arise in populations at high extinction risk. One piece of evidence that suggests that mutations with stochastic effects might be relevant to population extinction comes from microbial experimental evolution. It has been shown that small populations have reduced extinction risk if they overexpress genes encoding molecular chaperones that assist with protein folding [45]. These overexpressed chaperones presumably compensate for other mutations that cause increased rates of stochastic protein misfolding. Therefore, mutations responsible for an increased likelihood of protein misfolding may be an example of a class of mutations with a stochastic effect that enhance extinction risk. However, this is only speculation and further work is needed to determine if stochastic viability is a possible mechanism behind extinction risk.

The most prominent model of small population extinction is the mutational melt-down model [29, 30, 31], which argues that even intermediate-sized asexual and sexual populations (i.e., 10^3 individuals) can go extinct on the order of thousands of generations. In contrast to mutational meltdown models, only very small populations go extinct in Avida, and extinction occurs on a longer timescale. The difference between our results and previous results from mutational meltdown models is likely due to differences in the character of selection between the two models. Selection is hard in mutational meltdown models, and the accumulation of deleterious mutations directly increases the probability that offspring will be non-viable [31]. In Avida, selection on deleterious mutations is soft and the accumulation of deleterious mutations is unrelated to the likelihood of non-viable offspring. Without the positive feedback loop between deleterious mutation accumulation and population size, avidian populations only evolve a high rate of non-viable mutants if they evolve large genomes, thus explaining the trends we saw here.

These differences between extinction in hard selection models and the Avida selection model emphasizes the need to consider whether selection in biological populations is primarily hard or soft. Unfortunately, there has been little resolution on this question [42, 55]. There is some evidence that soft selection may be more prevalent than hard selection. For instance, soft selection has been invoked as an explanation for why humans are able to experience high rates of deleterious mutations per generation [8, 27]. Moreover, the persistence of small, isolated populations [6, 57, 43] suggests that not only is selection primarily soft in nature, but that the extinction dynamics we study here are relevant to a subset of biological populations. While large genome size may not be the factor that causes populations to decline, it could drive an already-reduced population to extinction.

In a previous study, we observed that small populations evolved the largest genomes, the greatest phenotypic complexity, and the greatest rates of extinction [23]. This result raised the question of whether greater biological complexity itself could increase a population's rate of extinction. Although we did not test whether increased phenotypic complexity had a role in extinction, we have shown that genome size did drive small-population extinction. While it is possible that phenotypic complexity also enhanced the likelihood of extinction, the Avida phenotypic traits likely do not increase the lethal mutation rate. Thus, both high extinction rates and increased phenotypic complexity arise due to the same mechanism: greater genome size. This result illustrates an evolutionary constraint for small populations. While weakened selection and stronger genetic drift can lead to increases in biological complexity, small populations must also evolve genetic architectures that reduce the risk of extinction [45, 52]. Otherwise, small populations cannot maintain greater complexity and their lethal mutational load inexorably drives them to extinction.

Acknowledgements We thank M. Zettlemoyer and C. Weisman for comments on previous drafts of the manuscript. This work was supported in part by Michigan State University through computational resources provided by the Institute for Cyber-Enabled Research.

References

1. Adami, C.: *Introduction to Artificial Life*. TELOS, Springer Verlag (1998)
2. Adami, C.: *Digital genetics: Unravelling the genetic basis of evolution*. Nature Reviews Genetics **7**(2), 109–118 (2006)
3. Adami, C., LaBar, T.: From entropy to information: Biased typewriters and the origin of life. In: S.I. Walker, P.C. Davies, C.F. Ellis (eds.) From Matter to Life: Information and Causality, chap. 7, pp. 130–154. Cambridge University Press, Cambridge (2017)
4. Barnosky, A.D., Matzke, N., Tomiya, S., Wogan, G.O., Swartz, B., Quental, T.B., Marshall, C., McGuire, J.L., Lindsey, E.L., Maguire, K.C., et al.: *Has the Earth's sixth mass extinction already arrived?* Nature **471**(7336), 51–57 (2011)
5. Batut, B., Parsons, D.P., Fischer, S., Beslon, G., Knibbe, C.: *In silico experimental evolution: A tool to test evolutionary scenarios*. BMC Bioinformatics **14**(Suppl 15), S11 (2013)
6. Benazzo, A., Trucchi, E., Cahill, J.A., Delser, P.M., Mona, S., Fumagalli, M., Bunnefeld, L., Cornetti, L., Ghirotto, S., Girardi, M., et al.: *Survival and divergence in a small group: The extraordinary genomic history of the endangered apennine brown bear stragglers*. Proceedings of the National Academy of Sciences **114**(45), E9589–E9597 (2017)
7. CG, N., LaBar, T., Hintze, A., Adami, C.: *Origin of life in a digital microcosm*. Philosophical Transactions of the Royal Society A: Mathematical, Physical, and Engineering Sciences **375**(2109), 20160,350 (2017)
8. Charlesworth, B.: *Why we are not dead one hundred times over*. Evolution **67**(11), 3354–3361 (2013)
9. Clavero, M., García-Berthou, E.: *Invasive species are a leading cause of animal extinctions*. Trends in Ecology and Evolution **20**(3), 110–110 (2005)
10. Covert, A.W., Lenski, R.E., Wilke, C.O., Ofria, C.: *Experiments on the role of deleterious mutations as stepping stones in adaptive evolution*. Proceedings of the National Academy of Sciences **110**(34), E3171–E3178 (2013)
11. Diss, G., Gagnon-Arsenault, I., Dion-Coté, A.M., Vignaud, H., Ascencio, D.I., Berger, C.M., Landry, C.R.: *Gene duplication can impart fragility, not robustness, in the yeast protein interaction network*. Science **355**(6325), 630–634 (2017)
12. Dunn, R.R., Harris, N.C., Colwell, R.K., Koh, L.P., Sodhi, N.S.: *The sixth mass coextinction: are most endangered species parasites and mutualists?* Proceedings of the Royal Society of London B: Biological Sciences **276**(1670), 3037–3045 (2009)
13. Elena, S.F., Wilke, C.O., Ofria, C., Lenski, R.E.: *Effects of population size and mutation rate on the evolution of mutational robustness*. Evolution **61**(3), 666–674 (2007)
14. Fry, E., Kim, S.K., Chigurapti, S., Mika, K.M., Ratan, A., Dammermann, A., Mitchell, B.J., Miller, W., Lynch, V.J.: *Accumulation and functional architecture of deleterious genetic variants during the extinction of Wrangel Island mammoths*. bioRxiv p. 137455 (2017)
15. Gabriel, W., Lynch, M., Burger, R.: *Muller's ratchet and mutational meltdowns*. Evolution **47**(6), 1744–1757 (1993)
16. Gu, Z., Steinmetz, L.M., Gu, X., Scharfe, C., Davis, R.W., Li, W.H.: *Role of duplicate genes in genetic robustness against null mutations*. Nature **421**(6918), 63–66 (2003)
17. Gupta, A., LaBar, T., Miyagi, M., Adami, C.: *Evolution of genome size in asexual digital organisms*. Scientific Reports **6**, 25,786 (2016)
18. Hindré, T., Knibbe, C., Beslon, G., Schneider, D.: *New insights into bacterial adaptation through in vivo and in silico experimental evolution*. Nature Reviews Microbiology **10**(5), 352–365 (2012)
19. Hunter, J.D.: *Matplotlib: A 2D graphics environment*. Computing In Science & Engineering **9**(3), 90–95 (2007)
20. Jablonski, D.: *Background and mass extinctions: The alternation of macroevolutionary regimes*. Science **231**(4734), 129–133 (1986)
21. Jones, E., Oliphant, T., Peterson, P., et al.: SciPy: Open source scientific tools for Python (2001–)

22. Kawecki, T.J., Lenski, R.E., Ebert, D., Hollis, B., Olivieri, I., Whitlock, M.C.: *Experimental evolution*. Trends in Ecology & Evolution **27**(10), 547–560 (2012)
23. LaBar, T., Adami, C.: *Different evolutionary paths to complexity for small and large populations of digital organisms*. PLoS Computational Biology **12**(12), e1005,066 (2016)
24. LaBar, T., Adami, C.: *Evolution of drift robustness in small populations*. Nature Communications **8**, 1012 (2017)
25. Lande, R.: *Risk of population extinction from fixation of new deleterious mutations*. Evolution **48**(5), 1460–1469 (1994)
26. Lenski, R.E., Ofria, C., Pennock, R.T., Adami, C.: *The evolutionary origin of complex features*. Nature **423**(6936), 139–144 (2003)
27. Lesecque, Y., Keightley, P.D., Eyre-Walker, A.: *A resolution of the mutation load paradox in humans*. Genetics **191**(4), 1321–1330 (2012)
28. Lindsey, H.A., Gallie, J., Taylor, S., Kerr, B.: *Evolutionary rescue from extinction is contingent on a lower rate of environmental change*. Nature **494**(7438), 463–467 (2013)
29. Lynch, M., Bürger, R., Butcher, D., Gabriel, W.: *The mutational meltdown in asexual populations*. Journal of Heredity **84**(5), 339–344 (1993)
30. Lynch, M., Conery, J., Burger, R.: *Mutation accumulation and the extinction of small populations*. American Naturalist **146**(4), 489–518 (1995)
31. Lynch, M., Gabriel, W.: *Mutation load and the survival of small populations*. Evolution **44**(7), 1725–1737 (1990)
32. Matthies, D., Bräuer, I., Maibom, W., Tscharntke, T.: *Population size and the risk of local extinction: empirical evidence from rare plants*. Oikos **105**(3), 481–488 (2004)
33. Maynard Smith, J.: *The causes of extinction*. Philosophical Transactions of the Royal Society of London B: Biological Sciences **325**(1228), 241–252 (1989)
34. McKinney, W.: *Python for Data Analysis: Data Wrangling with Pandas, NumPy, and IPython*. O'Reilly Media, Inc. (2012)
35. Melbourne, B.A., Hastings, A.: *Extinction risk depends strongly on factors contributing to stochasticity*. Nature **454**(7200), 100–103 (2008)
36. Misevic, D., Lenski, R.E., Ofria, C.: *Sexual reproduction and Muller's ratchet in digital organisms*. In: Ninth International Conference on Artificial Life, pp. 340–345 (2004)
37. Ofria, C., Bryson, D.M., Wilke, C.O.: *Avida: A software platform for research in computational evolutionary biology*. In: M. Komosinski, A. Adamatzky (eds.) Artificial Life Models in Software, pp. 3–35. Springer London (2009)
38. O'Grady, J.J., Brook, B.W., Reed, D.H., Ballou, J.D., Tonkyn, D.W., Frankham, R.: *Realistic levels of inbreeding depression strongly affect extinction risk in wild populations*. Biological Conservation **133**(1), 42–51 (2006)
39. Pedersen, A.B., Jones, K.E., Nunn, C.L., Altizer, S.: *Infectious diseases and extinction risk in wild mammals*. Conservation Biology **21**(5), 1269–1279 (2007)
40. Pennock, R.T.: *Models, simulations, instantiations, and evidence: the case of digital evolution*. Journal of Experimental & Theoretical Artificial Intelligence **19**(1), 29–42 (2007)
41. Raup, D.M.: *Extinction: Bad genes or bad luck?* WW Norton & Company, New York (1992)
42. Reznick, D.: *Hard and soft selection revisited: How evolution by natural selection works in the real world*. Journal of Heredity **107**(1), 3–14 (2015)
43. Robinson, J.A., Ortega-Del Vecchyo, D., Fan, Z., Kim, B.Y., Marsden, C.D., Lohmueller, K.E., Wayne, R.K., et al.: *Genomic flatlining in the endangered island fox*. Current Biology **26**(9), 1183–1189 (2016)
44. Rogers, R.L., Slatkin, M.: *Excess of genomic defects in a woolly mammoth on Wrangel Island*. PLoS Genetics **13**(3), e1006,601 (2017)
45. Sabater-Muñoz, B., Prats-Escriche, M., Montagud-Martínez, R., López-Cerdán, A., Toft, C., Aguilar-Rodríguez, J., Wagner, A., Fares, M.A.: *Fitness trade-offs determine the role of the molecular chaperonin GroEL in buffering mutations*. Molecular Biology and Evolution **32**(10), 2681–2693 (2015)
46. Singh, T., Hyun, M., Sniegowski, P.: *Evolution of mutation rates in hypermutable populations of Escherichia coli propagated at very small effective population size*. Biology Letters **13**(3), 20160,849 (2017)

47. Spielman, D., Brook, B.W., Frankham, R.: *Most species are not driven to extinction before genetic factors impact them.* Proceedings of the National Academy of Sciences **101**(42), 15,261–15,264 (2004)
48. Strona, G., Lafferty, K.D.: *Environmental change makes robust ecological networks fragile.* Nature Communications **7**, 12,462 (2016)
49. Turner, C.B., Blount, Z.D., Lenski, R.E.: *Replaying evolution to test the cause of extinction of one ecotype in an experimentally evolved population.* PloS One **10**(11), e0142,050 (2015)
50. Urban, M.C.: *Accelerating extinction risk from climate change.* Science **348**(6234), 571–573 (2015)
51. Van Der Walt, S., Colbert, S.C., Varoquaux, G.: *The NumPy array: A structure for efficient numerical computation.* Computing in Science & Engineering **13**(2), 22–30 (2011)
52. Vermeij, G.J., Grosberg, R.K.: *Rarity and persistence.* Ecology Letters **21**(1), 3–8 (2018)
53. Vinogradov, A.E.: *Selfish DNA is maladaptive: Evidence from the plant Red List.* Trends in Genetics **19**(11), 609–614 (2003)
54. Vinogradov, A.E.: *Genome size and extinction risk in vertebrates.* Proceedings of the Royal Society of London B: Biological Sciences **271**, 1701–1706 (2004)
55. Wallace, B.: *Fifty Years of Genetic Load: An Odyssey.* Cornell University Press (1991)
56. Whitlock, M.C., Bürger, R., Dieckmann, U.: Fixation of new mutations in small populations. In: R. Ferrire, U. Dieckmann, D. Couvet (eds.) Evolutionary Conservation Biology, chap. 9, pp. 155–170. Cambridge University Press, Cambridge (2004)
57. Xue, Y., Prado-Martinez, J., Sudmant, P.H., Narasimhan, V., Ayub, Q., Szpak, M., Frandsen, P., Chen, Y., Yngvadottir, B., Cooper, D.N., et al.: *Mountain gorilla genomes reveal the impact of long-term population decline and inbreeding.* Science **348**(6231), 242–245 (2015)
58. Yedid, G., Ofria, C., Lenski, R.: *Historical and contingent factors affect re-evolution of a complex feature lost during mass extinction in communities of digital organisms.* Journal of Evolutionary Biology **21**(5), 1335–1357 (2008)
59. Yedid, G., Ofria, C.A., Lenski, R.E.: *Selective press extinctions, but not random pulse extinctions, cause delayed ecological recovery in communities of digital organisms.* The American Naturalist **173**(4), E139–E154 (2009)
60. Yedid, G., Stredwick, J., Ofria, C.A., Agapow, P.M.: *A comparison of the effects of random and selective mass extinctions on erosion of evolutionary history in communities of digital organisms.* PloS One **7**(5), e37,233 (2012)
61. Zeyl, C., Mizesko, M., De Visser, J.: *Mutational meltdown in laboratory yeast populations.* Evolution **55**(5), 909–917 (2001)

Part III
Evolution of Behavior and Intelligence

Chapter 15
Temporal Niche Evolution and the Sensory Brain

Barbara Lundrigan, Andrea Morrow, Paul Meek and Laura Smale

Abstract Mammals that occupy divergent temporal niches are active in worlds that provide different sensory cues. In particular, activity during the day affords greater opportunity to use light to obtain information. But to capitalize on sensory cues, an animal must have a brain that can process them, and brain tissue is costly. Here, we present preliminary data evaluating the hypothesis that there have been trade-offs in investment among sensory regions of the brain at evolutionary transitions from one temporal niche to another. We compare the sizes of five brain structures (olfactory bulb, superior colliculus, lateral geniculate nucleus, inferior colliculus, medial geniculate nucleus) in four rodent species (southern flying squirrel and red squirrel, suborder Sciuromorpha; Australian bush rat and Nile grass rat, suborder Myomorpha), representing at least two independent evolutionary transitions from one temporal niche to another. Investment in olfactory bulbs and one visual structure (superior colliculus) was consistent with an evolutionary tradeoff, i.e. nocturnal representatives of each clade had a larger olfactory bulb and smaller superior colliculus. The other visual structure (lateral geniculate nucleus) was larger in the diur-

Barbara Lundrigan
Department of Integrative Biology; Program in Ecology, Evolutionary Biology and Behavior; BEACON Center for the Study of Evolution; and Michigan State University Museum; Michigan State University, East Lansing, MI 48824, e-mail: lundriga@msu.edu

Andrea Morrow
Department of Integrative Biology; Program in Ecology, Evolutionary Biology and Behavior; BEACON Center for the Study of Evolution; and Michigan State University Museum; Michigan State University, East Lansing, MI 48824, e-mail: Morrowa6@msu.edu

Paul Meek
NSW Dept Primary Industries, National Marine Science Centre, Bay Road, Coffs Harbour, NSW; University of New England, School of Environmental and Rural Science, Australia, e-mail: paul.meek@dpi.nsw.gov.au

Laura Smale
Department of Psychology; Department of Integrative Biology; Neuroscience Program; Program in Ecology, Evolutionary Biology and Behavior; BEACON Center for the Study of Evolution; Michigan State University, East Lansing, MI 48824, e-mail: smale@msu.edu

© Springer Nature Switzerland AG 2020
W. Banzhaf et al. (eds.), *Evolution in Action: Past, Present and Future*,
Genetic and Evolutionary Computation, https://doi.org/10.1007/978-3-030-39831-6_15

nal than the nocturnal rat, but did not differ significantly between the two squirrels. Moreover, the two auditory structures were larger in the nocturnal than the diurnal sciuromorph, but smaller in the nocturnal than the diurnal myomorph. We evaluate these data with respect to the hypothesis that temporal niche transitions have shifted the value of sensory brain regions and discuss what they suggest more generally about the mosaic nature of sensory brain evolution.

Key words: Diurnal, nocturnal, temporal niche, mosaic evolution

15.1 Introduction

Virtually every mammal faces major challenges that result from the earth spinning on its axis and the concomitant daily fluctuations in light and temperature. However, the ways in which species cope with these challenges can vary markedly: some concentrate their activity at night, others during the day or at dawn/dusk transitions, and still others adopt an arrhythmic pattern. In addition to variation in the characteristic timing of activity, species vary in the plasticity of activity rhythms, with some readily switching their most active period from one time of day to another, while others exhibit little or no flexibility. Evolutionary transitions from one temporal niche to another require substantial change. Mechanisms coordinating rhythms of a multitude of behavioral and physiological parameters must be transformed, and these must be accompanied by changes in other features (e.g. insulation) that support activity in a cold/dark night rather than a warm/bright day. Each of these transitions represents a shift from one extremely complex suite of adaptations to an equally complex but opposing one, via what would seem to be a deep fitness valley in which intermediate states could be deadly.

Despite this complexity, evolutionary transitions in temporal niche have occurred many times since mammals first appeared, approximately 220 MYA [21]. Both morphological evidence [14, 34] and historical reconstructions [2, 21] suggest that the first mammals were nocturnal, presumably an adaptation for avoiding the diurnal dinosaurs that dominated the landscape at that time (reviewed in [13]). Activity primarily at night appears to have characterized the group for its first 150 million years or so [2], a period in mammalian history sometimes referred to as the "nocturnal bottleneck" [13, 24]. But with the extinction of dinosaurs at the end of the Mesozoic, and subsequent adaptive radiation of mammals in the early Cenozoic, the range of mammalian activity patterns expanded to include greater diversity in temporal niche, including fully diurnal forms. The number of evolutionary transitions in temporal niche that have occurred within the mammalian radiation is not known, but our preliminary examination of temporal niche evolution in the largest extant mammalian order, Rodentia, identified a minimum of 20 transitions between nocturnal and diurnal activity patterns within that clade.

Many things must change when a diurnal animal evolves from a nocturnal one and vice versa, and transformations of sensory systems are among the most important, as sensory data are vital to survival, and the information available to support activity during the day differs from that available at night. Photic cues, in particular, are severely limited at night, but can carry considerable information during the day, assuming the animal has the visual systems necessary to extract that information. While many studies describe adaptations of eye structure and function to diurnal or nocturnal conditions (reviewed in [13, 17]), investment in brain tissues needed to obtain information from visual signals is less well studied. Consider, for example, the ability to discriminate light of different wavelengths, i.e. color vision, which depends not only on populations of receptors (cones) that respond differently to that light, but also on brain tissue that can disentangle the wavelengths and intensities of the signals that reach those receptors. Obtaining visual information about patterns, sizes, shapes, distances, and movements of objects similarly depends on brain tissue able to extract it. Although olfactory, auditory, and somatosensory cues vary in effectiveness to a lesser degree across the day/night cycle, they too rely for their interpretation on specialized sensory tissue in the brain.

A central goal of our research is to investigate the impact of temporal niche transitions on investment in the sensory brain. What happens, for example, when a nocturnal mammal gives rise to a diurnal one? Jerison [18] proposed that the evolution of large brains in the earliest mammals was driven by selection favoring enhancement of non-visual regions when these animals became more active in the dark. Perhaps transitions back to diurnality led to a further increase in brain size, with expansion of tissue able to extract more information from visual cues available during the day. Although this increase in visual areas (and thus overall brain size) would facilitate adaptation to the changed temporal niche, it would also engender significant energetic costs. Indeed, in terms of the energy it requires to operate, brain tissue is among the most "expensive" there is, second only to the muscles that keep the heart pumping [1, 26]. One approach to offsetting the cost of increased investment in the visual brain would be concurrently decreasing investment in other sensory brain regions (i.e. tradeoffs), especially where visual systems could provide more and/or redundant information, replacing to some extent the need for olfactory, auditory, and somatosensory tissue. However, both expansion of a single brain sensory region, and tradeoffs between regions, would depend on the ability of those regions to evolve independently of one another. Another consideration is that evolutionary change in the sensory brain regions might be developmentally constrained, such that a change in any one sensory region would be linked to coordinated change in others. This hypothesis follows from a seminal paper by Finlay and Darlington [12] that investigated the nature of evolutionary change in the relative sizes of 11 non-overlapping brain regions, together comprising almost all of the mammalian brain. Two models were considered: mosaic evolution, in which individual brain structures and circuits change in size independently of one another (as in the first two hypotheses above), and concerted evolution, in which developmental constraints result in different parts of the brain changing size in a coordinated manner. The Finlay and Darlington data [12] strongly support the latter. Interestingly, the only brain region in their data set that did not change size in a predictable way with respect to others was the olfactory bulb.

Although the relationship between sensory brain investment and temporal niche has not been systematically investigated in mammals, a number of studies have explored the influence of the sensory environment or temporal niche on particular regions of the sensory brain. It is, for example, well known that in species that spend their lives underground where there is effectively no information to be obtained from photic signals (e.g. mole rats), visual regions of the brain are all but non-existent [9]. Studies of primates have shown that diurnal species have a larger visual cortex relative to hindbrain volume than nocturnal species [3], and in rodents, the percent of dorsolateral cortex devoted to visual processing is higher, and that devoted to processing somatosensory data lower, in diurnal Nile grass rats (*Arvicanthis niloticus*) and sciurids (*Spermophilus beecheyi* and *Sciurus carolinesis*) than in nocturnal Norway rats, *Rattus norvegicus* [7, 8]. Moreover, the superior colliculus, a region of the midbrain that receives direct input from the retina, is estimated to be 10 times larger in the strictly diurnal 13-lined ground squirrel (*Spermophilus tridecemlineatus*) than in the nocturnal (and more malleable) Norway rat [19]. These findings contrast with those of Finlay et al. [11], who sampled 32 mammal species representing 7 orders and found no effect of temporal niche on the size of the lateral geniculate nucleus of the thalamus (a retinorecipient region of the brain involved in processing visual information). Together, these data suggest that some regions of the sensory brain have coevolved in a predictable way with temporal niche, whereas others have not. The pattern of evolutionary change among sensory brain regions remains largely unknown because most studies to date have measured only one sensory area (usually visual) and have not taken phylogeny into account.

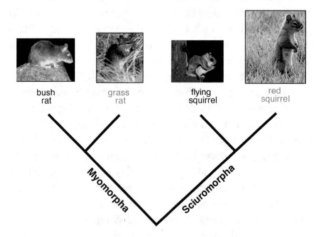

Fig. 15.1: Phylogenetic relationships, suborder, and temporal niche of species examined in this study: Australian bush rat (*Rattus fuscipes*), Nile grass rat (*Arvicanthis niloticus*), southern flying squirrel (*Glaucomys volans*), and red squirrel (*Tamiasciurus hudsonicus*). Black font indicates nocturnal species; green font indicates diurnal species.

Here, we present preliminary data comparing sensory brain investment (olfactory, visual, auditory) in four rodent species (two nocturnal and two diurnal) encompassing at least two independent transitions in temporal niche (Figure 15.1). These data speak to the general issue of whether brain tissue supporting extraction of information from photic cues expanded (or retracted) at these transitions, and if so, whether other sensory systems changed in the opposite directions, indicating tradeoffs in investment in the sensory brain. The species we focus on are ones that rigidly adhere to activity patterns that are either completely nocturnal or completely diurnal.

15.2 Materials and Methods

15.2.1 Animals

Brain data were collected from two nocturnal rodent species, the Australian bush rat (*Rattus fuscipes*) and southern flying squirrel (*Glaucomys volans*), and two diurnal rodent species, the Nile grass rat (*Arvicanthis niloticus*) and red squirrel (*Tamiasciurus hudsonicus*). These represent two suborders (Myomorpha and Sciuromorpha) whose lineages separated more than 65 million years ago ([10], Figure 15.1). The exact number of transitions in temporal niche that have occurred along these two lineages is unknown. However, phylogenetic reconstructions of temporal niche in order Rodentia indicate that the first rodents were nocturnal and that the transitions to diurnality represented here are historically independent [10, 21]. In addition, the nocturnal flying squirrel evolved from a diurnal sciurid ancestor, resulting in a minimum of three transitions in this rodent lineage (i.e. nocturnal ancestor, to diurnal squirrel, to nocturnal squirrel).

Southern flying squirrels and red squirrels were live-trapped in and around East Lansing, Michigan, during the summers of 2016 and 2017. Australian bush rats were live-trapped in New South Wales, Australia, in August of 2015. The Nile grass rats were obtained from a Michigan State University (MSU) laboratory colony established in 1993 from animals live-trapped in Kenya that year. To avoid the possibility that differences in age, sex, or reproductive condition might confound species differences, we examined only adult females (no males) and only individuals that were not pregnant or lactating; 2-3 animals were examined per species, resulting in a total sample size of 11 (Table 15.1).

All animals were handled according to protocols approved by the following institutional and regional authorities: American Society of Mammalogists [32], MSU Institutional Animal Care and Use Committee (protocol No.07/16-116-00), Office of Environment and Heritage (License No.SL100634), and NSW Department of Industry and Investment Animal Research Authority (ORA 14/17/009).

Table 15.1: Sample sizes (*n*), and species means and standard errors for body mass, brain mass, olfactory bulb mass (OB), and volumes of visual (SC, LNG) and auditory (IC, MGN) brain structures. SC = superior colliculus, LGN = lateral geniculate nucleus, IC = inferior colliculus, MGN = medial geniculate nucleus.

	n	Body mass [g]	Brain mass [mg]	OB mass [mg]	SC [mm^3]	LGN [mm^3]	IC [mm^3]	MGN [mm^3]
Bush rat	3	124 ± 16	$1,931 \pm 34$	112 ± 9	8.17 ± 0.12	4.14 ± 0.18	22.93 ± 0.20	3.35 ± 0.07
Grass rat	3	97 ± 4	896 ± 1	35 ± 0	7.41 ± 0.13	3.40 ± 0.03	17.52 ± 0.07	3.58 ± 0.02
Flying squirrel	3	68 ± 2	$1,892 \pm 83$	56 ± 1	16.27 ± 0.45	6.98 ± 0.76	20.45 ± 0.92	4.60 ± 0.39
Red squirrel	2	218 ± 24	$3,920 \pm 90$	81 ± 1	64.59 ± 3.86	15.65 ± 0.15	27.6 ± 1.20	7.47 ± 0.32

15.2.2 Brain Collection and Histology

Each animal was euthanized with a lethal dose of sodium pentobarbital, adminis-tered intraperitoneally. Immediately after death, the animal was weighed to the near-est gram and its brain was extracted, placed in powdered dry ice for 2-5 minutes, and transferred to a $-20°$ or a $-80°$ freezer until further processing. After removal from the freezer, each brain was trimmed immediately caudal to the medulla oblongata (to eliminate residual non-brain tissue) and weighed to the nearest milligram. Olfac-tory bulbs (OB) were then dissected from the brain, using an incision just anterior to the olfactory peduncles, and weighed to the nearest milligram. The portion of the brain extending from the anterior thalamus to just caudal to the tectum was coro-nally sectioned at 40μm thickness, and the resulting tissue sections were sorted into 3 alternating series, mounted onto superfrost slides, and placed in a $-80°$ freezer. The first two series are being held (in the -80) for future work. The third series (i.e. every third slice) was stained for acetylcholinesterase as follows: slides were incu-bated for 5 hours in a solution consisting of 0.0072% ethopropazine HCl, 0.075% glycine, 0.05% cupric sulfate, 0.12% acetylthiocholine iodide, and 0.68% sodium acetate (pH 5.0); rinsed 2 times (3 minutes each) with distilled H_2O; and developed in a 0.77% sodium sulfide solution (pH 7.8) for 45 minutes. Slides were then rinsed with 2 changes of distilled H_2O (3 minutes each), run through a series of ascending ethanol concentrations (70%, 95%, 100%, and 100%) for 1 minute each (to dehy-drate the adhering tissue), cleared through 2 changes of xylenes for 5 minutes each, and coverslipped using DPX mounting medium.

15.2.3 Data Collection and Regions of Interest

Photographs of stained sections were taken with a digital camera (MBF Bioscience CX9000) attached to a Zeiss light microscope (Carl Zeiss, Göttingen, Germany, 5x objective), using the 2D slide scanning module on Stereo Investigator 2017 (MBF

Bioscience). Volumetric measurements of four brain structures were collected from the photomicrographs (Figure 15.2), using the Cavelieri method (150 × 150 μm grid) in Stereo Investigator. These include two visual structures, the superior colliculus (SC) and lateral geniculate nucleus (LGN), and two auditory structures, the inferior colliculus (IC) and medial geniculate nucleus (MGN). LGN, IC, and MGN were measured in their entirety, whereas SC volume included only the 3 most superficial layers (i.e. the zonal layer, superficial gray layer, and optic nerve layer); these layers contain most of the retinal input to this structure (e.g. [20]). For each structure, only one side (left or right) was measured and that volume was doubled to obtain total volume. The rat brain atlas [28] was used as a reference guide in determining the boundaries of each brain region.

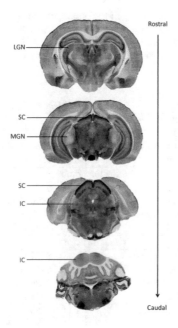

Fig. 15.2: Photomicrographs of grass rat brain sections showing the visual and auditory brain regions measured in this study: superior colliculus (SC), lateral geniculate nucleus (LGN), inferior colliculus (IC), and medial geniculate nucleus (MGN).

15.2.4 Analysis

To account for intra- and interspecific differences in overall brain size, each of the sensory tissue measurements (OB, SC, LGN, IC, MGN) was divided by the brain weight of the individual before analysis. The effects of temporal niche (nocturnal, diurnal) and suborder membership (Myomorpha, Sciuromorpha) on the size of each brain structure, relative to overall brain weight, was assessed using a 2-way ANOVA on the log transformed ratios.

15.3 Results

15.3.1 Brain Weight

Brain weights ranged from 2.8% of body weight in the smallest of the four species,

Table 15.2: Two-way ANOVA examining effects of temporal niche and suborder on total brain mass as a proportion of body mass; olfactory bulb mass as a proportion of total brain mass (OB); and volumes of the superior colliculus (SC), lateral geniculate nucleus (LGN), inferior colliculus (IC), and medial geniculate nucleus (MGN) as proportions of total brain mass.

Regions	Factors	Df	Mean Sq	F value	Pr(>F)
Total brain	Temporal Niche	1	0.1739	56.731	0.000284
	Suborder	1	0.18602	60.687	0.000236
	Temporal Niche*Suborder	1	0.00061	0.198	0.671913
	Residuals	6	0.00307		
Olfactory Bulbs					
OB	Temporal Niche	1	0.06325	33.96	0.000645
	Suborder	1	0.24889	133.64	8.17×10^{-6}
	Temporal Niche*Suborder	1	0.00032	0.17	0.692861
	Residuals	7	0.00186		
Visual Structures					
SC	Temporal Niche	1	0.18221	873.8	1.31×10^{-8}
	Suborder	1	0.25471	1221.48	4.07×10^{-9}
	Temporal Niche*Suborder	1	0.00004	0.21	0.66
	Residuals	7	0.00021		
LGN	Temporal Niche	1	0.05480	46.73	0.000245
	Suborder	1	0.05273	44.95	0.000276
	Temporal Niche*Suborder	1	0.0297	25.32	0.00151
	Residuals	7	0.00117		
Auditory Structures					
IC	Temporal Niche	1	0.01003	248	1.01×10^{-6}
	Suborder	1	0.13378	3308	1.26×10^{-10}
	Temporal Niche*Suborder	1	0.11084	2741	2.43×10^{-10}
	Residuals	7	0.00004		
MGN	Temporal Niche	1	0.07195	153.81	5.09×10^{-6}
	Suborder	1	0.01068	22.83	0.00202
	Temporal Niche*Suborder	1	0.14592	311.97	4.6×10^{-7}
	Residuals	7	0.00047		

the flying squirrel, to less than 1% of body weight in the next smallest, the Nile grass rat (Table 15.1). Suborder and temporal niche are both significant predictors of the brain weight to body weight ratio (Table 15.2), which is higher in the squirrels than the rats, and in the nocturnal than in the diurnal species.

15.3.2 Olfactory Bulbs

The rat species have significantly larger OBs, relative to brain weight, than the squirrels (Table 15.2, Figure 15.3). Temporal niche is also a significant contributor to OB weight relative to brain weight, with nocturnal species having larger values than diurnal species.

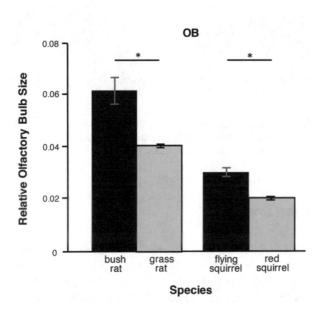

Fig. 15.3: Species means and standard errors for olfactory bulb size (mg) as a proportion of total brain mass (mg). Black indicates nocturnal species; gray indicates diurnal species. $P < 0.05^*, P < 0.01^{**}, P < 0.001^{***}$.

15.3.3 Visual Brain

Temporal niche and suborder both have a significant effect on volume of the SC relative to brain weight (Table 15.2). SC volume, relative to brain weight, is greater in the diurnal grass rat than in the nocturnal bush rat, and in the diurnal red squirrel than the nocturnal flying squirrel (Figure 15.4(a)).

The volume of the LGN, relative to brain weight, is significantly affected by temporal niche, suborder, and by an interaction between temporal niche and suborder (Table 15.2). Of the two rats, the diurnal species has a much larger LGN relative to brain weight, whereas for squirrels, the LGN to brain weight ratio is similar in nocturnal and diurnal species (Figure 15.4(b)).

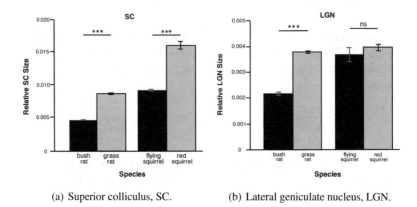

(a) Superior colliculus, SC.   (b) Lateral geniculate nucleus, LGN.

Fig. 15.4: Species means and standard errors for volumes of visual brain structures as a proportion of total brain mass. Black indicates nocturnal species; gray indicates diurnal species. $P < 0.05^*$, $P < 0.01^{**}$, $P < 0.001^{***}$.

15.3.4 Auditory Brain

Temporal niche, suborder, and their interaction are all significant predictors of the volumes of IC and MGN relative to brain weight (Table 15.2). However, the relationship between temporal niche and the size of these auditory structures is reversed in the two suborders (Figures 15.5(a) and 15.5(b)). In the myomorphs, the volumes of IC and MGN relative to brain weight are larger in the diurnal than in the nocturnal species, while for sciuromorphs, these ratios are larger in the nocturnal than in the diurnal species.

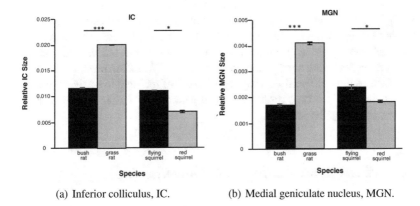

(a) Inferior colliculus, IC. (b) Medial geniculate nucleus, MGN.

Fig. 15.5: Species means and standard errors for volumes of auditory brain structures as a proportion of total brain mass. Black indicates nocturnal species; gray indicates diurnal species. $P < 0.05^*$, $P < 0.01^{**}$, $P < 0.001^{***}$.

15.4 Discussion

The data presented here suggest that some temporal niche transitions in order Rodentia have been associated with a mosaic pattern of change in the sensory brain that may have been driven by opportunities and costs associated with using olfactory and visual cues to navigate the world at night versus during the day. Interesting differences between species occupying the same temporal niche were also evident. We investigate these issues below, focusing on olfactory, visual, and auditory regions in turn, then more generally discuss caveats, conclusions, and the questions driving our continued work on this subject.

15.4.1 Olfactory Bulbs

Olfactory bulb weight relative to brain weight was significantly greater in the nocturnal than the diurnal representative of both of the suborders examined here (Figure 15.3). This is consistent with the hypothesis that the emergence of diurnality in the ancestors of *Arvicanthis* brought with it a decrease in reliance on, and investment in, olfactory information, whereas the emergence of nocturnality along the lineage leading to the flying squirrels did the reverse. We also saw evidence of a phylogenetic effect, in that investment in OB was substantially greater in the myomorphs than in the sciuromorphs. The information available in olfactory cues is different in a variety of ways from that contained in light or sound, thus tradeoffs favoring olfaction would have complex consequences for sensory abilities [33]. For example, olfactory cues can last for days, informing animals not only about what predators, potential mates, or other conspecifics are present in the immediate environment, but also about those present in the recent past. Visual cues, by contrast, provide an animal much more precise and immediate information about critical features of their environment, assuming sufficient light is available. Where the mechanisms to obtain visual information are well developed, the costs of maintaining extensive olfactory brain tissue could become prohibitive. The patterns we saw in the OB and SC are consistent with that hypothesis.

15.4.2 Visual Structure: Superior Colliculus (SC)

Our data suggest that the evolution of diurnality in the *Arvicanthis* lineage was accompanied by increased investment in the SC, whereas the evolutionary transition to nocturnality characteristic of *Glaucomys* was accompanied by decreased investment in it (Figure 15.4(a)). The SC has not previously been examined quantitatively in relation to temporal niche, but images of it in 13-lined ground squirrels (diurnal) and Norway rats (nocturnal) led Kaas and Collins [19] to suggest that it is 10 times larger in the former than in the latter. The chronotype-related differences observed in our study were in the same direction but considerably smaller, i.e. the SC (relative to brain size) of each diurnal rodent was roughly twice that of its nocturnal relative.

Effects of phylogeny were also apparent in these data, i.e. sciuomorphs had a larger SC (relative to brain size) than myomorphs. This might reflect, at least in part, the importance of midbrain visual tissue for navigating in, and in the case of flying squirrels, gliding through, the trees. The SC contains information about the location and nature of visual stimuli, and it links that information to motor regions, especially those that guide head and eye movements; it also contains cells that are active during presentation of a looming visual stimulus (reviewed in Seabrook et al. [30]). This general arboreality hypothesis predicts that the SC should be smaller in ground dwelling sciurids than in tree dwelling ones, an idea that, to our knowledge, has not yet been explored.

When considered together, the species differences we observed in the OB and the SC would appear to reflect an example of a mosaic pattern of brain evolution in which components of a functional circuit that have evolved together did so in a manner that was dissociated from other systems and structures [4, 25]. This model is generally presented as an alternative to Finlay and Darlington's conjecture [12] that developmental links between brain structures result in constrained patterns of evolutionary change. The latter hypothesis was supported by data from 10 of 11 brain regions, the one exception being the OB. Our results appear to provide another example of a dissociation between the OB and other structures with respect to patterns of evolutionary change. While our data for OB and SC represent an example of how brain evolution can proceed in a mosaic manner, they do not present a challenge to the overarching hypothesis of concerted evolution.

15.4.3 Visual Structure: Lateral Geniculate Nucleus (LGN)

Our data on the LGN do not parallel those for the SC; though larger in diurnal than nocturnal myomorphs, the LGN volume (relative to brain weight) was roughly the same in the two sciuromorph species. The latter finding is consistent with the conclusions that Finlay et al. [11] arrived at from their analysis of LGN volumes in 82 mammal species: that LGN volumes are unrelated to temporal niche. The sample of rodents was small in that study, consisting of six species, including four myomorphs, only one of which was diurnal. Paradoxically, our myomorph data are not in accord with theirs, as the grass rat LGN was substantially larger than that of the bush rat. Similarly, when Campi and Krubitzer [8] compared grass rats and Norway rats, they found that the visual cortex, the primary target of the LGN, was more expansive in the former than the latter species. The current data, together with those of [8] and [11], suggest that the transition to diurnality along the lineage leading to grass rats was unusual in that both the LGN and its target in the cortex expanded, something that did not occur at a number of other temporal niche transitions in an array of mammalian lineages. However, a more comprehensive sampling of species, especially of rodents, is needed to determine whether the pattern of change seen in the myomorph species sampled here is indeed unusual in mammals.

A second question raised by our LGN findings is why the LGN is roughly the same in the two sciuromoph species, whereas the SC is markedly larger in one than the other. This would seem to represent a counterexample to the concerted patterns of evolution thought to reflect factors that limit the decoupling of interwoven networks of brain structures [12]. Such interconnnectedness is considerable in the visual system of laboratory mice, in which 100% of the retinal projections to the LGN are collaterals of axons that go to the SC. In diurnal primates that figure is only 10% [30]. It would be interesting to know whether sciuromorphs are more like primates in this regard, as that could help to explain the dissociation suggested by our data in the pattern of evolutionary change in these two visual structures.

15.4.4 Auditory Structures: Inferior Colliculus (IC) and Medial Geniculate Nucleus (MGN)

The sizes of the auditory fore- and midbrain structures examined here were associated with temporal niche in the predicted direction in the sciuromorphs, but in a direction counter to prediction in the myomorphs, i.e. the diurnal Nile grass rat has a markedly larger auditory brain than the nocturnal bush rat. Though contrary to expectations, this finding is consistent with Campi et al. [7], who found that auditory cortex as a proportion of dorsolateral cortex is significantly larger in the Nile grass rat than in the Norway rat. These data suggest that if there is any influence of temporal niche on myomorph auditory brain structures it may be one that can be readily reversed by other factors. One can only speculate about ways in which auditory information could be of greater value to grass rats than bush rats, but one possibility is related to vocal communication. While both species undoubtedly communicate through sound, that form of signaling might be better developed in grass rats, which live in groups [27], than in the more solitary bush rats [22]. Evaluating the influence of vocal communication on the size of auditory regions of the sensory brain will require a much larger sample of species, ideally including extreme forms. Future work in our lab, for example, will examine the sensory brain of the long-tailed singing mouse, *Scotinomys xerampelinus*, a species whose social communication is highly developed [6].

15.4.5 Caveats

15.4.5.1 Grass rats were raised in captivity

The grass rats used in this study came from a breeding colony that had been maintained in captivity for 23 years. This is potentially problematic in that it has been clear for some time that the complexity of the environment in which a laboratory rodent is raised can influence brain development [31]. Indeed, sensory regions of the cortex are measurably different in wild caught Norway rats than in commercial ones, which have generally been selectively bred in captivity for over one hundred years [7]. Although it is possible that captivity had some influence on the sensory brain of our grass rats, we believe that this is unlikely. Great care has been taken to prevent inbreeding in the MSU grass rat colony and our subject animals were raised in a relatively rich environment. They were housed in large cages with both parents, siblings, substantial nesting material, and objects to play with; their cages were kept in a colony room where individuals could see, hear, and smell animals beyond those in their own cages.

15.4.5.2 Volumes and weights as indicators of "investment"

For our measure of investment in olfaction, we used the weights of the OBs relative to the weight of the whole brain, but for investment in visual and auditory regions, we used the volumes of fore- and midbrain structures relative to the weight of the whole brain. We do not see this as problematic because the key analyses were aimed at determining whether and how the species differ with respect to investment in each sensory structure, not how these structures differ from one another. Of greater concern is that little is known about actual energy expenditure in different regions of the sensory brain [26], and neuron packing density (the ratio of the number of neurons to the mass of the structure in which they are contained) decreases with brain size at different rates in different structures [15]. It is thus not entirely clear what species differences in the size of a given sensory brain region, relative to the brain as a whole, tell us about the costs of maintaining those tissues. Nor can we assess the influence of differences in overall brain size on regional energy investment. The brain of a red squirrel, for example, weighs approximately twice that of a flying squirrel, which makes it likely that the density and size of neurons within a given region are not the same in the two species [16]. Clearly, additional data are needed to precisely characterize species differences in the costs of these sensory regions. However, we believe it unlikely that such data would change the direction of the species differences reported here.

15.5 Conclusions and Questions

The data presented here, though preliminary, suggest that evolutionary transitions from one temporal niche to another have involved a mosaic pattern of evolutionary change that might reflect tradeoffs in investment between some forebrain and midbrain tissues that process olfactory and visual information. Specifically, these results suggest that the evolution of diurnality in the *Arvicanthis* lineage was accompanied by increases in at least one brain structure that processes visual information (the SC) and a compensatory decrease in one that processes olfactory information (the OB), whereas a return from diurnality to nocturnality in the *Glaucomys* lineage did the reverse. Species differences in other sensory structures were not as clearly associated with temporal niche, but they too reveal some dissociation in the degree and direction of change in different regions of the brain. For example, the size of the SC (relative to brain size) in flying squirrels is less than half that of red squirrels, whereas LGN (relative to brain size) is roughly the same in the two species (Figure 15.4). Although these data suggest a dissociated pattern of change in some midbrain and forebrain structures, they do not represent a serious challenge to the overarching principle of concerted evolution laid out by Finlay and Darlington [12], which was proposed as a general model to account for large scale patterns of change as the brain evolved. Those authors also noted variance around the regression lines representing the sizes of each brain structure relative to brain weight. Our results suggest that if such a regression line were developed for sensory regions of the brain, temporal niche might help explain some of the variance around it. Another important discovery of Finlay and Darlington [12] was that that the OB has not changed in concert with other brain structures; it is not clear why that is the case. The explanation likely involves differences in the dissociability of mechanisms that guide the development of the OB and other brain structures, and our data suggest the possibility that this may have enabled selection pressures associated with temporal niche to play a role in shaping its evolution.

In this study, we focused on species that are completely nocturnal or completely diurnal and whose behavioral rhythms appear to be obligate ones. If the differences we saw in the OB and SC have truly been driven by changes in temporal niche then they should be smaller, or perhaps even nonexistent, in species whose rhythms are more cathemeral, i.e. that include activity both day and night, as is the case in many rodents [5, 17, 29]. An interesting comparison, for example, could be made between the nocturnal Australian bush rat examined here, and the closely related Australian swamp rat, which is active during both day and night in some parts of its range, and is almost completely diurnal when its home range overlaps that of the bush rat [23]. If the patterns found here represent sensory tradeoffs, our expectation is that the swamp rat would have a smaller OB and larger SC than the bush rat, but with more modest differences between these species than between grass rats and bush rats. A deeper understanding of these general issues will require both analysis of sensory structures in a broad array of species representing multiple independent transitions from one temporal niche to another, and inclusion of species that have intermediate activity patterns and/or are more plastic than those of the species examined here.

Acknowledgements We thank Lily Yan for her advice on laboratory procedures. We are also indebted to Michigan State University Museum for loaning traps and trapping gear. This research was supported by NSF award DBI-0939454.

References

1. Aiello, L.C., Wheeler, P.: *The expensive-tissue hypothesis: The brain and the digestive system in human and primate evolution.* Current Anthropology **36**(2), 199–221 (1995)
2. Anderson, S.R., Wiens, J.J.: *Out of the dark: 350 million years of conservatism and evolution in diel activity patterns in vertebrates.* Evolution **71**(8), 1944–1959 (2017)
3. Barton, R.: *Evolutionary specialization in mammalian cortical structure.* Journal of Evolutionary Biology **20**(4), 1504–1511 (2007)
4. Barton, R.A., Harvey, P.H.: *Mosaic evolution of brain structure in mammals.* Nature **405**(6790), 1055–1058 (2000)
5. Bennie, J.J., Duffy, J.P., Inger, R., Gaston, K.J.: *Biogeography of time partitioning in mammals.* Proceedings of the National Academy of Sciences **111**(38), 13,727–13,732 (2014)
6. Blondel, D.V., Pino, J., Phelps, S.M.: *Space use and social structure of long-tailed singing mice (Scotinomys xerampelinus).* Journal of Mammalogy **90**(3), 715–723 (2009)
7. Campi, K.L., Collins, C.E., Todd, W.D., Kaas, J., Krubitzer, L.: *Comparison of area 17 cellular composition in laboratory and wild-caught rats including diurnal and nocturnal species.* Brain, Behavior and Evolution **77**(2), 116–130 (2011)
8. Campi, K.L., Krubitzer, L.: *Comparative studies of diurnal and nocturnal rodents: Differences in lifestyle result in alterations in cortical field size and number.* Journal of Comparative Neurology **518**(22), 4491–4512 (2010)
9. Cooper, H.M., Herbin, M., Nevo, E.: *Visual system of a naturally microphthalmic mammal: The blind mole rat, Spalax ehrenbergi.* Journal of Comparative Neurology **328**(3), 313–350 (1993)
10. Fabre, P.H., Hautier, L., Dimitrov, D., Douzery, E.J.: *A glimpse on the pattern of rodent diversification: A phylogenetic approach.* BMC Evolutionary Biology **12**(1), 88 (2012)
11. Finlay, B.L., Charvet, C.J., Bastille, I., Cheung, D.T., Muniz, J.A.P., de Lima Silveira, L.C.: *Scaling the primate lateral geniculate nucleus: Niche and neurodevelopment in the regulation of magnocellular and parvocellular cell number and nucleus volume.* Journal of Comparative Neurology **522**(8), 1839–1857 (2014)
12. Finlay, B.L., Darlington, R.B.: *Linked regularities in the development and evolution of mammalian brains.* Science **268**(5217), 1578–1584 (1995)
13. Gerkema, M.P., Davies, W.I., Foster, R.G., Menaker, M., Hut, R.A.: *The nocturnal bottleneck and the evolution of activity patterns in mammals.* Proc. R. Soc. B **280**(1765), 20130508 (2013)

14. Hall, M.I., Kamilar, J.M., Kirk, E.C.: *Eye shape and the nocturnal bottleneck of mammals.* Proceedings of the Royal Society of London B: Biological Sciences p. rspb20122258 (2012)
15. Herculano-Houzel, S.: *Scaling of brain metabolism with a fixed energy budget per neuron: Implications for neuronal activity, plasticity and evolution.* PLoS One **6**(3), e17514 (2011)
16. Herculano-Houzel, S., Manger, P.R., Kaas, J.H.: *Brain scaling in mammalian evolution as a consequence of concerted and mosaic changes in numbers of neurons and average neuronal cell size.* Frontiers in Neuroanatomy **8**(77), 1–28 (2014)
17. Hut, R.A., Kronfeld-Schor, N., van der Vinne, V., De la Iglesia, H.: *In search of a temporal niche: Environmental factors.* Progress in Brain Research **199**, 281–304 (2012)
18. Jerison, H.J.: On theory in comparative psychology. In: R. Sternberg, J. Kaufman (eds.) The Evolution of Intelligence, pp. 251–288. Lawrence Erlbaum Associate, Mahwah, NJ (2002)
19. Kaas, J.H., Collins, C.E.: *Variability in the sizes of brain parts.* Behavioral and Brain Sciences **24**(2), 288–290 (2001)
20. Major, D., Rodman, H., Libedinsky, C., Karten, H.: *Pattern of retinal projections in the California ground squirrel (Spermophilus beecheyi): Anterograde tracing study using cholera toxin.* Comparative Neurology **463**(3), 317–340 (2003)
21. Maor, R., Dayan, T., Ferguson-Gow, H., Jones, K.E.: *Temporal niche expansion in mammals from a nocturnal ancestor after dinosaur extinction.* Nature Ecology & Evolution **1**(12), 1889–1895 (2017)
22. Meek, P.D.: Pers. obs.
23. Meek, P.D., Zewe, F., Falzon, G.: *Temporal activity patterns of the swamp rat (*Rattus lutreolus*) and other rodents in north-eastern new south wales, australia.* Australian Mammalogy **34**(2), 223–233 (2012)
24. Menaker, M., Tosini, G.: *The evolution of vertebrate circadian system.* In: K. Honma, S. Honma (eds.) Sixth Sapporo Symp. On Biological Rhythms: Circadian organization and oscillatory coupling, pp. 39–52. Hokkaido University Press, Sapporo, Japan (1996)
25. Montgomery, S.H., Mundy, N.I., Barton, R.A.: *Brain evolution and development: Adaptation, allometry and constraint.* Proc. R. Soc. B **283**(1838), 20160433 (2016)
26. Niven, J.E., Laughlin, S.B.: *Energy limitation as a selective pressure on the evolution of sensory systems.* Journal of Experimental Biology **211**(11), 1792–1804 (2008)
27. Packer, C.: *Demographic changes in a colony of nile grassrats (*Arvicanthis niloticus*) in tanzania.* Journal of Mammalogy **64**(1), 159–161 (1983)
28. Paxinos, G., Watson, C.: *The rat nervous system.* Academic Press (2014)
29. Refinetti, R.: *The diversity of temporal niches in mammals.* Biological Rhythm Research **39**(3), 173–192 (2008)
30. Seabrook, T.A., Burbridge, T.J., Crair, M.C., Huberman, A.D.: *Architecture, function, and assembly of the mouse visual system.* Annual Review of Neuroscience **40**, 499–538 (2017)
31. Sharpless, S.K.: *Reorganization of function in the nervous system – use and disuse.* Annual Review of Physiology **26**(1), 357–388 (1964)
32. Sikes, R. and the Animal Care and Use Committee of the American Society of Mammalogists: *2016 Guidelines of the American Society of Mammalogists for the use of wild mammals in research and education.* Journal of Mammalogy **97**, 663–688 (2016)
33. Striedter, G.: *Principles of brain evolution.* Sinauer Assoc., Sunderland, MA (2005)
34. Walls, G.L.: *The vertebrate eye and its adaptive radiation.* Cranbrook Institute of Science, Bloomfield Hills, MI (1942)

Chapter 16
Time Makes You Older, Parasites Make You Bolder — Toxoplasma Gondii Infections Predict Hyena Boldness toward Definitive Lion Hosts

Eben Gering†, Zachary Laubach†, Patricia Weber, Gisela Hussey, Julie Turner, Kenna Lehmann, Tracy Montgomery, Kay Holekamp and Thomas Getty

Abstract There is growing interest in the alteration of host behaviors by parasites, yet crucial gaps remain in our understanding of its ecological and evolutionary significance. Here, we present the first evidence that the enhanced boldness of infected intermediate hosts of *Toxoplasma gondii* can increase their risk of mortality by the parasite's definitive feline hosts. In a long-term study of hyenas in Kenya's Masai Mara region, we found that 65% of hyenas were seropositive for *T. gondii* in ELISA IgG assays. Seropositive hyenas approached lions more closely than uninfected counterparts, and also showed longer latencies to approach a simulated conspecific territorial intruder. Lastly, although not significant, the ratio of mortalities caused by lions (vs. other sources) was higher for hyenas that were infected by *T. gondii*. These results accord with a long-standing hypothesis that the manipulation of host boldness and/or ailurophilia evolved to enhance disease transmission. Since hyenas are rarely consumed by lions, however, elevating their boldness toward lions may not be adaptive for *T. gondii*. Instead, it may reflect "collateral manipulation" that evolved to influence homologous mechanisms underlying behaviors of alternative hosts (e.g. rodents). This model is often invoked to explain *T. gondii's* many effects in humans, but is virtually unexplored in natural settings. For *T. gondii*, these effects could feasibly impact both behavior and fitness in a vast array, and significant proportion, of earth's mammals and birds. In addition to characterizing behavioral covariates of infection, we examined spatial and temporal patterns of *T. gondii* prevalence within the Mara landscape. Contrary to our predictions, disease prevalence did not differ 1) at a protected vs. disturbed locality, or 2) over three decades of increasing human activity within the disturbed locality.

Eben Gering, Zachary Laubach, Julie Turner, Kenna Lehmann, Tracy Montgomery, Kay Holekamp, and Thomas Getty
Department of Integrative Biology, and Ecology, Evolutionary Biology, and Behavior Program, Michigan State University, East Lansing, MI 48824, USA

Patricia Weber, Gisela Hussey
College of Veterinary Medicine, Michigan State University, East Lansing, MI 48824, USA

† Equally contributing first authors

© Springer Nature Switzerland AG 2020
W. Banzhaf et al. (eds.), *Evolution in Action: Past, Present and Future*,
Genetic and Evolutionary Computation, https://doi.org/10.1007/978-3-030-39831-6_16

16.1 Introduction

Parasites often influence their hosts' behaviors, and these changes have important consequences for health and survival in wildlife, livestock, and humans. Behavioral changes can arise within infected hosts via numerous mechanisms including 1) effects of host resource depletion, e.g. lethargy, that are not under natural selection to benefit hosts or parasites. 2) host adaptations that reduce infection costs, e.g. grooming, behavioral fever, and self-medication [14]. 3) "host manipulation" that evolved in parasites to facilitate transmission from infected hosts (e.g. [25]). 4) "collateral manipulation" that evolved to manipulate a subset of hosts, but does not increase transmission form a focal, "collateral" host taxon. This last form of behavioral alteration is hypothesized to be a common byproduct of homologies between the hosts in which extended parasite phenotypes evolve, and alternative "collateral" hosts in which parasite-induced behaviors do not promote disease transmission. Of the four mechanisms described above, the first three are well-studied in the biomedical, ecological, and evolutionary literatures. Collateral manipulation, in contrast has attracted very little research interest apart from the many known effects of zoonotic parasites on humans [26]. Among the most important questions concerning collateral manipulation are 1) does it produce similar behavioral phenotypes in 'targeted' and collateral hosts? 2) how does it impact host fitness? 3) what environmental factors control its frequency and significance in nature?

Since the recent discovery of its life cycle in 1970 [6], *Toxoplasma gondii* has become an infamous example of putatively adaptive host manipulation by a parasite. In several studied hosts, including humans, *T. gondii* infections are associated with reduced aversion towards, or even affinities for, the odor of feline urine (e.g. [10, 31]). This "fatal attraction" to an indirect cue of feline presence is hypothesized to have evolved via natural selection on *T. gondii* to facilitate trophic (prey to predator) transmission from a subset of the parasite's non-definitive hosts (e.g. certain rodents), into definitive feline hosts. Only within these definitive feline hosts can *T. gondii* undergo sexual reproduction and produce recombinant, environmentally stable spores called oocysts [7].

Impressively, *T. gondii* can also produce other changes in many of its non-definitive hosts, including the enhancement of behavioral boldness. In rodents and other small mammals and birds, this boldness may further promote trophic (i.e. prey to predator) transmission of the parasite [34]. In humans, which are typically dead-end *T. gondii* hosts, enhancement of risky behavior appears to be non-adaptive for *T. gondii*. Instead, it is usually regarded as a by-product of homologies between human hosts and other mammals in which manipulation has evolved [7]. This view is supported by experimental evidence (esp. from rodents, see [30]) that shows several neural and hormonal processes controlling human behavior are also modified by *T. gondii* infection of other hosts. Correlations between human behavioral phenotypes and infection severity further support this view, altogether indicating *T. gondii* is likely causal of (vs. correlated with) changes in complex behaviors of human hosts. Again, however, we do not currently know if, or how, this type of collateral manipulation affects fitness-related behavior in non-human *T. gondii* hosts.

Among *T. gondii*'s many hosts worldwide, there is also surprisingly little evidence to link behavioral covariates of infection with risks of mortality by feline definitive hosts. This gap reflects the fact that infection-related behaviors, such as attraction to feline urine, have been chiefly investigated in laboratory models and/or in human subjects that are isolated from the risk of mortality by felines. More limited work with wild hosts, e.g. rats and chimpanzees, shows that *T. gondii* can modulate host responses to highly specific olfactory cues of local felines [4, 24]. While these results are highly suggestive of potential fitness costs of reduced ailurophobia (fear of cats), their connection with mortality risks have never been demonstrated in the wild ecosystems where *T. gondii* co-evolves with intermediate and definitive hosts. This is also the case for other infection-associated traits, such as increased host boldness and reduced host neophobia, that could similarly influence host mortality and/or fitness in nature [7]. The first aim of this study (Table 16.1) is therefore to test for relationships between fitness-related behavioral phenotypes and naturally-occurring *T. gondii* infections in wild non-human hosts.

Here, we test for behavioral consequences of *T. gondii* infection in a long lived and highly social host, the spotted hyena (*Crocuta crocuta*), within a natural setting in Kenya where intermediate hosts frequently interact with lions (*Panthera leo*). Several prior studies indicate lions are an important source of hyena mortality. In Namibia, for example, 71% of hyena mortalities were found to result from lions [29]. Lions were also the leading source of mortalities with known causes in an earlier study of Mara hyenas [33]. Despite these apparent costs, hyenas can also benefit from engaging with lions under some conditions - including interactions that function to defend territories, protect relatives, and/or steal food. Tension between benefits and costs of lion interactions may underlie a previous observation of stabilizing selection favoring hyenas with intermediate boldness in the presence of lions [36]. By focusing our study on hyenas, we can therefore characterize relationships between infection status and behaviors with established fitness consequences. Further, because hyenas are somewhat unlikely to transmit *T. gondii* to lions (see discussion), our study can also shed light on the ecological and evolutionary implications of collateral host manipulation.

In addition to examining behavioral covariates of infection, we also explored relationships between focal hyenas' social ranks and *T. gondii* serostatuses. In a subset of its hosts, *T. gondii* causes significant declines in overall condition [7]. If this were the case for hyenas, we would predict infection to be associated with lower social rank.

The second overarching aim of our study was to ascertain if, and how, environmental factors modulate the frequency of *T. gondii* infections in wild populations (Table 16.1). Earlier studies have shown this generalist parasite has long been both geographically and taxonomically widespread in warm blooded African vertebrates (e.g. [2, 27]). There are, however, several ways the continent's growing human population might further increase *T. gondii* prevalence and/or exacerbate its ecological consequences for African ecosystems. Agricultural communities, which are rapidly growing within our study area, represent a large potential reservoir for *T. gondii* due to the high density of vertebrate hosts (livestock, rodents, and humans) that can facilitate horizontal transmission from prey to predator [12]. Feral and domestic cats are commonly fed potentially infectious scraps of meat in African households to encourage rodent control [22]. Like lions and many other felids, housecats can serve as definitive hosts that propagate *T. gondii* by shedding oocysts into local water and soil (e.g. [20]). In light of these anthropogenic sources of infection, we predicted *T. gondii* prevalence would be positively correlated with human presence in the Mara landscape - both in space and in time.

Table 16.1: Predicted covariates of *T. gondii* infection in spotted hyenas of Kenya's Masai Mara region

H1: *T. gondii* infections alter hyena behavior behavior in a context-dependent manner (predictions 1 & 2), and also reduce overall host condition (prediction 3)
Prediction 1: *T. gondii*+hyenas approach lions more closely than uninfected hyenas
Prediction 2: *T. gondii*+hyenas' behavior towards lions differs from their responses toward conspecifics
Prediction 3: *T. gondii*+hyenas have lower social ranks than uninfected hyenas
H2: Human activities influence local *T. gondii* prevalence
Prediction 4: *T. gondii* prevalence is lowest within a protected wildlife area[1]
Prediction 5: *T. gondii* prevalence is increasing over time in an area undergoing human development[2]
[1] Serena (Eastern side of Masai Mara Reserve) [2] Talek (Western side of Masai Mara Reserve near Talek town)

The results and discussion provided below are somewhat preliminary; they are based on our pilot studies of the data available when an accompanying Festschrift volume was being assembled. As noted in the addendum, we are now compiling and analyzing a much larger dataset of spotted hyena behaviors, life history data, and disease diagnostics. The results will be presented in a forthcoming manuscript.

16.2 Methods

16.2.1 Study Sites in Protected and Developing Regions of the Masai Mara National Reserve

This research uses data and samples from the Mara Hyena Project, a long-term field study of individually known spotted hyenas that have been continuously observed since 1988. Study hyenas are monitored daily and behavioral, demographic, and ecological data are systematically collected and entered into a database. Here, we used data from 4 different hyena groups, called clans, as well as historic information about ecological conditions in the Masai Mara National Reserve. Taking advantage of a preexisting natural experiment we compared *T. gondii* infection between hyenas living near high versus low human presence by exploiting differing ecological conditions that resulted from two management strategies in the reserve, and a growing human population along one of its eastern borders. More specifically, we classified hyenas from the Serena North, Serena South, and Happy Zebra clans as members of the Serena low human presence group because they live in the remote western side of the Reserve. Here, isolation from pastoralist villages and a strict ban on livestock grazing reduce hyenas' exposures to domestic animals and humans, which are known reservoirs for *T. gondii*. It should also be noted that darting at the Serena site was only recently incorporated into the Masai Mara Hyena project, so all samples from this low human presence location were sampled after 2012. On the eastern side of the reserve are pastoralist villages that have seen increased human population growth, especially around the burgeoning Talek community [16]. We classified hyenas from the Talek clan as experiencing high or low human presence based on livestock counts we conducted inside the reserve. These counts have shown a marked increase in the number of livestock within the reserve beginning in 2000, followed by another livestock increase between 2009 and 2013. These changes coincided with parallel shifts in hyena demography and wildlife community composition [13]. We therefore classified Talek hyenas sampled before 2000 as part of the Talek low human presence group, and after 2012 as part of the Talek high human presence group.

16.2.2 Collection of Demographic, Behavioral, and Biological Samples from Mara Hyenas

We maintained detailed records on the demographics of our study population, including sex, age, and the dates of key life-history milestones such as birth, weaning and death. Spotted hyenas live in matriarchal societies, each structured by a linear social hierarchy. We calculated each hyena's rank within the clan based on wins and losses in repeated agonistic interactions with other clan mates. Rank was updated annually to account for changes in clan demography, and we standardized it as a relative score between -1, the lowest-ranking individual in a clan, and 1, the highest-ranking individual, to allow rank comparisons between clans that vary in size [8].

Hyena and lion interactions occur because of fierce competition over food and considerable overlap in their prey resources [19]. Over the course of our study, we have documented 339 hyena and lion interactions involving individual hyenas for which we also obtained *T. gondii* diagnoses. Because our serodiagnostics were obtained only for the date individuals were darted (and plasma was taken), we cannot know infection statuses of hyenas at the time of each observed hyena-lion interaction. We controlled for this by filtering our data set to include only hyena-lion interactions in which a focal hyena's serostatus was known. For individuals testing negative, we analyzed only observations that occurred prior to the date of plasma sampling, thus ensuring these represented behaviors of uninfected hyenas. We similarly restricted our analysis of hyenas that tested positive for *T. gondii* to observations that occurred after the date of plasma collection for *T. gondii* diagnosis. To correct for developmental changes in behavior toward lions that are independent of *T. gondii* infection status, these analyses were adjusted for individual hyena ages on the date(s) they were observed interacting with lions (see statistical analyses). For individual hyenas observed multiple times with lions (where serostatus was also known), we analyzed the average of both recorded behaviors (described below) and hyena age (in months) on the dates the behaviors were recorded.

As part of daily behavioral observations, anytime we saw hyena-lion interactions we recorded data about which hyenas were present, their behaviors, their approach distances towards lions [36], and their ranked distance among all hyenas in the observed hyena group. As with social rank, we standardized these ranked distances to permit meaningful comparisons among ranked distances of focal animals in groups of varying sizes. We also recorded information about the group sizes, ages, and sexes of the lions involved in observed hyena-lion interaction [13]. Finally, as part of our long-term data collection, we record sources of mortality whenever we find dead hyenas that we can identify. As part of our necropsy protocol we determine the cause of death when known. Here we were particularly interested in deaths due to lions. All other sources of mortality were binned into a second category (i.e. death by other causes) for analyses in this manuscript.

In addition to collecting passive observations, the Mara Hyena Project also conducts behavioral experiments in the field. In the pre-sent study, we analyzed responses from an earlier experiment in which focal subadult and adult individuals with known serodiagnostics ($n = 31$) were presented with a commercially-available model of a full-grown spotted hyena to simulate territorial intrusion by an unfamiliar individual. Intruding hyenas are a common threat that hyenas experience frequently as immigrant males attempt to join new clans. Simulated intrusions therefore permit us to assess a naturally-occurring risk-taking behavior that hyenas would experience in both high- and low-disturbance areas [18]. To simulate territory intrusion, we blocked a focal animal's view of the hyena model with a vehicle while deploying it in the field, then drove approximately 30m away and parked parallel to the simulated intruder. We then recorded behavioral responses with video recorders, commencing when mock intruders were first noticed by focal hyenas as determined by a startle reflect and accompanying pause in other activities. Trials ended when the focal individual either walked > 50m from the model, or lay down for at least five minutes within 50m of the model. We terminated four sessions when focal hyenas attacked the simulated intruder, and a fifth session when other wildlife interfered with a focal animal during the trial. Further details and results from this study will be presented in a forthcoming paper. In the present study, we present data from focal individual's latency to approach the target, and also their minimum distance from the target throughout a given trial.

Finally, we routinely dart the study animals in order to collect biological samples and take morphological measurements. Here we immobilized hyenas using 6.5

mg/kg of tiletamine-zolazepam (Telazol®) in a pressurized dart that is fired from a CO_2 powered rifle. We then drew blood from the sedated hyena's jugular vein into ethylene-diaminetetraacetic acid (EDTA) coated vacuum tubes. After the hyena was secured in a safe place to recover from the anesthesia, we took the samples back to camp where a portion of the collected blood was spun in a centrifuge to separate red blood and white blood cells from plasma. Plasma was aliquoted into multiple cryogenic vials. Immediately, the blood derivatives, including plasma, were flash frozen in liquid nitrogen where they remained until they were delivered to a -80 °C freezer in the U.S.

16.2.3 ELISA Assays of Toxoplasma Gondii Exposure

We diagnosed individual hyenas as seropositive, seronegative, or seroambiguous using the ID-Vet multi-species Toxoplasma gondii ELISA kit. In previously published validation studies, IgG-based ELISAs have yielded concordant results to other diagnostic approach-es in both hyenas [32] and other mammals [1]. This kit contains wells that are precoated with *T. gondii*'s P-30 antigen. If present in tested plasma or serum, IgG anti-bodies to *T. gondii* P-30 bind to the wells. After rinsing the well, bound antibodies are further complexed by the kit conjugate, a cocktail of molecules that target IgGs from a wide array of mammals. These conjugated antibodies produce a color-changing reaction when exposed to the kit's substrate. After a 30-minute incubation, further reaction between the conjugate and substrate are blocked, and a plate read-er is used to measure the color intensity in each well. Colorimetric values are used to infer the presence of conjugated antibody, and to calculate SP ratios (described below) that are used to diagnose prior *T. gondii* exposure.

The SP ratio is the ratio of the colorimetric signal from a sample (S) divided by the positive control (P) after subtracting out the background signal for the plate (i.e. a negative control) from both S and P. The kit manufacturer's criteria for interpreting S/P are: <40% = negative result, 40%-50% = doubtful result, >50% = positive result. As described in the results section, only a small fraction of hyenas were seroambiguous, falling in the "doubtful" range of ELISA S/P values. We treated these individuals as negative in our analyses, but will also be confirming that excluding them would not have impacted our findings and conclusions using sensitivity analyses.

16.2.4 Quality Control of ELISA Testing

We used a well validated commercial kit (multi-species ID Screen® Toxoplasmosis Indirect, IDVET, Montpellier) for our ELISA assays. Duplicate assays of each individual hyena yielded highly consistent SP ratios and, consequently, a repeatable *T. gondii* diagnosis. Among all tested hyenas, the average pairwise differences in SP ratios (indicative of antibody-bound *T. gondii* antigen) was $0.039 + 0.039$sd. This minor, inter-replicate noise would be unlikely to impact categorical *T. gondii* diagnostics of individual hyenas, given the much larger ranges for the kit's positive $(S/P > 50.000)$ and negative $(S/P < 40.000)$ windows, and the distributions of our observed SP values (Figure 16.1). We also re-tested one focal individual a) on each of our five ELISA plates, and b) in a series of serial dilutions on a single plate. This individual (#653) produced similar SP ratios across all five plates (Table 16.2). In the dilution series, dilution volumes were also strongly predictive of observed SP ratios (results not shown). Altogether, these results suggest that the ELISA kit provided a repeatable and informative index of immunological reactivity to *T. gondii*'s P30 antigen.

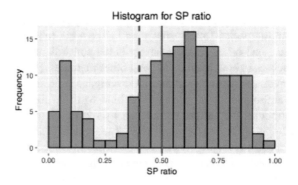

Fig. 16.1: Distribution of SP ratios (from ELISA assays) among spotted hyenas sampled in or near Kenya's Masai Mara National Reserve. The dashed red line is the upper SP ratio cutoff for seronegative and the solid red line is the lower cutoff for serpositive. In between the red lines corresponds to the SP ratio range for seroambiguous.

16.2.5 *Statistical Analyses*

We selected a subset of hyenas for this study from the Masai Mara Hyena project, with the objective of maximizing power to test our key predictions (Table 16.1). Our final data selection (Table 16.3) reflects an effort to balance representation of hyena age classes and sampling localities, while also selecting individuals for which behavioral observations were available. In our final data selection (Table 16.3), we had 33 samples collected during or after 2012 from Talek hyenas (exposed to high human presence), 21 samples from hyenas living in the remote Serena region of the Mara since 2012 (exposed to low human presence), and 118 samples which were collected from Talek hyenas prior to 2000 (also exposed to low human presence).

Table 16.2: Toxoplasma gondii ELISA plate control. One individual hyena (#653) was tested on all five ELISA plates, along with a positive control of porcine serum included by the test kit manufacturer. Both this individual (#653) and the manufacturer's positive control yielded consistent S/P ratios and, consequently, repeatable *T. gondii* diagnostics across all five plates.

plate_ID	Individual #653 replicate 1	Individual #653 replicate 2	+control replicate 1	+control replicate 2	S/P	Diagnosis
1	0.735	0.836	0.877	0.985	0.844	positive
2	0.864	0.679	1.122	1.062	0.707	positive
3	0.813	0.763	0.969	1.063	0.775	positive
4	0.891	0.864	1.013	0.996	0.874	positive
5-upper	0.907	0.839	1.093	1.087	0.801	positive
5-lower	0.877	0.879	1.093	1.087	0.806	positive

We assessed the background characteristics of this dataset including checking the distributions and reporting raw means and standard deviations for continuous variables (SP ratio, lifetime average minimum approach distance towards lions, lifetime average standardized rank, minimum approach distance toward lions during lion-hyena interactions, minimum approach distance towards a model hyena intruder, latency to approach to the minimum distance towards a model hyena intruder, and cause of mortality. For all nominal categorical variables, including infection status, sex, categorical age, standardized social rank, and human presence, we calculated within group sample sizes and frequency ratios (Table 16.3). Finally, we conducted bivariate analyses to assess for potential confounding variables.

Our formal analyses were carried out in two steps using simple and multiple variable linear regression when the outcome of interest was continuous, and logistic regression when we had a binary outcome. First, in unadjusted linear regression, we modeled the separate effects of standardized social rank, and human presence as potential determinants of toxoplasma ELISA SP ratios, a continuous variable that serves as a proxy for *T. gondii* antibody titers. In our adjusted models we controlled for the potential confounding effect of continuous age (in months). Associations from adjusted models were interpreted as a 1 unit change in SP ratio for each unit of change in standardized social rank or as the difference in SP ratio between each level of human presence. Using the same set of independent variables but replacing the outcome of SP ratio with our binary metric of toxoplasma serostatus (seropositive vs. nonseropositive), we used logistic regression to identify independent variables that predict the odds of infection. Here again, we first examined unadjusted models, then recalculated parameter estimates in adjusted models that controlled for age (in months). The estimates from logistic regression models were exponentiated to transform them from the log odds to the odds scale.

We also used linear regression to assess the effects of *T. gondii* exposure on hyena boldness behaviors obtained from observational and experimfental data. To do this, we fit two sets of models that used ELISA results as predictor variables, expressed categorically (seropositive vs. non-seropositive) vs. continuously (as SP ratios). For

Table 16.3: Background characteristics and *T. gondii* infection of 168 spotted hyenas from Masai Mara, Kenya

Measure of *T. gondii* infection	%(N)[a] and Mean ± SD	
Seroprevalence		
Uninfected	35% (58)	
Infected	65% (110)	
SP ratio (n = 168)	0.55 ± 0.26	
Population characteristics: Determination of *T. gondii*	**Uninfected**	**Infected**
Social Rank	17% (8)	83% (39)
Relative rank scale (-1:1)	0.20 ± 0.70	0.18 ± 0.71
Human Presence		
Talek (High; post 2012)	4% (7)	16% (26)
Talek (Low; pre 2000)	28% (47)	39 % (63)
Serena (Low; post 2012)	2% (4)	10% (17)
Population characteristics: Consequences *T. gondii*	**Uninfected**	**Infected**
Boldness towards lions	10% (8)	90% (69)
Lifetime average minimum approach distance (m)	77.81 ± 45.83	37.83 ± 40.78
Lifetime average standardized rank minimum distances	0.24 ± 0.14	0.33 ± 0.19
Boldness towards model hyena intruder	23% (7)	77% (24)
Minimum approach distance (m)	22.00 ± 18.88	32.80 ± 34.10
Latency to approach (min)	1.77 ± 0.67	4.58 ± 2.40
Mortality Source		
Other mortality sources	18% (6)	39% (13)
Lion mortalities	6% (2)	36% (12)

[a] Percentages and Ns may not add up to 100% and 168 individuals respectively, due to missing values.

hyena-lion observations in which interactions were limited to focal individuals with known serostatus (see methods), our response variables were: minimum approach distance to lions (averaged across observations when a hyena interacted with lions on multiple occasions) and standardized ranked distance to lions (similarly averaged when there were multiple measurements per hyena). In our adjusted models we controlled for the focal hyena's age in months at the time of darting because age may affect how hyenas behave towards lions. For the experimental dataset, we modelled focal individuals' minimum approach distance and latency to approach the simulated intruder. Again, because we found a significant age structure in our ELISA assays of *T. gondii* serostatus (see results), we refit each model to adjust for confounding effects of hyena ages (in months) on the date their tested plasma sample was collected.

Finally, we used a logistic regression model to compare the odds of mortality due to lions versus any other known cause between infected and uninfected hyenas. This model was underpowered due to small sample size, and was not adjusted for any confounding variables.

16.3 Results

16.3.1 T. gondii infection Is Common In Spotted Hyenas, And Increases With Age

110 of 168 hyenas (65%) tested positive for IgG antibodies to *T. gondii* by ELISA, indicating prior exposures to the parasite. 37 individuals (22%) tested negative, and 21 hyenas (13%) yielded inconclusive results within the "doubtful/uncertain" range of SP ratios (Figure 16.1). Bivariate analyses revealed that female and male hyenas did not differ in their seroprevalence (OR 0.76 [95% CI: 0.40, 1.46]; $\chi^2 = 0.67, df = 1, p = 0.41$). Hyena cubs had a signifcantly lower seroprevalence than subadults (OR 5.32 [95% CI: 2.18, 13.58) and adults (OR 7.70 [95% CI: 3.52, 17.69]; $\chi^2 = 27.0, df = 2, p < 0.0$).

16.3.2 Seropositive Hyenas Show More Boldness Towards Lions

There was a significant relationship between our *T. gondii* diagnostics and the distances within which hyenas approached lions over the course of their lifetimes. After filtering our observational dataset to include lion interactions where the focal hyena's serostatus was known, and also controlling for age at the time of diagnosis, we found that infected hyenas approached lions more closely (approaching 36.17 meters closer [95% CI: 0.17, 72.16]; t = -1.96, df = 62, p = 0.05). At the group level, we did not detect relationships between *T. gondii* diagnosis and ranked distances from lions (Table 16.4). Lastly, we observed that hyenas which are infected with *T. gondii* had a higher odds, 2.77 (95% CI: 0.52, 21.49); $^2 = 1.3$, df = 1, p = 0.26 of dying due to lions vs. other sources of mortality. This result was not significant, and also likely underpowered due to the small sample size.

Table 16.4: Associations of *T. gondii* infection with boldness behaviors towards lions and lion related mortality among 77 spotted hyenas.

Measures of T. gondii inf.	N	Lifetime average minimum approach distance (m) β (95% CI)		Lifetime average standardized rank minimum distances β (95% CI)		N	Lion versus other mortality causes
		Unadjusted Models	Adjusted Models[a]	Unadjusted Models	Adjusted Models[a]		Mortality Sources OR (95% CI)[b]
Seropreval.							
Uninfected	8	0.00 (Reference)	0.00 (Reference)	0.00 (Reference)	0.00 (Reference)	8	0.00 (Reference)
Infected	69	-39.98 (-70.20, -9.77)	-37.17 (-72.16, -0.17)	0.09 (-0.05, 0.23)	0.09 (-0.04, 0.22)	25	2.77 (0.52, 21.49)
P		0.01	0.05	0.19	0.18		0.26
SP ratio	77	-22.05 (-68.49, 24.40)	-9.33 (-65.63, 45.96)	0.06 (-0.14, 0.26)	0.08 (-0.12, 0.28)	33	3.84 (0.35, 58.98)
P		0.36	0.75	0.56	0.46		0.29

[a] Adjusted models are controlled for hyena age in months when diagnosed for *T. gondii*.
There may be fewer N in adjusted models due to missing data.

[b] Mortality was assessed as being caused by lions or other when the cause of death was not lions.

16.3.3 T. Gondii Seropositive Hyenas Are Less Responsive To Simulated Territorial Intruders

Seronegative hyenas approached a simulated hyena intruder approximately 1.6 times more rapidly (Table 16.5; 2.89 [95% CI: 0.91, 4.86]; t = 2.86, df = 23, p = 0.01). Both seropositive and seronegative individuals ultimately approached within the same minimum distance of the simulated intruder.

Table 16.5: Associations of *T. gondii* infection with boldness behaviors towards a model hyena intruder among 31 spotted hyenas.

Measures of *T. gondii* infection	N	Minimum approach distance (m) β (95%CI)		Latency to approach (min) β (95% CI)	
		Unadjusted Models	Adjusted Models[a]	Unadjusted Models	Adjusted Models[a]
Streovalence					
Uninfected	7	0.00 (Reference)	0.00 (Reference)	0.00 (Reference)	0.00 (Reference)
Infected	24	10.80 (-15.77, 37.38)	3.73 (-25.01, 32.48)	**2.81 (0.99, 4.62)**	**2.89 (0.91, 4.86)**
P		0.43	0.80	0.01	0.01
SP ratio	77	14.37 (-49.67, 78.40)	-8.52 (-76.67, 59.62)	**6.88 (2.56, 11.19)**	**6.73 (2.01, 11.44)**
P		0.66	0.81	<0.01	0.01

[a] Adjusted models are controlled for hyena age in months when diagnosed for *T. gondii*.
There may be fewer N in adjusted models due to missing data.

16.3.4 T. Gondii Prevalence Is Not Predicted By Spatial Or Temporal Variation In Human Presence

Among 19 adult hyenas sampled from the protected Serena part of the reserve, 15 (78.9%) tested positive for *T. gondii*. Among 44 adult hyenas sampled in Talek before 2000, predating subsequent expansions of the local human population, 36 (81.8%) tested positive. Among adult hyenas sampled in Talek after 2012, 12 of 15 (80%) adults tested positive (Figure 16.2). While prevalence initially appeared to change over time at Talek, this effect was eliminated after adjustment for age at diagnosis. Thus, hyenas in our Talek low human presence sample (animals darted pre 2000) had similar odds of infection as compared to the Talek high human presence group (post 2012; see Table 1.5), and the difference between them was not significant. The Serena low human presence group (sampled post 2012) also did not differ from the low disturbance Talek sample, which was collected over the same (pre-2000) time interval (0.51 [95% CI: 0.08, 4.25]; $\chi^2 = 2.5, df = 1, p = 0.12$).

Our sampling design did not allow for meaningful assessments of *T. gondii* prevalence in subadults or cubs from Serena ($n = 0$ and $n = 2$, respectively), nor in cubs from Talek post-2012 ($n = 0$). Thus, while prevalence increases with age, we are presently unable to compare the age structure in prevalence among localities or

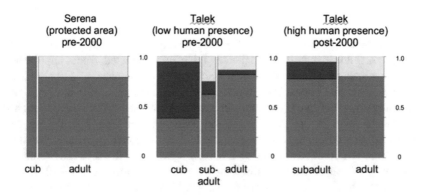

Fig. 16.2: Serodiagnostics for spotted hyenas from a protected area (Serena), and an area that has
experienced increasing human disturbance since 2000 (Talek). Red indicates
seropositive hyenas, blue indicates seronegative hyenas, and light blue indicates
seroambigous "doubtful" diganoses. Serodiagnostics for spotted hyenas from a
protected area (Serena), and an area that has experienced increasing human disturbance
since 2000 (Talek). Red indicates seropositive hyenas, blue indicates seronegative
hyenas, and light blue indicates seroambigous "doubtful" diganoses.

time strata using data from cubs. Among the subadults sampled from Talek, 10
of 16 sampled before 2000 were seropositive (62.5%), and 14 of 18 (77.8%) sampled after 2012 were seropositive. Across these time strata (and concomitant human
density changes at Talek), the small difference in subadult seroprevalence did not
approach statistical significance (z=-0.9759, p=0.33).

Table 16.6: Social and ecological determinants of *T. gondii* infection in 168 spotted hyenas.

	N	Unadjusted Models		Adjusted Models[c]	
		β (95% CI) SP Ratio[a]	OR (95% CI) Seroprevalence[b]	β (95% CI) SP Ratio[a]	OR (95% CI) Seroprevalence[b]
Determinants of *T. gondii* infection					
Social Rank					
Relative rank scale (-1:1)	47	0.00 (-0.07, 0.07)	0.96 (0.30, 2.84)	0.02 (-0.06, 0.10)	1.39 (0.34, 5.91)
P		0.93	0.94	0.84	0.64
Human Presence					
Talek (High; post 2012)	33	0.00 (Reference)	1.00 (Reference)	0.00 (Reference)	1.00 (Reference)
Talek(Low; pre 2000)	111	**-0.10 (-0.20, 0.00)**	**0.37 (0.14, 0.88)**	**-0.10 (-0.20, 0.00)**	0.45 (0.16, 1.19)
Serena (Low; post 2012)	21	0.02 (-0.12, 0.16)	1.14 (0.30, 4.92)	-0.01 (-0.19, 0.18)	0.51 (0.08, 4.25)
P		0.04	0.03	0.01	0.12

[a]Significance is based on type I sums of squares F-test or t-test.
[b]Significance is based on Wald Chi-squared test.
[c]Adjusted models are controlled for hyena age in months when diagnosed for *T. gondii*

16.4 Discussion

This is the first study to show that *T. gondii* infected hosts incur greater risks of mortality via felids in nature. As predicted (Table 16.3), infected hyenas approached lions more closely than their uninfected counterparts. This behavior is likely costly for hyenas given that lions readily attack and kill them (e.g. [18, 29]), and represent the primary source of hyena mortality in many populations [33]. A lion is roughly four times as large as a spotted hyena, and a single swipe from a lion's paw can maim or kill an adult hyena. As a definitive host for *T. gondii*, lions can also shed recombinant, environmentally stable *T. gondii* spores into the local environment. These "oocysts" can infect a multitude of warm-blooded host species, inclusive of hyenas, that inhabit the Masai Mara landscape. Nonetheless, hyenas are relatively unlikely to transmit *T. gondii* to lions because lions rarely consume hyenas after killing them [15]. Because this makes lions unlikely to ingest infectious hyena tissues, hyena boldness towards lions most likely reflects "collateral manipulation" by *T. gondii*. Here, we introduce this term to describe extended parasite phenotypes that evolved to manipulate a subset of hosts, but which are not under positive selection (i.e. do not enhance transmission) in a focal alternative host.

Effects of collateral manipulation on host behavior are virtually unexplored in wild hosts. There is, however, an array of indirect evidence of their ubiquity and significance in mammalian hosts of *T. gondii*. In humans, which are dead end hosts for the parasite, infections are correlated with changes to neural, hormonal, and behavioral phenotypes that are also seen in experimentally-infected rodent models despite these hosts being separated by $\sim$100 million years of evolutionary divergence [7]. In sea otters, encephalitis caused by *T. gondii* is also associated with elevated risks of shark predation [17]. It is unclear, however, if this stems from the manipulation of otter boldness, or merely declines in condition that impede shark evasion and/or force otters to engage in riskier foraging. For hyenas, however, declines in condition are unlikely to explain behavioral covariates of *T. gondii* infections. Infected hyenas have shown otherwise normal behavior in our extensive field observations. And unlike sea otters, which can also die from *T. gondii*-induced encephalitis, disease was rarely determined to be the cause of death in extensive necropsies of Mara hyenas. Given the high prevalence of *T. gondii* in Mara hyenas (> 80% of tested adult hyenas), this suggests the parasite causes low morbidity for the focal population. As described below, we also failed to see any relationship between hyena ranks and infection statuses, which we predicted would occur if infections reduce overall condition (Table 16.3). Finally, a prior study of spotted hyenas found that latent *T. gondii* infections were common in a zoo population, yet did not involve any symptoms of ill health [32].

We also did not find a relationship between *T. gondii* serostatus and the ranked distance of individual hyenas within groups that we observed interacting with lions. This may reflect our inability to control for the serostatuses of other individuals within these groups, and/or additional factors that influence a focal individual's boldness within a group. Another limitation of the present study is that we could only obtain hyena serostatus on the date when individuals were tranquilized to collect plasma samples. This limits the power of our behavioral comparisons, but also makes them conservative. Thus, effects of *T. gondii* on hyena behavior might be even more pronounced and/or multifaceted than those suggested by our present dataset. In the near future, we hope to pursue serial diagnostics of focal individuals to pinpoint sources, timing, and consequences of parasitic infections of wildlife in the Masai Mara National Reserve.

Because only 33 of our focal individuals have died of known causes at the time of this writing, we are currently unable to draw informative comparisons between mortality rates and/or causes in uninfected vs. infected hyenas. Nonetheless, we noted a non-significant pattern of higher mortality by lion (vs. other sources) in hyenas that were seropositive for *T. gondii*. This suggests it will be fruitful to diagnose a larger panel of individuals with known mortality causes. Meanwhile, given previously-established links between hyena boldness toward lions and fitness [36], the present study provides the clearest evidence to date that enhanced feline proximity can incur fitness costs for an intermediate *T. gondii* host.

As predicted (Table 16.3), and similar to findings from other hosts, *T. gondii*'s effects on hyena boldness are also context-dependent. In the absence of lions, we found that seropositive hyenas were slower to respond toward a simulated territorial intruder. In these trials, and in contrast to the pattern found for lion interactions, infected hyenas and uninfected hyenas did not differ in how closely they approached the behavioral stimulus (here, a model intruder).

Absent further studies, it is difficult to ascertain the reason(s) for the observed differences in approach latencies of infected vs. uninfected hyenas during simulated territorial intrusion. A simulated intruder would likely be regarded by focal animals as an unfamiliar individual. Longer response latencies might, therefore, reflect reduced neophobia in the presence of a novel threat. Reduced neophobia has also been correlated with infection of other hosts, e.g. rodents, that are commonly predated by felines [34]. This same effect could explain infected hyenas' closer approaches to lions, though we cannot exclude olfactory modulation by the parasite, nor other mechanisms of reducing ailurophobia, without further experimental studies.

Regardless of their proximate causes(s), the different behavioral outputs of seropositive hyenas in the presence of lions (i.e. closer approaches) vs. toward simulated intruding conspecifics (i.e. slower approach to within an equal distance) reveals an intriguing context-specificity in *T. gondii*-associated host traits. This echoes findings from other social mammals, in which *T. gondii*'s highly specific modulation of responses to olfactory cues imply fine-tuned manipulation mechanisms (e.g. [30]). For example, in other intermediate hosts, *T. gondii* selectively modulates response to urine odors of wild (vs. domestic), and/or to local (vs. allopatric) felines. These effects have been interpreted by previous authors as evidence of adaptive host manipulation that facilitates local transmission (e.g. [24, 28]). If hyenas are indeed collaterally manipulated by *T. gondii*, our results reveal remarkable similarities in the extended phenotypes of targeted and collateral hosts, despite these hosts' striking evolutionary distances and ecological dissimilarities.

Our ELISA tests revealed that a high proportion of adult hyenas in the Masai Mara have been exposed to *T. gondii*. Our finding of age structure in disease prevalence is not surprising since, with very few known exceptions, infected hosts remain seropositive throughout their lives ([23] outlines possible exceptions). Observed *T. gondii* prevalences in our study are very similar to findings from a recent, independent study of infection prevalence in a smaller number of spotted hyenas at nearby field sites [9]. Altogether, these observations likely reflect the aggregate influence of several non-exclusive routes of exposure within African landscapes including 1) *T. gondii*'s prevalence within the diverse wild prey hyenas consume, 2) a rich community of indigenous felids that can serve as definitive hosts, 3) increasing presence of domesticated non-definitive hosts (e.g. sheep, goats and swine) [13], which can act as reservoirs and/or sources of trophic infection, and 4) the popularity of housecats (*Felis domesticus*) within Kenya for household rodent control [22].

While the relative role(s) of these potential infection sources for hyenas are not entirely clear, the current study and earlier work offer several informative insights. Since hyenas rarely migrate between Talek and Serena, we conclude *T. gondii* is currently well-distributed throughout the Mara. We further suspect that infection often originates from ingested oocysts in this ecosystem, given the high prevalence of *T. gondii* in Africa's native and domesticated herbivorous mammals [27], which are unlikely to consume infected animal tissues. Trophic transmission from wild and/or domesticated prey is also likely contributing to *T. gondii* prevalence in Mara hyenas, given that these and other generalist carnivores have higher infection rates than sympatric herbivores, and also that grazing herbivores would, most likely, ingest oocysts at equal or higher rates than carnivores [2].

Counter to our predictions (Table 16.3), temporal increase in human presence was not associated with rising seroprevalence in Talek hyenas. Human activities may still be impacting the parasite's contemporary evolution (e.g. by facilitating strain recombination), but do not appear to be necessary to produce high prevalence of *T. gondii* in wild hyena hosts.

Finally, while several non-definitive hosts can also sustain *T. gondii* through congenital infection of their offspring [7], this does not appear to be the main source of hyena infections. We found that seroprevalence increases in hyenas after weaning, when subadults transition to carnivorous diets. This again suggests that environmental routes of infection, i.e. oocysts and/or trophic transmission, are probably common sources of infection in the Mara's spotted hyenas.

Further discoveries lie ahead concerning the direct and indirect costs of latent *T. gondii* infections for hyenas in nature. Though non-significant, our observed pattern of hyena mortality sources (lion vs. other; Table 16.3) hints at possible fitness costs of collateral manipulation by *T. gondii*. Territorial behavior, which covaries significantly with *T. gondii* infection (Table 16.6), also plays a critical role in the organization of hyena societies. In view of the parasite's multiplicative effects on fitness-related behaviors, we suggest collateral manipulation by this parasite can have important and dynamic consequences for the fitness of Mara hyenas. For example, hyenas at Talek compete more intensely with lions for food [33], and this may change both the costs and benefits of boldness in lion presence. However, these costs and benefits may also be shifting rapidly, since Talek's lions are more sensitive than hyenas to the growing human population in the area [13]. Clearly, further work will be needed to understand the dynamic feedbacks between human activities, disease prevalence, and the behavior and fitness of interacting *T. gondii* hosts.

Our finding of high *T. gondii* prevalence in hyenas should be of general concern to conservationists of Africa's remarkable and threatened biodiversity. Acute *T. gondii* infections are contributing to declines of many endangered species (e.g. [35]), and there is growing concern that increasing *T. gondii* abundance and connectivity facilitates recombination between strains that differ in infectiousness and virulence [5]. Circulating strains are also more diverse and virulent in South America and Africa than other parts of the world, and Kenya has the second highest recorded *T. gondii* diversity among African nations [11].

T. gondii's presence in areas of both human and wildlife activity, such as the Masai Mara, are further concerning in light of rapidly increasing HIV infections in Africa's human population. Latent *T. gondii* infections are often benign in healthy humans, though infection is a risk factor for a growing list of physical and psychological diseas-es [21]. In contrast, *T. gondii* often causes debilitating or lethal symptoms in individuals living with AIDS or other immunodeficiencies [3]. Preventing human *T. gondii* infections is thus an especially important public health objective in Africa, where ~25 million people live with HIV; it also requires a detailed understanding of the parasite's local mechanisms of transmission. Prior studies have suggested only limited exchange between the sylvatic (wildlife) and anthropogenic (urban) populations of *T. gondii* [12], but this may differ in the Mara where wildlife, livestock, and humans interact with atypically high frequency. These considerations suggest that both public health and wildlife conservation in Kenya will benefit from further research into the behavioral and ecological drivers of local transmission between humans, domestic animals, and wildlife. And lastly, further studies of host manipulation by parasites will continue to boldly advance our understanding of animal behavior, ecology, and evolution.

Acknowledgements This material is based in part upon work supported by the National Science Foundation under Cooperative Agreement No. DBI-0939454. Any opinions, findings, and conclusions or recommendations expressed in this material are those of the author(s) and do not necessarily reflect the views of the National Science Foundation. The authors wish to thank Chiara Bowen for help in the lab. Kathryn Fiedler, Kay Chalkowski, Dan Nunez, and Whitman the cat have helped to inspire their interest in Toxoplama gondii. Countless MSU undergraduates, Holekamp lab graduate students, and Masai field workers contributed to collection of plasma and behavioral data included in this study. Getty and Holekamp lab members also contributed to the development of project aims, and offered feedback on the analyses and writing of this manuscript. Finally, we thank Erik Goodman, director of the NSF-BEACON center, along with many BEACON staff, for facilitating the collaborative boldness needed to embark on this work.

ADDENDUM

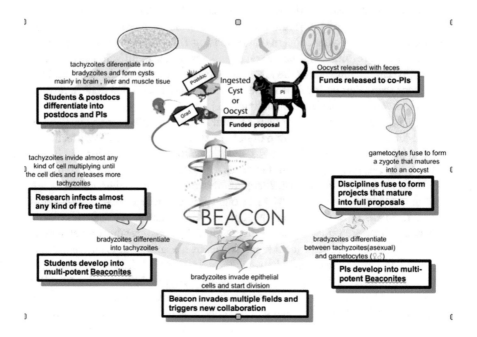

Similarities between life cycles of *T. gondii*, the protist which causes toxoplamsosis, and *B. eacon*, which causes beaconosmosis.

It has not escaped our notice that Toxoplama gondii exhibits several biological similarities with a newly discovered superorganism: *Boldnessbuilding eacon*. This is somewhat surprising, since *T. gondii* is a carbon-based apicomplexan protist whose life cycle was discovered in 1988. In contrast, *B. eacon* is an MSU-based research center directed by Dr. Erik Goodman that invaded East Lansing in 2010. Despite these phylogenetic, biological, and historical distinctions, both organisms have remarkably similar, potentially fitness-altering impacts on their many hosts. Specifically, individuals infected with either organism exhibit greater boldness in the face of unknown and/or formidable challenges.

In the near future, we will begin experimental computational research, supported by two beaconosmosis experts (Hintze and Adami), to test whether simulated *T. gondii* infections promote risky cooperation at the group level in co-evolving digital hosts. The execution of the research reported above has already shown that this is true for *B. eacon*, which significantly emboldened the co-authors of the brain-altering manuscript you have now almost finished reading.

Our studies of Kenyan hyenas document, for the first time, how parasitically-enhanced boldness elevates risk of predation by felids in *T. gondii*-infected hosts.

It remains to be seen whether *B. eacon*-enhanced boldness will incur similar risks, e.g. attacks from lionesque reviewers. Regardless, we are confident many readers will eventually acquire beaconosmosis through consumption of the infectious ideas embedded herein. This need not concern anyone, however, since *B. eacon* is only known to have enhancing effects on the intellects of its many grateful hosts.

References

1. Al-Adhami, B., Gajadhar, A.A.: *A new multi-host species indirect ELISA using protein A/G conjugate for detection of anti-Toxoplasma gondii IgG antibodies with comparison to ELISA-IgG, agglutination assay and Western blot.* Veterinary Parasitology **200**(1-2), 66–73 (2014)
2. Bakal, P.M., Karstad, L., Veld, N.I.T.: *Serologic evidence of toxoplasmosis in captive and free-living wild mammals in Kenya.* Journal of Wildlife Diseases **16**, 559–564 (1980)
3. Basavaraju, A.: *Toxoplasmosis in HIV infection: An overview.* Tropical Parasitology **6**(2), 129–135 (2016)
4. Berdoy, M., Webster, J.: *Fatal attraction in toxoplasma-infected rats: A case of parasite manipulation of its mammalian host.* Proc R Soc B. **267**(1452), 1591–1594 (2000)
5. Dardé, M.L.: *Toxoplasma gondii "new" genotypes and virulence.* Parasite **15**(3), 366–371 (2008)
6. Dubey, J.P.: *History of the discovery of the life cycle of toxoplasma gondii.* International Journal for Parasitology **39**(8), 877–882 (2009)
7. Dubey, J.P.: *Toxoplasmosis of animals and humans.* CRC Press (2016)
8. Engh, A.L., Esch, K., Smale, L., Holekamp, K.E.: *Mechanisms of maternal rank "inheritance" in the spotted hyaena, Crocuta crocuta.* Animal Behaviour **60**, 323–332 (2000)
9. Ferreira, S.C.M., et al.: *High prevalence of anti-Toxoplasma gondii antibodies in free-ranging and captive African carnivores.* International Journal for Parasitology: Parasites and Wildlife **8**, 111–117 (2018)
10. Flegr, J., Lenochoyá, P., Hodný, Z., Vondrová, M.: *Fatal attraction phenomenon in humans-cat odour attractiveness increased for toxoplasma-infected men while decreased for infected women.* PLoS Neglected Tropical Diseases **5**(11), e1389 (2011)
11. Galal, L., Ajzenberg, D., Hamidović, A., Durieux, M.F., Dardé, M.L., Mercier, A.: *Toxoplasma and Africa: One Parasite, Two Opposite Population Structures.* Trends in Parasitology **34**(2), 140–154 (2017)
12. Gilot-Fromont, E., et al.: The life cycle of *Toxoplasma gondii* in the natural environment. In: O. Djurkovic-Djakovic (ed.) Toxoplasmosis — Recent Advances, pp. 1–36 (2012)
13. Green, D.S., Johnson-Ulrich, L., Couraud, H.E., Holekamp, K.E.: *Anthropogenic disturbance induces opposing population trends in spotted hyenas and African lions.* Biodiversity and Conservation **27**(4), 871–889 (2018)
14. Hart, B.L.: *Behavioral adaptations to pathogens and parasites: Five strategies.* Neuroscience & Biobehavioral Reviews **14**, 273–294 (1990)
15. Holekamp, K.: Personal communication
16. Kolowski, J.M., Holekamp, K.E.: *Spatial, temporal, and physical characteristics of livestock depredations by large carnivores along a Kenyan reserve border.* Biological Conservation **128**(4), 529–541 (2006)
17. Kreuder, C., et al.: *Patterns of mortality in southern sea otters (Enhydra lutris nereis) from 1998-2001.* Journal of Wildlife Diseases **39**, 495–509 (2003)
18. Kruuk, H.: *The spotted hyena: A study of predation and social behavior.* University of Chicago Press, Chicago (1972)
19. Lehmann, K.D., et al.: *Lions, hyenas and mobs (oh my!).* Current Zoology **63**(3), 313–322 (2017)

20. Miró, G., Montoya, A., Jimnez, S., Frisuelos, C., Mateo, M., Fuentes, I.: *Prevalence of antibodies to Toxoplasma gondii and intestinal parasites in stray, farm and household cats in Spain.* Veterinary Parasitology **126**, 249–255 (2004)

21. Ngô, H., Zhou, Y., Lorenzi, H., Wang, K., Kim, T., Zhou, Y., El Bissati, K., Mui, E., Fraczek, L., Rajagopala, S., Roberts, C.: *Toxoplasma modulates signature pathways of human epilepsy, neurodegeneration and cancer.* Scientific Reports **7**, 11,496 (2017)

22. Ogendi, E., Maina, N., Kagira, J., Ngotho, M., Mbugua, G., Karanja, S.: *Questionnaire survey on the occurrence of risk factors for Toxoplasma gondii infection amongst farmers in Thika District, Kenya.* Journal of the South African Veterinary Association **84**, 00–00 (2013)

23. Opsteegh, M., Swart, A., Fonville, M., Dekkers, L., Van Der Giessen, J.: *Age-related Toxoplasma gondii seroprevalence in Dutch wild boar inconsistent with lifelong persistence of antibodies.* PLoS One **6**, e16,240 (2011)

24. Poirotte, C., Kappeler, P.M., Ngoubangoye, B., Bourgeois, S., Moussodji, M., Charpentier, M.J.: *Morbid attraction to leopard urine in Toxoplasma-infected chimpanzees.* Current Biology **26**(3), R98–R99 (2016)

25. Poulin, R.: *Parasite manipulation of host behavior: An update and frequently asked questions.* In: H. Brockmann (ed.) Advances in the Study of Behavior, vol. 41, pp. 151–186. Academic Press, Burlington (2010)

26. Poulin, R., Levri, E.P.: *Applied aspects of host manipulation by parasites.* In: D. Hughes, J. Brodeur, F. Thomas (eds.) Host Manipulation by Parasites, pp. 172–194. Oxford University Press, New York (2012)

27. Riemann, H.P., Burridge, M.J., Behymer, D.E., Franti, C.E.: *Toxoplasma gondii antibodies in free-living african mammals.* Journal of Wildlife Diseases **11**(4), 529–533 (1975)

28. Tenter, A.M., Heckeroth, A.R., Weiss, L.M.: *Toxoplasma gondii: From animals to humans.* International Journal for Parasitology **30**(12-13), 1217–1258 (2000)

29. Trinkel, M., Kastberger, G.: *Competitive interactions between spotted hyenas and lions in the Etosha National Park, Namibia.* African Journal of Ecology **43**(3), 220–224 (2005)

30. Vyas, A.: *Mechanisms of host behavioral change in Toxoplasma gondii rodent association.* PLoS Pathogens **11**(7), e1004,935 (2015)

31. Vyas, A., Kim, S.K., Giacomini, N., Boothroyd, J.C., Sapolsky, R.M.: *Behavioral changes induced by Toxoplasma infection of rodents are highly specific to aversion of cat odors.* Proceedings of the National Academy of Sciences **104**(15), 6442–6447 (2007)

32. Wait, L.F., Srour, A., Smith, I.G., Cassey, P., Sims, S.K., McAllister, M.M.: *A comparison of antiserum and protein A as secondary reagents to assess Toxoplasma gondii antibody titers in cats and spotted hyenas.* The Journal of Parasitology **101**(3), 390–392 (2015)

33. Watts, H.E., Holekamp, K.E.: *Ecological determinants of survival and reproduction in the spotted hyena.* Journal of Mammalogy **90**(2), 461–471 (2009)

34. Webster, J.P.: *The effect of Toxoplasma gondii on animal behavior: Playing cat and mouse.* Schizophrenia bulletin **33**(3), 752–756 (2007)

35. Work, T.M., Massey, J.G., Rideout, B.A., Gardiner, C.H., Ledig, D.B., Kwok, O.C.H., Dubey, J.P.: *Fatal toxoplasmosis in free-ranging endangered 'Alala from Hawaii.* Journal of Wildlife Diseases **36**(2), 205–212 (2000)

36. Yoshida, K., Van Meter, P., Holekamp, K.: *Variation among free-living spotted hyenas in three personality traits.* Behaviour **153**, 1665–1722 (2016)

Chapter 17
Behavioral Strategy Chases Promote the Evolution of Prey Intelligence*

Aaron P. Wagner, Luis Zaman, Ian Dworkin and Charles Ofria

Abstract Predator-prey coevolution is commonly thought to result in reciprocal arms races that produce increasingly extreme and complex traits. However, such directional change is not inevitable. Here, we provide evidence for a previously un-demonstrated dynamic that we call 'strategy chases,' wherein populations explore strategies with similar levels of complexity, but differing behaviorally. Indeed, in populations of evolving digital organisms, as prey evolved more effective predator-avoidance strategies, they explored a wider range of behavioral strategies in addition to exhibiting increased levels of behavioral complexity. Furthermore, coevolved prey became more adept in foraging, evidently through coopting components of explored sense-and-flee avoidance strategies into sense-and-retrieve foraging strategies. Specifically, we demonstrate that coevolution induced non-escalating exploration of behavioral space, corresponding with significant evolutionary advancements, including increasingly intelligent behavioral strategies.

Aaron P. Wagner
Metron Inc, Reston, VA 20190, USA

Luis Zaman
Department of Ecology and Evolutionary Biology, University of Michigan, Ann Arbor, MI, 48109, USA

Ian Dworkin
Department of Biology, McMaster University, 1280 Main St. West, Hamilton, Ontario L8S 4L8, Canada and BEACON Center for the Study of Evolution in Action, Michigan State University, East Lansing, MI 48824, USA

Charles Ofria
BEACON Center for the Study of Evolution in Action, and Department of Computer Science and Engineering, and Program in Ecology, Evolutionary Biology, and Behavior, Michigan State University, East Lansing, MI 48824, USA

* This paper was externally peer-reviewed.

© Springer Nature Switzerland AG 2020
W. Banzhaf et al. (eds.), *Evolution in Action: Past, Present and Future*,
Genetic and Evolutionary Computation, https://doi.org/10.1007/978-3-030-39831-6_17

17.1 Introduction

Dawkins and Krebs [13] famously proposed that Red Queen Dynamics [37] in antagonistic systems should produce reciprocal evolutionary arms races [10]. This hypothesis predicts that the interacting species coevolve traits in a tit-for-tat exchange of increasingly extreme adaptations and counter-adaptations. I.e. "swords get sharper, so shields get thicker, so swords get sharper still". While variations of this arms race interpretation often dominate popular explanations and scientific expectations [2], there is also support for alternative and non-escalating coevolutionary mechanisms, including trait cycling [14] and defense-preference alternation [12, 33]. However, the potential importance of non-escalating coevolutionary exploration of behavioral strategies remains largely unconsidered and untested.

In order to understand how expectations for behavioral phenotypes may differ from other traits, consider the common expectations under arms race models. Arms races are typically couched in terms of effects on the complexity [4, 5] of an individual aspect of morphology or behavior [2, 7, 13, 15, 16, 38, 39]. E.g., stronger claws vs. thicker shells or speed of chase vs. speed of flight. In such a model, directional selection is predicted to drive increased complexity in both players over evolutionary time. To test for this dynamic, traits can be evaluated in terms of how much information they incorporate about the environment (e.g., shell thickness reflecting predator capabilities for crushing; after [24]. Such a directional model requires the (unrealistic) assumption that potential evolutionary responses fall along very limited axes. Given this constraint, the antagonistic nature of predator-prey interactions would ensure that only one direction of travel along an axis is viable. For example, thicker shells are the only viable evolutionary response to increased predator crushing strength when no phenotypic alternatives are available to be explored. It is also important to note that phenotypes of equivalent behavioral complexity can carry different fitness effects: a grey moth with an expressed behavioral preference for perching on grey trees is likely safer, but no more behaviorally complex, than a grey moth expressing a preference for perching on black trees. Of course, behavior is not defined by single, isolated actions, but a series of interrelated actions. Given the numbers and combinations of potential actions, the dimensionality of options for even simple behavioral strategies can be vast. For example, while the complexity of prey responses to coursing predators could increase over evolutionary time, viable alternative flight behaviors could include zig-zagging, hiding, or sudden stops and redirections, as well as variations of each. For most definitions of behavioral complexity, these strategies could reasonably be considered to be of comparable complexity.

Since equally complex strategies are unlikely to be uniformly effective against a given predator, we should expect evolution to produce exploratory "chases" through behavioral option space as often as producing arms races for increasing complexity. While a number of studies [2, 6, 7, 9, 10, 12, 14, 15, 16, 17, 20, 33, 37, 38] have discussed escalating arms races and non-escalating alternatives in antagonistic interactions, we are not aware of any that have examined the relative importance of evolutionary behavioral strategy exploration in defining the outcomes of predator-prey coevolution. A major constraint on testing for these processes is the inherent difficulty of the simultaneous, detailed, and prolonged experimental study of behavior in predators and prey [31], particularly over evolutionary time. However, computational systems permit this sort of inquiry. Specifically, the experimental evolution software Avida [34] carries all of the benefits of evolutionary simulations (e.g., rapid generation times and full control over experimental environments), without incorporating explicit fitness functions to artificially select individuals for reproduction. Importantly, Avida does not merely simulate evolution [35], nor does it carry the assumptions inherent to selection regimes and other mathematical models [34]: a digital organism in Avida has a genome subject to random mutations that are inherited by its offspring, as well as a fitness determined by realized competitive abilities to survive, collect needed resources, and produce offspring. Uniquely among computational systems, this combination allows for unrestricted, unsupervised, and undetermined evolution via natural selection, and direct testing of biological hypotheses [34]. Here we use the Avida system to show that coevolution among predators and prey produces both escalating arms races and non-escalating chases through behavioral strategies.

Avida populations exhibit a rich range of evolutionary dynamics and have been used to understand many factors behind the evolution of complexity [24, 30], including its emergence as a consequence of antagonistic host-parasite interactions [40]. The genomes of the digital organisms consist of low-level computational instructions, including those for environmental sensing, controlling the order and conditions of instruction execution, and for reproduction (at the cost of consumed resources). During reproduction, mutations can occur, producing genetic differences between parent and offspring genotypes. We modified Avida to include a predation instruction [19] that, if mutated into a genome, makes the carrier capable of killing and consuming non-predator organisms. An organism is classified as a predator if it makes a successful kill using the predation instruction. All organisms were required to consume enough resources to meet a threshold for reproduction. Accordingly, prey needed to locate and consume food in the environment, while predators needed to locate, successfully attack, and consume multiple prey. As such, predators are simply organisms that evolved to eat other organisms, sharing a common genetic instruction set with prey and interacting in the same ways with their environment. As in nature, it is only evolved changes in genetic sequences and behaviors that differentiate predators from their prey (see Fig. S1 and Movie S1).

We initialized all evolutionary trials with prey that randomly moved about the environment, indiscriminately attempting to consume resources and reproduce. Among potential adaptive targets, evolution could refine these simple behaviors via adaptations for sensing and responding to objects (i.e. food, organisms, barriers) and more controlled navigation or avoidance strategies. We performed evolutionary trials conducted with (*Pred+*) and without (*Pred−*) the possibility of predator coevolution, and monitored both frequency of sensor use (Fig. 17.1) and behavioral intelligence and complexity, defined as the proportion of genetic actions (decisions) that relied on sensory information.

17.2 Results and Discussion

After two million evolutionary time steps ($\approx 19,500$ prey generations), termed updates, sensor use was higher for prey populations evolved with predators (*mean* =
0.027, 95% CI: 0.019,0.033) vs. those evolved in the absence of predators (*mean* =
0.015, 95% CI: 0.012,0.018; Kruskal-Wallis $p = 0.033$). Likewise, behavioral intelligence and complexity, was also higher in prey populations evolved with predators (*Pred+*: *mean* = 0.094, 95% CI: 0.070,0.120; *Pred−*: *mean* = 0.050, 95% CI:
0.038,0.061; Kruskal-Wallis $p = 0.005$). In contrast, behavioral intelligence and
complexity did not change in response to more complex abiotic environments: dis-
tributing barriers (Fig. S2) throughout the environment had no detectable effect on
evolved levels of behavioral complexity (Fig. S3).

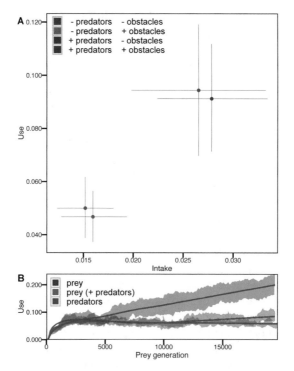

Fig. 17.1: **Coevolution promotes increased prey behavioral intelligence and complexity. (A)**
Coevolution with predators significantly increases both the rate of sensory intake and
the rate of information use (realized behavioral intelligence), while abiotic
environmental complexity (obstacles) has little effect. Data shown are from the final
evolutionary time-point. **(B)** While predator behavioral intelligence increases linearly
over the course of evolution, coevolving prey evolve to use sensory information later
and at a lower rate. Lines are LOESS fits. Mean prey generation times were 102.21
updates (± 0.11 se), with predator-to-prey generation time ratios of 2.49:1 (± 0.06 se).
Shaded regions and error bars are bootstrapped 95% confidence intervals.

As predicted, prey coevolving with predators explored a greater area of behavioral space, as described by executed rates of moves and turns (the only two possible "physical" behavioral actions for prey): *Pred+* prey populations made frequent forays into new areas of move-turn behavioral space, while *Pred−* prey remained in a much smaller sub-area (Fig. 17.2a vs. 17.2b, see also Fig. S4). As a consequence of this behavioral exploration, 27 of the 30 prey populations coevolving with predators discovered, moved to, and then remained in, an area of behavioral space clearly separated from that used by naïve populations (i.e. *Pred−* populations and evolutionarily young *Pred+* populations). As a measure of their true and realized extent of exploration, cumulative lengths of the paths connecting observation points in this move-turn behavioral space were substantially longer for prey coevolving with predators than in counterpart populations (*Pred+*: median path length=11,222 steps,

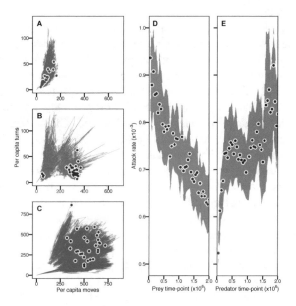

Fig. 17.2: **Coevolving populations explore more behavioral strategies while improving performance.** Shown are mean number of turns and moves taken in each population of (**A**) *Pred−* prey, (**B**) *Pred+* prey, and (**C**) predators (note change in scale) over evolutionary time. Points denote final behaviors in each of 30 trials. Even when returning to the low-movement and low-turn behavioral strategies nearer to that of the naïve ancestor (at the origin), *Pred+* prey populations explore parts of behavioral space never investigated by *Pred−* prey. For all but three *Pred+* prey populations, that exploration leads to a behavioral transition, allowing them into an area defined by high movement rates. (**D**) Mean attack rates when prey from different time-points are reintroduced with predators from the middle of the same evolutionary timeline are highest when predators face the most naïve prey and lowest when facing the most fully evolved prey. (**E**) Likewise, attack rates increase when predators from each time-point face prey from the middle of their evolutionary timeline (**E**). X-axes in (**D**) and (**E**) indicate the time-point (update) from which the indicated populations were drawn; shaded regions are bootstrapped 95% confidence intervals.

range: 6026-26,323; *Pred—: median* $= 6,399$, range: 3,387-9,079), and even longer for predators themselves (*median*=74,243, range: 49,270-92,223) (Fig. S5).

In addition to exploring more of behavioral space and taking in and using more information in making decisions, prey coevolving with predators exhibited increasingly effective predator avoidance strategies (Fig. 17.2 and Movie S1): attack rates decreased for predators that were reintroduced into time-shift trials [22, 21] with the prey from earlier vs. later in their evolutionary history. Likewise, hunting performance of predators clearly improved over time, as measured by presenting predator populations along each evolutionary timeline with the prey from the middle of that timeline. Additionally, attack rates on evolving prey declined at a constant rate (*mean* $= 0.937$, 95% CI: 0.818,1.057, at the first sample, declining to *mean* $= 0.636$, 95% CI: 0.571,0.702, at the final sample), even while use of sensory information exhibited minimal change (e.g., the second quarter of the evolutionary timelines, Fig. 17.1), indicating that prey continued to explore new and more effective anti-predator behavioral strategies even in the absence of increased behavioral intelligence and complexity. Similarly, there was no indication of a movement arms race: while *Pred*+ prey settled in an area of behavioral space defined by relatively high rates of movement, final movement rates for coevolving species were below explored maxima (Fig. 17.2, Fig. S4). Furthermore, in behavioral assays, removal of predators resulted in similar declines in prey movement (a proxy for length of flight responses) over most of evolutionary time (*mean* $= 15.453$ %, 95% CI: -45.030,15.180, movement decline with predators removed at update 50,000 vs. *mean* $= 21.120$%, 95% CI: -41.322, -2.367, decline if removed at the final update; Fig. S6).

We hypothesized that the evolution of behaviorally intelligent traits improving predator avoidance would also result in increased use of sensory data by prey for foraging. Indeed, prey coevolved with predators demonstrated a substantial reliance on information about their environment in making foraging decisions, and increasingly so over evolutionary time (Fig. 17.3): in additional behavioral assays, the 'blinding' of prey to food resources resulted in a mean fitness (the quotient of lifetime food intake by replication time) decline of 0.968% (95% CI: -0.380,2.277) for populations tested at update 50,000, and a decline of 9.812% (95% CI: -15.926,-4.058) for fully evolved populations. In contrast, the blinding of prey evolved in the absence of predators decreased their fitness only slightly, and with little change in the magnitude of that effect over time (initial mean=-1.478%, 95% CI: -3.212,0.200, decline vs. 0.608, 95% CI: -1.000,2.248, decline at the final update). Hence greater evolved use of information about the environment contributed significantly to prey fitness, beyond its importance for predator avoidance.

Finally, the three coevolutionary effects on prey (increased information intake, use of information in decision making, and broader behavioral strategy exploration) also increased prey competitiveness. Specifically, we competed all *Pred*+ prey populations against all *Pred*— populations in new, predator-free environments. At the end of competition, the descendants of prey coevolved with predators represented the majority in most populations (Fig. 17.4; *Pred*+: 23.5 median in-majority counts, 95% CI: 21.487-25.953; *Pred—: median* 7, 95% CI: 3.827-10.440; Kruskal-Wallis $p < 0.001$). The competitive performance of prey coevolved with predators was

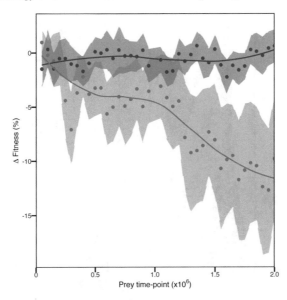

Fig. 17.3: **Coevolution promotes intelligent use of sensory data in foraging decisions.** Shown are mean per-population changes in fitness (reflecting foraging success and gestation time) for prey before and after being blinded to resources. Prey evolved without predators (top, purple) exhibit limited declines in foraging success, indicating a lack of reliance on sensor information. In contrast, prey coevolved with predators (red) experience significant and increasing declines in success over evolutionary time. Shaded regions are bootstrapped 95% confidence intervals. Lines are LOESS fits.

further pronounced in additional trials with a 75% reduction in resource regrowth rates. Thus, prey evolved with predators proved to be more adept and competitive in foraging than prey evolved without predators, including in the very environments one would otherwise expect the latter, not the former, to be more closely adapted. This result appears to be a consequence of a reciprocal evolutionary relationship in which, as prey become better at sensing and reacting to predators, they more readily evolve to become better at sensing and reacting to resources, which further increases evolutionary discovery of adaptations for responsiveness to predators (Fig. S7).

Coevolution with predators produced more behaviorally complex and behaviorally intelligent prey. However, prey performance continued to improve even when complexity indicators did not. Instead, we observed an ongoing exploration of equally complex behaviors. Unlike pure arms races, such exploration of behavioral options need not be directional, nor is it as directly and tightly constrained as are physical traits (e.g., as in [9, 29]). While the extent of reciprocity in this process remains unexamined [1, 2, 23, 37], we have demonstrated that such chases do produce significant evolutionary advancements, including early forms of behavioral intelligence producing more fit and competitive populations. We expect additional examination of the interplay between ecological interactions [25, 28, 32] and the

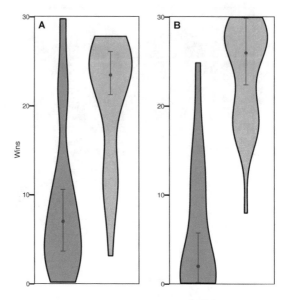

Fig. 17.4: **Coevolution enhances prey competitiveness and behavioral flexibility. (A)** Results
of competition between prey evolved with and without predators; shown are numbers of
predator free competitions in which *Pred−* prey (purple) and *Pred+* prey (red)
dominated at the end of competition (30 competitions per population). **(B)** Results of
competition in a more extreme environment in which resource regrowth rates were 25%
of that used in the evolutionary trials. In both environments, prey coevolved with
predators dominated most competitions, with their competitiveness enhanced in the
novel reduced growth environment. Points indicate medians. Error bars are
bootstrapped 95% confidence intervals. Shaded areas show the full distribution of
per-population per-treatment wins.

exploration of behavioral strategy spaces will further highlight its importance in the
evolutionary discovery of key innovations.

17.2.1 Conclusions

In nature, predation often occurs non-randomly, with predators preferring low condition individuals [11, 18, 36]. This preference, along with a myriad of other factors
are known to influence the evolution of both predator and prey populations (reviewed in [27]. However most studies of coevolutionary dynamics have been limited to a small number of traits, which may underestimate the evolutionary potential
of behavior [3]. In this study we observe many of the expected dynamics of populations under risk of predation such as increased use of sensory information and
movement (Fig. 17.1 & 17.2). More importantly, the evolved expansion of behavioral repertoires when faced with the risk of predation enabled new evolutionary

opportunities, such as improved foraging behaviors. This interaction between selective pressures resulted in a general increase in fitness and competitiveness, even in the absence of predation risk (Fig. 17.3 & 17.4).

These findings are important not only for biological studies, but also for computational problem solving using evolutionary algorithms. Given that a combination of distinct selective pressures (in this case the need to simultaneously forage for food and avoid predators) results in each evolved behavior becoming more effective than if they were selected for individually, we should be able to create a similar dynamic in applied evolution. More exploration is needed to understand this effect, but our observations have led us to hypothesize that behavioral traits effective for one goal (i.e., spotting food) can be co-opted for another (spotting a predator before it gets too close). We hypothesize that the coevolutionary pressures that accelerate evolution on one axis, can also accelerate evolution of entangled behaviors. Many additional studies will be needed to disentangle and isolate the key components of selection leading to such improvements, and to apply those results toward automated problem solving.

17.3 Methods

17.3.1 Environment

We initialized each evolutionary trial with nine simple, identical prey organisms that randomly moved about the environment, blindly attempting to collect resources and reproduce. Each of the 30 evolutionary trials (of each treatment) was conducted for 2 million updates ($\approx$19,500 prey generations). Updates are the unit of time in Avida and one update is defined as the time required for each organism, on average, to execute 30 instructions.

All experiments were conducted in bounded grid-worlds of 251 by 251 cells. Each cell could contain up to one unit of food. When a prey fed from a cell, the prey consumed that full unit and the resource would then regenerate at a rate of 0.01 units per update. Thus any particular cell could be fed from no more than once per 100 updates. Organisms were required to consume ten units of resources from the environment before they could reproduce.

In the treatments that included barriers (which block movement), we created 25 pairs of barriers (Fig. S2). Within each pair, one barrier extended north to south, and the other barrier extended east to west, intersecting at their north and west ends, respectively. Intersection points were separated by 50 cells on each axis, leaving 30 cells between the end of a barrier in one pair and the intersection point of the next pair. Pairs were placed in five columns and five rows, with the northern-most row along the northern boundary of the grid (so that, for this row, the north-south barrier extended downward from the northern boundary, while the east-west barrier lay along the boundary) and the western-most column along the western boundary.

17.3.2 Reproduction

Provided that an organism had consumed sufficient resources and was old enough (minimum = 100 updates), reproduction occurred when an organism executed a single reproduction instruction (i.e. organisms used a composite instruction and were not required to copy individual instructions as in traditional configurations of Avida). For each new offspring genome, there was a 25% chance of a single substitution mutation occurring, and 5% chances for single insertion and deletion mutations occurring, independently. Genome lengths were unrestricted. New genetic mutations were suppressed in all reintroduction and competition experiments. Whereas most instructions took 1/30th of an update to execute (= 1 "cycle"), reproduction required a full update to complete.

New organisms were born into the cell faced by their parent. To limit population-size artifacts, populations evolving in the absence of predators were limited to 700 organisms. When a new birth would have caused the population level to exceed this cap, a random organism (other than the parent) was removed from the existing population. Organisms older than 500 updates were also removed.

17.3.3 Predation

Predators attack prey via the execution of a single instruction. If there is a prey in front of the predator and a kill is made, predators consume the prey with a conversion efficiency of 10%; that is, a predator would gain one unit of resource from consuming a prey that had eaten ten units of environmental resources. Reproduction for both predators and prey is limited by resource consumption: the faster an organism gathers food, the sooner it will be able to reproduce. For predators, mutations allowing for more effective location, pursuit, and capture of prey will therefor provide evolutionary advantages. Likewise, mutations in prey that improve foraging efficiency or predator avoidance provide selective advantages. In previous work in Avida, each cell in the world could hold only one organism. Here, however, there is no limit on the number of organisms per cell. Consequently, populations are limited from the bottom-up by resources, by setting explicit population caps, or, for prey, by top-down predation pressure. Because the experimental systems were effectively closed, and in order to allow for consistency in prey densities across trials and treatments, we prevented predator attacks from succeeding if the prey population fell below a minimum threshold of 700. In practice, this means repeated attacks could be required to make a kill. For each failed attack, the targeted prey was "injured" via a 10% reduction of their consumed, stored resource.

17.3.4 Forager Types

All organisms were born in a neutral "juvenile" state. Organisms could then alter their classification to become a predator or prey by executing a specific classifier instruction with the appropriate predator or prey value identifier in the modifying register. Organisms had to be classified as prey to consume environmental resources. Organisms were always classified as predators as soon as they had successfully executed any attack on a prey. Alternatively, organisms could adopt the forager classification of their parent if the parent had executed a "teach" instruction and the (offspring) organism executed a "learn" instruction. While prey could become predatory, predators were prevented from being reclassified as prey during their lifetime. In practice, success in the former was rare once predator and prey behaviors diverged significantly (which occurred early in the evolutionary timelines, see Fig. 17.2 and Fig. S1) and each became more efficient in its own niche.

17.3.5 Sensors

Organisms could evolve the use and control of environmental sensors capable of providing information about objects. Each organism"s area of vision was limited to its front octant out to a distance of 10 cells. Objects in the environment included other organisms, food and, in specified treatments, barriers blocking movement. Walls were also placed around the outer perimeter of the grid world, making the boundary detectable by organisms.

The capacities of the sensors were designed to allow organisms the ability to evolve extensive capacities for sight. Dependent on their evolved behavior, organisms could set and use integer values in four of their internal registers to query sensors for information, specifying: 1) the type of object they were looking for, including predators vs. prey, 2) the maximum distance to look to, 3) whether they were looking for the closest object of that type, or a count of all objects of that type in their visual field, 4) any specific instance of the type of object sought (e.g., a particular known organism). Eight integer outputs were returned by every use of a sensor: 1) the type of object searched for, 2) the distance to the object, or distance used if nothing was visible, 3) whether the closest object or a count of objects was sought, 4) the specific instance of an object that was sought, if specified in the input controls, 5) the count of objects of the correct type seen, 6) the values of the objects seen, 7) the identity of the object seen, 8) in a search for organisms, the type of organism seen (predator vs. prey). In essence, the sensors could become perfect eyes. However, they are useless (and potentially detrimental) unless organisms evolve mechanisms for controlling what information is processed from visual inputs. A complete list of sensor default behaviors is available in the Avida documentation.

17.3.6 Hardware

In Avida, the virtual hardware defines critical aspects of an organism"s construction, e.g., memory registers, potential instructions, and genome execution rules. We used the EX hardware [8], modified to include the eight registers needed for organisms to control their sensors and to allow up to four parallel execution threads. Threads were created if an organism executed a "fork" instruction. Any instruction occurring in the genome between the fork and an "end-thread" instruction were effectively copied to a second genome execution stream. Each thread also maintained its own complement of registers and a single stack (there was also one stack common to all threads). Each cycle, the current instruction for each thread was executed, in the order that the threads were created. Additional instructions were also available for threads to pause their own execution until certain values appeared in the registers of other threads. Beyond new instructions for predation and thread control, the instruction set also included instructions for detecting an organism"s heading (i.e. a compass), rotating multiple times, and rotating until a specific organism (detected via sensors and remembered) came back into view.

17.3.7 Complexity and Intelligence

We measured potential complexity as the mean proportion of per-capita, total lifetime instructions executed that were sensing instruction executions, i.e. the level of information intake [4, 5]. We measured realized behavioral complexity and behavioral intelligence as the mean proportion of instructions that used data originating from the sensors as regulatory or modifying inputs, i.e. the extent to which information was used and incorporated into decisions and actions [26].

17.3.8 Behavioral Exploration

We measured behavioral exploration in an x-y plane of per-capita moves and turns, recorded every 1,000 updates for every population. The travel distance between recorded points was calculated (using the Pythagorean theorem) as the square root of the sum of the squared difference in per-capita turns and the squared difference in per-capita moves. Total explored distance, or path length, for each population was the sum of these distances over the two million updates of evolution (sum of 2,000 distances per population).

17.3.9 Time-shifts

We saved complete records of the genomes and birth locations of all living organisms every 50,000 updates during each evolutionary trial. To limit any potential artifacts related to location within the grid-world or age and developmental state, for all reintroductions, organisms were placed at their original birth location and with all internal states (e.g., memory) reset, as it is in new births, but retaining information about the organisms" parents (e.g., whether the parent had executed a "teach" instruction).

To evaluate changes in prey abilities to avoid predators, each predator population from one million updates was reintroduced, in turn, along with the prey from each time point of the same trial. Likewise, to evaluate changes in predator abilities to catch prey, we reintroduced each saved predator population with the prey population from the middle of their evolutionary timeline. We then measured attack rates as the proportion of all lifetime instructions that were successful attacks for the parents of the predators alive at 1,000 updates post-reintroduction (data from parents are used to allow evaluation over complete lifetimes).

17.3.10 Foraging Decisions

We reintroduced each saved prey population, evolved with and without predators, into predator-free environments and measured mean fitness at 1,000 updates. Fitness in Avida is calculated as lifetime food intake divided by gestation time (in updates). We then altered the sensors so that they would always return signals indicating the equivalent of "no food seen" in response to an organism"s attempts to look for food, and again evaluated fitness in new reintroduction trials of the same source prey populations. Because the only variable changed across these two assays was the ability to acquire and respond to visual information about resources, we used the per-population changes in fitness as our measure of the importance of that knowledge in informing foraging decisions.

17.3.11 Foraging Competitions

To compete the prey coevolved with predators against the prey evolved without predators, each of the final prey populations from the evolutionary trials was paired once with each of the 30 final prey populations from the opposing treatment and reintroduced into a new environment. For each population, we then counted the number of competitions in which its descendants constituted the majority of the final total composite population after 200 generations of competition in environments with 100% and 25% of the resource regrowth rates used in the evolutionary trials.

17.3.12 Software

We used Avida version 2.12 for all experiments. Data were post-processed using Python 2.7.1. Statistical analyses and plotting were conducted in R version 2.15.2 using the ggplot2 and boot libraries.

17.3.13 Author Contributions

A.P.W conceived and conducted the study and prepared the manuscript. L.Z. helped inspire the work and devise experiments. I.D. and C.O. were involved in the study design, implementation, and analyses. All authors discussed the methods and results and edited the manuscript.

17.3.14 Author Information

Avida configuration files, datasets, and analysis scripts have been deposited in the Dryad database. Full results produced by these configuration files (approximately 20 GB) are available upon request. Reprints and permissions information is available at www.nature.com/reprints. The authors declare no competing financial interests. Correspondence and requests for materials should be addressed to C.O. (ofria@msu.edu).

Acknowledgements We thank D.M. Bryson and G. Wright for their assistance in developing the experimental system. This work was supported by the BEACON Center for the Study of Evolution in Action (NSF Cooperative Agreement DBI-0939454), NSF Grant DEB-1655715, and the Michigan State University Institute for Cyber Enabled Research. We especially thank Erik Goodman whose leadership of the BEACON Center has inspired us all.

References

1. Abrams, P.A.: *Adaptive responses of predators to prey and prey to predators: the failure of the arms-race analogy.* Evolution **40**, 1229–1247 (1986)
2. Abrams, P.A.: *Is predator-prey coevolution an arms race?* TREE **1**, 108–110 (1986)
3. Abrams, P.A.: *The evolution of predator-prey interactions: Theory and evidence.* Annual Review of Ecology and Systematics **31**, 79–105 (2000)
4. Adami, C.: *What is complexity?* BioEssays **24**, 1085–1094 (2002)
5. Adami, C., Ofria, C., Collier, T.C.: *Evolution of biological complexity.* PNAS **97**, 4463–4468 (2000)
6. Brodie Jr., E.D., Ridenhour, B.J., Brodie, E.D.I.: *The evolutionary response of predators to dangerous prey: Hotspots and coldspots in the geographic mosaic of coevolution between garter snakes and newts.* Evolution **56**, 2067–2082 (2002)

7. Brodie, E.D.I., Brodie Jr., E.D.: *Predator-prey arms races.* Bioscience **49**, 557–568 (1999)
8. Bryson, D.M., Ofria, C.: (2013). Understanding Evolutionary Potential in Virtual CPU Instruction Set Architectures
9. Buckling, A., Rainey, P.B.: *Antagonistic coevolution between a bacterium and a bacteriophage.* Proc. R. Soc. Lond. B **269**, 931–936 (2002)
10. Cott, H.B.: *Adaptive Coloration in Animals.* Methuen, London (1940)
11. Curio, E.: *The ethology of predation.* Springer Verlag, Berlin (1976)
12. Davies, N.B., Brooke, M.D.J.: *An experimental study of co-evolution between the cuckoo, Cuculus canorus, and its hosts. I. Host egg discrimination.* Animal Ecol. **58**, 207–224 (1989)
13. Dawkins, R., Krebs, J.R.: *Arms races between and within species.* Proc. R. Soc. Lond. B **205**, 489–511 (1979)
14. Dieckmann, U., Marrow, P., Law, R.: *Evolutionary cycles in predator-prey interactions: Population dynamics and the Red Queen.* J. Theor. Biol. **176**, 91–102 (1995)
15. Dietl, G.P.: *Coevolution of a marine gastropod predators and its dangerous bivalve prey.* Biol. J. Linnean Soc. **80**, 409–436 (2003)
16. Dietl, G.P., Kelley, P.H.: *The fossil record of predator-prey arms races: Coevolution and escalation hypotheses.* Paleontological Soc. Papers **8**, 353–374 (2002)
17. Endler, J.A.: *Defense against predators.* In: M.E. Feder, G.V. Lauder (eds.) Predator-Prey Relationships, pp. 109–134. The University of Chicago Press, Chicago (1986)
18. Errington, P.L.: *Predation and vertebrate populations.* Q Rev Biol **21**, 144–177 (1946)
19. Fortuna, M.A., Zaman, L., Wagner. A.P. & Ofria, C.: *Evolving digital ecological networks.* PLoS Comp. Biol. **9**, e1002928 (2013)
20. Franceschi, V.R., Krokene, P., Christiansen, E., Krekling, T.: *Anatomoical and chemical defenses of conifer bark against bark beetles and other pests.* New Phytologist **167**, 353–376 (2005)
21. Gandon, S., Buckling, A., Decaestecker, T., Day, T.: *Host-parasite coevolution and patterns of adaptation across time and space. arms races give way to flutuating selection.* J. Evol. Biol **21**, 1861–1866 (2008)
22. Hall, A.R., Scanlan, P.D., Morgan, A.D., Buckling, A.: *Host-parasite coevolutionary arms races give way to fluctuating selection.* Ecol. Letters **14**, 635–642 (2011)
23. Harper, E.M.: *Dissecting post-Palaeozoic arms races.* Palaeogeogr. Palaeoclimatol. Palaeoecol **232**, 322–343 (2006)
24. Hazen, R.M., Griffin, P.L., Carothers, J.M., Szostak, J.W.: *Functional information and the emergence of biocomplexity.* PNAS **104**, 8574–8581 (2007)
25. Johnson, M.T.J., Stinchombe, J.R.: *An emerging synthesis between community ecology and evolutionary biology.* TREE **22**, 250–257 (2007)
26. Kamil, A.C.A.: *Synthetic approach to the study of animal intelligence.* Nebraska Symposium on Motivation **7**, 257–308 (1987)
27. Langerhans, R.B.: Evolutionary consequences of predation: Avoidance, escape, reproduction, and diversification. In: A.M.T. Elewa (ed.) Predation in Organisms, pp. 177–220. Springer, Berlin, Heidelberg (2007)
28. Lawrence, D., et al.: *Species interactions alter evolutionary responses to a novel environment.* PLoS Biol **10**, e1001330 (2012)
29. Lenski, R.E.: *Coevolution of bacteria and phage: Are the endless cycles of bacterial defenses and phage counterdefenses?* J. Theor. Biol **108**, 319–325 (1984)
30. Lenski, R.E., Ofria, C., Pennock, R.T., Adami, C.: *The evolutionary origin of complex features.* Nature **423**, 139–144 (2003)
31. Lima, S.L.: *Putting predators back into behavioral predator-prey interactions.* TREE **17**, 70–75 (2002)
32. Meyer, J.R., et al.: *Repeatability and contingency in the evolution of a key innovation in phage lambda.* Science **335**, 428–432 (2012)
33. Nuismer, S.L., Thompson, J.N.: *Coevolutionary alternation in antagonistic interactions.* Evolution **60**, 2207–2217 (2006)
34. Ofria, C., Wilke, C.O.A.: *A software platform for research in computational evolutionary biology.* Artificial Life **10**, 191–229 (2004)

35. Pennock, R.T.M.: *Simulations, instantiations, and evidence: The case of digital evolution.* J. Exp. Theor. Art. Int. **19**, 29–42 (2007)
36. Temple, S.A.: *Do predators always capture substandard individuals disproportionately from prey populations?* Ecology **68**, 669–674 (1987)
37. van Valen, L.: *A new evolutionary law.* Evol. Theory **1**, 1–30 (1973)
38. Vermeij, G.J.: *Evolution in the consumer age: Predators and the history of life.* Paleontological Soc. Papers **8**, 375–393 (2002)
39. Vermeij, J.: *The evolutionary interaction among species: Selection, escalation, and coevolution.* Ann. Rev. Ecol. Syst. **25**, 219–236 (1994)
40. Zaman, L., Devangam, S., Ofria, C.: Rapid host-parasite coevolution drives the production and maintenance of diversity in digital organisms. In: N. Krasnogor, P.L. Lanzi (eds.) Proceedings of the 13th Annual Conference on Genetic and Evolutionary Computation (GECCO-2011), pp. 219–226. ACM Press (2011)

17.4 Supplementary Figures

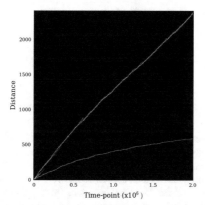

Fig. S1. Predators and prey diverge genotypically as well as behaviorally. Shown is a plot from a sample population of evolving predators and prey. X-axis indicates evolutionary time (in updates). Y-axis indicates the mutational distance for every genotype in the population to the common ancestor. Color indicates number of organisms at that depth and time. Top lineage corresponds with prey. Bottom cluster corresponds with predators. Over the course of 2 million updates of evolution, mutations created significant divergence in predators and prey genetics, as well as "behavioral speciation" (e.g., as in Fig. 2 and Movie S1). Mutations tend to accumulate slower in established predator lineages because foraging inefficiencies across trophic levels slow reproductive output and generation times.

Fig. S2. Predators and prey coevolved in bounded, cell-based grid-worlds. Shown is a sample evolved population of predators (red) and prey (blue) in their 251 X 251 grid-cell environment. Black lines indicate barriers, included in some treatments (as specified in the main text), that block movement (shown here, full sized = 20 cells long on each axis, and one cell wide). Grey to white background illustrates prey forage levels by cell (grey = edible, white = consumed and regrowing). Maximum sight distance was 10 cells.

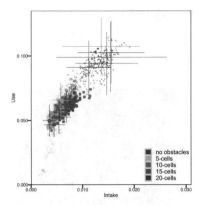

Fig. S3. Behavioral intelligence and complexity did not scale with complexity of the abiotic environment. The complexity of the abiotic environment was adjusted by adding obstacles 5, 10, 15, and 20 cells long to the environment (see Fig. S2). Sensory intake was measured as the ratio of lifetime sensor information intake to total actions taken. Behavioral intelligence was measured as the mean proportion of lifetime decisions that used sensory data about the environment. Error bars illustrate bootstrapped 95% confidence intervals around the means for the final time-point of the evolutionary trials. Points indicate mean within-treatment values at 20,000 time-point (update) intervals. Shading indicates update sampled (lighter = older). Circles indicate means for prey populations evolving without predators (none of which escaped the bottom cluster of low rates of information intake and use). Triangles represent data for prey populations coevolving with predators (all of which reached the top cluster of high complexity and intelligence). While there was no clear pattern of environmental complexity driving the evolution of prey behavioral complexity and intelligence, coevolution with predators consistently increased both measures, both within and across environmental treatments.

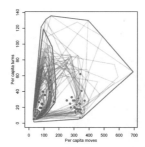

Fig. S4. Prey populations coevolving with predators explore a larger area of behavioral space. While traveling greater distances (Fig. 2), prey coevolving with predators explore larger and more varied areas of behavioral strategy space, with most settling in an area of relatively low turn rates, but high movement rates. Populations coevolving with predators are shown in red. Prey populations evolving in the absence of predators are shown in purple. Points indicate final per-capita move-turn rates for each of the prey populations under each treatment. Dark outlines indicate cumulative convex hulls for all populations of each treatment. Lighter outlines indicate convex hulls for the areas explored by each individual population. Note that three populations coevolving with predators did not escape the low-movement behavioral space, never exploring beyond, or successfully crossing, an apparent behavioral valley bordering the area in which all nave populations started and in which all populations evolved without predators remained (see also Fig. 2).

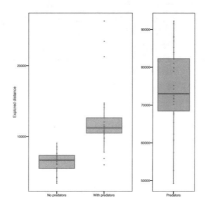

Fig. S5. Predators and their prey travel farther in their explorations of behavioral space than prey populations evolving alone. For each population, the explored distance was measured as the cumulative distance traveled over the plane defined by per-capita executions of moves and turns. Points indicate total explored distance over the full two million updates of evolution. Boxes extend from first to third quartiles. Whiskers extend from the first/third quartiles out to the highest/lowest values within one and half times the distance between the first and third quartiles. Travel distance for prey populations evolving alone and those coevolving with predators are show on the left. Predator exploration of behavioral space, at a different scale, is shown on the right.

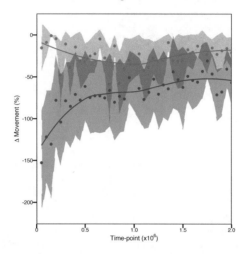

Fig. S6. Prey evolve to respond to predation pressure by increasing movement, but do not increase that response over evolutionary time. Shown (in red) are the mean per-population changes in number of steps taken by prey in environments with predators removed relative to their level of movement in environments with predators included. When predators are removed, prey consistently respond by reducing motility. However, the level of decreased movement does not change over evolutionary time. Therefore, improved anti-predator success (Fig. 2) could not have been reliant on a chase-flee movement arms race. At the same time, predators (blue) reduce levels of movement when introduced into test environments with prey removed, with the magnitude of the change stabilizing over evolutionary time. Populations were drawn from the source populations at time-point intervals of 50,000, and tested with and without the competing species removed. Data shown are from update 1,000 in the test environment. Shaded regions are bootstrapped 95% confidence intervals. Lines are LOESS fits.

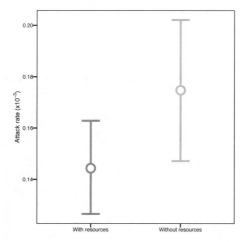

Fig. S7. Prey adaptive use of sensors for finding food improves evolved predator-avoidance skills. Prey evolved in environments without resources are less able to avoid predators, as indicated by attack rates, than those evolved in the presence of resources. After two million updates of evolution, mean predator attack rates on prey evolved in the presence of resource was 0.144 (x10^{-3}), 95% CI: 0.126,0.062, compared to a mean of 0.175 (x10^{-3}), 95% CI: 0.148,0.202, on prey evolved in the absence of resources. Given that prey evolved in the absence of predators are less successful in foraging than prey evolved in the presence of predators (Fig. 4), there appears to be a reciprocal evolutionary relationship in which, as prey become better at sensing and reacting to predators, they also evolve to better sense and react to resources, which further enhances evolving responsiveness to predators. Here, organisms were required to consume 10 units of resource (and prevented from consuming more). Exclusively for these treatments, for every unit collected, the metabolic rate of organisms was increased by an additive factor of one. In the resource environment, as in the main experimental treatments, prey consumed resource by feeding from cells. For the resource free environment used here, "resources" were "consumed" simply by moving, but no resources were removed from the environment. Thus a prey in the resource environment would increase fitness by avoiding predators while also finding and consuming resources. Contrastingly, a prey in the resource-free environment could improve its fitness simply by moving and avoiding predators. Because resources were unlimited in these two environments, prey populations were capped at 800 individuals, and the full population at 1,000. Thus, in order to allow removal of spatially distributed prey food resources, this test required substantial changes to the evolutionary environment and this particular result should be viewed with some caution. Vertical bars indicate bootstrapped 95% confidence intervals around the means (points) for the 30 trials.

17.5 Supplementary Videos

Movie S1. Predators and prey evolve complex and intelligent processes for taking in, processing, and responding to information about their environment. Shown are clones of an evolved predator (red) and prey (blue) pulled from a larger population. For this example, the predators are prevented from killing and eating the prey. While predators have evolved to look for, identify, orient toward, target, chase, and attack individual prey, prey have evolved to consume the food resources they need (grey background; white = consumed) while also avoiding predators. Neither predators nor prey can see behind them and so prey escape from predators, as in nature, is aided by frequent changes in movement directions. Sight distance is limited to 10 cells (steps), so the predators can and do lose sight of prey. Video available via Figshare: https://doi.org/10.6084/m9.figshare.7210355.v1

Chapter 18
A Hologenomic Approach to Animal Behavior*

Kevin R. Theis, Danielle J. Whittaker and Connie A. Rojas

Abstract The hologenome concept of evolution posits that animals and their symbiotic microbes are emergent individuals, or holobionts, exhibiting synergistic phenotypes that are subject to evolutionary forces. Its premises are that interactions between animals and their microbes affect the fitness of holobionts, in both beneficial and deleterious ways, and that microbes and their functional genes can persist across animal host generations with fidelity. Covariance between the host genome and the microbiome can thus be maintained and holobiont phenotypes encoded by the microbiome can be subject to selection and drift within holobiont populations. Animal behavior researchers have historically underappreciated the beneficial effects of symbiotic microbes on animals' behavioral phenotypes, but this is changing. Symbiotic microbes can protect their host animals from predators, increase their foraging efficiencies and reproductive outputs, and mediate their chemical communication systems. The objectives of this chapter are to introduce the hologenome concept of evolution and, within the framework of the concept's premises, to present highlights of the current understanding of how symbiotic microbes contribute to animals' behavioral phenotypes and how animals facilitate transmission of beneficial microbes to their offspring and kin. The chapter concludes with a discussion of how behavioral ecologists, in particular, are well-positioned to evaluate the explanatory value of host-microbial evolutionary models, such as the hologenome concept.

Kevin R. Theis
Department of Biochemistry, Microbiology and Immunology, Wayne State University, Detroit, MI,
e-mail: ktheis@med.wayne.edu

Danielle J. Whittaker · Connie A. Rojas
BEACON Center for the Study of Evolution in Action, Michigan State University, East Lansing, MI

Connie A. Rojas
Department of Integrative Biology, Michigan State University, East Lansing, MI

* This paper was externally peer-reviewed.

© Springer Nature Switzerland AG 2020   247
W. Banzhaf et al. (eds.), *Evolution in Action: Past, Present and Future*,
Genetic and Evolutionary Computation, https://doi.org/10.1007/978-3-030-39831-6_18

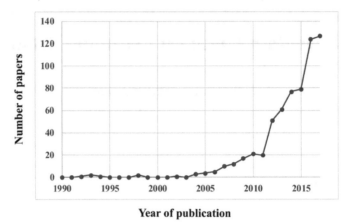

Year of publication

Fig. 18.1: There has been a marked increase in the use of the terms holobiont and hologenome in
the biological literature over the past decade. The frequency of papers published
between 1990 and 2017 with holobiont and/or hologenome as keywords. All papers
were surveyed by searching for the keywords (i.e. topic) holobiont★ or hologenom★
using Thomson Reuters' Web of Science.

18.1 The Hologenome Concept Redefines That Which Constitutes An Individual In Animal Biology

Over the last decade, advances in "omics" technologies have provided unprece-
dented insight into the diverse and dynamic associations that animals have with
symbiotic (i.e., resident) microbes [11, 16, 67]. Most, if not all, exposed anatomical niches of animals (i.e., eyes, ears, nose, skin, and the epithelia of respiratory,
gastrointestinal, urinary and reproductive tracts) are populated by microbes, and it
has been estimated that at least half of the cells and the vast majority of the unique
genes associated with animals' bodies are of microbial origin [88]. Thus, animals'
phenotypes are products of multitudinous biomolecular interactions between host
genomes, the genomes of resident microbial communities, and their shared environments. As a consequence, biologists and philosophers of biology are increasingly
viewing animals as emergent individuals, or holobionts, rather than as autonomous
entities [10, 83, 92]. A *holobiont* is defined as the emergent phenotype of an animal host and its symbiotic microbes at a given point in time, and a *hologenome* is
defined as the genetic content of the host and its microbiome.

The use of the term holobiont was originally applied to corals and their microbial symbionts. Indeed, it was observations of the disease ecology of coral holobionts that led Eugene Rosenberg, Ilana Zilber-Rosenberg, and colleagues to propose the hologenome concept of evolution [80]. However, between 2008 and 2010,
Rosenberg and Zilber-Rosenberg published three essays developing and extending
the hologenome concept to animals and plants in general [81, 82, 101]. Since publication of these essays, the use of the terms holobiont and hologenome in the sci-

entific literature have greatly increased (Figure 18.1). The terms are now widely used in contemporary host biology [10, 83], and international research conferences and workshops on hologenomics have been established (e.g., International Conference on Holobionts, Paris, France in April, 2017 [2]; Interdisciplinary Workshop on Holobionts [Bordeaux, France in November, 2017 [1]).

The hologenome concept is a hypothesis explaining the evolution of animals (and plants) in the context of the microbiome. Specifically, the concept posits that selection can act on transgenerational associations between hosts and their symbiotic microbes, that these networked associations are more numerous and dynamic than has been appreciated previously, and that progress in elucidating the underlying etiologies of holobiont phenotypes may be best achieved through integrated consideration of host, microbial, and environmental features [92]. The hologenome concept of evolution has four premises [81, 101]. First, all animals (at least during their post-embryonic life stages) are populated by communities of microbial symbionts and these associations are often non-random [12]. Microbial symbionts can be permanent or transient and can act in a context-dependent manner as harmful, harmless, or helpful to the host [92]. Second, these associations have fitness consequences for the holobiont. The holobiont may or may not be a primary unit of selection, but selection can nevertheless act on these host-microbial associations [10, 91, 92]. Third, hologenomes can be transgenerational, facilitated by vertical transmission of symbiotic microbes between parents and offspring, horizontal transmission of microbes among kin and social partners, and highly specific host-microbial cross talk that promotes transgenerational coassociation of host and microbial alleles even when individuals must obtain these microbial alleles during development from the environment anew each generation. Fourth, variation in hologenomes can arise through changes in host genomes or in microbiomes (i.e., horizontal gene transfer, microbial immigration / emigration, and microbial replication). Changes to the microbiome can be rapid, thereby enabling holobiont populations to adjust to swiftly changing environments and for newly acquired traits to potentially be passed on to offspring [82].

Evolution via selection is inevitable when heritable phenotypic variation results in differential reproductive success within populations. The hologenome concept is a hypothesis explaining the evolution of animals given that their phenotypes are necessarily products of interactions between their genomes and their microbiomes. In this sense, the hypothesis is host-centric, and this has been a criticism of the hypothesis. Specifically, host-associated microbial populations can often also reside in habitats beyond the host (e.g. *Vibrio fischeri* lives both within the light organ of the bobtail squid, *Euprymna scolopes*, and the water column). Microbial populations are therefore often subject to selective pressures in multiple environments and thus the hologenome concept may be more useful for understanding the evolution of hosts than their microbial populations [27]. Another criticism of the hologenome concept is that there are many different selective pressures acting upon the diverse microbial populations within holobionts and that, without concordance of selective interests among microbial symbionts and between symbionts and their hosts, selection is unlikely to act at the level of the holobiont [27, 70].

The debatable and testable crux of the hypothesis is whether the interactions between host and microbial alleles are of sufficient fitness consequence, and are heritable and/or persistent enough across generations, for selection to act on holobiont populations. The development and testing of quantitative predictions is a critical next step in the refinement and evaluation of the hologenome concept [46, 54, 73, 85]. For example, Kopac and Klassen [54] introduced a model assessing whether microbial symbionts increased their host's realized ecological niche and vice versa, and whether the respective realized niches of hosts and microbial symbionts were dependent upon one another. Similarly, Huitzil et al [46] introduced a model to assess the potential of the evolution of holobionts and observed predicted behaviors, such as reduced conflict of interest facilitated by division of labor. Moving forward, behavioral ecology laboratories, in particular, are well-positioned for testing predictions of the hologenome concept [91]. Specifically, symbiotic microbes can substantially contribute to animals' behavioral phenotypes, protecting their hosts from predators, increasing their foraging efficiencies and reproductive outputs, and mediating their chemical communication systems [5, 33]. These behavioral phenotypes can be clearly defined and are quantifiable, and behavioral ecology investigations are often long-term projects, spanning multiple generations within natural animal populations, wherein metadata are collected on genetic relatedness, parentage, and relative reproductive success [91]. Collection of microbiome data should become a staple of these investigations as well.

18.2 Symbiotic Microbes Can Beneficially Contribute To Animals' Behavioral Phenotypes

Behavior is a primary means for animals to mediate their circumstances within complex physical and social environments, and symbiotic microbes beneficially contribute to animals' behavioral phenotypes in a broad range of animal taxa, including invertebrates, fish, amphibians, reptiles, birds and mammals [5, 21, 33]. Symbiotic microbes have even been implicated in human behavior [25, 96]. Here we highlight examples of how microbes can protect their host animals from predators, increase their foraging efficiencies and reproductive outputs, and mediate their chemical communication systems.

Defense: Symbiotic microbes play pivotal roles in priming their host animals' immune systems, and they routinely protect their hosts from pathogens and parasites through competitive exclusion and the production of antimicrobial compounds [3, 31, 44]. Symbiotic microbes can also directly contribute to their host animals' antipredator behaviors [35].

Many marine animals use counterillumination to avoid predation [47]. For the Hawaiian bobtail squid, *Euprymna scolopes*, counterillumination is provided by a bioluminescent symbiotic bacterium, *Vibrio fischeri* [65, 66]. Squids sequester *V. fischeri* from surrounding seawater into a specialized light organ within their mantle cavity. Importantly, they can do so while effectively excluding other environmental bacteria [65, 66]. As a squid forages in the water column in the evening, its bacterial symbionts luminesce, emitting light downward with the same intensity as the downwelling light from the stars above [48, 65, 66]. This illumination reduces the squid's silhouette, enabling it to forage undetected and thus uninterrupted by predatory fish below.

In contrast, symbiotic microbes afford some coral reef isopods (*Santia* spp.) conspicuous warning coloration [58]. These isopods are covered with a dense lawn of fluorescent red cyanobacteria. Reef fish find these isopods unpalatable, allowing them to forage on sponge colonies in broad daylight. If the isopods are attacked by fish, they are quickly released, a process which they survive. However, if their cyanobacterial lawns are experimentally depleted, these isopods are readily eaten by reef fish, as are members of a sympatric cryptic isopod species that is inhabited by a different, non-fluorescent cyanobacterium. The fluorescent red cyanobacteria thus appear to make their animal host both aposematic and unpalatable [58]. Notably, as with the squid-*Vibrio* example, the symbiotic microbe not only affords its animal host protection from predators, it also increases its foraging efficiency.

Feeding: Most animals harbor symbiotic microbes that enable them to effectively extract nutrients from food [67]. Symbiotic microbes can also contribute to their host animals' foraging and predatory behaviors.

Some animals, most notably fungus-growing ants, termites and beetles, actively cultivate populations of environmental microbes as food sources [71, 90]. Other animals farm the microbes that inhabit their own bodies. In the previous section, we discussed how fluorescent red cyanobacteria protect their host isopods from fish predation [58]. These isopods also use the resident cyanobacteria populations as a food source. By foraging in shallow coral reef waters during the day, the isopods provide their cyanobacteria with ideal conditions for photosynthetic growth. The isopods capitalize on this growth, occasionally scooping the cyanobacterial lawns off their back and antennae and eating them [58].

In addition to affording their host animals protection from predators, bioluminescent bacteria can also directly contribute to their hosts' predatory behavior. Many species of marine fish are bacterially luminous, and some use their luminescence to attract and illuminate prey [28, 41]. The sea urchin cardinalfish, *Siphamia versicolor*, harbors a bioluminescent bacterium, *Photobacterium mandapamensis*, in its specialized ventral light organ [28, 30]. The cardinalfish hunts for zooplankton during twilight, a period of low predation risk for itself when neither diurnal nor nocturnal predators are typically active. The symbiotic bacteria enable the cardinalfish to hunt at twilight by illuminating the areas below and around the fish, thereby attracting zooplankton prey and making them visible to the cardinalfish [29, 30]. Interestingly, two other *Siphamia* species, *S. permutata* and *S. cephalotes*, possess a second light organ in their mouths [34]. These organs are also densely populated by bioluminescent bacteria, and they, like the bacterially luminous light organs of anglerfish, appear to function as lures for prey [34, 41].

Reproduction: Microbes can provide cues that regulate animals' reproductive behaviors. For example, when making settlement decisions, marine planktonic larvae commonly rely on chemical cues from microbial biofilms that cover stable structures on the ocean floor and thus represent reliable environments for larval development [42, 98]. For the gregarious acorn barnacle, *Balanus amphitrite*, larval settlement cues originate from microbes inhabiting the shells of adult conspecifics [8]. These barnacle-associated microbial biofilms differ from those on surrounding rocks, and the barnacles' symbiotic bacteria promote larval settlement whereas rock-associated bacteria do not [8]. To successfully reproduce, acorn barnacles must settle near conspecifics, technically within extensible penile-reach of them, so heeding these symbiotic microbial cues critically contributes to a barnacle's reproductive success [18].

Oviposition cues, which provide information about the quality of potential egg-laying sites to females, are common among insects [21]. Female house flies, *Musca domestica*, are iterative and socially-facilitated ovipositors, and asynchronous oviposition can result in larval age disparity and cannibalism [55]. The bacterium *Klebsiella oxytoca* is a rapidly growing symbiont of house fly eggs, and experimentally increasing its abundance on fresh eggs inhibits oviposition. By paying attention to *Klebsiella*'s chemical cues, and avoiding ovipositing where they are high, female flies can ensure that their offspring develop in environments surrounded only by same-age conspecifics and, therefore, are not cannibalized [55].

The protective effects of symbiotic microbes can increase offspring survival, and many animals exhibit behaviors facilitating those effects. Most birds have pronounced preen glands whose secretions have a number of functions, including pathogen defense [77]. Prior to egg-laying, the preen glands of female European hoopoes, *Upupa epops*, increase greatly in size and begin producing copious amounts of dark and pungent secretion [63]. Females only produce dark secretions while incubating eggs and brooding nestlings; during the rest of the year, their preen gland secretions are white, as are those of male hoopoes [63]. One of the primary members of nesting female hoopoes' preen gland bacterial communities is the bacterium *Enterococcus faecalis*, which produces a bacteriocin, a powerful and broadly effective antimicrobial peptide [61, 62]. While nesting, female hoopoes coat their eggs with preen gland secretions, turning the eggs from pale blue to brown, and thereby protect the eggs and hatchlings from pathogen infection [61, 62].

Female solitary wasps in the genera *Philanthus*, *Trachypus* and *Philanthinus* harbor a bacterium, *Streptomyces philanthi*, in specialized antennal glands [51, 52]. Females deposit the *Streptomyces*-rich secretions from these glands into their subterranean brood cells, wherein their larvae consume and incorporate the microbial symbionts into their own cocoons. *S. philanthi* produces a suite of antimicrobial compounds, affording the cocooning larvae protection from bacterial and fungal pathogens in the soil [50, 52]. As newly adult females emerge from their cocoons, they inoculate their antennal glands with *S. philanthi* from the cocoon itself and the generational process repeats itself [49, 52].

Communication: The potential roles of symbiotic microbes in animal chemical communication have been broadly studied [5, 15, 21]. The fermentation hypothesis for animal chemical communication posits that some components of animals' chemical signals are products of symbiotic microbial metabolism, and that variation in these signals, both among and within species, is due in part to underlying variation in the structure of animals' odor-producing microbial communities. Symbiotic microbes have been suggested to contribute odors signaling a great deal of information about their host animals, including their species and individual identity, group membership, sex, reproductive state, and general quality [5, 21]. The explanatory potential of the hypothesis is actually limited only by the capacity of animals' physiological and social circumstances to modulate their microbial communities in ways that predictably affect their odor profiles. The hypothesis predicts that 1) animals' scent organs should harbor odor-producing microbes, 2) these microbial communities should vary with the host traits being signaled, 3) scent organs' odor profiles should vary with the same traits, 4) the microbial and odor profiles of scent organs should covary, 5) manipulating scent organs' microbial communities should change the organs' odor profiles, and 6) the consequent changes in odor should alter receivers' behavioral responses to the organs' chemical signals [5, 21, 26]. To date, in evaluating the hypothesis, the primary focus has been on aggregating insects and scent-marking mammals.

Desert locusts, *Schistocerca gregaria*, are notorious for congregating into massive swarms as they attempt to locate new food patches [4, 99]. A key component of the aggregation pheromone behind these swarms is guaiacol, a volatile compound that is produced by the locusts' gut bacteria [22, 24, 23]. The fecal pellets of adult wild-type locusts emit guaiacol, but the pellets of germ-free adults do not. Furthermore, when locust gut bacteria, including *Pantoea agglomerans*, *Klebsiella pneumoniae* and *Enterobacter cloacae*, are cultivated on germ-free fecal pellets they produce guaiacol. Given that the production of guaiacol by these bacteria requires plant-derived vanillic acid as a substrate, the aggregation pheromone is a reliable indicator of the presence of conspecifics that have successfully located suitable food patches [22, 24, 23].

African, *Loxodonta africana*, and Asian, *Elephas maximus*, male elephants exhibit periods of heightened aggression and reproductive activity, called musth [17, 38]. During musth, males continuously dribble pungent smelling urine which effectively signals their musth status to male competitors who investigate it [38, 43]. This signaling appears to be microbially-mediated [40]. The urine of males in musth contains higher abundances of alkan-2-ones and alkan-2-ols than urine of non-musth males. While the abundance of these compounds increases in musth urine incubated for several days at ambient temperatures (mimicking dribbled urine marks), they do not change over time if the musth urine is first filter-sterilized to remove urinary microbes. Notably, alkan-2-ones and alkan-2-ols can be produced by bacteria in other environments [87]. Collectively, these data suggest that bacteria in male urine are responsible for signaling musth, and potentially other chemical signals, among elephants [40, 86].

Like many mammals, spotted, *Crocuta crocuta*, and striped, *Hyaena hyaena*, hyenas scent mark with secretions from specialized skin glands [94, 95] (Figure 18.2). Hyenas' subcaudal scent pouches are warm, moist, nutrient-rich, largely anaerobic environments, and are thus highly conducive to the proliferation of symbiotic microbes. Their scent secretions, called "pastes" are populated by diverse communities of fermentative bacteria from well-documented odor-producing clades [94, 95]. The bacterial and volatile fatty acid profiles of pastes differ between

Fig. 18.2: A spotted hyena scent marking (i.e., "pasting") with secretions from its subcaudal scent gland on a grass stalk in the Masai Mara National Reserve, Kenya. (Photo credit: Andy Flies, previously of the Holekamp Laboratory at Michigan State University)

the two hyena species, and among spotted hyenas the two profiles also vary with group membership and with sex and female reproductive state within a social group [14, 94, 95].

Furthermore, the bacterial and odor profiles of hyena pastes consistently covary [95]. These empirical data suggest that symbiotic bacteria mediate hyena chemical signaling behavior. To definitively demonstrate that they do, researchers must next show that manipulating hyenas' paste bacterial communities results in changes to their odor profiles, and that the consequent changes in odor alter receivers' behavioral responses to pastes [5, 26, 95].

18.3 Animals Exhibit Behaviors Promoting The Transmission Of Beneficial Symbiotic Microbes Across Host Generations With Fidelity

For it to have widespread explanatory value, the hologenome concept of evolution requires the coassociation of cooperative symbiotic microbial partners across animal host generations. Transgenerational coassociation can be accomplished through vertical transmission (i.e. from a parent), horizontal transmission (i.e. from kin or another conspecific), or through persistent environmental acquisition. Animal-microbial transmission has been broadly reviewed elsewhere [60, 84], so we will here focus on behaviorally-relevant examples.

Vertical transmission of symbiotic microbes from mother to offspring, via sex cells or inoculation during the egg-laying or birthing process, is ubiquitous among animals [37, 84]. As reviewed above, female hoopoes coat their eggs with microbially laden preen gland secretions, and they continue to produce copious amounts of secretion while brooding and preening their nestlings [61, 62]. In another bird, the dark-eyed junco (*Junco hyemalis*), nestlings share highly similar preen gland and cloacal microbial communities with their mothers, which brooded and fed them, and less similar communities with their fathers, which only fed them, highlighting the role of close physical contact in vertical transmission of microbes [100]. Also recall that female solitary wasps deposit *Streptomyces*-rich secretions into their offspring's brood cells so that the larvae can consume the secretions and incorporate the defensive microbes into the walls of their cocoons [49, 50, 52]. Female bull-headed dung beetles, *Onthophagus taurus*, similarly deposit beneficial microbes in their offspring's brood chambers; their offspring are effectively inoculated as they feed on the brood chamber walls during development [32]. Also, many animals, including insects, reptiles and mammals, engage in coprophagy, whereby offspring inoculate themselves with their parents' beneficial gut microbes by consuming their parents' feces [84].

Horizontal transmission of microbes within and across generations is also common. In scent-marking mammals, juveniles frequently overmark the scent marks of adults, thereby potentially inoculating themselves with scent gland-specific microbes [5, 93]. Adults can also share microbes through overmarking or by marking each other, as in the allomarking behavior of European badgers, *Meles meles* [13]. In multiple non-human primate species, social networks shape gut microbiomes, such that individuals that groom each other more frequently have more similar microbiomes than those that interact less often [69, 75, 97]. Sociality itself has even been suggested as an adaptation for promoting the horizontal transmission of beneficial microbes [59]. For example, female four-toed salamanders, *Hemidactylum scutatum*, can build solitary nests or they can nest communally with other females [9]. Those that nest communally increase the likelihood that their embryos become exposed to and are populated by antifungal symbiotic bacteria [9]. Microbes have even been proposed to facilitate the evolution of host altruistic behavior: if microbes can manipulate their host's behavior, it would be advantageous to manipulate the host into increasing the survival and reproduction of another host recipient of vertically or horizontally transmitted related microbes [57]. This effect has been demonstrated recently through computational population genetics modeling – the models indicated that microbe-induced altruism can evolve within mixed host populations even without repeated interactions or individual recognition among hosts [57]. These results provide intriguing possibilities for empirical study.

Environmental transmission entails the acquisition of symbiotic microbes anew through a conduit each generation. The microbially-luminous bobtail squid and cardinalfish are excellent examples. Adult bobtail squid "vent" their *Vibrio* symbionts into the water column daily. Juvenile squid have morphological and immunological adaptations enabling them to sequester these specific *Vibrio*, while excluding other environmental microbes, from the seawater [65, 66, 68]. Juvenile sea urchin cardinalfish likewise selectively acquire their *Photobacterium* symbionts from seawater rather than from their mothers or their mouth-brooding fathers [29]. Notably, in the context of the hologenome concept, microbial populations transmitted between hosts through the environment, as opposed to those limited to and directly transmitted between hosts' bodies, are less likely to be coassociated with fidelity across host generations and are more likely to replicate independent of hosts (thus being influenced by selective pressures external to the host). Therefore, if selection is to operate on the holobiont as a unit, it is most likely to occur in cases where the microbiome is largely vertically inherited and least likely to occur in cases where the microbiome is obtained from the broader environment [27, 54].

18.4 Animal Behavior Studies Provide Opportune Forums For Evaluating Animal-microbial Evolutionary Models, Such As The Hologenome Concept

We have reviewed how microbes can protect their host animals from predators, increase their foraging efficiencies and reproductive outputs, and mediate their chemical communication systems, and we have discussed animal behavioral mechanisms that ensure beneficial symbiotic microbes are transmitted across generations with fidelity. Nevertheless, historically, these phenomena have not been considered by the animal behavior research community (Table 18.1). Of the more than 25,000 original data papers published in behavioral journals over the last three decades, less than 1% considered the influence of symbiotic microbes on animal behavior. Of those that did, almost all (93.1%) considered how animals avoid being infected with pathogens, cope once they are infected, or make mate choice decisions based on their potential partners' infection and immune status (Table 18.1). Only nine of the more than 25,000 papers surveyed considered the potential beneficial contributions of symbiotic microbes to animals' behavioral phenotypes or the behaviors animals engage in to facilitate the transmission of beneficial symbiotic microbes among social partners and kin [13, 36, 45, 53, 55, 56, 64, 74, 78].

Fortunately, the animal behavior research community has begun showing increasing interest in their subjects' symbiotic microbes, including the beneficial ones [5, 33]. In fact, the nine papers from behavior journals that considered beneficial symbiotic microbes were published since the year 2003. This increase coincides with a growing interest in beneficial microbes in general [11, 67], and it is fueled in large part by rapidly developing tools and analytical approaches for effectively characterizing animals' microbial communities, including their metagenomes and metabolomes, in different developmental and social contexts both within the laboratory and in the host animals' natural habitats [16, 7]. As these tools and approaches become more accessible to the animal behavior research community, the scope and depth of our collective understanding of symbiotic microbial contributions to animal behavioral phenotypes will certainly improve [5, 33].

Table 18.1: The beneficial contributions of symbiotic microbes to animals' behavioral phenotypes remain largely unexplored.

⋆Manuscripts do not include reviews, perspectives, non-data commentaries, or papers validating methodology (e.g. statistical approaches). As our specific interest was the endo- and ecto-microbial symbionts of animal hosts, we did not include manuscripts on leaf-cutting ants or other fungal gardening insects, unless the study addressed the transmission of microbial symbionts via ant bodies or insect pathogen exposure. All journal issues were surveyed by searching for the keywords virus⋆, bacteri⋆, archae⋆, fung⋆, protist⋆, and microb⋆ using Thomson Reuters' Web of Science. The abstracts of identified manuscripts were then manually reviewed to determine if the study addressed host-microbial symbiosis.

† *Acta Ethologica* and *Behavioral Ecology* began publishing in 1999 and 1990, respectively.

‡ Symbiont here refers to resident microbes regardless the interaction type (e.g. antagonism, commensalism, mutualism).

Behavioral journal	Years (manuscripts⋆) surveyed†	Effects of symbiotic‡ microbes on animal behavior	Contributions of symbiotic microbes to animal behavior
Acta Ethologica	1999 – 2017 (351)	0.85 % (3)	0.00 % (0)
Animal Behaviour	1988 – 2017 (8391)	0.36 % (30)	0.02 % (2)
Behaviour	1988 – 2017 (8391)	0.13 % (3)	0.04 % (1)
Behavioral Ecology	1990 – 2017 (3273)	0.79 % (26)	0.03 % (1)
Behavioral Ecology and Sociobiology	1988 – 2017 (4216)	1.02 % (43)	0.05 % (2)
Behavioural Processes	1988 – 2017 (2936)	0.34 % (10)	0.10 % (3)
Ethology	1988 – 2017 (2775)	0.40 %(11)	0.00 % (0)
Journal of Ethology	1988 – 2017 (925)	0.43 % (4)	0.00 % (0)
TOTAL	**1988 – 2017 (25137)**	**0.52 % (130)**	**0.04 % (9)**

Moving forward, there are several noteworthy opportunities for additional behaviorally-relevant, pioneering animal-microbial research. The first opportunity is elucidating the potential benefits of symbiotic microbes to animal learning. One of the most exciting lines of inquiry in animal biology is the emerging concept of the "microbiota-gut-brain axis" which posits that variation in animals' gut microbes can affect brain development and behavior, potentially even the capacity for learning [19, 20]. For example, consuming *Mycobacterium vaccae* improves performance in spatial learning tasks by laboratory mice [64]. A second opportunity is determining the degree to which beneficial microbes inform animals' mate choice decisions [5]. Behavioral researchers have considered whether pathogens provide animals with cues about the infection and immune status of prospective mates [79], but new data suggest that animals can also heed cues from beneficial microbes in making mate choice decisions [89]. Another opportunity is further pursuing two phenomena at the core of the hologenome concept of evolution: cooperation and the mitigation of cheating [39, 84, 6]. These phenomena have been extensively discussed by behavioral scientists [76], but the complexity of animal-microbial functional integration should catalyze debate on the evolution of interkingdom partnerships and the mitigation of cheating within them. Several potential mechanisms for regulating cheating within holobionts have been proposed [84], but one particularly germane to animal behavior, and for which some preliminary evidence exists, is the exercise of partner choice by the host animal through its immune system [51, 66, 72, 84]. This hypothesis should be tested using an array of animal taxa.

18.5 Conclusion

Most animal biologists consider the individual to be the unit of selection in animal evolution. The hologenome concept of evolution redefines that which constitutes an individual, by positing that animals and their symbiotic microbes are emergent phenotypes, or holobionts, upon which selection can act. Holobionts exhibit complex functional integration, with symbiotic microbes critically contributing to animal nutrition and immune health, the proper development and functioning of animal tissues, and animals' behavioral phenotypes. We have highlighted how symbiotic microbes can protect their host animals from predators, increase their foraging efficiencies and reproductive outputs, and mediate their chemical communication systems. We also described behavioral mechanisms that animals use to ensure the effective transmission of beneficial microbes from themselves to their offspring, and that juvenile animals employ when their beneficial microbes need to be acquired from environmental sources. Although the beneficial contributions of symbiotic microbes to animal behavior have been largely unconsidered, behavioral researchers have begun to show an increased interest in the functional and evolutionary significance of beneficial microbes, catalyzed by rapidly developing tools and analytical approaches for effectively studying microbial communities in their natural habitats, including animals' bodies. The beneficial contributions of symbiotic microbes to animal behavior are potentially widespread and dynamic. Moving forward, animal behavior researchers will be in a prime position to elucidate animal-microbial evolutionary processes and to evaluate the explanatory potential of the hologenome concept.

Acknowledgements This work was funded by the National Science Foundation (IOS0920505) and the BEACON Center for the Study of Evolution in Action (DBI0939454).

References

1. Interdisciplinary workshop on holobionts. https://www.immuconcept.org/erc-idem/holobiont-workshop
2. International conference on holobionts. https://symposium.inra.fr/holobiont-paris2017
3. Abt, M.C., Artis, D.: *The dynamic influence of commensal bacteria on the immune response to pathogens.* Current Opinion in Microbiology **16**(1), 4–9 (2013)
4. Anstey, M.L., Rogers, S.M., Ott, S.R., Burrows, M., Simpson, S.J.: *Serotonin mediates behavioral gregarization underlying swarm formation in desert locusts.* Science **323**(5914), 627–630 (2009)
5. Archie, E.A., Theis, K.R.: *Animal behaviour meets microbial ecology.* Animal Behaviour **82**(3), 425–436 (2011)
6. van Baalen M: The unit of adaptation, the emergence of individuality, and the loss of evolutionary sovereignty. In: F. Bouchard and P. Huneman (eds.) From Groups to Individuals: Evolution and Emerging Individuality, pp. 117–140. MIT Press (2013)
7. van Baarlen, P., Kleerebezem, M., Wells, J.M.: *Omics approaches to study host–microbiota interactions.* Current Opinion in Microbiology **16**(3), 270–277 (2013)
8. Bacchetti De Gregoris, T., Khandeparker, L., Anil, A., Mesbahi, E., Burgess, J., Clare, A.: *Characterisation of the bacteria associated with barnacle, Balanus amphitrite, shell and their role in gregarious settlement of cypris larvae.* Journal of Experimental Marine Biology and Ecology **413**, 7–12 (2012)
9. Banning, J.L., Weddle, A.L., III, G.W.W., Simon, M.A., Lauer, A., Walters, R.L., Harris, R.N.: *Antifungal skin bacteria, embryonic survival, and communal nesting in four-toed salamanders, hemidactylium scutatum.* Oecologia **156**(2), 423–429 (2008)
10. Bordenstein, S.R., Theis, K.R.: *Host biology in light of the microbiome: Ten principles of holobionts and hologenomes.* PLOS Biology **13**(8), e1002226 (2015)
11. Bosch, T.C., McFall-Ngai, M.J.: *Metaorganisms as the new frontier.* Zoology **114**(4), 185–190 (2011)
12. Brooks, A.W., Kohl, K.D., Brucker, R.M., van Opstal, E.J., Bordenstein, S.R.: *Phylosymbiosis: Relationships and functional effects of microbial communities across host evolutionary history.* PLOS Biology **14**(11), e2000225 (2016)
13. Buesching CD Stopka P, M.D.: *The social function of allo-marking in the European badger (Meles meles).* Behaviour **140**(8-9), 965–980 (2003)
14. Burgener, N., East, M.L., Hofer, H., Dehnhard, M.: Do spotted hyena scent marks code for clan membership? In: Chemical Signals in Vertebrates 11, pp. 169–177. Springer New York (2008)
15. Carthey, A.J., Gillings, M.R., Blumstein, D.T.: *The extended genotype: Microbially mediated olfactory communication.* Trends in Ecology & Evolution **33**(11), 885–894 (2018)
16. Chaston, J., Douglas, A.: *Making the most of "omics" for symbiosis research.* The Biological Bulletin **223**(1), 21–29 (2012)
17. Chelliah, K., Sukumar, R.: *The role of tusks, musth and body size in male–male competition among Asian elephants, Elephas maximus.* Animal Behaviour **86**(6), 1207–1214 (2013)
18. Clare, A.S.: Toward a characterization of the chemical cue to barnacle gregariousness. In: Chemical Communication in Crustaceans, pp. 431–450. Springer New York (2010)
19. Collins, S.M., Surette, M., Bercik, P.: *The interplay between the intestinal microbiota and the brain.* Nature Reviews Microbiology **10**(11), 735–742 (2012)
20. Cryan, J.F., Dinan, T.G.: *Mind-altering microorganisms: The impact of the gut microbiota on brain and behaviour.* Nature Reviews Neuroscience **13**(10), 701–712 (2012)
21. Davis, T.S., Crippen, T.L., Hofstetter, R.W., Tomberlin, J.K.: *Microbial volatile emissions as insect semiochemicals.* Journal of Chemical Ecology **39**(7), 840–859 (2013)
22. Dillon, R., Charnley, K.: *Mutualism between the desert locust Schistocerca gregaria and its gut microbiota.* Research in Microbiology **153**(8), 503–509 (2002)
23. Dillon, R., Vennard, C., Charnley, A.: *A note: Gut bacteria produce components of a locust cohesion pheromone.* Journal of Applied Microbiology **92**(4), 759–763 (2002)

24. Dillon, R.J., Vennard, C.T., Charnley, A.K.: *Exploitation of gut bacteria in the locust.* Nature **403**(6772), 851–851 (2000)

25. Dinan, T.G., Stilling, R.M., Stanton, C., Cryan, J.F.: *Collective unconscious: How gut microbes shape human behavior.* Journal of Psychiatric Research **63**, 1–9 (2015)

26. Douglas, A.E., Dobson, A.J.: *New synthesis: Animal communication mediated by microbes: Fact or fantasy?* Journal of Chemical Ecology **39**(9), 1149–1149 (2013)

27. Douglas, A.E., Werren, J.H.: *Holes in the hologenome: Why host-microbe symbioses are not holobionts.* mBio **7**(2), 7 (2016)

28. Dunlap, P.V., Ast, J.C., Kimura, S., Fukui, A., Yoshino, T., Endo, H.: *Phylogenetic analysis of host-symbiont specificity and codivergence in bioluminescent symbioses.* Cladistics **23**(5), 507–532 (2007)

29. Dunlap, P.V., Gould, A.L., Wittenrich, M.L., Nakamura, M.: *Symbiosis initiation in the bacterially luminous sea urchin cardinalfish Siphamia versicolor.* Journal of Fish Biology **81**(4), 1340–1356 (2012)

30. Dunlap, P.V., Nakamura, M.: *Functional morphology of the luminescence system of Siphamia versicolor (Perciformes: Apogonidae), a bacterially luminous coral reef fish.* Journal of Morphology **272**(8), 897–909 (2011)

31. Erturk-Hasdemir, D., Kasper, D.L.: *Resident commensals shaping immunity.* Current Opinion in Immunology **25**(4), 450–455 (2013)

32. Estes, A.M., Hearn, D.J., Snell-Rood, E.C., Feindler, M., Feeser, K., Abebe, T., Hotopp, J.C.D., Moczek, A.P.: *Brood ball-mediated transmission of microbiome members in the dung beetle, Onthophagus taurus (Coleoptera: Scarabaeidae).* PLoS ONE **8**(11), e79061 (2013)

33. Ezenwa, V.O., Gerardo, N.M., Inouye, D.W., Medina, M., Xavier, J.B.: *Animal behavior and the microbiome.* Science **338**(6104), 198–199 (2012)

34. Fishelson, L., Gon, O., Goren, M., Ben-David-Zaslow, R.: *The oral cavity and bioluminescent organs of the cardinal fish species Siphamia permutata and S. cephalotes (Perciformes, Apogonidae).* Marine Biology **147**(3), 603–609 (2005)

35. Flórez, L.V., Biedermann, P.H.W., Engl, T., Kaltenpoth, M.: *Defensive symbioses of animals with prokaryotic and eukaryotic microorganisms.* Natural Product Reports **32**(7), 904–936 (2015)

36. Foltan, P., Puza, V.: *To complete their life cycle, pathogenic nematode–bacteria complexes deter scavengers from feeding on their host cadaver.* Behavioural Processes **80**(1), 76–79 (2009)

37. Funkhouser, L.J., Bordenstein, S.R.: *Mom knows best: The universality of maternal microbial transmission.* PLoS Biology **11**(8), e1001631 (2013)

38. Ganswindt, A., Rasmussen, H.B., Heistermann, M., Hodges, J.K.: *The sexually active states of free-ranging male African elephants (Loxodonta africana): Defining musth and non-musth using endocrinology, physical signals, and behavior.* Hormones and Behavior **47**(1), 83–91 (2005)

39. Gardner, A.: Adaptation of individuals and groups. In: F. Bouchard and P. Huneman (eds.) From Groups to Individuals: Evolution and Emerging Individuality, pp. 99–116. MIT Press (2013)

40. Goodwin, T.E., Broederdorf, L.J., Burkert, B.A., Hirwa, I.H., Mark, D.B., Waldrip, Z.J., Kopper, R.A., Sutherland, M.V., Freeman, E.W., Hollister-Smith, J.A., Schulte, B.A.: *Chemical signals of elephant musth: Temporal aspects of microbially-mediated modifications.* Journal of Chemical Ecology **38**(1), 81–87 (2012)

41. Haddock, S.H., Moline, M.A., Case, J.F.: *Bioluminescence in the sea.* Annual Review of Marine Science **2**(1), 443–493 (2010)

42. Hadfield, M.G.: *Biofilms and marine invertebrate larvae: What bacteria produce that larvae use to choose settlement sites.* Annual Review of Marine Science **3**(1), 453–470 (2011)

43. Hollister-Smith, J.A., Alberts, S.C., Rasmussen, L.: *Do male African elephants, Loxodonta africana, signal musth via urine dribbling?* Animal Behaviour **76**(6), 1829–1841 (2008)

44. Hooper, L.V., Littman, D.R., Macpherson, A.J.: *Interactions between the microbiota and the immune system.* Science **336**(6086), 1268–1273 (2012)

45. Hosokawa, T., Hironaka, M., Mukai, H., Inadomi, K., Suzuki, N., Fukatsu, T.: *Mothers never miss the moment: A fine-tuned mechanism for vertical symbiont transmission in a subsocial insect.* Animal Behaviour **83**(1), 293–300 (2012)
46. Huitzil, S., Sandoval-Motta, S., Frank, A., Aldana, M.: *Modeling the role of the microbiome in evolution.* Frontiers in Physiology **9**, 1836 (2018)
47. Johnsen, S., Widder, E.A., Mobley, C.D.: *Propagation and perception of bioluminescence: Factors affecting counterillumination as a cryptic strategy.* The Biological Bulletin **207**(1), 1–16 (2004)
48. Jones, B.W., Nishiguchi, M.K.: *Counterillumination in the Hawaiian bobtail squid, Euprymna scolopes Berry (Mollusca: Cephalopoda).* Marine Biology **144**(6), 1151–1155 (2004)
49. Kaltenpoth, M., Goettler, W., Koehler, S., Strohm, E.: *Life cycle and population dynamics of a protective insect symbiont reveal severe bottlenecks during vertical transmission.* Evolutionary Ecology **24**(2), 463–477 (2009)
50. Kaltenpoth, M., Gttler, W., Herzner, G., Strohm, E.: *Symbiotic bacteria protect wasp larvae from fungal infestation.* Current Biology **15**(5), 475–479 (2005)
51. Kaltenpoth, M., Roeser-Mueller, K., Koehler, S., Peterson, A., Nechitaylo, T.Y., Stubblefield, J.W., Herzner, G., Seger, J., Strohm, E.: *Partner choice and fidelity stabilize coevolution in a cretaceous-age defensive symbiosis.* Proceedings of the National Academy of Sciences **111**(17), 6359–6364 (2014)
52. Kaltenpoth, M., Yildirim, E., Gürbüz, M.F., Herzner, G., Strohm, E.: *Refining the roots of the beewolf-streptomyces symbiosis: Antennal symbionts in the rare genus Philanthinus (Hymenoptera, Crabronidae).* Applied and Environmental Microbiology **78**(3), 822–827 (2011)
53. Kohoutová, D., Rubešová, A., Havlíček, J.: *Shaving of axillary hair has only a transient effect on perceived body odor pleasantness.* Behavioral Ecology and Sociobiology **66**(4), 569–581 (2011)
54. Kopac, S.M., Klassen, J.L.: *Can they make it on their own? Hosts, microbes, and the holobiont niche.* Frontiers in Microbiology **7**, 6 (2016)
55. Lam, K., Babor, D., Duthie, B., Babor, E.M., Moore, M., Gries, G.: *Proliferating bacterial symbionts on house fly eggs affect oviposition behaviour of adult flies.* Animal Behaviour **74**(1), 81–92 (2007)
56. Leclaire, S., Nielsen, J.F., Drea, C.M.: *Bacterial communities in meerkat anal scent secretions vary with host sex, age, and group membership.* Behavioral Ecology **25**(4), 996–1004 (2014)
57. Lewin-Epstein, O., Aharonov, R., Hadany, L.: *Microbes can help explain the evolution of host altruism.* Nature Communications **8**, 14040 (2017)
58. Lindquist, N., Barber, P.H., Weisz, J.B.: *Episymbiotic microbes as food and defence for marine isopods: Unique symbioses in a hostile environment.* Proceedings of the Royal Society B: Biological Sciences **272**(1569), 1209–1216 (2005)
59. Lombardo, M.P.: *Access to mutualistic endosymbiotic microbes: An underappreciated benefit of group living.* Behavioral Ecology and Sociobiology **62**(4), 479–497 (2007)
60. Mandel, M.J.: *Models and approaches to dissect host–symbiont specificity.* Trends in Microbiology **18**(11), 504–511 (2010)
61. Martin-Platero, A.M., Valdivia, E., Ruiz-Rodriguez, M., Soler, J.J., Martin-Vivaldi, M., Maqueda, M., Martinez-Bueno, M.: *Characterization of antimicrobial substances produced by Enterococcus faecalis MRR 10-3, isolated from the uropygial gland of the hoopoe (Upupa epops).* Applied and Environmental Microbiology **72**(6), 4245–4249 (2006)
62. Martín-Vivaldi, M., Peña, A., Peralta-Sánchez, J.M., Sánchez, L., Ananou, S., Ruiz-Rodríguez, M., Soler, J.J.: *Antimicrobial chemicals in hoopoe preen secretions are produced by symbiotic bacteria.* Proceedings of the Royal Society B: Biological Sciences **277**(1678), 123–130 (2010)
63. Martín-Vivaldi, M., Ruiz-Rodríguez, M., Soler, J.J., Peralta-Sánchez, J.M., Méndez, M., Valdivia, E., Martín-Platero, A.M., Martínez-Bueno, M.: *Seasonal, sexual and developmental differences in hoopoe Upupa epops preen gland morphology and secretions: Evidence for a role of bacteria.* Journal of Avian Biology **40**(2), 191–205 (2009)

64. Matthews, D.M., Jenks, S.M.: *Ingestion of Mycobacterium vaccae decreases anxiety-related behavior and improves learning in mice*. Behavioural Processes **96**, 27–35 (2013)

65. McFall-Ngai, M.: Host-microbe symbiosis: The squid-*vibrio* association—a naturally occurring, experimental model of animal/bacterial partnerships. In: Advances in Experimental Medicine and Biology, pp. 102–112. Springer New York (2008)

66. McFall-Ngai, M.: *Divining the essence of symbiosis: Insights from the squid-Vibrio model.* PLoS Biology **12**(2), e1001783 (2014)

67. McFall-Ngai, M., Hadfield, M.G., Bosch, T.C.G., Carey, H.V., Domazet-Lošo, T., Douglas, A.E., Dubilier, N., Eberl, G., Fukami, T., Gilbert, S.F., Hentschel, U., King, N., Kjelleberg, S., Knoll, A.H., Kremer, N., Mazmanian, S.K., Metcalf, J.L., Nealson, K., Pierce, N.E., Rawls, J.F., Reid, A., Ruby, E.G., Rumpho, M., Sanders, J.G., Tautz, D., Wernegreen, J.J.: *Animals in a bacterial world, a new imperative for the life sciences*. Proceedings of the National Academy of Sciences **110**(9), 3229–3236 (2013)

68. McFall-Ngai, M., Heath-Heckman, E.A., Gillette, A.A., Peyer, S.M., Harvie, E.A.: *The secret languages of coevolved symbioses: Insights from the Euprymna scolopes–Vibrio fischeri symbiosis*. Seminars in Immunology **24**(1), 3–8 (2012)

69. Moeller, A.H., Foerster, S., Wilson, M.L., Pusey, A.E., Hahn, B.H., Ochman, H.: *Social behavior shapes the chimpanzee pan-microbiome*. Science Advances **2**(1), e1500997 (2016)

70. Moran, N.A., Sloan, D.B.: *The hologenome concept: Helpful or hollow?* PLoS Biology **13**(12), e1002311 (2015)

71. Mueller, U.G., Gerardo, N.M., Aanen, D.K., Six, D.L., Schultz, T.R.: *The evolution of agriculture in insects*. Annual Review of Ecology, Evolution, and Systematics **36**(1), 563–595 (2005)

72. Nyholm, S.V., Graf, J.: *Knowing your friends: Invertebrate innate immunity fosters beneficial bacterial symbioses*. Nature Reviews Microbiology **10**(12), 815–827 (2012)

73. Osmanovic, D., Kessler, D.A., Rabin, Y., Soen, Y.: *Darwinian selection of host and bacteria supports emergence of Lamarckian-like adaptation of the system as a whole*. Biology Direct **13**(1), 13 (2018)

74. Pechova, H., Foltan, P.: *The parasitic nematode Phasmarhabditis hermaphrodita defends its slug host from being predated or scavenged by manipulating host spatial behaviour*. Behavioural Processes **78**(3), 416–420 (2008)

75. Perofsky, A.C., Lewis, R.J., Abondano, L.A., Fiore, A.D., Meyers, L.A.: *Hierarchical social networks shape gut microbial composition in wild Verreaux's sifaka*. Proceedings of the Royal Society B: Biological Sciences **284**(1868), 20172274 (2017)

76. Raihani, N.J., Thornton, A., Bshary, R.: *Punishment and cooperation in nature*. Trends in Ecology & Evolution **27**(5), 288–295 (2012)

77. Rajchard, J.: *Biologically active substances of bird skin: A review*. Veterinarni Medicina **55**, 413–421 (2010)

78. Rittschof, C.C., Pattanaik, S., Johnson, L., Matos, L.F., Brusini, J., Wayne, M.L.: *Sigma virus and male reproductive success in Drosophila melanogaster*. Behavioral Ecology and Sociobiology **67**(4), 529–540 (2012)

79. Roberts, M., Buchanan, K., Evans, M.: *Testing the immunocompetence handicap hypothesis: A review of the evidence*. Animal Behaviour **68**(2), 227–239 (2004)

80. Rosenberg, E., Koren, O., Reshef, L., Efrony, R., Zilber-Rosenberg, I.: *The role of microorganisms in coral health, disease and evolution*. Nature Reviews Microbiology **5**(5), 355–362 (2007)

81. Rosenberg, E., Sharon, G., Atad, I., Zilber-Rosenberg, I.: *The evolution of animals and plants via symbiosis with microorganisms*. Environmental Microbiology Reports **2**(4), 500–506 (2010)

82. Rosenberg, E., Sharon, G., Zilber-Rosenberg, I.: *The hologenome theory of evolution contains Lamarckian aspects within a Darwinian framework*. Environmental Microbiology **11**(12), 2959–2962 (2009)

83. Rosenberg, E., Zilber-Rosenberg, I.: *The hologenome concept of evolution after 10 years.* Microbiome **6**(1), 14 (2018)

84. Rosenberg E, Zilber-Rosenberg, I.: *The Hologenome Concept: Human, Animal and Plant Microbiota*. Springer, Switzerland (2013)

85. Roughgarden, J., Gilbert, S.F., Rosenberg, E., Zilber-Rosenberg, I., Lloyd, E.A.: *Holobionts as units of selection and a model of their population dynamics and evolution*. Biological Theory **13**(1), 44–65 (2017)

86. Schulte, B.A., Freeman, E.W., Goodwin, T.E., Hollister-Smith, J., Rasmussen, L.E.L.: *Honest signalling through chemicals by elephants with applications for care and conservation*. Applied Animal Behaviour Science **102**(3-4), 344–363 (2007)

87. Schulz, S., Dickschat, J.S.: *Bacterial volatiles: The smell of small organisms*. Natural Product Reports **24**(4), 814 (2007)

88. Sender, R., Fuchs, S., Milo, R.: *Revised estimates for the number of human and bacteria cells in the body*. PLoS Biology **14**(8), e1002533 (2016)

89. Sharon, G., Segal, D., Ringo, J.M., Hefetz, A., Zilber-Rosenberg, I., Rosenberg, E.: *Commensal bacteria play a role in mating preference of Drosophila melanogaster*. Proceedings of the National Academy of Sciences **107**(46), 20051–20056 (2010)

90. Six, D.L.: *The bark beetle holobiont: Why microbes matter*. Journal of Chemical Ecology **39**(7), 989–1002 (2013)

91. Theis, K.R.: *Hologenomics: Systems-level host biology*. mSystems **3**(2), 5 (2018)

92. Theis, K.R., Dheilly, N.M., Klassen, J.L., Brucker, R.M., Baines, J.F., Bosch, T.C.G., Cryan, J.F., Gilbert, S.F., Goodnight, C.J., Lloyd, E.A., Sapp, J., Vandenkoornhuyse, P., Zilber-Rosenberg, I., Rosenberg, E., Bordenstein, S.R.: *Getting the hologenome concept right: An eco-evolutionary framework for hosts and their microbiomes*. mSystems **1**(2), 6 (2016)

93. Theis, K.R., Heckla, A.L., Verge, J.R., Holekamp, K.E.: The ontogeny of pasting behavior in free-living spotted hyenas, *Crocuta crocuta*. In: J. Hurst, R. Beynon, S. Roberts, T. Wyatt (eds.) Chemical Signals in Vertebrates 11, pp. 179–187. Springer New York (2008)

94. Theis, K.R., Schmidt, T.M., Holekamp, K.E.: *Evidence for a bacterial mechanism for group-specific social odors among hyenas*. Scientific Reports **2**(1), 615 (2012)

95. Theis, K.R., Venkataraman, A., Dycus, J.A., Koonter, K.D., Schmitt-Matzen, E.N., Wagner, A.P., Holekamp, K.E., Schmidt, T.M.: *Symbiotic bacteria appear to mediate hyena social odors*. Proceedings of the National Academy of Sciences **110**(49), 19832–19837 (2013)

96. Tillisch, K., Mayer, E.A., Gupta, A., Gill, Z., Brazeilles, R., Nevé, B.L., van Hylckama Vlieg, J.E., Guyonnet, D., Derrien, M., Labus, J.S.: *Brain structure and response to emotional stimuli as related to gut microbial profiles in healthy women*. Psychosomatic Medicine **79**(8), 905–913 (2017)

97. Tung, J., Barreiro, L.B., Burns, M.B., Grenier, J.C., Lynch, J., Grieneisen, L.E., Altmann, J., Alberts, S.C., Blekhman, R., Archie, E.A.: *Social networks predict gut microbiome composition in wild baboons*. eLife **4**, 05224 (2015)

98. Wahl, M., Goecke, F., Labes, A., Dobretsov, S., Weinberger, F.: *The second skin: Ecological role of epibiotic biofilms on marine organisms*. Frontiers in Microbiology **3**, 292 (2012)

99. Wertheim, B., van Baalen, E.J.A., Dicke, M., Vet, L.E.: *Pheromone-mediated aggregation in nonsocial arthropods: An evolutionary ecological perspective*. Annual Review of Entomology **50**(1), 321–346 (2005)

100. Whittaker, D.J., Gerlach, N.M., Slowinski, S.P., Corcoran, K.P., Winters, A.D., Soini, H.A., Novotny, M.V., Ketterson, E.D., Theis, K.R.: *Social environment has a primary influence on the microbial and odor profiles of a chemically signaling songbird*. Frontiers in Ecology and Evolution **4**, 90 (2016)

101. Zilber-Rosenberg, I., Rosenberg, E.: *Role of microorganisms in the evolution of animals and plants: The hologenome theory of evolution*. FEMS Microbiology Reviews **32**(5), 723–735 (2008)

Chapter 19
Creative AI Through Evolutionary Computation

Risto Miikkulainen

Abstract The main power of artificial intelligence is not in modeling what we already know, but in creating solutions that are new. Such solutions exist in extremely large, high-dimensional, and complex search spaces. Population-based search techniques, i.e. variants of evolutionary computation, are well suited to finding them. These techniques are also well positioned to take advantage of large-scale parallel computing resources, making creative AI through evolutionary computation the likely "next deep learning".

In the last decade or so we have seen tremendous progress in Artificial Intelligence (AI). AI is now in the real world, powering applications that have a large practical impact. Most of it is based on modeling, i.e. machine learning of statistical models that make it possible to predict what the right decision might be in future situations. For example, we now have object recognition, speech recognition, game playing, language understanding, and machine translation systems that rival human performance, and in many cases exceed it [7, 8, 20]. In each of these cases, massive amounts of supervised data exists, specifying the right answer to each input case. With the massive amounts of computation that is now available, it is possible to train neural networks to take advantage of the data. Therefore, AI works great in tasks where we already know what needs to be done.

The next step for AI is machine creativity. Beyond modeling there is a large number of tasks where the correct, or even good, solutions are not known, but need to be discovered. For instance designing engineering solutions that perform well at low costs, or web pages that serve the users well, or even growth recipes for agriculture in controlled greenhouses are all tasks where human expertise is scarce and good solutions difficult to come by [4, 12, 10, 11, 18]. Methods for machine

Risto Miikkulainen
The University of Texas at Austin and Cognizant, Inc., USA
e-mail: risto@cs.utexas.edu

© Springer Nature Switzerland AG 2020   265
W. Banzhaf et al. (eds.), *Evolution in Action: Past, Present and Future*,
Genetic and Evolutionary Computation, https://doi.org/10.1007/978-3-030-39831-6_19

creativity have existed for decades. I believe we are now in a similar situation as deep learning was a few years ago: with the million-fold increase in computational power, those methods can now be used to scale up to real-world tasks.

Evolutionary computation is in a unique position to take advantage of that power, and become the next deep learning. To see why, let us consider how humans tackle a creative task, such as engineering design. A typical process starts with an existing design, perhaps an earlier one that needs to be improved or extended, or a design for a related task. The designer then makes changes to this solution and evaluates them. S/he keeps those changes that work well and discards those that do not, and iterates. It terminates when a desired level of performance is met, or when no better solutions can be found—at which point the process may be started again from a different initial solution. Such a process can be described as a hill-climbing process (Figure 19.1a). With good initial insight it is possible to find good solutions, but much of the space remains unexplored and many good solutions may be missed.

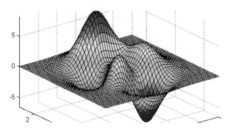
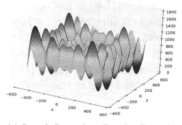

(a) Search Space Appropriate for Hill-Climbing   (a) Search Space in a Creative Domain

Fig. 19.1: Challenge of Creative Problem Solving. Human design process as well as deep learning and reinforcement learning can be seen as hill-climbing processes. They work well as long as the search space is relatively small, low-dimensional, and well behaved. However, creative problems where solutions are not known may require search in a large, high-dimensional spaces with many local optima. Population-based search through evolutionary computation is well-suited for such problems: it discovers and utilizes partial solutions, searches along multiple objectives, and novelty. (Image credit: http://deap.readthedocs.io/en/latest/api/benchmarks.html)

Interestingly, current machine learning methods are also based on hill-climbing. Neural networks and deep learning follow a gradient that is computed based on known examples of desired behavior [14, 22]. The gradient specifies how the neural network should be adjusted to make it perform slightly better, but it also does not have a global view of the landscape, i.e. where to start and which hill to climb. Similarly, reinforcement learning starts with an individual solution and then explores modifications around that solution, in order to estimate the gradient [21, 26]. With large enough networks and datasets and computing power, these methods have achieved remarkable successes in recent years.

However, the search landscape in creative tasks is likely to be less amenable to hill climbing (Figure 19.1b). There are three challenges: (1) The space is large, consisting of too many possible solutions to be explored fully, even with multiple restarts; (2) the space is high-dimensional, requiring that good values are found for many variables at once; and (3) the space is deceptive, consisting of multiple peaks and valleys, making it difficult to make progress through local search.

Evolutionary computation, as a population-based search technique, is in a unique position to meet these challenges. First, it makes it possible to explore many areas of the search space at once. In effect, evolution performs multiple parallel searches, not a single hill-climb. By itself such parallel search would result in only a linear improvement, however, the main advantage is that the searches interact: if there is a good partial solution found in one of the searches, the others can immediately take advantage of it as well. That is, evolution finds building blocks, or schemas, or stepping stones, that are then combined to form better comprehensive solutions [5, 9, 17].

This approach can be highly effective, as shown e.g. in the multiplexer benchmark problem [13]. Multiplexers are easy to design algorithmically: the task is to output the bit (among 2^n choices) specified by an n-bit address. However, as a search problem in the space of logical operations they grow very quickly, as $2^{2^{n+2^n}}$. There is, however, structure in that space that evolution can discover and utilize effectively. It turns out that evolution can discover solutions in extremely large such cases, including the 70-bit multiplexer (i.e. $n = 6$) with a search space of at least $2^{2^{70}}$ states. It is hard to conceptualize a number that large, but to give an idea, imagine having the number printed on a 10pt font on a piece of paper. It would take light 95 years to traverse from the beginning to the end of that number.

Second, population-based search makes it possible to find solutions in extremely high-dimensional search spaces as well. Whereas it is very difficult to build a model with high-order interactions beyond pairs or triples, the population represents such interactions implicitly, as the collection of actual combinations of values that exist in the good solutions in the population. Recombination of those solutions then makes it possible to collect good values for a large number of dimensions at once.

As an example, consider the problem of designing an optimal schedule for metal casting [3]. There are variables for number of each type of object to be made in each heat (i.e. melting process). The number of objects and heats can be grown from a few dozen, which can be solved with standard methods, to tens of thousands, resulting in billion variables. Yet, utilizing an initialization process and operators customized to exploit the structure in the problem, it is possible to find good combinations for them, i.e. find near-optimal solutions in a billion-dimensional space. Given that most search and optimization methods are limited to six orders of magnitude fewer variables, this scale-up makes it possible to apply optimization to entire new category of problems.

Third, population-based search can be adapted naturally to problems that are highly deceptive. One approach is to utilize multiple objectives [2]: if search gets stuck in one dimension, it is possible to make progress among other dimensions, and thereby get around deception. Another approach is to emphasize novelty, or diversity, of solutions in search [25]. The search does not simply try to maximize fitness, but also favors solutions that are different from those that already exist.

Novelty can be expressed as part of fitness, or a separate objective, or serve as a minimum criterion for selection, or as a criterion for mate selection and survival [1, 6, 15, 16, 19].

For instance, in the composite novelty method [23], different objectives are defined for different aspects of performance, and combined so that they specify an area of search space with useful tradeoffs. Novelty is then used as the basis for selection and survival within this area. This method was illustrated in the problem of designing minimal sorting networks, which have to sort a set of n numbers correctly, but also consist of as few comparator elements as possible (which swap two numbers), and as few layers as possible (where comparisons can be performed in parallel). The search space is highly deceptive in that often the network structure needs to be changed substantially to make it smaller. Combining multiple objectives and novelty results in better solutions, and finds them faster, than traditional evolution, multi-objective evolution, and novelty search alone. The approach has already found a new minimal network for 20 inputs [24], and is now being extended to larger networks.

To conclude, evolutionary computation is an AI technology that is on the verge of a breakthrough, as a way to take machine creativity to the real world. Like deep learning, it can take advantage of the large amount of compute that is now becoming available. Because it is a population-based search method, it can scale with compute better than other machine learning approaches, which are largely based on hill-climbing. With evolution, we should see many applications in the near future where human creativity is augmented by evolutionary search in discovering complex solutions, such as those in engineering, healthcare, agriculture, financial technology, biotechnology, and e-commerce, resulting in more complex and more powerful solutions than are currently possible.

References

1. Cuccu, G., Gomez, F.: *When Novelty is Not Enough*. In: Proceedings of the 2011 International Conference on Applications of Evolutionary Computation - Volume Part I, pp. 234–243. Springer-Verlag, Berlin, Heidelberg (2011)
2. Deb, K., Agrawal, S., Pratab, A., Meyarivan, T.: *A Fast Elitist Non-Dominated Sorting Genetic Algorithm for Multi-Objective Optimization: NSGA-II*. In: Proceedings of Parallel Problem Solving from Nature: PPSN VI, 6th International Conference, pp. 849–858 (2000)
3. Deb, K., Myburgh, C.: *A population-based fast algorithm for a billion-dimensional resource allocation problem with integer variables: Breaking the Billion-Variable Barrier in Real-World*. European Journal of Operational Research **261**, 460–474 (2017)
4. Dupuis, J.F., Fan, Z., Goodman, E.: *Evolutionary design of discrete controllers for hybrid mechatronic systems*. International Journal of Systems Science **46**, 303–316 (2015)
5. Forrest, S., Mitchell, M.: *Relative Building-Block Fitness and the Building-Block Hypothesis*. In: L.D. Whitley (ed.) Foundations of Genetic Algorithms, pp. 109–126. Kaufmann (1993)
6. Gomes, J., Mariano, P., Christensen, A.L.: *Devising Effective Novelty Search Algorithms: A Comprehensive Empirical Study*. In: Proceedings of the 2015 Annual Conference on Genetic and Evolutionary Computation, pp. 943–950. ACM, New York, NY (2015)

7. Hassan Awadalla, H., Aue, A., Chen, C., Chowdhary, V., Clark, J., Federmann, C., Huang, X., Junczys-Dowmunt, M., Lewis, W., Li, M., Liu, S., Liu, T.Y., Luo, R., Menezes, A., Qin, T., Seide, F., Tan, X., Tian, F., Wu, L., Wu, S., Xia, Y., Zhang, D., Zhang, Z., Zhou, M.: *Achieving Human Parity on Automatic Chinese to English News Translation.* Tech. rep., Microsoft Research (2018)
8. Hessel, M., Modayil, J., van Hasselt, H., Schaul, T., Ostrovski, G., Dabney, W., Horgan, D., Piot, B., Azar, M.G., Silver, D.: *Rainbow: Combining Improvements in Deep Reinforcement Learning.* CoRR **abs/1710.02298** (2017)
9. Holland, J.H.: *Adaptation in Natural and Artificial Systems: An Introductory Analysis with Applications to Biology, Control and Artificial Intelligence.* University of Michigan Press, Ann Arbor, MI (1975)
10. Hu, J., Goodman, E.D., Li, S., Rosenberg, R.C.: *Automated synthesis of mechanical vibration absorbers using genetic programming.* Artificial Intelligence in Engineering Design and Manufacturing **22**, 207–217 (2008)
11. Ishida Lab: The N700 Series Shinkansen (Bullet Train) (2018). Retrieved 9/29/2018
12. Johnson, A.J., Meyerson, E., de la Parra, J., Savas, T.L., Miikkulainen, R., Harper, C.B.: *Flavor-Cyber-Agriculture: Optimization of plant metabolites in an open-source control environment through surrogate modeling.* PLoS ONE **14**, e0213,918 (2019)
13. Koza, J.R.: A hierarchical approach to learning the boolean multiplexer function. In: G.J.E. Rawlins (ed.) Foundations of Genetic Algorithms, pp. 171–192. Morgan Kaufmann (1991)
14. LeCun, Y., Bengio, Y., Hinton, G.: *Deep Learning.* Nature **521**, 436–444 (2015)
15. Lehman, J., Stanley, K.O.: *Revising the Evolutionary Computation Abstraction: Minimal Criteria Novelty Search.* In: Proceedings of the Genetic and Evolutionary Computation Conference (GECCO-2010), pp. 103–110. ACM Press, New York (2010)
16. McQuesten, P.: Cultural Enhancement of Neuroevolution. Ph.D. thesis, Department of Computer Sciences, The University of Texas at Austin
17. Meyerson, E., Miikkulainen, R.: *Discovering Evolutionary Stepping Stones through Behavior Domination.* In: Proceedings of the Genetic and Evolutionary Computation Conference (GECCO 2017), Berlin, Germany, pp. 139–146. ACM Press, New York (2017)
18. Miikkulainen, R., Iscoe, N., Shagrin, A., Rapp, R., Nazari, S., McGrath, P., Schoolland, C., Achkar, E., Brundage, M., Miller, J., Epstein, J., Lamba, G.: *Sentient Ascend: AI-Based Massively Multivariate Conversion Rate Optimization.* In: Proceedings of the Thirtieth Innovative Applications of Artificial Intelligence Conference. AAAI (2018)
19. Mouret, J.B., Doncieux, S.: *Encouraging behavioral diversity in evolutionary robotics: An empirical study.* Evolutionary Computation **20**, 91–133 (2012)
20. Russakovsky, O., Deng, J., Su, H., Krause, J., Satheesh, S., Ma, S., Huang, Z., Karpathy, A., Khosla, A., Bernstein, M.S., Berg, A.C., Li, F.: *ImageNet Large Scale Visual Recognition Challenge.* CoRR **abs/1409.0575** (2014)
21. Salimans, T., Ho, J., Chen, X., Sutskever, I.: *Evolution Strategies as a Scalable Alternative to Reinforcement Learning.* CoRR **abs/1703.03864** (2017)
22. Schmidhuber, J.: *Deep Learning in Neural Networks: An Overview.* Neural Networks **61**, 85–117 (2015)
23. Shahrzad, H., Fink, D., Miikkulainen, R.: *Enhanced Optimization with Composite Objectives and Novelty Selection.* In: Proceedings of the 2018 Conference on Artificial Life, Tokyo, Japan, pp. 616–622. MIT Press, Cambridge, MA (2018)
24. Shahrzad, H., Hodjat, B., Dolle, C., Denissov, A., Lau, S., Goodhew, D., Dyer, J., Miikkulainen, R.: *Enhanced Optimization with Composite Objectives and Novelty Pulsation.* In: W. Banzhaf, E. Goodman, L. Sheneman, L. Trujillo, B. Worzel (eds.) Genetic Programming Theory and Practice XVII. Springer, New York (2020, in press)
25. Stanley, K.O., Lehman, J.: *Why Greatness Cannot Be Planned: The Myth of the Objective.* Springer, New York (2015)
26. Zhang, X., Clune, J., Stanley, K.O.: *On the Relationship Between the OpenAI Evolution Strategy and Stochastic Gradient Descent.* arXiv:1712.06564 (2017)

Part IV
Evolution of Communities and Collective Dynamics

Chapter 20
Subtle Environmental Differences have Cascading Effects on the Ecology and Evolution of a Model Microbial Community*

Justin R. Meyer and Richard E. Lenski

Abstract Predicting ecological and evolutionary dynamics is challenging because the phenomena of interest emerge from complex nonlinear interactions between genomes, organisms, and environments. Complexity theory predicts that small changes in a basal element of an ecosystem can impact higher-order features such as population dynamics and biodiversity. Here we use a simple two-species laboratory system to demonstrate how slight alterations to the environment can have cascading effects on the ecology and evolution of that system. We cultured the bacterium *Escherichia coli* and a virus, phage λ, together in a carbon-limited medium. We varied the carbon source by supplying one of three similar sugars: glucose, maltose, or maltotriose. These sugars were chosen because we predicted they would impose varying degrees of constraint on the potential for the bacteria to evolve resistance to the phage. The sugars have different routes into the cell: both maltodextrins rely on the outer-membrane pore LamB, whereas glucose does not. LamB also serves as the receptor for λ attachment to the cell surface, and mutations that alter its structure or reduce its expression can confer resistance to λ. By varying the sugar and thereby the bacteria's reliance on LamB, we predicted they would evolve different types of resistance and engage in different coevolutionary trajectories with λ. We saw even more striking effects than expected. This simple resource manipulation caused differences in the bacteria's cost of resistance, which in turn affected population dynamics, community composition, coexistence, and coevolution. This cascade has important implications for predicting ecology and evolution. On the one

Justin R. Meyer
Division of Biological Sciences, University of California, San Diego, La Jolla, CA 92093, USA
e-mail: jrmeyer@ucsd.edu

Richard E. Lenski
BEACON Center for the Study of Evolution in Action, Michigan State University, East Lansing, MI 48824, USA
Department of Microbiology and Molecular Genetics, Michigan State University, East Lansing, MI 48824, USA

* This paper was externally peer-reviewed.

© Springer Nature Switzerland AG 2020   273
W. Banzhaf et al. (eds.), *Evolution in Action: Past, Present and Future*,
Genetic and Evolutionary Computation, https://doi.org/10.1007/978-3-030-39831-6_20

hand, it reveals that even subtle environmental differences can have large and complex effects, making predictions difficult. On the other hand, important features of the environment (here, the specific limiting resource) can sometimes be identified *a priori* given sufficient knowledge of the molecular biology and physiology of the organisms.

Key words: coevolution, experimental evolution, parasites, phage lambda, trade-offs

20.1 Introduction

Charles Darwin famously expressed the complexity of the living world in the closing paragraph of *The Origin of Species* [13]: "It is interesting to contemplate an entangled bank, clothed with many plants of many kinds, with birds singing on the bushes, with various insects flitting about, and with worms crawling through the damp earth, and to reflect that these elaborately constructed forms, so different from each other, and dependent on each other in so complex a manner, have all been produced by laws acting around us." He envisioned that a handful of basic processes—reproduction, inheritance, variation, and the struggle for survival—could produce such complexity.

Today, the idea that interactions among a few simple processes can generate complex systems is the subject of complexity theory [31]. One implication of this theory is that subtle changes in the system can cause major transformations. If this sounds familiar, the idea was popularized in the movie Jurassic Park as the "butterfly effect." Applied to biology, one can imagine a scenario in which slight changes in the environment, say in average temperature, could impact the metabolism of organisms, alter the growth potential of populations, destabilize key predator-prey interactions, and cause ecosystem collapse. Here we report on a set of experiments that examine the idea that subtle changes to an environment can cause large changes in the ecology and evolution of interacting species.

Sensitivity to minor perturbations can occur in complex systems when interactions between components are strong, nonlinear, and produce feedbacks [31]. Viruses and their hosts offer a good example of such interactions, and they are pervasive throughout nature. They often have strong antagonistic effects, where the survival of one depends on the death or morbidity of the other. Their population dynamics are characterized by nonlinear functions [24], and their interactions readily generate feedbacks. One potential feedback is a coevolutionary arms race, in which the host evolves resistance and the virus evolves counter-defenses [24]. In line with complexity theory, viral-host coevolution is thought to contribute to many emergent properties of biological systems, including the production and maintenance of biodiversity [4, 6, 40], the evolution of more evolvable systems [30, 43], and the emergence of sexual reproduction [19]. Also consistent with complexity theory, coevolutionary trajectories are often highly sensitive to other environmental variables [18, 26].

Here we report results from experiments with a simple microbial community comprising just two species, a virus and its bacterial host. We manipulated the environment in a subtle way and documented how that manipulation altered the ecology and evolution of the community. The host is *Eschierichia coli* and the virus is a strictly lytic variant of phage λ. Previous studies have shown that *E. coli* and λ undergo an arms race in the laboratory [2, 5, 7, 21, 28, 32, 35, 41]. Resistance to λ typically evolves by mutations affecting the expression or amino-acid sequence of the receptor, LamB [5, 38]. *E. coli* expresses LamB to consume maltodextrins [9], which are chains of glucose molecules, such as maltose (two glucose units) and maltotriose (three units) (Fig. 20.1). By contrast, glucose usually enters the cell by another porin, OmpF, which wild-type λ cannot use. OmpF has a low affinity for maltose, and even lower for maltotriose [12]. We predicted that varying which of these three sugars we supplied would impose different constraints on the ability of *E. coli* to evolve resistance and thereby impact its coevolution with λ (see also [17, 24, 40]).

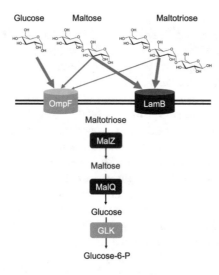

Fig. 20.1: Transport of the three sugars used in this study across the outer membrane (horizontal double line) via two different pore proteins and their initial metabolic conversions once in the cytoplasm. The thickness of the gray arrows represents schematically the relative importance of each pore type for each sugar's transport. The proteins shown in black are activated by MalT, whereas those shown in gray are not.

20.2 Materials and Methods

20.2.1 Strains

We used as starting hosts *E. coli* B strains REL606 and REL607 because their evolution in the laboratory has been well documented [3, 25, 37, 42]. Moreover, these strains readily coevolve with phage in a receptor-focused manner because they lack other defenses including restriction-modification systems, a functional CRISPR system, and the capacity to produce a broadly resistant mucoid phenotype [35, 36]. REL606 and REL607 differ by a mutation that prevents REL606 from growing on arabinose (Ara$^-$), which provides a marker that is selectively neutral [25, 39]. The λ strain that we used is cI26, an obligatorily lytic mutant provided by Donald Court (National Cancer Institute). This strain has a single nucleotide deletion that causes a frameshift in the *cI* repressor gene [28] that otherwise maintains lysogeny [20].

20.2.2 Experimental Procedures and Culture Conditions

Populations of phage λ and *E. coli* were cultured together in a minimal medium in which the growth of the *E. coli* population is limited by the available carbon. Three treatments were imposed by supplying one of three sugars: glucose, maltose, or maltotriose. For each community, we inoculated $\sim$100 phage and $\sim$1,000 cells in 10 mL of modified Davis Medium (DM) [25] with 10x the standard concentration of magnesium sulfate (1 μg mL^{-1}) and 125 μg mL^{-1} of sugar. Because of the similar structure of these sugars, supplying the same concentration yields effectively the same maximum density of bacteria. Six replicate flask communities were started for each sugar treatment. The bacteria were preconditioned during the prior day in the same medium with the sugar corresponding to their experimental treatment. Half of the flasks in each treatment were inoculated with REL606 and half with REL607, in an alternating design, so that any inadvertent cross-contamination of bacteria between the communities would likely be discovered by the difference in the arabinose-utilization marker. Small initial population sizes were used to ensure that mutations for resistance and host-range expansions evolved *de novo*. All cultures were kept at 37°C with shaking at 120 rpm. After 24 h, a random volumetric sample of 1% of the community was transferred to a flask with fresh medium and allowed to regrow. At the end of each 24-h period, two 1-mL samples of the flasks were preserved frozen at -80°C with 15% v/v glycerol. Population densities were determined each day by counting bacterial colonies on LB (Luria-Bertani) agar plates [34], and by counting phage plaques on lawns of REL606 [1]. Serial transfers and sampling were repeated for 25 days. Some populations stopped producing plaques, which we confirmed resulted from extinction of the phage by quantitative polymerase chain reaction (PCR) [33].

E. coli clones were isolated by picking individual colonies from the plates described above, streaking them on a fresh LB agar plate, allowing new colonies to form, and repeating the procedure once more to eliminate any possible phage carriage. Colonies were then selected from the second streak and grown in liquid LB overnight at 37°C and 120 rpm, and a sample was frozen in 1 mL with glycerol as described above. Phages were isolated by picking individual plaques and culturing them on REL606 using standard methods [1]. Phage isolates were stored at 4°C with 4% *v/v* chloroform added to maintain sterility.

20.2.3 Sequencing Resistance Genes

Candidate genes from clones sampled on days 3, 10, and 25 were sequenced to discover any resistance mutations. We targeted two loci: *lamB*, which encodes the receptor LamB; and *malT*, which encodes MalT, a transcriptional activator for *lamB* [15]. We examined one clone from each time-point from three of the six replicate populations for each treatment. We chose to use three replicates because only three phage populations survived for the full 25 days in glucose and maltose. We sequenced DNA fragments amplified by PCR and purified with GFX columns. An ABI sequencer run by Michigan State University's Research Technology Support Facility was used for sequencing. The primer sequences used to amplify *lamB* were 5' TTCCCGGTAATGTGGAGATGC 3' and 5' AATGTTTGCCGGGACGCTGTA 3' positioned at 1,398 bases upstream and 504 bases downstream of the protein-coding sequence. The primer sequences for amplifying *malT* were 5' CACCG-GTTTGGCGAATGG 3' and 5' GCGGCGGTGGGGGAATA 3' at 424 bases upstream and 212 bases downstream of the coding sequence.

20.2.4 Competition Experiments to Determine Fitness Costs of Resistance

To determine whether maltodextrins led to the evolution of more costly resistance than did glucose, we competed a resistant clone isolated from each community against the ancestor with the opposite arabinose-utilization marker. The resistant clones were sampled on day three of the experiment. We chose this early time-point because preliminary experiments indicated that resistance had already evolved by then, and it was too soon for the bacteria to evolve additional mutations that might ameliorate the cost. Competitions were performed in the absence of phage by inoculating a flask with $\sim 1.25 \times 10^6$ preconditioned cells each of the resistant clone and the ancestor, allowing them to grow and compete, and quantifying their relative fitness as the ratio of their realized growth rates [25, 42]. We measured each clone's fitness in the same sugar environment in which it evolved, with each assay running for three days (three 1:100 serial transfers), and with three replicate competition assays for every clone.

20.2.5 Experimental Test of Tradeoff-mediated Coexistence

We performed an experiment to test whether a tradeoff between resistance to phage and competitiveness for limiting resources was responsible for maintaining genetic variation in the bacterial populations, and if this variation in turn allowed phage λ to persist in the maltotriose environment. We selected two clones that evolved from REL606 (Ara$^-$ marker state) with maltotriose as the limiting resource. Both were previously shown to have high levels of resistance to λ [16], but one of them, clone 19a, suffered a cost for this resistance in terms of a reduced growth rate on maltotriose, whereas the other one, clone 56a, did not [29]. We constructed communities with phage cI26, sensitive host strain REL607 (Ara$^+$ marker state), and one of the two resistant clones. If the tradeoff model for phage-mediated coexistence applies, then the sensitive host REL607 should decline when the resistant clone 56a is present, because the sensitive host has no growth advantage and yet can be infected by phage. The declining REL607 population should, in turn, cause the phage population to decline. By contrast, those communities that include the resistant clone 19a, which has reduced competitive fitness, should maintain populations of both the sensitive host REL607 and phage λ. Communities were initiated with $\sim 1,000$ particles of λ and ~ 500 cells each of the sensitive and resistant E. coli strains. We constructed six replicates with each resistant clone, and propagated the communities by serial transfer for four days in modified DM with maltotriose as the limiting sugar. Population densities were sampled at the end of each day. Phage populations were enumerated as before. The total bacterial populations (resistant and sensitive combined) were enumerated by counting colonies on LB agar plates; sensitive cells only were enumerated by counting colonies on minimal arabinose agar plates, where only the sensitive Ara$^+$ REL607 progenitors could form colonies.

Before running this experiment, we performed two additional checks. First, we confirmed that the differences in growth rate led to differences in competitive fitness between the two resistant hosts by running the previously described competition protocol using clones 19a and 56a, with a control in which the marked sensitive clones REL606 and REL607 competed against one another. Second, we analyzed the full genomes of the two resistant clones to determine if they had other mutations. To that end, the Research Technology Support Facility at Michigan State University sequenced the genomes using an Illumina Genome Analyzer IIx. To obtain genomic samples, we revived frozen bacteria in LB medium, grew them overnight, and isolated DNA from several mL of culture with Qiagen genome tips. The DNA samples were fragmented by sonication, labeled with bar-coded attachments, and run as multiplexed samples over four lanes. We called mutations from the resulting 75-base single-end reads using *breseq* version 0.13 [14] and using the REL606 genome [22] as the reference (GenBank accession: NC_012967.1).

20.3 Results

20.3.1 Population Dynamics Vary by Environment

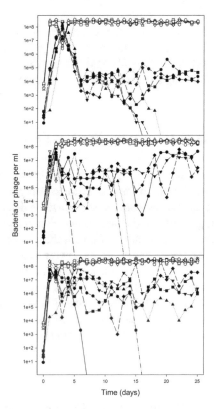

Fig. 20.2: Phage population dynamics (filled symbols) vary by treatment (top, glucose; middle, maltose; bottom, maltotriose), whereas bacterial dynamics (open symbols) do not. Each panel shows six replicate communities. The limit of phage detection was ∼3 per mL; the lines that cross the x-axis indicate extinctions. Except for day 0, densities were measured at the end of each 24-h cycle, just before 1% of the culture volume was transferred to a flask with fresh medium.

The phage populations went extinct in almost half of the communities by the end of the 25-day experiment (Fig. 20.2). Contrary to our expectation, the extinctions were not treatment specific; instead, phage survived in half of the populations in glucose, half in maltose, and two-thirds in maltotriose. We had predicted that λ would not survive in the glucose treatment because *E. coli* could evolve mutations that suppress expression of the LamB receptor, and do so without fitness costs because glucose enters cells by another pore. This absence of a tradeoff would allow resistant mutants to fix in the bacterial population and cause λ to decline to extinction.

Instead, however, a small but stable phage population of $\sim 10^4 \text{mL}^{-1}$ persisted for an extended period (Fig. 20.2, top panel). The phage were likely maintained as a consequence of so-called "leaky" resistance, in which a subpopulation of sensitive cells is continuously generated from largely resistant cells, probably by occasional spontaneous induction of LamB production, despite mutations in the gene that encodes the MalT activator [10, 28]. The phage populations, prior to extinction in some cases, were generally larger and experienced greater fluctuations in the two maltodextrin treatments (Fig. 20.2, middle and bottom panels). By contrast, the bacterial populations remained high and stable in all populations in all three treatments.

20.3.2 Fitness Costs of Resistance Vary by Environment

In the glucose environment without λ present, the six evolved resistant clones showed little or no loss of competitive fitness, with no change on average (Fig. 20.3). In maltose, by contrast, all six resistant clones showed reductions in their competitive fitness. The average fitness reduction in the maltotriose-evolved clones was similar to those that evolved in maltose, although the clones from the maltotriose environment showed greater variance in fitness. Owing to this heterogeneity in maltotriose, the differences among the three sugar treatments were only marginally significant (Kruskal-Wallis test, $p < 0.05$), with that result driven largely by the difference between the glucose and maltose treatments (Mann-Whitney test, $p < 0.01$). On balance these data imply that resistance typically imposed a substantial fitness cost when the bacteria evolved on maltodextrins, but not on glucose, consistent with our expectations.

20.3.3 Resistance Mutations Vary by Environment

We sequenced two candidate genes, *lamB* and *malT*, to find mutations responsible for the resistance to phage λ that evolved in all three of the sugar environments (Table 20.1). In the glucose treatment, we found many mutations in *malT*, which encodes MalT, a transcriptional activator of genes involved in maltodextrin uptake and metabolism, including LamB, which is the receptor for phage λ (Fig. 20.1). These gene products are not useful for growth on glucose, and so the *E. coli* in this treatment were able to achieve high levels of resistance without an associated fitness cost by disrupting this activator and turning off expression of LamB. Also, the *malT* gene is about three times as long as *lamB*, which exposes more sequence to mutations and may thus explain why it was mutated more often than *lamB*. Additionally,

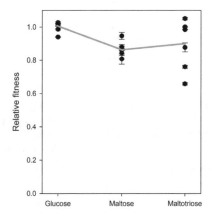

Fig. 20.3: Fitness of λ-resistant *E. coli* clones that evolved in three resource environments. Each
       resistant clone competed in the environment where it evolved against the sensitive
       ancestor with the opposite arabinose-utilization marker state. Each point shows the
       mean of three competition assays; error bars indicate standard errors, which in some
       cases are hidden by the symbols. One clone was examined from each independently
       evolved community (six from each treatment). The gray line connects the grand mean
       fitness values for the clones from each environment.

malT has an enigmatic 25-bp DNA sequence that has an elevated rate of duplication
[10, 28].

In the maltose and maltotriose treatments, mutations did not evolve in the *malT*
gene. Such mutations would reduce or eliminate expression not only of LamB, but
also other proteins used to metabolize these two sugars after they enter the cell
(Fig. 20.1). Instead, mutations evolved repeatedly in *lamB* in these two environments (Table 20.1). In maltose, all of the observed *lamB* mutations were frameshift
or nonsense mutations, which presumably eliminated any functional LamB protein
and thereby conferred high-level resistance to phage λ. These mutations are costly
in maltose because LamB is the main pore protein used to move maltose across the
outer cell envelope. However, such loss-of-function mutations are not lethal because
the OmpF pore protein provides an alternative, though less efficient, route.

In the maltotriose treatment, by contrast, most mutations in *lamB* were point
mutations, which changed the amino-acid sequence of LamB. Five of the six point
mutations affected protein loops and probably interfered with λ binding (Table 20.1;
position 856 is the exception). These mutations affected different loops, which
might explain why maltotriose-evolved clones vary so much in their fitness costs
(Fig. 20.3). The LamB pore is also larger than the OmpF pore, and maltotriose is
larger than maltose, which may explain why loss-of-function mutations were less
common in the maltotriose environment. Two of the *lamB* alleles sampled in the
maltotriose treatment at the end of the experiment (day 25) had two separate muta-

tions each, suggesting a multi-step arms race between phage λ and the *lamB* gene in the *E. coli* hosts.

In all three sugar treatments, some clones that were resistant to phage λ had no mutations in either *lamB* or *malT* (Table 20.1). These clones presumably had mutations in other genes that can confer resistance [8, 27, 28].

Table 20.1: Mutations found by sequencing two candidate genes, *malT* and *lamB*, in λ-resistant clones sampled after 3, 10, and 25 days in three replicate communities of each treatment. The notation "unknown" indicates that no mutations were found; however, these clones presumably had mutations in other genes that conferred resistance.

Condition	Replicate population	Day	Gene mutated	Nucleotide position*	Mutation	Effect
Glucose	-3	3	*malT*	992	25 base duplication	frameshift
Glucose	+1	3	*malT*	912	△1 base duplication	frameshift
Glucose	+3	3	unknown			
Glucose	-3	10	unknown			
Glucose	+1	10	unknown			
Glucose	+3	10	*malT*	992	25 base duplication	frameshift
Glucose	-3	25	*malT*	2681	$T \longrightarrow C$	Leu $\longrightarrow$ Pro
Glucose	+1	25	unknown			
Glucose	+3	25	*malT*	2177	$A \longrightarrow T$	Asp $\longrightarrow$ Val
Maltose	-1	3	*lamB*	1162	△29 base duplication	frameshift
Maltose	-2	3	*lamB*	1162	△11 base duplication	frameshift
Maltose	+1	3	*lamB*	308	△11 base duplication	frameshift
Maltose	-1	10	unknown			
Maltose	-2	10	*lamB*	966	△11 base duplication	frameshift
Maltose	+1	10	*lamB*	429	$C \longrightarrow G$	stop
Maltose	-1	25	unknown			
Maltose	-2	25	*lamB*	966	△11 base duplication	frameshift
Maltose	+1	25	*lamB*	429	$C \longrightarrow G$	stop
Maltotriose	-2	3	*lamB*	697	△11 base duplication	frameshift
Maltotriose	-3	3	unknown			
Maltotriose	+1	3	*lamB*	1128	$G \longrightarrow T$	Trp $\longrightarrow$ Cys
Maltotriose	-2	10	*lamB*	709	$G \longrightarrow T$	Gly $\longrightarrow$ Cys
				856	$T \longrightarrow A$	Leu$\longrightarrow$Gln
Maltotriose	-3	10	*lamB*	509	$G \longrightarrow C$	Arg $\longrightarrow$ Pro
Maltotriose	+1	10	unknown			
Maltotriose	-2	25	*lamB*	716	$T \longrightarrow A$	stop
Maltotriose	-3	25	*lamB*	796	+18 bases	+6 amino acids
				1128	$G \longrightarrow T$	Leu$\longrightarrow$Pro
Maltotriose	+1	25	unknown			

20.3.4 Experimental Test of Tradeoff-mediated Coexistence

There are multiple hypotheses for how populations of bacteria and lytic phage can coexist, despite their strong antagonism. Previous work has shown that λ persists in the glucose condition, despite *E. coli*'s ability to evolve high levels of resistance, because that resistance is often leaky [10, 28]. In particular, resistant cells with mutations in *malT* experience occasional spontaneous induction of LamB expression, thereby generating a small population of phenotypically sensitive cells that can sustain a small phage population. However, phage populations were much larger in the two maltodextrin treatments (Fig. 20.2) and, moreover, the evolved bacterial resistance in those communities was not caused by *malT* mutations (Table 20.1). These facts suggest that a different mechanism supports the λ population in those treatments. The coexistence hypothesis that has received the most attention in the literature is based on a tradeoff in bacteria between resistance and competitive fitness [4, 23, 24]. Such tradeoffs can prevent resistant mutants from sweeping to fixation, thus allowing the maintenance of a genetically sensitive host population, which in turn can support the phage.

Given the evidence for this tradeoff in maltodextrins (Fig. 20.2), we tested this hypothesis as follows. We constructed two types of synthetic communities in the maltotriose environment, each with three players: the ancestral λ, the sensitive *E. coli* ancestor, and an evolved resistant clone. The two types of community differed, however, in the identity of the resistant clone. Both clones had mutations in the *lamB* gene (Fig. 20.4A). One of them had a nonsense mutation in *lamB* and suffered a severe reduction in fitness in the absence of phage; the other resistant clone had an insertion mutation in that gene and showed no measurable fitness cost (Fig. 20.4B). The total bacterial population size at the end of this four-day experiment was unaffected by which resistant clone was used in these synthetic communities (Fig. 20.4C). However, as predicted by the tradeoff hypothesis, both the sensitive bacterial population (Fig. 20.4D) and the phage population (Fig. 20.4E) were significantly larger in the communities with the costly resistance than in those where the resistant bacteria were as good as their sensitive counterparts in competing for resources.

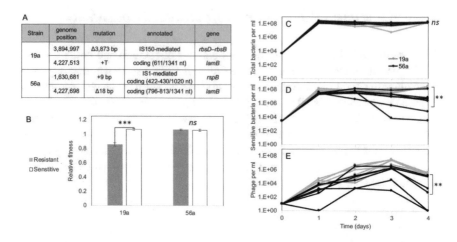

Fig. 20.4: Coexistence of phage λ and *E. coli* in the maltotriose environment depends on the tradeoff between resistance and competitive fitness in the absence of phage. (A) Mutations in two resistant clones, based on whole-genome sequencing. (B) Clone 19a has lower fitness than its sensitive ancestor, whereas clone 56a does not. (C) Total bacterial density is similar whether the communities include the 19a (gray lines) or 56a (black lines) resistant clone. (D) The final density of the sensitive ancestral strain is lower in the presence of the high-fitness resistant clone 56a (black lines) than in presence of the low-fitness clone 19a (gray lines). (E) The final density of phage λ is also lower in the presence of 56a (black lines) than 19a (gray lines); a density of 1 indicates the phage population was below the limit of detection. Significance levels are based on *t* tests in (B) and Mann-Whitney tests in (C), (D), and (E), with results shown as follows: ns, not significant; $P < 0.01^{**}$; $P < 0.001^{***}$.

Genome sequencing of these resistant clones revealed that each had another mutation in addition to the one in *lamB*. These mutations were an insertion and a deletion, both caused by IS elements (Fig. 20.4A). Neither affected gene is known to confer resistance to λ [27]. This type of mutation occurs at a high rate in many bacteria including this strain of *E. coli* [10, 11, 37], and these mutations were probably inconsequential hitchhikers without major phenotypic effects.

20.4 Discussion

We examined the effects at several levels – from genomes to community structure – of a subtle change in the environment. In particular, we provided otherwise identical model communities of one lytic phage and one bacterial species with "glucose" in three forms: glucose, maltose, and maltotriose. Maltose and maltotriose are sugars with two and three linked glucose moieties, respectively, each involving one additional covalent bond.

Despite their fundamental similarity, these treatments generated distinct eco-evolutionary dynamics. These differences follow from the fact that *E. coli* expresses a different pore protein, LamB, which allows the larger maltose and maltotriose molecules to cross its outer membrane more efficiently than does the smaller-diameter pore protein, OmpF, which suffices for glucose (Fig. 20.1). In glucose, the populations of *E. coli* typically evolved resistance by mutations in a regulatory gene called *malT*, severely reducing the expression of LamB, which phage λ also uses to enter host cells. These *malT* mutations had no fitness costs for the bacteria when growing on glucose, and the phage populations declined following their initial expansion. The phage also appeared unable to readily evolve counter-defenses. Nonetheless, half of the phage populations persisted because this mode of resistance was phenotypically leaky, such that LamB production was spontaneously induced in a small fraction of the resistant mutants.

In the maltose and maltotriose treatments, by contrast, mutations in *malT* that inactivate LamB expression would be extremely deleterious to the bacteria. Such mutations would not only reduce the protein used for efficient acquisition of these larger sugars, they would also reduce the enzymes used in the first steps of metabolizing these sugars (Fig. 20.1). Instead, in these treatments the bacteria typically evolved resistance via mutations in *lamB*, the gene that encodes the LamB pore and receptor used by phage λ. In the case of the maltose environment, many of the *lamB* mutations were frameshift and nonsense mutations, which presumably eliminated any functional LamB pores. These mutations were quite costly to the bacteria in terms of competitive fitness for maltose, but some intermediate-sized maltose molecules could nonetheless enter cells via the smaller OmpF pores (Fig. 20.1). Evidently, reduced competitiveness was worth it when faced with the lethal consequences of phage infection. As a result of this tradeoff, a minority population of genetically sensitive cells (as opposed to leaky resistant cells) could persist and, in turn, sustain a phage population.

In the maltotriose environment, the OmpF pores are of little use in allowing such large molecules to enter the cell, and the LamB protein is even more important than it is in maltose. Mutations that destroy LamB were not tolerated in this treatment. Instead, we typically saw point mutations that altered the structure of the protein in ways that interfered with the phage's ability to adsorb to the altered protein, but evidently still allowed maltotriose to enter the cell. These mutations had variable effects on competitive fitness, which in turn affected whether sensitive cells and phage could persist (Fig. 20.4). Moreover, the subtle changes in LamB structure appear to have led to an evolutionary response by phage λ, which in turn sometimes led to further changes in LamB, indicating a multi-step coevolutionary arms race.

20.5 Coda

Has the reader ever left an ecology seminar discouraged because the presenter identified yet another variable that matters for some ecosystem property or dynamics?

For readers who are not ecologists, this feeling is similar to the exhaustion experienced after seeing yet another headline describing supposed health benefits from some dietary supplement. An overarching lesson from much of life-sciences research is that subtle perturbations often matter for the health of biological systems, whether individual organisms or whole ecosystems. Complexity theory provides an explanation for why living systems are so sensitive, and therefore why it is difficult to make predictions in biology. Our results, however, should not be interpreted as just another example of identifying a variable that makes it difficult to predict ecological and evolutionary dynamics. Instead, our study offers a ray of hope for prediction in the face of biocomplexity. By leveraging information gained from microbial genetics and molecular biology, we were able to identify in advance the proverbial "needle in a haystack" of chemical bonds that would significantly impact the ecology and evolution of an experimental community. This work shows that integrating molecular knowledge with population-biology models can improve the predictive power of biological theories.

Acknowledgements We thank Erik Goodman for his outstanding leadership of BEACON. This work was supported, in part, by the BEACON Center for the Study of Evolution in Action (National Science Foundation Cooperative Agreement DBI-0939454) and the Defense Advanced Research Projects Agency (HR0011-09-1-0055). We thank Anurag Agrawal, Jeff Barrick, Tom Ferenci, Mike Wiser, and Luis Zaman for helpful discussions, Neerja Hajela for help with lab work, and Donald Court for sharing phage cI26. J.R.M. will make the evolved strains available to qualified recipients, subject to completion of a material transfer agreement (http://blink.ucsd.edu/research/conducting-research/mta/index.html).

References

1. Adams, M.H.: *Bacteriophages*. New York: Interscience (1959)
2. Appleyard, R., McGregor, J., Baird, K.: *Mutation to extended host range and the occurrence of phenotypic mixing in the temperate coliphage Lambda*. Virology **2**(4), 565–574 (1956)
3. Barrick, J.E., Yu, D.S., Yoon, S.H., Jeong, H., Oh, T.K., Schneider, D., Lenski, R.E., Kim, J.F.: *Genome evolution and adaptation in a long-term experiment with Escherichia coli*. Nature **461**(7268), 1243–1274 (2009)
4. Bohannan, B.J., Lenski, R.E.: *Linking genetic change to community evolution: Insights from studies of bacteria and bacteriophage*. Ecol Lett **3**(4), 362–377 (2000)
5. Braun-Breton, C., Hofnung, M.: *In vivo and in vitro functional alterations of the bacteriophage lambda receptor in lamB missense mutants of Escherichia coli K-12*. J Bacteriol **148**(3), 845–852 (1981)
6. Brown, J.S., Vincent, T.L.: *Organization of predator-prey communities as an evolutionary game*. Evolution **46**(5), 1269–1283 (1992)
7. Burmeister, A.R., Lenski, R.E., Meyer, J.R.: *Host coevolution alters the adaptive landscape of a virus*. Proc. R. Soc. B **283**(1839), 20161528 (2016)
8. Burmeister, A.R., Sullivan, R.M., Lenski, R.E.: Fitness costs and benefits of resistance to phage Lambda in experimentally evolved *Escherichia coli*. In: W. Banzhaf, B. Cheng, K. Deb, K. Holekamp, R. E. Lenski, C. Ofria, R. Pennock, W. Punch, and D. Whittaker, editors. Evolution in Action: Past, Present, Future., pp. 123–142. Springer (2019)
9. Charbit, A.: *Maltodextrin transport through LamB*. Front Biosci **8**, 265–274 (2003)

10. Chaudhry, W.N., Pleška, M., Shah, N.N., Weiss, H., McCall, I.C., Meyer, J.R., Gupta, A., Guet, C.C., Levin, B.R.: *Leaky resistance and the conditions for the existence of lytic bacteriophage*. PLoS Biology **16**(8), e2005971 (2018)

11. Cooper, V.S., Schneider, D., Blot, M., Lenski, R.E.: *Mechanisms causing rapid and parallel losses of ribose catabolism in evolving populations of E. coli B*. J Bacteriol **183**(9), 2834–2841 (2001)

12. Dargent, B., Rosenbusch, J., Pattus, F.: *Selectivity for maltose and maltodextrins of maltoporin, a pore-forming protein of E. coli outer membrane*. FEBS Letters **220**(1), 136–142 (1987)

13. Darwin, C.: *On the Origin of Species*. Murray, London (1859)

14. Deatherage, D.E., Barrick, J.E.: *Identification of mutations in laboratory-evolved microbes from next-generation sequencing data using breseq*. Methods Mol Biol **1151**, 165–188 (2014)

15. Dippel, R., Boos, W.: *The maltodextrin system of Escherichia coli: Metabolism and transport*. J Bacteriol **187**(24), 8322–8331 (2005)

16. Flores, C.O., Meyer, J.R., Valverde, S., Farr, L., Weitz, J.S.: *Statistical structure of host-phage interactions*. Proc Natl Acad Sci **108**(28), E288–E297 (2011)

17. Forde, S.E., Thompson, J.N., Bohannan, B.J.: *Adaptation varies through space and time in a coevolving host-parasitoid interaction*. Nature **431**(7010), 841–844 (2004)

18. Gómez, P., Buckling, A.: *Bacteria-phage antagonistic coevolution in soil*. Science **332**(6025), 106–109 (2011)

19. Hamilton, W.D.: *Sex versus non-sex versus parasite*. Oikos **35**, 282–290 (1980)

20. Hendrix, R.W.: *Lambda II*. Cold Spring Harbor Laboratory Press, Cold Spring Harbor, NY (1983)

21. Hofnung, M., Jezierska, A., Braun-Breton, C.: *lamB mutations in E. coli K12: Growth of Lambda host-range mutants and effect of nonsense suppressors*. Mol Gen Genet **145**(2), 207–213 (1976)

22. Jeong, H., Barbe, V., Lee, C.H., Vallenet, D., Yu, D.S., Choi, S.H., Couloux, A., Lee, S.W., Yoon, S.H., Cattolico, L., et al.: *Genome sequences of Escherichia coli B strains REL606 and BL21 (DE3)*. J Mol Biol **394**(4), 644–652 (2009)

23. Koskella, B.: *Resistance gained, resistance lost: An explanation for host-parasite coexistence*. PLoS Biology **16**(9), e3000013 (2018)

24. Lenski, R.E., Levin, B.R.: *Constraints on the coevolution of bacteria and virulent phage: A model, some experiments, and predictions for natural communities*. Am Nat **125**(4), 585–602 (1985)

25. Lenski, R.E., Rose, M.R., Simpson, S.C., Tadler, S.C.: *Long-term experimental evolution in Escherichia coli. I. Adaptation and divergence during 2,000 generations*. Am Nat **138**(6), 1315–1341 (1991)

26. Lopez-Pascua, L., Buckling, A.: *Increasing productivity accelerates host-parasite coevolution*. J Evol Biol **21**(3), 853–860 (2008)

27. Maynard, N.D., Birch, E.W., Sanghvi, J.C., Chen, L., Gutschow, M.V., Covert, M.W.: *A forward-genetic screen and dynamic analysis of Lambda phage host-dependencies reveals an extensive interaction network and a new anti-viral strategy*. PLoS Genetics **6**(7), e1001017 (2010)

28. Meyer, J.R., Dobias, D.T., Weitz, J.S., Barrick, J.E., Quick, R.T., Lenski, R.E.: *Repeatability and contingency in the evolution of a key innovation in phage Lambda*. Science **335**(6067), 428–432 (2012)

29. Meyer, J.R., Gudelj, I., Beardmore, R.: *Biophysical mechanisms that maintain biodiversity through trade-offs*. Nature Comm **6**, 6278 (2015)

30. Moxon, E.R., Rainey, P.B., Nowak, M.A., Lenski, R.E.: *Adaptive evolution of highly mutable loci in pathogenic bacteria*. Curr Biol **4**(1), 24–33 (1994)

31. Murphy, J.T.: *Complexity Theory*. Oxford Bibliographies, Oxford (2017)

32. Petrie, K.L., Palmer, N.D., Johnson, D.T., Medina, S.J., Yan, S.J., Li, V., Burmeister, A.R., Meyer, J.R.: *Destabilizing mutations encode nongenetic variation that drives evolutionary innovation*. Science **359**(6383), 1542–1545 (2018)

33. Refardt, D., Rainey, P.B.: *Tuning a genetic switch: Experimental evolution and natural varia-tion of prophage induction.* Evolution **64**(4), 1086–1097 (2010)
34. Sambrook, J., Russell, D.W.: *Molecular Cloning, 3rd ed.* Cold Spring Harbor Laboratory Press, Cold Spring Harbor, NY (2001)
35. Spanakis, E., Horne, M.: *Co-adaptation of Escherichia coli and coliphage λ-vir in continuous culture.* J Gen Microbiol **133**(2), 353–360 (1987)
36. Studier, F.W., Daegelen, P., Lenski, R.E., Maslov, S., Kim, J.F.: *Understanding the differences between genome sequences of Escherichia coli B strains REL606 and BL21 (DE3) and com-parison of the E. coli B and K-12 genomes.* J Mol Biol **394**(4), 653–680 (2009)
37. Tenaillon, O., Barrick, J.E., Ribeck, N., Deatherage, D.E., Blanchard, J.L., Dasgupta, A., Wu, G.C., Wielgoss, S., Cruveiller, S., Médigue, C., et al.: *Tempo and mode of genome evolution in a 50,000-generation experiment.* Nature **536**(7615), 165–170 (2016)
38. Thirion, J., Hofnung, M.: *On some genetic aspects of phage Lambda-resistance in E. coli K12.* Genetics **71**(2), 207–216 (1972)
39. Travisano, M., Mongold, J.A., Bennett, A.F., Lenski, R.E.: *Experimental tests of the roles of adaptation, chance, and history in evolution.* Science **267**(5194), 87–90 (1995)
40. Weitz, J.S., Hartman, H., Levin, S.A.: *Coevolutionary arms races between bacteria and bac-teriophage.* Proc Natl Acad Sci USA **102**(27), 9535–9540 (2005)
41. Werts, C., Michel, V., Hofnung, M., Charbit, A.: *Adsorption of bacteriophage lambda on the LamB protein of Escherichia coli K-12: Point mutations in gene J of λ responsible for extended host range.* J Bacteriol **176**(4), 941–947 (1994)
42. Wiser, M.J., Ribeck, N., Lenski, R.E.: *Long-term dynamics of adaptation in asexual popula-tions.* Science **342**, 1364–1367 (2013)
43. Zaman, L., Meyer, J.R., Devangam, S., Bryson, D.M., Lenski, R.E., Ofria, C.: *Coevolu-tion drives the emergence of complex traits and promotes evolvability.* PLoS Biol **12**(12), e1002023 (2014)

Chapter 21
Ecological Context Influences Evolution in Host-Parasite Interactions: Insights from the *Daphnia*-Parasite Model System*

Katherine D. McLean and Meghan A. Duffy

Abstract Parasites exert strong selective pressure on their hosts, and many hosts can evolve rapidly in response. As such, host-parasite interactions have a special place in the study of contemporary evolution. However, these interactions are often considered in isolation from the ecological contexts in which they occur. Here we review different ways in which the ecological context of host-parasite interactions can modulate their evolutionary outcomes in important and sometimes unexpected ways. Specifically, we highlight how predation, competition, and abiotic factors change the outcome of contemporary evolution for both hosts and parasites. In doing so, we focus on insights gained from the *Daphnia*-microparasite system. This system has emerged as a model system for understanding the ecology and evolution of host-parasite interactions, and has provided important insights into how ecological context influences contemporary evolution.

21.1 Introduction

"I want to suggest that the struggle against disease, and particularly infectious disease, has been a very important evolutionary agent, and that some of its results have been rather unlike those of the struggle against natural forces, hunger, and predators, or with members of the same species." —JBS Haldane (1949)

Parasites are ubiquitous, and the outcomes of host-parasite interactions can often be measured in terms of life or death. Thus, it is not surprising that in the 70 years since Haldane postulated the importance of parasites as selective agents, studies of host-parasite interactions have provided striking examples of evolution in action

---

Katherine D. McLean · Meghan A. Duffy
Department of Ecology & Evolutionary Biology, University of Michigan, Ann Arbor, MI 48109
e-mail: kdmclean@umich.edu

* This paper was externally peer-reviewed.

W. Banzhaf et al. (eds.), *Evolution in Action: Past, Present and Future*,
Genetic and Evolutionary Computation, https://doi.org/10.1007/978-3-030-39831-6_21

[4, 11, 16, 17, 35, 38, 46, 51, 55, 60, 114, 115]. Moreover, we now realize that the ecological context—the "natural forces, hunger, and predators" and "members of the same species" to which Haldane referred—modulates the evolutionary outcomes of infectious disease in important and sometimes unexpected ways. Here, we review recent studies that demonstrate that predators, competitors, and the abiotic environment strongly influence the evolutionary dynamics of host-parasite interactions.

Host-parasite interactions are often considered in isolation, but the larger ecological context matters, too. To give just two examples: excluding large vertebrate herbivores increased the prevalence of viruses in plants by increasing the abundance of highly competent hosts [12]. Similarly, increasing nutrient inputs to ponds elevated levels of disease in frogs by increasing algal abundance which, in turn, increased the abundance of snails, who are intermediate hosts for the parasite [73]. These parasites strongly impact their hosts: the plant virus reduces plant longevity, growth, and seed production and the frog parasite causes severe limb deformities. Therefore, it does not require a large leap to imagine that these alterations to ecological context might alter parasite-mediated selection.

Human activities are strongly impacting the ecological context in which host-parasite interactions are embedded. Humans are changing abiotic factors in terrestrial and aquatic habitats, including nutrient levels, precipitation regimes, temperature, and pH [25, 24, 128]. Human activities are also strongly impacting species assemblages via environmental disturbance, climate change, and the introduction and extirpation of different species, including parasites and predators [14, 40, 101, 111, 125]. Because ecological context influences the prevalence and severity of disease, human-driven changes in abiotic factors and species assemblages can have dramatic consequences for evolution in host-parasite systems.

In this review, we highlight some of the ways in which ecological context, including human-driven changes to ecosystems, can influence evolution in host-parasite interactions. We also touch on some ways in which contemporary evolution may change ecological dynamics (i.e. eco-evolutionary feedbacks [68, 122]). In doing so, we focus in on one particular study system that has yielded key insights: *Daphnia* and their microparasites. *Daphnia* is a genus of planktonic microcrustacean found in various aquatic environments. *Daphnia* are ecologically important and experimentally tractable, and have emerged as a model system for understanding the ecology and evolution of host-parasite interactions [21, 47]. We first introduce this system, then review studies demonstrating the importance of predators, competitors, and the abiotic environment in altering evolution in host-parasite interactions.

21.2 The *Daphnia*-Microparasite Study System

Ecologists and evolutionary biologists have long studied *Daphnia*, both because of their ecological importance and because of their tractability as a study organism [48, 77]. *Daphnia* are dominant herbivores in many temperate aquatic ecosystems and serve as important links between primary producers (the phytoplankton they consume from the water column) and consumers (the small fish and predatory invertebrates that feed on *Daphnia*). In addition, their small size and rapid generation time make it possible to work with them in the laboratory and in field studies, allowing scientists to test possible mechanisms underlying patterns observed in nature—an important bridge between the laboratory and the natural world that is not easily crossed in many study systems.

The reproductive system of *Daphnia* also helps explain why they have emerged as an important study system. Most *Daphnia* are cyclical parthenogens, meaning they can reproduce sexually and asexually. Asexual reproduction makes it possible to propagate isofemale (i.e., clonal) lines under standardized laboratory conditions, allowing researchers to differentiate genetic and environmental effects on phenotypic traits. At the same time, the sexually produced offspring are enclosed in long-lived dormant eggs that accumulate in sediments, allowing studies that "resurrect" genotypes from earlier populations so that scientists may understand how populations have changed on scales from decades to centuries [35, 57, 64, 107].

Another advantage of the *Daphnia* system comes from the ability to study multiple replicate lakes or ponds that have well-defined boundaries; this means it is possible to study multiple populations (essential for evolutionary studies, where population is the unit of replication) and to do so across ecological gradients (e.g., in predation or productivity).

In the past few decades, *Daphnia* and their microparasites have emerged as a model system for understanding infectious diseases [21, 47, 48, 78, 80]. A number of parasites including viruses, bacteria, fungi, oomycetes, microsporidians and protozoa regularly infect *Daphnia* [47, 61, 124]. These parasites have diverse infection dynamics (horizontal vs. vertical transmission, obligate killers vs. continuous transmission) and exert a wide range of effects on their hosts (including early death, castration, and even gigantism [47]).

As is true for all organisms, any particular *Daphnia*-parasite interaction is embedded within a much larger, richer ecological context [89]. When thinking about this ecological context, we need to consider not only the types of interactions that have traditionally been the focus of ecological studies (such as resource levels and predation regimes), but also that pathogens are likely to be infecting multiple members of the food web, and that any one member of the food web is likely to be infected by multiple pathogens.

We begin by reviewing the impact of predation on evolution in host-parasite systems. Next, we consider the potential for species interactions within the same trophic level (especially the presence of multiple host species or multiple parasite species) to alter host-parasite evolution. Finally, we review some ways the abiotic environment can alter evolution of hosts and parasites. In each case, evidence from *Daphnia*-parasite interactions demonstrates that the ecological context impacts the evolution of both the host and the parasite.

21.3 Predation Alters Evolution in Host-Parasite Systems

Predators should alter evolution in host-parasite systems in multiple ways, including by altering the amount of disease in a focal host population. Predation is often thought of as reducing infection prevalence in hosts, especially in cases where predators selectively remove infected hosts [70, 92, 95]. However, predators can also increase disease in their host populations via a variety of mechanisms (reviewed in [42]), including by changing prey community composition so that high-quality hosts dominate [12], inducing behavioral changes in the prey that increase the risk of infection [91], or by spreading transmission stages while feeding [20].

Predators can also alter evolution in host-parasite systems by introducing trade-offs that alter the selective landscape. As one example, hosts may face trade-offs between anti-predator defenses and mounting an effective immune response [90, 106, 119]. Predators can also influence parasite evolution via impacts on trade-offs, most notably those related to virulence. Virulence is generally defined as the reduction in host fitness caused by infection, usually due to changes in fecundity or lifespan [34]. Overall, much theory related to the evolution of parasite virulence has focused on the influence of parasites on the host's instantaneous mortality rate [5, 34]. Under this framework, parasite fitness increases with transmission rate and also with the length of time that a host is infectious; it is generally assumed that higher replication rates of parasites increase transmission but reduce the length of time a host is infectious (e.g., by killing it or triggering an immune response), driving an intermediate optimal virulence [79, 109]. Because of the nature of this trade-off, increases in mortality rates from sources other than infection (including predators) are expected to increase parasite virulence, since the cost the parasite pays for killing the host quickly is reduced [79], though other outcomes are also possible depending on the specifics of the interaction [26, 34, 69].

The impacts of predation on host evolution have been the focus of several studies in the *Daphnia dentifera-Metschnikowia bicuspidata* system. In this system, predators influence the overall amount of disease from the common fungal parasite *Metschnikowia*: fish preferentially feed on infected *Daphnia*, strongly reducing infection prevalence [41, 43], whereas invertebrate predators are "predator spreaders" that increase infection prevalence [20]. There are also important trade-offs between predation risk, resistance to disease, and fecundity. Larger bodied *Daphnia* are more susceptible to fish predation [15] and to *Metschnikowia* [66], but less susceptible to predation by the common, voracious, gape-limited invertebrate predator *Chaoborus* [96]. In addition, there is also a trade-off between resistance to *Metschnikowia* and fecundity, with larger animals being more fecund but less resistant [66], though some populations contain animals with high fecundity and high resistance [9]. These trade-offs—combined with variation among lakes in vertebrate and invertebrate predation rates, resource levels, and host genetic variation—likely explain different evolutionary responses of populations to disease outbreaks, with some populations evolving increased resistance to disease, some evolving increased susceptibility, and some experiencing disruptive selection on resistance [41, 44, 45].

Work in the *Daphnia dentifera-Metschnikowia bicuspidata* system has focused on evolution of the host but not the parasite because the parasite shows surprisingly little variation and limited evolutionary potential [8, 45, 117]. However, predators seem likely to drive evolution in other *Daphnia* parasites, including the common bacterial parasite *Pasteuria ramosa*. In an artificial selection experiment using *Daphnia dentifera* hosts and *Pasteuria*, parasites that were selected in an environment that simulated high predation evolved to produce more spores in that environment; however, that came at the cost of reduced performance in low predation environments (Figure 21.1, [8]). These results suggest that *Pasteuria* collected from lakes with high predation might be more virulent than those from low predation lakes.

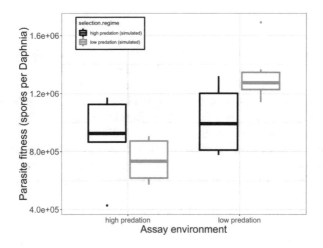

Fig. 21.1: The bacterial parasite *Pasteuria ramosa* evolved higher fitness in high predation environments, but this came at the cost of lower fitness is low predation environments. *Pasteuria* was selected in environments that simulated high predation (shorter host life span; shown with black bars) or environments that simulated low predation (longer host life span; shown with gray bars). Parasite fitness was then assayed in two environments, one simulating high predation (shorter host lifespan) and one simulating low predation (longer host life span). High predation selection lines produced significantly more spores in high predation assay environments than did low predation selection lines when assayed in high predation environments (compare the gray and black bars on the left; planned contrast: $z = -3.07$, $p = 0.0021$). When assayed in low predation environments, however, low predation selection lines produced more spores (compare the gray and black bars on the right; planned contrast: $z = 2.70$, $p = 0.0070$). Data are replotted from [8].

Overall, studies on *Daphnia* and their microparasites demonstrate that predators can strongly alter parasite-mediated selection on host populations, trade-offs faced by hosts and by parasites, and selection on parasite traits.

21.4 Multihost, Multiparasite Interactions: Influences of Competition on Evolution in Host-Parasite Systems

Most parasites can infect multiple hosts, and most hosts are infected by multiple parasites [56, 81]. While this is the rule rather than the exception in nature, the majority of research on host-parasite evolution is based around a one-host one-parasite model [103, 105]. However, there is an expanding field of research uncovering the complex ways in which interspecific competition can change disease dynamics and evolutionary outcomes.

21.4.1 Multiple Hosts

High biodiversity in hosts can dilute [71, 74, 75, 82, 93] or amplify [116, 102, 121, 131] the prevalence of disease. The dilution effect arises when species-rich communities contain lower quality (that is, less competent) hosts that slow the spread of the parasite and therefore protect competent focal hosts from infection. While many studies have documented the dilution effect in wild systems [27], there is still vigorous debate about how common dilution is [102, 112, 131].

One major shortcoming of dilution effect theory is that it generally ignores competition between diluter and focal hosts, despite coexistence theory showing interspecific host competition has a strong impact on host-parasite dynamics [13, 62, 63, 110]. Altering the number of host species changes not only the amount of disease in the system, but also the amount of interspecific competition a focal host experiences, with potentially complex effects on focal host density and disease prevalence [19]. If adding a host species increases total host density, it could potentially drive an increase in disease in a focal host (amplification), even if the additional host is less competent than the focal host [116].

Thus, when considering how multiple hosts might alter evolution in host-parasite systems, it is important to recognize that selection will occur both via changes in the amount of disease and via changes in host density, mediated by interspecific competition. A recent study tested the joint influence of infectious disease and competition on eco-evolutionary dynamics in the *Daphnia-Metschnikowia* system [122]. The additional host species, *Ceriodaphnia*, is more resistant to *Metschnikowia* than the focal host, *Daphnia dentifera*, but also a competitor for resources. The expectation was that the combination of a virulent parasite and strong interspecific competition from *Ceriodaphnia* might drive populations of *Daphnia dentifera* to extinction [122]. Indeed, in populations where *Daphnia dentifera* had little genetic diversity (and thus low evolutionary potential), the combination of parasitism and interspecific competition resulted in very low densities of the focal host. However, in populations where *Daphnia dentifera* had high diversity (and thus high evolutionary potential), the populations thrived. Surprisingly, this rescue effect arose because hosts evolved increased competitive ability, but not increased resistance. Evolution rescued the focal host from the negative impacts of competition, but also drove larger disease out-breaks (as compared to populations with low evolutionary potential). This demonstrates that introducing a diluter host to curb an epidemic may have unexpected results if we ignore the potential for competition—and rapid evolution—between focal and diluter hosts.

At present, we know that interactions between host species can change transmission dynamics and drive evolution in unexpected ways, but the eco-evolutionary effects of parasitism and competition on a focal host remain difficult to predict. However, by integrating a mechanistic understanding of the types of host-host and host-parasite interactions that occur [85, 116, 121], we can better understand how multihost systems can impact host fitness, change parasite transmission dynamics, and ultimately drive rapid evolution in hosts and parasites.

21.4.2 Multiple Parasites

When multiple parasites coexist within a host population they have the potential to influence each other directly (via competition or facilitation within coinfected hosts) or indirectly (e.g., via altering host lifespan or population density). As a result, the addition of a new parasite has the potential to alter selection on existing parasites in the system. Coinfections between helminths (including nematodes) and microparasites have been a particular focus of study, in part as a result of influences of helminths on host immune systems [53]. Work on African buffalo, nematodes, and bovine tuberculosis has demonstrated how coinfecting parasites can influence one another, and also the importance of tests in real world situations. Nematodes suppress the response of the Th1 arm of the immune system in buffalo hosts; Th1 cells protect against microparasites, so the nematode-induced suppression of this part of the immune system should facilitate the invasion of tuberculosis in buffalo [52]. Those results suggest that removing helminths should decrease microparasite fitness. However, treating African buffalo with anthelminthics actually promoted the spread of bovine tuberculosis: anthelminthic treatment did not influence the likelihood of infection with tuberculosis, but did increase survival after infection, increasing transmission opportunities [54]. Such contrasting impacts of coinfection at the within-host scale vs. the host population scale is not unique to macroparasite-microparasite coinfections. As discussed more below, recent work motivated by the *Daphnia-microparasite* system found that priority effects (where the order of infection determines the impacts parasite species have on each other's fitness) can drive scenarios where parasite competition within a host can actually promote coexistence at the population scale [29].

21.4.3 Host Mortality

One way in which multi-parasite infections may alter the evolution of one or more of the coexisting parasites is by changing the lifespan of the host. As discussed in the predation section, shortening the lifespan of a host generally selects for the evolution of higher virulence, as the optimal virulence of a parasite is thought to reflect a trade-off between transmission rate and host mortality [18]. If a single host individual is coinfected—that is, simultaneously infected by two or more parasite strains or species—that has the potential to alter evolutionary outcomes. In particular, if a coinfecting parasite is virulent (increasing mortality rate on the host), that should select for higher virulence in the other parasite [87]. However, both in theory and in practice, coinfections often yield results that are more complicated than might initially be predicted (as reviewed in [3]). For example, in a rodent malaria system, immunopathology leads to additional costs associated with parasite virulence, with the potential to drive negative virulence-transmission relationships [84]. As a result, competition between genotypes coinfecting a single host individual can have major impacts on parasite evolution, increasing or decreasing virulence [84, 88].

Work in the *Daphnia*-parasite system has also demonstrated that interactions between competing parasites can sometimes drive initially counterintuitive results. In an experiment using *Daphnia magna* and the gut microsporidian *Glugoides intestinalis*, treatments with low host mortality rates resulted in the evolution of higher virulence [49]. This pattern arose due to competition between coinfecting strains of the parasite; lower host mortality rates increased the amount of time parasites spent competing amongst themselves within hosts, driving the evolution of faster parasite growth and therefore higher virulence [59]. This underscores the need to understand the mechanisms of within-host interactions in order to predict parasite evolution.

21.4.4 Order of Infection

While much theory on the evolution of virulence focuses on the impacts of changes in host mortality rate, other factors can also influence virulence evolution. Increasingly, scientists are recognizing that the order in which parasites arrive in a host can influence both host and parasite fitness and that those impacts can vary between genotypes [108, 2, 86, 99].

In the *Daphnia-Pasteuria* system, a study found that virulence was influenced not only by infections consisting of multiple strains of a parasite, but also by the order of infection [10]. In simultaneous coinfections or sequential infections where a more virulent parasite strain arrived first, virulence (host mortality rate) and parasite fitness (spore production) matched that of the more virulent strain. However, when the less virulent parasite infected first, virulence resembled an average between single infections of the two strains. Additionally, both parasites suffered lower fitness, likely due to interactions akin to scramble competition. Surprisingly, these mixed-strain infections also led to higher host fecundity than did single infections

(*Pasteuria* has dramatic effects on fecundity [47]), suggesting coinfections may be less harmful to hosts than single infections in the short term. Overall, the authors concluded that high rates of coinfection would select for virulent parasites, which outcompete less-virulent strains [10].

Studies of *Daphnia* infected with multiple parasite species (rather than multiple strains of the same species) also have found that the order of infection is important to host and parasite fitness. A study of *Daphnia galeata*, the fungus *Metschnikowia*, and the ichthyosporean *Caullerya mesnili* found that simultaneous coinfections were significantly more virulent (in terms of host lifespan and fecundity) than were single infections or sequential coinfections [83]. They found that *Caullerya* had higher fitness when it arrived first in sequential coinfections, whereas *Metschnikowia* had higher fitness if it arrived second. A new study on *Daphnia dentifera*, *Pasteuria*, and *Metschnikowia* also found *Metschnikowia* benefitted from second arrival [29]. However, in this case, *Pasteuria* fitness was highest in single infections and low in coinfections, regardless of whether it arrived first or second, likely due to the shortened host lifespan of coinfected hosts. Overall, priority effects can influence parasite prevalence and coexistence, changing pathogen community structure [28, 29], which underscores the importance of linking within- and between-host processes to understand host-multiparasite dynamics.

In the case of the interactions between *Pasteuria* and *Metschnikowia*, it is interesting to note that the dominant driver of low fitness for *Pasteuria* in coinfections seems to be shortened host lifespans driven by *Metschnikowia* [29]. *Pasteuria* is a parasitic castrator, with a relatively slow life history compared to *Metschnikowia* [8]. However, as discussed above in the predation section, experimental evolution studies have demonstrated that *Pasteuria* can evolve to increase its fitness in high mortality environments [8]. In the future, it would be interesting to use experimental evolution to explore the potential of *Pasteuria* to evolve to better compete with *Metschnikowia* and other coinfecting parasites.

21.5 The Influence of the Abiotic Environment on Evolution in Host-Parasite Systems

Humans are dramatically altering the abiotic environment in which host-parasite interactions take place. Perhaps most obviously, climate change is altering mean environmental temperature, as well as the duration and variation of temperature extremes [25], which can strongly influence the outcome of host-parasite interactions [76]. However, climate change also alters precipitation regimes, with consequences for water clarity in aquatic systems [130]. Human activities also drastically alter nutrient levels in natural ecosystems (which drives changes in primary producer communities) and add pesticides and other novel chemicals to environments [23, 120]. Our understanding of evolution in action developed from the *Daphnia*-parasite system makes it clear that these anthropogenic alterations to the abiotic environment should influence evolutionary dynamics of hosts and parasites (Figure 21.2).

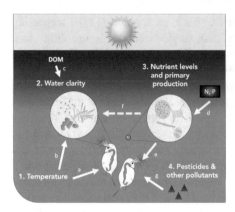

Fig. 21.2: Humans are dramatically altering the abiotic environment, with consequences for
evolution of hosts and parasites. 1) Human activities are increasing average
temperatures as well as the duration of temperature extremes; (a) this has direct effects
on *Daphnia* (by impacting their feeding rate), and (b) parasites (by impacting their
development rates). 2) Human activities are also altering precipitation regimes, which
increases the amount of dissolved organic matter (DOM) arriving in lakes, making
water darker and (c) potentially reducing degradation of parasites by sunlight. 3)
Humans are also altering nutrient levels and therefore (d) primary production, which (e)
changes *Daphnia* feeding rates (and therefore growth and infection rates). (f) The
prevalence of primary producers indirectly affects parasites through changes to host
feeding rates, plus the nutritional and medicinal quality of different phytoplankton
mediate host resistance and tolerance. Finally, 4) agricultural practices and other human
activities are adding pesticides and other novel chemicals to the environment, which (g)
can impact the wellbeing of hosts by reducing their tolerance to pathogens. All of those
changes should have impacts on host-parasite interactions, as discussed in the main text.

21.5.1 Temperature

Climate change is altering mean temperatures as well as variability in temperature in
ecosystems worldwide [25, 33]. Temperature can influence the likelihood of a host
encountering and/or being infected by a pathogen (e.g., [50, 67]), parasite development rates and transmission stage production (e.g., [100]), host thermal stress (with
impacts on things such as immune function [39]), and the fitness impacts of infection on hosts (e.g., [126]). Thus, temperature should strongly influence evolution in
host-parasite systems.

Research on the *Daphnia*-parasite system has helped us understand how altered
temperatures might influence the amount of disease and how hosts evolve in response to disease outbreaks. Recent research on the *Daphnia-Metschnikowia* system
suggests that a warmer world will be a sicker world [118]. A mesocosm experiment
showed that warmer temperatures resulted in larger epidemics, primarily because of
temperature dependence in transmission rates. Temperature-dependent transmission
arose because the host encounters fungal spores while foraging, and foraging rate
(and, therefore, parasite exposure rate) increased with temperature.

An experimental study of the *Daphnia-Pasteuria* system shows that these alterations in disease levels can alter evolutionary outcomes [7]. The timing and magnitude of disease outbreaks depended on mean temperature and temperature variability, as did parasite-driven evolution of the host populations. An increase of 3C drove much larger disease outbreaks that were associated with strong parasite-driven selection and associated reductions in host diversity. Interestingly, this study also looked at a second aspect of environmental variation—the impact of spatial structure on host-parasite populations. The study used physical mixing to homogenize populations, while no mixing allowed populations to form and retain spatial structure. As with temperature, the size of the epidemic and the tempo and mode of evolution were influenced by the mixing treatment. Furthermore, a follow up study found that mixing influenced patterns of adaptation and (co)evolution in the host and parasite [6]. This has interesting potential links with climate change as well, as increasing severity of storms might change mixing regimes in aquatic habitats.

21.5.2 Water Clarity

An underappreciated component of climate change is that increases in heavy precipitation bring more dissolved organic matter into aquatic systems, making surface waters darker and increasing turbidity in the water column [129, 130]. This means that climate change is leading to reduced light penetration in surface waters. This can reduce prey visibility, which will change the rate of predation and its impact on host-parasite dynamics. Notably, one study found that selective predation on infected *Daphnia* was eliminated in high dissolved organic matter conditions [72], so the ability of fish predators to reduce disease in *Daphnia* hosts might be eliminated in darker waters.

Darker surface waters may also reduce the likelihood that waterborne parasites will be killed by sunlight. For example, *Metschnikowia* is highly sensitive to light, and darker lakes generally have larger disease outbreaks [94]. Thus, changes in lake light environments should alter the size of disease outbreaks and the parasite-mediated selection associated with them. Moveover, it might drive selection on the parasite, if parasite genotypes vary in their sensitivity to light. An exciting potential avenue for future research would be to take advantage of spore banks [35, 36] to look for evolutionary change over time in the parasite's ability to tolerate light associated with changes in light penetration.

21.5.3 Nutrient Levels and Primary Production

Humans strongly alter nutrient levels, greatly increasing the amount of bioavailable nitrogen and phosphorus in the environment. This increases primary productivity, which can increase the amount of disease a focal host experiences, especially due to increases in host density [73].

Work on several different *Daphnia*-parasite systems has explored links between nutrient levels, disease, and host fitness or evolution. In the *Daphnia-Metschnikowia* system, more productive lakes had larger disease outbreaks during which hosts evolved greater resistance to infection, whereas lakes with lower productivity had smaller disease outbreaks during which hosts evolved greater susceptibility to infection [44]. In the *Daphnia*-White Fat Cell Disease system, nutrient enrichment increased infection prevalence and intensity [37] but also led to less efficient nutrient assimilation in *Daphnia*, resulting in lower disease tolerance [104]. A study on a natural lake population of *Daphnia longispina* found that the seasonal influx of environmental nutrients increased algal food quality, driving higher prevalence of two gut endoparasites; however, this also drove a decrease in the prevalence of an epibiont and overall parasite species richness [1]. These contradictory effects are likely due to species-specific stoichiometric demands of parasites and hosts [1]. Finally, a laboratory study using the *Daphnia-Pasteuria* system demonstrated that the nutrient content (C:P ratio) of the food a host consumes influences parasite virulence [58].

Increased nutrient levels can also strongly influence the community of primary producers; in lakes, high nutrient levels are typically associated with dominance by cyanobacteria. Work on *Daphnia* and their parasites suggests that cyanobacteria alter host susceptibility, though the specific effects vary across parasites [30, 113, 123]. An interesting avenue for future research will be understanding how human-driven changes in phytoplankton communities alter parasite-driven evolution of *Daphnia* populations.

21.5.4 Pesticides

Pesticides are widely used human-made chemicals, trailing only fertilizers in terms of the extent and amount of use [120]. Work on other systems shows that pesticide use can strongly influence host-parasite interactions. Sub-lethal pesticide exposure has been shown to increase susceptibility of the European honeybee *Apis mellifera* to a gut pathogen, the fungus *Nosema spp.* [98, 132], increase the within-host density of the fungus [97], and even increase the mortality rate of bees already infected with the pathogen [127].

Research on *Daphnia* also shows that pesticides can alter the virulence of their parasites. The virulence of *Pasteuria* on *Daphnia magna* increased with increasing concentrations of the pesticide carbaryl, including higher levels of early mortality and earlier castration of infected hosts [32], even with just short term exposure [31]. Notably, increased virulence was also seen with a second parasite, the microsporidian *Flabelliforma magnivora* [32]. Thus, the presence of pesticides in lakes and ponds could alter the virulence of parasites, which should alter the nature of the transmission-mortality trade-off (and, thus, the evolution of virulence), as well as alter selection on *Daphnia* populations. In future research, it would also be interesting to focus on the impact of other anthropogenic pollutants, including road salts [22], on *Daphnia*-parasite interactions.

21.6 Conclusions

In the 70 years since Haldane [65] suggested that parasites might be especially important drivers of evolution, it has become abundantly clear that parasitism is, indeed, a major selective force. Haldane contrasted the impacts of parasites with those of other "natural forces, hunger, and predators, or with members of the same species". However, we now know that populations are not influenced by parasites *or* by other food web members—rather, they all interact. Thus, when studying evolution in host-parasite interactions, we need to consider that the amount of disease and the nature and tempo of evolution will be modulated by the biotic and abiotic context in which the host-parasite interaction is embedded.

Acknowledgements We thank Patrick Clay, Camden Gowler, Clara Shaw, and four anonymous reviewers for feedback on earlier drafts of this manuscript. We acknowledge support from the National Science Foundation (DEB-1655856 and DEB-1748729). We also want to acknowledge the innovative and interdisciplinary research carried out through the BEACON Center, and to thank Erik Goodman for his excellent leadership and vision.

References

1. Aalto, S., Ketola, T., Pulkkinen, K.: *No uniform associations between parasite prevalence and environmental nutrients.* Ecology **95**, 2558–2568 (2014)
2. Al-Naimi, F., Garrett, K., Bockus, W.: *Competition, facilitation, and niche differentiation in two foliar pathogens.* Oecologia **143**, 449–457 (2005)
3. Alizon, S., de Roode, J., Michalakis, Y.: *Multiple infections and the evolution of virulence.* Ecology Letters **16**, 556–567 (2013)
4. Allison, A.: *Protection afforded by sickle-cell trait against subtertian malareal infection.* Br Med J **1**, 290–294 (1954)
5. Anderson, RM and, M.R.: *Coevolution of hosts and parasites.* Parasitology **85**, 411 (1982)
6. Auld, S., Brand, J.: *Environmental variation causes different (co) evolutionary routes to the same adaptive destination across parasite populations.* Evolution Letters **1**, 245–254 (2017)
7. Auld, S., Brand, J.: *Simulated climate change, epidemic size, and host evolution across host-parasite populations.* Global Change Biology **23**, 5045–5053 (2017)
8. Auld, S., Hall, S., Ochs, J., Sebastian, M., Duffy, M.: *Predators and patterns of within-host growth can mediate both among-host competition and evolution of transmission potential of parasites.* American Naturalist **18**, S77–S90 (2014)
9. Auld, S., Penczykowski, R., Ochs, J., Grippi, D., Hall, S., Duffy, M.: *Variation in costs of parasite resistance among natural host populations.* Journal of Evolutionary Biology **26**, 2479–2486 (2013)
10. Ben-Ami, F., Mouton, L., Ebert, D.: *The effects of multiple infections on the expression and evolution of virulence in a Daphnia-endoparasite system.* Evolution **62**, 1700–1711 (2008)

11. Boots, M., Hudson, P., Sasaki, A.: *Large shifts in pathogen virulence relate to host population structure.* Science **303**, 842–844 (2004)
12. Borer, E., Mitchell, C., Power, A., Seabloom, E.: *Consumers indirectly increase infection risk in grassland food webs.* Proceedings of the National Academy of Sciences **106**, 503–506 (2009)
13. Bowers, R., Turner, J.: *Community structure and the interplay between interspecific infection and competition.* Journal of Theoretical Biology **187**, 95–109 (1997)
14. Britton, J.: *Introduced parasites in food webs: new species, shifting structures?* Trends in Ecology & Evolution **28**, 93–99 (2013)
15. Brooks, J., Dodson, S.: *Predation, body size, and composition of plankton.* Science **150**, 28–35
16. Buckling, A., Rainey, P.: *The role of parasites in sympatric and allopatric host diversification.* Nature **420**, 496–499 (2002)
17. Buckling, A., Taylor, L., Jane, M., Read, A.: *Adaptive changes in plasmodium transmission strategies following chloroquine chemotherapy.* Proceedings of the Royal Society of London. Series B: Biological Sciences **264**, 553–559 (1997)
18. Bull, J., Lauring, A.: *Theory and empiricism in virulence evolution.* PLoS Pathogens **10**(e1004387) (2014)
19. Caceres, C., Davis, G., Duple, S., Hall, S., Koss, A., Lee, P., Rapti, Z.: *Complex Daphnia interactions with parasites and competitors.* Mathematical Biosciences **258**, 148–161 (2014)
20. Caceres, C., Knight, C., Hall, S.: *Predator spreaders: Predation can enhance parasite success in a planktonic host-parasite system.* Ecology **90**, 2850–2858 (2009)
21. Caceres, C., Tessier, A., Duffy, M., Hall, S.: *Disease in freshwater zooplankton: What have we learned and where are we going?* Journal of Plankton Research **36**, 326–333 (2014)
22. Cando-Arguelles, M., Kefford, B., Schaefer, R.: *Salt in freshwaters: Causes, effects and prospects - Introduction to the theme issue.* Philosophical Transactions of the Royal Society B: Biological Sciences **374**, 20180,002 (2019)
23. Carpenter, S.: *Phosphorus control is critical to mitigating eutrophication.* Proceedings of the National Academy of Sciences **105**, 11,039–11,040 (2008)
24. Carpenter, S., Caraco, N., Correll, D., Howarth, R., Sharpley, A., Smith, V.: *Nonpoint pollution of surface waters with phosphorus and nitrogen.* Ecological Applications **8**, 559–568 (1998)
25. CB, F., et al.: *Managing the risks of extreme events and disasters to advance climate change adaptation: Special report of the Intergovernmental Panel on Climate Change (IPCC).* Cambridge University Press, Cambridge, UK (2012)
26. Choo, K., Williams, P., Day, T.: *Host mortality, predation and the evolution of parasite virulence.* Ecology Letters **6**, 310–315 (2003)
27. Civitello, D., et al.: *Biodiversity inhibits parasites: Broad evidence for the dilution effect.* Proceedings of the National Academy of Sciences **112**, 8667–8671 (2015)
28. Clay, P., Cortez, M., Duffy, M., Rudolf, V.: *Priority effects within co-infected hosts can drive unexpected population-scale patterns of parasite prevalence.* Oikos **128**, 571–583 (2019)
29. Clay, P., Dhir, K., Rudolf, V., Duffy, M.: *Within host priority effects systematically alter pathogen coexistence.* American Naturalist **193**, 187–199 (2019)
30. Coopman, M., Muylaert, K., Lange, B., Reyserhove, L., Decaestecker, E.: *Context dependency of infectious disease: the cyanobacterium Microcystis aeruginosa decreases white bacterial disease in Daphnia magna.* Freshwater Biology **59**, 714–723 (2014)
31. Coors, A., De Meester, L.: *Fitness and virulence of a bacterial endoparasite in an environmentally stressed crustacean host.* Parasitology **138**, 122–131 (2011)
32. Coors, A., Decaestecker, E., Jansen, M., Meester, L.: *Pesticide exposure strongly enhances parasite virulence in an invertebrate host model.* Oikos **117**, 1840–1846 (2008)
33. Coumou, D., Rahmstorf, S.: *A decade of weather extremes.* Nature Climate Change **2**, 491 (2012)
34. Day, T.: *On the evolution of virulence and the relationship between various measures of mortality.* Proceedings of the Royal Society of London, B **269**, 1317–1323 (2002)

35. Decaestecker, E., Gaba, S., Raeymaekers, J., Stoks, R., Van Kerckhoven, L., Ebert, D., De Meester, L.: *Host-parasite 'Red Queen' dynamics archived in pond sediment.* Nature **450**, 870–873 (2007)
36. Decaestecker, E., Lefever, C., De Meester, L.: *Haunted by the past: Evidence for dormant stage banks of microparasites and epibionts of Daphnia.* Limnology and Oceanography **49**, 1355–1364 (2004)
37. Decaestecker, E., Verreydt, D., Meester, L.D., Declerck, S.A.J.: *Parasite and nutrient enrichment effects onDaphnia interspecific competition.* Ecology **96**(5), 1421–1430 (2015)
38. Deom, C.M., Caton, A.J., Schulze, I.T.: *Host cell-mediated selection of a mutant influenza A virus that has lost a complex oligosaccharide from the tip of the hemagglutinin.* Proceedings of the National Academy of Sciences **83**(11), 3771–3775 (1986)
39. Dittmar, J., Janssen, H., Kuske, A., Kurtz, J., Scharsack, J.P.: *Heat and immunity: An experimental heat wave alters immune functions in three-spined sticklebacks (Gasterosteus aculeatus).* Journal of Animal Ecology **83**(4), 744–757 (2014)
40. Doherty, T.S., Glen, A.S., Nimmo, D.G., Ritchie, E.G., Dickman, C.R.: *Invasive predators and global biodiversity loss.* Proceedings of the National Academy of Sciences **113**(40), 11,261–11,265 (2016)
41. Duffy, M., Brassil, C., Hall, S., Tessier, A., Caceres, C., Conner, J.: *Parasite mediated disruptive selection in a natural Daphnia population.* BMC Evolutionary Biology **8**, 80 (2008)
42. Duffy, M., Caceres, C., Hall, S.: *Healthy herds or predator spreaders? Insights from the plankton into how predators suppress and spread disease.* In: Wildlife Disease Ecology: Linking Theory to Data and Application, p. 458-479. Cambridge University Press, Cambridge, UK (2019)
43. Duffy, M., Hall, S., Tessier, A., Huebner, M.: *Selective predators and their parasitized prey: Are epidemics in zooplankton under top-down control?* Limnology and Oceanography **50**, 412–420 (2005)
44. Duffy, M., Ochs, J., Penczykowski, R., Civitello, D., Klausmeier, C., Hall, S.: *Ecological context influences epidemic size and parasite-mediated selection.* Science **335**, 1636–1638 (2012)
45. Duffy, M., Sivars-Becker, L.: *Rapid evolution and ecological host-parasite dynamics.* Ecology Letters **10**, 44–53 (2007)
46. Dybdahl, M., Lively, C.: *Host-parasite coevolution: Evidence for rare advantage and time-lagged selection in a natural population.* Evolution **52**, 1057–1066 (1998)
47. Ebert, D.: *Ecology, epidemiology and evolution of parasitism in Daphnia.* National Library of Medicine (US), National Center for Biotechnology Information, Bethesda, MD (2005)
48. Ebert, D.: *A genome for the environment.* Science **331**(6017), 539–540 (2011)
49. Ebert, D., Mangin, K.: *The influence of host demography on the evolution of virulence of a microsporidian gut parasite.* Evolution **51**, 1828–1837 (1997)
50. Elderd, B.D., Reilly, J.R.: *Warmer temperatures increase disease transmission and outbreak intensity in a host-pathogen system.* Journal of Animal Ecology **83**(4), 838–849 (2014)
51. Epstein, B., Jones, M., Hamede, R., Hendricks, S., McCallum, H., Murchison, E.P., Schönfeld, B., Wiench, C., Hohenlohe, P., Storfer, A.: *Rapid evolutionary response to a transmissible cancer in Tasmanian devils.* Nature Communications **7**(1), 12,684 (2016)
52. Ezenwa, V., Rampal, E., Gordon, L., Beja-Pereira, A., Jolles, A.: *Hidden consequences of living in a wormy world: Nematode-induced immune suppression facilitates tuberculosis invasion in African buffalo.* The American Naturalist **176**, 613–624 (2010)
53. Ezenwa, V.O.: *Helminth-microparasite co-infection in wildlife: Lessons from ruminants, rodents and rabbits.* Parasite Immunology **38**(9), 527–534 (2016)
54. Ezenwa, V.O., Jolles, A.E.: *Opposite effects of anthelmintic treatment on microbial infection at individual versus population scales.* Science **347**(6218), 175–177 (2015)
55. Fenner, F., Fantini, B.: *Biological control of vertebrate pests: The history of Myxomatosis, an experiment in evolution.* CABI Publishing, Wallingford, Oxon, UK (1999)
56. Fenton, A., Pedersen, A.: *Community epidemiology framework for classifying disease threats.* Emerging Infectious Diseases **11**, 1815–1821 (2005)

57. Frisch, D., Morton, P.K., Chowdhury, P.R., Culver, B.W., Colbourne, J.K., Weider, L.J., Jeyasingh, P.D.: *A millennial-scale chronicle of evolutionary responses to cultural eutrophication in Daphnia.* Ecology Letters **17**(3), 360–368 (2014)

58. Frost, P., Ebert, D., Smith, V.: *Responses of a bacterial pathogen to phosphorus limitation of its aquatic invertebrate host.* Ecology **89**, 313–318 (2008)

59. Gandon, S., Jansen, V.A.A., Baalen, M.V.: *Host life history and the evolution of parasite virulence.* Evolution **55**(5), 1056–1062 (2001)

60. Gibson, A.K., Delph, L.F., Vergara, D., Lively, C.M.: *Periodic, parasite-mediated selection for and against sex.* The American Naturalist **192**(5), 537–551 (2018)

61. Green, J.: *Parasites and epibionts of Cladocera.* Transactions of the Zoological Society of London **32**, 417–515 (1974)

62. Greenman, J., Hudson, P.: *Parasite-mediated and direct competition in a two-host shared macroparasite system.* Theoretical Population Biology **57**(1), 13–34 (2000)

63. Gyllenberg, M., Liu, X., Yan, P.: *An eco-epidemiological model in two competing species.* Differential Equations & Applications (4), 495–519 (2012)

64. Hairston NG, J., et al.: *Rapid evolution revealed by dormant eggs.* Nature **401**, 446 (1999)

65. Haldane, J.: *Disease and evolution.* La Ricerca Scientifica Supplemento **19**, 1–11 (1949)

66. Hall, S., Becker, C., Duffy, M., Caceres, C.: *Genetic variation in resource acquisition and use among hosts creates key epidemiological tradeoffs.* American Naturalist **176**, 557–565 (2010)

67. Hall, S., Tessier, A., Duffy, M., Huebner, M., Caceres, C.: *Warmer does not have to mean sicker: Temperature and predators can jointly drive timing of epidemics.* Ecology **87**, 1684–1695 (2006)

68. Hendry, A.: *Eco-evolutionary dynamics.* Princeton University Press, Princeton, NJ (2016)

69. Houwenhuyse, S., Macke, E., Reyserhove, L., Bulteel, L., Decaestecker, E.: *Back to the future in a petri dish: Origin and impact of resurrected microbes in natural populations.* Evolutionary Applications **11**(1), 29–41 (2017)

70. Hudson, P., Dobson, A., Newborn, D.: *Do parasites make prey vulnerable to predation? Red grouse and parasites.* Journal of Animal Ecology **61**, 681–692 (1992)

71. Johnson, P., et al.: *Biodiversity decreases disease through predictable changes in host community competence.* Nature **494**, 230–233 (2013)

72. Johnson, P., Stanton, D., Preu, E., Forshay, K., Carpenter, S.: *Dining on disease: How interactions between parasite infection and environmental conditions affect host predation risk.* Ecology **87**, 1973–1980 (2006)

73. Johnson, P.T.J., Chase, J.M., Dosch, K.L., Hartson, R.B., Gross, J.A., Larson, D.J., Sutherland, D.R., Carpenter, S.R.: *Aquatic eutrophication promotes pathogenic infection in amphibians.* Proceedings of the National Academy of Sciences **104**(40), 15,781–15,786 (2007)

74. Keesing, F., et al.: *Impacts of biodiversity on the emergence and transmission of infectious diseases.* Nature **468**, 647–652 (2010)

75. Keesing, F., Holt, R., Ostfeld, R.: *Effects of species diversity on disease risk.* Ecology Letters **9**, 485–498 (2006)

76. Lafferty, K.D.: *The ecology of climate change and infectious diseases.* Ecology **90**(4), 888–900 (2009)

77. Lampert, W.: *Daphnia: Model herbivore, predator and prey.* Polish Journal of Ecology **54**, 607–620 (2006)

78. Lampert, W.: Parasitism. In: Daphnia: Development of a model organism in ecology and evolution, pp. 113–189. Excellence in Ecology Series. International Ecology Institute, Olendorf/Luhe, Germany (2011)

79. Lenski, R., May, R.: *The evolution of virulence in parasites and pathogens - reconciliation between 2 competing hypotheses.* Journal of Theoretical Biology **169**, 253–265 (1994)

80. Little, T., Ebert, D.: *Evolutionary dynamics of Daphnia and their microparasites.* Dronamraju K (ed) Infectious Disease: Host-pathogen evolution pp. 222–240 (2004)

81. Lively, C.M., de Roode, J.C., Duffy, M.A., Graham, A.L., Koskella, B.: *Interesting open questions in disease ecology and evolution.* The American Naturalist **184**(S1), S1–S8 (2014)

82. LoGiudice, K., et al.: *The ecology of infectious disease: Effects of host diversity and community composition on lyme disease risk.* Proceedings of the National Academy of Sciences **100**, 567–571 (2003)

83. Lohr, J.N., Yin, M., Wolinska, J.: *Prior residency does not always pay off – co-infections in Daphnia.* Parasitology **137**(10), 1493–1500 (2010)

84. Long, G.H., Graham, A.L.: *Consequences of immunopathology for pathogen virulence evolution and public health: Malaria as a case study.* Evolutionary Applications **4**(2), 278–291 (2011)

85. Luis, A.D., Kuenzi, A.J., Mills, J.N.: *Species diversity concurrently dilutes and amplifies transmission in a zoonotic host–pathogen system through competing mechanisms.* Proceedings of the National Academy of Sciences **115**(31), 7979–7984 (2018)

86. Marchetto, K.M., Power, A.G.: *Coinfection timing drives host population dynamics through changes in virulence.* The American Naturalist **191**(2), 173–183 (2018)

87. May, R., Nowak, M.: *Coinfection and the evolution of parasite virulence.* Proceedings of the Royal Society of London. Series B: Biological Sciences **261**(1361), 209–215 (1995)

88. Mideo, N.: *Parasite adaptations to within-host competition.* Trends in Parasitology **25**(6), 261–268 (2009)

89. Miner, B.E., Meester, L.D., Pfrender, M.E., Lampert, W., Hairston, N.G.: *Linking genes to communities and ecosystems: Daphnia as an ecogenomic model.* Proceedings of the Royal Society B: Biological Sciences **279**(1735), 1873–1882 (2012)

90. Navarro, C.: *Predation risk, host immune response, and parasitism.* Behavioral Ecology **15**(4), 629–635 (2004)

91. Orlofske, S.A., Jadin, R.C., Hoverman, J.T., Johnson, P.T.J.: *Predation and disease: Understanding the effects of predators at several trophic levels on pathogen transmission.* Freshwater Biology **59**(5), 1064–1075 (2014)

92. Ostfeld, R., Holt, R.: *Are predators good for your health? Evaluating evidence for top-down regulation of zoonotic disease reservoirs.* Frontiers in Ecology and the Environment **2**, 13–20 (2004)

93. Ostfeld, R., Keesing, F.: *Effects of host diversity on infectious disease.* Annual Review of Ecology, Evolution, and Systematics **43**, 157–182 (2012)

94. Overholt, E., Hall, S., Williamson, C., Meikle, C., Duffy, M., Caceres, C.: *Solar radiation decreases parasitism in Daphnia.* Ecology Letters **15**, 47–54 (2012)

95. Packer, C., Holt, R., Hudson, P., Lafferty, K., Dobson, A.: *Keeping the herds healthy and alert: Implications of predator control for infectious disease.* Ecology Letters **6**, 797–802 (2003)

96. Pastorok, R.: *Prey vulnerability and size selection by Chaoborus larvae.* Ecology **62**, 1311–1324 (1981)

97. Pettis, J.S., van Engelsdorp, D., Johnson, J., Dively, G.: *Pesticide exposure in honey bees results in increased levels of the gut pathogen nosema.* Naturwissenschaften **99**(2), 153–158 (2012)

98. Pettis, J.S., Lichtenberg, E.M., Andree, M., Stitzinger, J., Rose, R., van Engelsdorp, D.: *Crop pollination exposes honey bees to pesticides which alters their susceptibility to the gut pathogen Nosema ceranae.* PLoS ONE **8**(7), e70,182 (2013)

99. Pollitt, L.C., Bram, J.T., Blanford, S., Jones, M.J., Read, A.F.: *Existing infection facilitates establishment and density of malaria parasites in their mosquito vector.* PLoS Pathogens **11**(7), e1005,003 (2015)

100. Poulin, R.: *Global warming and temperature-mediated increases in cercarial emergence in trematode parasites.* Parasitology **132**(01), 143–151 (2006)

101. Prugh, L.R., Stoner, C.J., Epps, C.W., Bean, W.T., Ripple, W.J., Laliberte, A.S., Brashares, J.S.: *The rise of the mesopredator.* BioScience **59**(9), 779–791 (2009)

102. Randolph, S.E., Dobson, A.D.M.: *Pangloss revisited: A critique of the dilution effect and the biodiversity-buffers-disease paradigm.* Parasitology **139**(07), 847–863 (2012)

103. Read, A., Taylor, L.: *The ecology of genetically diverse infections.* Science **292**, 1099–1102

104. Reyserhove, L., Samaey, G., Muylaert, K., Coppé, V., Colen, W.V., Decaestecker, E.: *A historical perspective of nutrient change impact on an infectious disease in Daphnia*. Ecology **98**(11), 2784–2798 (2017)

105. Rigaud, T., Perrot-Minnot, M.J., Brown, M.J.F.: *Parasite and host assemblages: Embracing the reality will improve our knowledge of parasite transmission and virulence*. Proceedings of the Royal Society B: Biological Sciences **277**(1701), 3693–3702 (2010)

106. Rigby, M.C., Jokela, J.: *Predator avoidance and immune defence: Costs and trade–offs in snails*. Proceedings of the Royal Society of London. Series B: Biological Sciences **267**(1439), 171–176 (2000)

107. Rogalski, M.A.: *Maladaptation to acute metal exposure in resurrected Daphnia ambigua clones after decades of increasing contamination*. The American Naturalist **189**(4), 443–452 (2017)

108. de Roode, J., et al.: *Virulence and competitive ability in genetically diverse malaria infections*. Proceedings of the National Academy of Sciences of the United States of America **102**, 7624–7628 (2005)

109. de Roode, J., Yates, A., Altizer, S.: *Virulence-transmission trade-offs and population divergence in virulence in a naturally occurring butterfly parasite*. Proceedings of the National Academy of Sciences **105**, 7489–7494 (2008)

110. Saenz, R., Hethcote, H.: *Competing species models with an infectious disease*. Math Biosci Eng **3**, 219–235 (2006)

111. Sala, O.E.: *Global biodiversity scenarios for the year 2100*. Science **287**(5459), 1770–1774 (2000)

112. Salkeld, D.J., Padgett, K.A., Jones, J.H.: *A meta-analysis suggesting that the relationship between biodiversity and risk of zoonotic pathogen transmission is idiosyncratic*. Ecology Letters **16**(5), 679–686 (2013)

113. Sanchez, K., Huntley, N., Duffy, M., Hunter, M.: *Toxins or medicines? Algal diets mediate parasitism in a freshwater host-parasite system*. in press

114. Schiebelhut, L.M., Puritz, J.B., Dawson, M.N.: *Decimation by sea star wasting disease and rapid genetic change in a keystone species, Pisaster ochraceus*. Proceedings of the National Academy of Sciences **115**(27), 7069–7074 (2018)

115. Schild, G.C., Oxford, J.S., de Jong, J.C., Webster, R.G.: *Evidence for host-cell selection of influenza virus antigenic variants*. Nature **303**(5919), 706–709 (1983)

116. Searle, C., et al.: *Population density, not host competence, drives patterns of disease in an invaded community*. American Naturalist **188**, 554–566 (2016)

117. Searle, C.L., Ochs, J.H., Cáceres, C.E., Chiang, S.L., Gerardo, N.M., Hall, S.R., Duffy, M.A.: *Plasticity, not genetic variation, drives infection success of a fungal parasite*. Parasitology **142**(06), 839–848 (2015)

118. Shocket, M., et al.: *Warmer is sicker in a zooplankton-fungus system: a trait-driven approach points to higher transmission via host foraging*. American Naturalist **191**, 435–451 (2018)

119. Stoks, R., Block, M.D., Slos, S., Doorslaer, W.V., Rolff, J.: *Time constraints mediate predator-induced plasticity in immune function, condition and life history*. Ecology **87**(4), 809–815 (2006)

120. Stokstad, E., Grulln, G.: *Infographic: Pesticide planet*. Science **341**, 730–731

121. Strauss, A.T., Civitello, D.J., Cáceres, C.E., Hall, S.R.: *Success, failure and ambiguity of the dilution effect among competitors*. Ecology Letters **18**(9), 916–926 (2015)

122. Strauss, A.T., Hite, J.L., Shocket, M.S., Cáceres, C.E., Duffy, M.A., Hall, S.R.: *Rapid evolution rescues hosts from competition and disease but—despite a dilution effect—increases the density of infected hosts*. Proceedings of the Royal Society B: Biological Sciences **284**(1868), 20171,970 (2017)

123. Tellenbach, C., Tardent, N., Pomati, F., Keller, B., Hairston, N.G., Wolinska, J., Spaak, P.: *Cyanobacteria facilitate parasite epidemics in Daphnia*. Ecology **97**(12), 3422–3432 (2016)

124. Toenshoff, E.R., Fields, P.D., Bourgeois, Y.X., Ebert, D.: *The end of a 60-year riddle: Identification and genomic characterization of an iridovirus, the causative agent of white fat cell disease in zooplankton*. G3: Genes|Genomes|Genetics p. g3.300429.2017 (2018)

125. Urban, M.C.: *Accelerating extinction risk from climate change.* Science **348**(6234), 571–573 (2015)
126. Vale, P., Stjernman, M., Little, T.: *Temperature-dependent costs of parasitism and maintenance of polymorphism under genotype-by-environment interactions.* Journal of Evolutionary Biology **21**, 1418–1427 (2008)
127. Vidau, C., Diogon, M., Aufauvre, J., Fontbonne, R., Viguès, B., Brunet, J.L., Texier, C., Biron, D.G., Blot, N., Alaoui, H.E., Belzunces, L.P., Delbac, F.: *Exposure to sublethal doses of fipronil and thiacloprid highly increases mortality of honeybees previously infected by Nosema ceranae.* PLoS ONE **6**(6), e21,550 (2011)
128. Weiss, L.C., Pötter, L., Steiger, A., Kruppert, S., Frost, U., Tollrian, R.: *Rising pCO$_2$ in freshwater ecosystems has the potential to negatively affect predator-induced defenses in Daphnia.* Current Biology **28**(2), 327–332.e3 (2018)
129. Williamson, C.E., Madronich, S., Lal, A., Zepp, R.G., Lucas, R.M., Overholt, E.P., Rose, K.C., Schladow, S.G., Lee-Taylor, J.: *Climate change-induced increases in precipitation are reducing the potential for solar ultraviolet radiation to inactivate pathogens in surface waters.* Scientific Reports **7**(1), 13,033 (2017)
130. Williamson, C.E., Overholt, E.P., Pilla, R.M., Leach, T.H., Brentrup, J.A., Knoll, L.B., Mette, E.M., Moeller, R.E.: *Ecological consequences of long-term browning in lakes.* Scientific Reports **5**(1), 18,666 (2015)
131. Wood, C.L., Lafferty, K.D., DeLeo, G., Young, H.S., Hudson, P.J., Kuris, A.M.: *Does biodiversity protect humans against infectious disease?* Ecology **95**(4), 817–832 (2014)
132. Wu, J.Y., Smart, M.D., Anelli, C.M., Sheppard, W.S.: *Honey bees (Apis mellifera) reared in brood combs containing high levels of pesticide residues exhibit increased susceptibility to Nosema (Microsporidia) infection.* Journal of Invertebrate Pathology **109**(3), 326–329 (2012)

Chapter 22
Toward a Model of Investigating "Coordinated Stasis" Through Habitat Tracking in Communities of Digital Organisms

Zaki Ahmad Khan, Faraz Hasan, Matthew R. Rupp, Jian-long Zhu, Tian-tong Luo and Gabriel Yedid

Abstract The idea of stable biotic communities that are resilient to disturbance and invasion, and in which evolutionary change may be impeded by ecological interactions, has garnered considerable attention in both theoretical ecology and paleobiology. In the latter field, stable communities are often discussed in terms of "coordinated stasis", where entire biotic communities may display the canonical evolutionary stasis of punctuated equilibria for significant periods of time. A prevailing hypothesis that has been advanced to explain coordinated stasis is habitat tracking, in which species evolve little as long as they can migrate to regions of habitat containing their optimal living conditions. In this contribution, we describe how we have modified Avida, a proven digital evolution platform, for investigating the habitat tracking hypothesis in a rigorous, experimental manner. We model habitats consisting of spatially localized fields of resources and fixed landmarks, and modified Avida to give its digital organisms capabilities for movement, vision, resource sensing, and resource consumption. Our preliminary results show that while the movement and ecological subsystems we implemented each produce positive outcomes in isolation, they conflict with each other when combined and result in these outcomes occurring infrequently. Nonetheless, the results validate the effectiveness

Zaki Ahmad Khan · Jian-long Zhu · Tian-tong Luo · Gabriel Yedid
Nanjing Agricultural University
Department of Zoology
College of Life Sciences
Weigang, Nanjing, Jiangsu, 210095, China e-mail: gyedid02@gmail.com

Faraz Hasan
Department of Computer Applications
B.S. Abdur Rahman Crescent Institute of Science & Technology
Chennai, Tamil Nadu, 600048, India

Matthew R. Rupp
Michigan State University
BEACON Center For The Study of Evolution In Action
East Lansing, Michigan, 48824, U.S.A.

© Springer Nature Switzerland AG 2020
W. Banzhaf et al. (eds.), *Evolution in Action: Past, Present and Future*,
Genetic and Evolutionary Computation, https://doi.org/10.1007/978-3-030-39831-6_22

of these additions and set the stage for further habitat relocation experiments, with clear predictions about expected responses.

22.1 Introduction

22.1.1 Community Stability in Ecology and Evolution

The idea of "stable" ecological communities that are resistant to various kinds of disturbances and invasion has been extensively examined and debated in the ecological literature for more than half a century, particularly over whether there is a correlation between high diversity and high stability (see [36] for review). Recent theoretical research with food web models that account not just for web topology, but also interaction strengths and types, have demonstrated that more diverse communities may indeed be more stable if they include both antagonistic and mutualistic interactions [41, 42, 53], and also if much of their diversity contributes to weak interactions representing slow energy channels in the food web [37, 39, 52]. Such studies usually focus on the multispecies community level, but consideration of factors at the intraspecific and whole-ecosystem levels is also required for understanding the diversity/stability relationship [46].

Despite the traditional focus on community stability from a purely ecological perspective, the issue is also of interest from an evolutionary standpoint, as it raises the question of the relationship between ecological interactions and evolutionary responses to change [28, 31, 43, 55]. There is some evidence that greater diversity in a community may inhibit evolutionary responses [2, 3, 10, 35, 51], and the spatial extent of interactions and presence of non-transitivity can also contribute to evolutionary stability [29, 50]. Yet, a comprehensive theory that accounts for diversity and stability in an evolutionary context remains elusive. Consideration of community stability may also be encountered in the paleobiological realm [4, 11, 15, 58], particularly in the context of "coordinated stasis".

22.1.2 The Problem of Coordinated Stasis

"Coordinated stasis" refers to the idea that entire multispecies paleo-communities may show morphological, taxonomic and ecological persistence over geologically significant time periods, interrupted by episodes of abrupt mass turnover [7, 14, 57]. Such communities have been suggested to have the following properties:

1. the types of species within a community are stable, as is the geological context in which they occur
2. long-lasting species composition and community organization as measured by species list, species richness, rank order, abundance distribution, etc.

3. the component species themselves show long-lasting morphological invariance (i.e. the canonical stasis of punctuated equilibria [17];
4. resistance to minor environmental disturbances and invasion (both external from other communities and internal from in situ evolved species); and
5. pulsed episodes of extinction and origination, with episodes of rapid turnover and change that are short compared to the preceding and succeeding intervals of stasis. The new communities that emerge then behave in a like manner until the next such event.

Two major hypotheses have been advanced to explain coordinated stasis. One is habitat tracking, which posits that organisms track patches of preferred habitat as a response to changing environmental conditions; evolutionary change is only occurs when suitable habitat is no longer accessible [9, 18, 25]. The alternative is "ecological locking", in which a highly interconnected web of ecological interactions actually hinders evolutionary change. Evolution can proceed only if those interactions are broken through a massive environmental disruption [40], often involving extinction of key community members. Habitat tracking has recently been most accepted and emphasized by researchers in the field, given the lack of empirical support for (and difficulty of testing) ecological locking [6]. Leaving causal explanations aside, most recent research concerning coordinated stasis has focused on understanding its universality and evolutionary significance [8, 9, 24, 26, 48, 51].

However, coordinated stasis has proven difficult to study because of problems posed by the uneven preservation of the fossil record. Due to geological issues affecting record quality, many reported paleobiologic patterns may be partially or wholly artifactual [23, 30]. Initial conclusions may be overturned based on new and re-analyzed data even from the same geological sources (compare [5] vs. [9] and [27]. Incorrect identification of taxa, due to post-turnover external invasions [6] or undetected convergent evolution [25] may lead to erroneous conclusions about stasis of particular lineages within communities. Beyond these problems, there is a lack of repeatability and control; one can never know how evolution might have proceeded under different conditions.

22.1.3 Why Use Digital Evolution to Study Coordinated Stasis?

Given the aforementioned problems with studying such evolutionary issues only from fossil data, it becomes desirable to have an experimental platform with which questions about these phenomena can be addressed without the complications of the geological record. Digital evolution offers some compelling advantages for studying macro-evolutionary phenomena including coordinated stasis. First, and most importantly, it permits a truly manipulative, experimental approach to the problem, particularly the ability to "replay life's tape." With digital evolution, one can make changes at specific time points in what would otherwise be identical experimental replicates. For example, how would evolution unfold differently in the presence or absence of a particular macro-ecological treatment? Second, data can be recorded in far greater detail than is possible with the fossil record, without preservational artifacts that complicate quantitative analysis of fossil data. Although not all of the subtleties and complexity of the geological record can be recreated with digital evolution, it still provides a tractable system that allows one to focus on key data most appropriate for investigating the problem at hand, and to design appropriate follow-up experiments. Third, because digital evolution is an actual instantiation of Darwinian evolutionary processes, it allows the problem to be addressed with greater realism of population-dynamic, ecological, and evolutionary factors, as compared to other types of evolutionary simulations used in paleobiology (e.g., [47, 54].

22.2 Experimental Platform and Methodology

22.2.1 Experimental Platform

We used a development version (v.2.14.4) of the digital evolution software Avida [1, 45] as our experimental platform, allowing a complete record of the course of evolution. Avida is a system that features variation in digital organisms called *Avidians* through mutation, inheritance, and competition for limited resources, which are the core ingredients necessary for Darwinian evolution. Avida has been used to examine questions as diverse as mutational robustness and evolvability, evolution of complex, multipart features from simpler precursors, emergence of stable ecosystems through density-dependent selection, host-parasite coevolution, recovery from mass extinction, and more [13, 19, 22, 32, 33, 38, 49, 60, 62, 65].

The mechanics of Avida have been detailed at length elsewhere (see especially [45], but some details of our particular implementation are salient. First, the ecosystem has a finite population size, introducing an element of drift as new Avidians displace the old. Second, the environment features multiple depletable resources, which are associated with certain computations—called *tasks*—where a low concentration of a resource reduces the benefit gained by performing the associated task [12, 13, 59]. This feature leads to density-dependent competition for resources, favouring Avidians that more efficiently consume particular resources or target underutilized ones. Resources are spatially localized in the environment and are available only to Avidians in a local neighbourhood (Figure 22.1, left panel). Only a limited number of these resources are supplied exogenously, while the remainder are generated as by-products by Avidians when they successfully complete certain associated tasks [61, 63]. This spatially-localized cross-feeding introduces ecological interdependence, habitat heterogeneity, and niche construction [44] into the population dynamics, which are features of real communities.

Even with such a tractable system, it remains quite unclear how to contrive settings that would reliably result in ecological locking. However, the features available in Avida are conducive to examining the habitat tracking hypothesis, as in its current form Avida already includes capabilities for resources that are both depletable [13]

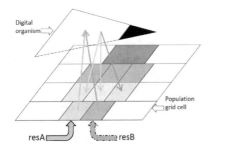
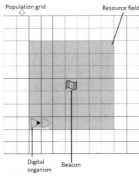

Fig. 22.1: Left panel: Spatial localization of cross-feeding in an Avida population cell. Each cell contains a "well" for each resource. Resources resA and resB are supplied into the cell. An Avidian (shown floating above the cell; black tip indicates facing) may consume either of these resources (green arrows) and deposit by-products (pink arrows) into the appropriate wells. Right panel: Schematic representation of the habitat model, consisting of a resource field (blue square) and beacon (orange banner). An Avidian (indicated by black-tipped triangle) can move one grid cell along its current facing (direction indicated by yellow arrow, destination indicated by grey-tipped Avidian).

and spatially localized [16, 20], and allows movement by the Avidians [20, 21]. In this contribution, we describe how we can use and expand on these capabilities for examining the habitat tracking hypothesis of coordinated stasis.

22.2.2 Modeling Habitats in Avida

In order to examine habitat tracking, we developed a simple model for habitats in Avida as follows:

1. We introduced the creation and removal of *resource fields*, a defined set of cells within the population grid that can hold resources and into which resources can flow into and out of in a spatially explicit way. For simplicity, fields are created as square or rectangular areas in the main population grid. Each cell in the field can hold differing amounts of any user-defined resources. Upon creation, each cell is set to receive a certain amount of a basal resource flowing into it every *update* (the basic Avida time step), and a certain percentage that flows out every update, allowing resource equilibration in the absence of consumption. Each field is composed of a single resource, but multiple potentially overlapping fields of different resources can be created. (Diffusion of resources into surrounding cells is not permitted in this study, but is an available option.) When a field is removed, every cell in the field has its inflow set to zero, while

any resource remaining in the former field's cells continues to flow out at the same rate until depletion.

2. We also implemented the capability to create and remove a *beacon*, a landmark-type feature that can be placed in a specific population cell and which can be detected by the Avidians as a target to move toward. In the present context, the beacon is placed at the center of the resource field, serving as a way of advertising the presence of resources and attracting Avidians onto the field. For the work described here, only a single beacon can be present at any time; an existing beacon must be removed before a new one is placed.

A *habitat* is thus a resource field with an associated beacon (Figure22.1, right panel). Habitats are static entities; in order to allow relocation, the habitat must be removed from one location and a new one created elsewhere.

In order to exploit these features, the Avidians require sensory capabilities for detecting both the beacon and resources, and capabilities for movement and resource consumption.

22.2.3 Environmental Configuration

The key features of our environmental configuration have been described above, but some additional details are relevant. As stated, resources have been spatially local-ized into fields made up of resource-containing cells. Each cell has its own inflow and outflow parameters for each resource defined in the environmental configuration file, only two of which are supplied into each cell in the field every update. When any of these resources is consumed by an Avidian successfully executing an associ-ated task, a specific amount of a by-product is then placed into the cell (Figure 22.1; details of the stoichiometry are user-configurable). By-products may then either ac-cumulate in the cell (though subject to outflow) or be consumed by any Avidian that both occupies that cell and has the ability to use that resource.

In our experiments, we used an additive reward model. The energy gained for per-forming a rewarded function was added directly to the Avidian's current execution rate, and could be applied to multiple executions of that function over an Avidian's life cycle. An Avidian could use 75% of a needed resource in a cell, provided there were at least 0.1 units of that resource available. Mutations occurred during repli-cation at a rate 0.0075 per instruction copied, and insertion/deletion mutations at a rate of 0.05 per division, permitting changes in genome sequence and size.

22.2.4 New Instructions and Tasks

In order to use the aforementioned environmental features, we implemented sev-eral new Avida instructions and tasks. These instructions effectively provide the

Avidians with senses of vision, smell, and taste. The *movement subsystem* has the following features:

1. The *pg-sense-beacon* instruction implements the vision model of Vickerstaff and DiPaolo [56]. It models a theoretical insect with an eye on either side of its head and an almost 360° field of vision. Each eye has a specific range of view represented by "activation energy". To obtain these values, we first calculate the difference in x and y coordinates between the Avidian and the beacon:

$$X_t = beacon_x - Avidian_x \qquad Y_t = beacon_y - Avidian_y$$

Each facing for the Avidian has a corresponding facing angle (A_F), represented in radians, which can be used to determine the rotation angle. First apply the rotation transformation:

$$\begin{bmatrix} X_{ba} \\ Y_{ba} \end{bmatrix} = \begin{bmatrix} X_t \\ Y_t \end{bmatrix} \times \begin{bmatrix} \cos(A_F) & \sin(A_F) \\ -\sin(A_F) & \cos(A_F) \end{bmatrix}$$

Then determine angle $\theta = \tan^{-1}\left(\frac{Y_{ba}}{X_{ba}}\right)$

Finally, calculate the left-side and right-side activation energies:

$$Left = 1000 \times \left(\frac{\cos(\theta - \frac{\pi}{2})}{2} + 0.5 \right) \qquad Right = 1000 \times \left(\frac{\cos(\theta + \frac{\pi}{2})}{2} + 0.5 \right)$$

These two values are stored in pre-specified, fixed registers in the Avidian's CPU.

2. The *pg-sense-distance* instruction calculates the squared Euclidean distance between the Avidian and the beacon. This value is stored in a single, *nop*-modifiable CPU register.

3. The *pg-sense-faced-distance-difference* instruction emulates the *sense-faced-resource* instruction of Elsberry et al. [20]. Rather than sensing differences in resource amounts, this instruction first calculates the distance from the Avidian to the beacon as described above, then does the same for the cell currently faced by the Avidian, taking the difference between the two distances. This value, too, is stored in a single, *nop*-modifiable CPU register.

4. The *pg-beacon-move* instruction is modified from the move instruction used in Elsberry et al. [20]. This modified instruction moves the Avidian one cell in the direction of its current facing. If the destination cell is already occupied, the two Avidians simply swap places. An associated task that is called upon completion of the move then rewards the Avidian. The reward has an associated *quality value* that scales with distance from the beacon and modifies the user-specified base reward value. The functional form of this quality value is specified by parameters in the master configuration file, effectively providing a fitness function for movement. We implemented and examined the performance of two quality functions, as follows:

- PERMISSIVE, where the quality declines as a constant linear function in small increments with increasing distance from the beacon, allowing substantially high movement rewards in a wide area around the beacon. The quality is calculated by:

$$Q = (1 - (F_D \times \sqrt{D_{SD}^2})) \times R_{D1}$$

where F_D is the value of the decay factor, D_{SD}^2 is the squared Euclidean distance to the beacon (as with *pg-sense-distance*), and R_{D1} is an additional reduction factor applied so that a distance of one cell from the beacon does not produce a larger value than when the distance is zero (typically set to 0.99). If the quality value is ever negative, it is set to zero.

- STRINGENT, where the quality declines proportionally to the inverse of distance from the beacon. This creates a much tighter zone around the beacon where Avidians may accrue high movement rewards. Here the quality is calculated as:

$$Q = \frac{\alpha}{\sqrt{D_{SD}^2}} \times R_{D1}$$

where the parameter α modulates how sharp the decline in reward is with distance from the beacon, and D_{SD}^2 and R_{D1} are as defined previously. This quality function does not produce negative values.

Avidians may change facing using the previously-implemented *rotate-l*, *rotate-r*, and *tumble* instructions [20]. The first two rotate the Avidian's facing one cell to the left or right, while *tumble* changes the facing randomly. Facings are usually preserved across moves, but they are altered if the new facing is not legal in the new grid cell (e.g. at the grid boundaries).

The *ecological subsystem* contains the following additional features:

5. The *smell-mod-resource* instruction allows an Avidian to compare the level of a particular resource in the currently occupied cell with those in the immediately-surrounding cells. If one of the surrounding cells has a greater amount of resource, then the Avidian rotates to face that cell and moves into it. If there are ties, the Avidian picks the cell needing the fewest rotations. If the occupied cell is still best, the Avidian stays put. The desired resource is determined by *nop*-templating in the genome immediately following this instruction; if no such template is present, the instruction fails.

6. The *eat-mod-resource* instruction first specifies a desired resource (again based on *nop*-templating in the genome, as described above), and stores a value in a CPU register containing the ID of that resource. It then triggers a series of tasks that check the ID value and allow consumption of that resource if present in the cell above a threshold limit (specified in the environmental configuration file). Each one of these tasks will fail if there is an ID mismatch or if the resource is below the user-specified minimum. Each resource available for consumption requires a separately implemented task; for this study, we made nine such tasks,

one for each of the resources used in the experiments. No correlation between smell and consumption of resources was explicitly enforced.

22.2.5 Experimental Methodology

22.2.5.1 Movement Subsystem-Only Experiments

We checked whether or not our implementation of movement and beacon-related instructions and tasks would produce results comparable to those observed previously. To that end, we performed a series of 100 replicates for each of the PERMISSIVE (with F_D = 0.015) and STRINGENT (with α = 2.0) environments. We added the aforementioned beacon-sensing and movement-related instructions to the basic Avida instruction set and enabled the movement reward task. Following Elsberry et al. [20], we used a bounded grid model and a relatively prime grid size (127 columns x 113 rows, 14351 cells in total) to minimize the possibility that Avidians might exploit some periodicity implicit in the grid size. The population size was capped at 720 organisms, leaving about 95% of the grid cells unoccupied at any time step during an experiment. The beacon was placed as close to the center of the grid as possible and remained static during the experiment. Offspring placement was random upon birth, to avoid the possibility of the Avidians exploiting birth location to accumulate high rewards. In these experiments, rewards could accrue only through movement, but unlike Elsberry et al [20], we did not impose an explicit cost for executing the *pg-beacon-move* instruction. However, the maximum reward available (prior to quality adjustment) was set to be low, only 0.1 per execution for PERMISSIVE and 0.5 per execution for STRINGENT (i.e., less than a full unit of merit would be added to the Avidian's current bonus on each move). As we did not use an environmentally explicit resource gradient here, movement used the default "infinite" resource. Runs lasted for 500,000 updates, with each update representing on average 30 instructions executed per Avidian; using fewer instructions per update (as in [20]) produced only negative results. Information about the Avidian population, including a tally of visits by Avidians to each grid cell, was kept from each run, with summaries logged every 1,000 updates. We categorized the results in a manner similar to [20] based on the form of the *cumulative cell visits* (CCV) heat map. Following this initial duration some runs were re-run for an additional 250,000 updates if the form of the CCV heat map at 500,000 updates was equivocal.

22.2.5.2 Movement Assays

From a selection of successful and partially successful runs (see Results) we examined the movement abilities of successful Avidian genotypes that were present in the end-experiment populations, based on one of several criteria: maximum fitness in the population at that time, most living individuals present at that time, or historically most common over the last 1000 updates before the experiment ended. We assessed these Avidians' abilities to locate and move to the beacon through a series of movement assays. Each assayed genotype was sterilized (i.e., rendered unable to replicate) placed in one of the four corners of the population grid, and given 2000 updates to see whether or not it could move to and remain in proximity to the beacon, with its position at each update saved to a file for later visualization. Each assayed genotype was tested four times, with the same random seed but different starting positions (i.e. the four grid corners).

22.2.5.3 Functional Genomics Tests

Having identified a number of Avidians with successful movement strategies, we also performed functional genomics assays on some of them to determine which (and how many) of the beacon- and distance-sensing instructions in their genomes were providing useful positional information. We picked one successful movement assay (particular genotype, random seed, and starting position) with a known outcome and replaced each occurrence of the *pg-sense-X* instructions with a null instruction to see if the outcome would be altered. An instance of the instruction was considered to provide information to the Avidian if the knockout test visibly altered the outcome, particularly if it abolished the Avidian's ability to locate the beacon.

22.2.5.4 Full-Habitat Experiments

We next investigated whether the Avidians could evolve what we term *habitat fidelity*, or the ability to stick tightly to the resource field through effective use of resource sensing. For these experiments, we enabled all components for the ecological and movement subsystems. Each run was seeded with a naive Avidian capable only of self-replication, placed into a population grid containing two resource fields corresponding to the two basal resources (Figure 22.1, right panel). Each field was of size 38 x 38, centered on the approximate population grid center (which also contained the beacon) and was always present. The environment was configured to enable spatially explicit cross-feeding as described previously (Figure 22.1, left panel). However, a key difference between these experiments and the movement-only sets was that offspring placement was local rather than random throughout the grid, because we think it is usually more biologically realistic that offspring are born near their parents. Also, the population carrying capacity was increased to 1,440 Avidians, as too small a population size might result in some resources going unexploited due to genotypic changes resulting from genetic drift. Each experiment ran for 250,000 updates. We again performed 100 runs under these conditions, saving the same data as for the movement-only experiments, along with grid data representing the locations and ecological profiles of each Avidian in the population, as well how often each reaction (successful execution of a task resulting in use of a resource) was triggered. From the saved files, we could scan the Avidians' genomes for signatures of effective use of smell and the number of *eat* tasks each Avidian could perform. For the experiments described here, we used the parameters for the STRINGENT configuration to assign movement reward quality (a sharp decrease with increasing distance from the beacon). The stoichiometry and reward values for resource use were those of Yedid et al. [63], except that each resource was associated with a separate *eat* task, rather than the logic functions commonly used in Avida research. A minimum of 0.1 units of a resource had to be present in a cell at the time of consumption in order for the corresponding *eat* task to be processed successfully.

22.3 Results

22.3.1 Movement Subsystem-Only Results

22.3.1.1 Cumulative Cell Visits (CCV) Categorization

A movement strategy found at the end of the experiment should demonstrate that evolved Avidians can in most cases recognize and track the beacon, which indicates a rudimentary form of intelligence. However, the variety of results requires categorization. Following the classification based on CCV heat maps of [20], we observed the emergence of three main movement strategies during our evolutionary runs: *Cockroach, Ziggurat,* and *Climber* (Figure 22.2. Summary results for each environment type are shown in Table 22.1. If a particular run failed to break even in terms of the population average Avida-fitness (the ratio of total rewards accumulated to total gestation time), it was designated a failure. Of particular note were the high number of failures for the PERMISSIVE environment (about half the data set, 22.1).

Table 22.1: Summary results for cumulative cell visits (CCV) categorization of movement-only runs

	PERMISSIVE	STRINGENT
Fitness failure	49	31
Cockroach	4	35
Ziggurat	12	0
Super-climber	9	16
Suboptimal climber	19	18
Crown	4	0

All totals are out of 100 replicates for each environment type

In *Cockroach* movement, the movement of Avidians was primarily along the grid boundaries, with occasional crossing of the grid diagonals; this pattern was observed

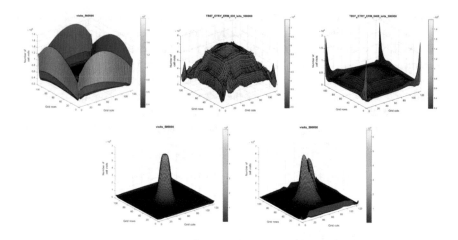

Fig. 22.2: Cumulative cell visits (CCV) heat maps in three dimensions showing major categories of movement subsystem results. In all panels, the x and y axes are the population grid columns and rows respectively, while the z-axis is the cumulative number of visits to each cell. All panels show the end-experiment state. Top row (L to R): *Cockroach*, *Ziggurat*, and *Crown*. Bottom row (L to R): *super-climber* and *suboptimal climber*.

for both PERMISSIVE and STRINGENT environments, but especially in the latter. We observed two types that occurred only in the PERMISSIVE environment. Populations in which *Ziggurat* movement was prevalent (15% of runs) indicate cell visits in the form of a stepped pyramid, whose top could be either flat or have a slight peak or dome. We also observed a minor type we called *Crown*, where CCVs form an increasing ridge along each boundary edge, reaching a maximum in each grid corner. *Crown* was more often observed as an intermediate state before the emergence of one of the other main states, occurring as an end-state in only 4% of PERMISSIVE replicates. Finally, the *Climber* movement strategy shows at least one clear peak in terms of cumulative cell visits performed by Avidians. We further sub-categorized these results into *super-climbers*, which showed a single peak centered over the beacon, and *suboptimal* climbers, where the peak was usually not centered over the beacon and there were occasionally multiple peaks present. Both of these types were found for both environmental configurations, and there were somewhat more suboptimal than super-climbers. We did not observe CCV results corresponding to the *Drunkard*, *Cake*, *Row*, *Column*, *Square*, or *Shark* strategies seen by Elsberry et al [20].

Despite the wider zone around the beacon where Avidians could accrue high-quality rewards for movement, the PERMISSIVE environment had a higher failure rate (nearly 50% of all runs) and produced relatively fewer good-quality outcomes than STRINGENT. However, PERMISSIVE did produce a greater diversity of CCV patterns.

22.3.1.2 Movement Assays

We sought to determine what kind of Avidians produced the observed CCV results, particularly for the *Climber*-type populations, as these have the strongest ability to recognize and move around the beacon. Our test Avidians, as selected by the criteria mentioned in Methods, demonstrate that our evolutionary runs produced a variety of outcomes with varying degrees of effectiveness. The *super-climber* runs produced Avidians with the best ability to locate and remain in the beacon's vicinity regardless of starting position (Figure 22.3). The Avidians whose movement traces are shown in Figure 22.3 show two *super-climber* outcomes illustrating an apparent trade-off between efficiency of beacon search and effective orbital behavior. The first Avidian (Figure 22.3, panels a-d) has a somewhat inefficient, triangular loop-based search but remains in a tight orbit around the beacon once in its vicinity. The second one (Figure22.3, panels e-h) has a much more direct search, moving straight approximately halfway along the grid edges before reorienting towards the beacon, followed by chaotic orbital behavior in which it wanders some distance from the beacon before heading back.

In contrast to *super-climber* Avidians, those from *suboptimal climber* runs often showed less consistent solutions: depending on the starting point and/or value of the random seed used, they successfully located the beacon some but not all of the time. Hence, these Avidians represent solutions that are only partially effective.

22.3.1.3 Functional Genomics Tests

The functional genomics tests revealed which beacon-related instructions contributed to the phenotype of the obtained Avidians. Despite the fact that the evolved Avidians usually contained multiple instances of each instruction in their genome, these tests showed that typically only one instance contributed critical positional information to the Avidian, and that this instance was located in or near the evolved copy loop. An example using the "chaotic orbiter" (Figure 22.4, panel a) shows that removal of only the red-highlighted *pg-sense-distance* abolishes the Avidian's ability to detect the beacon (Figure 22.4, panel b). These tests were also useful for showing that the red-highlighted *tumble* instruction is responsible for the eccentric orbit, as removing that instruction produces movement behavior that would lead to a *Cockroach*-type result (Figure 22.4, panel c). While our search was by no means exhaustive, we found one case where more than one beacon-related instruction was needed to produce the beacon-tracking phenotype, albeit from a *suboptimal climber* replicate (data not shown).

The tests also showed that Avidians producing the different CCV result types tended to use different instructions to obtain positional information. Avidians from *super-climber* runs invariably used the *pg-sense-distance* instruction to obtain critical information. In contrast, Avidians from *suboptimal climber* runs used either *pg-sense-distance-difference* or, more rarely, *pg-sense-beacon* to obtain that information. In general, these latter two instructions did not appear to provide sufficient

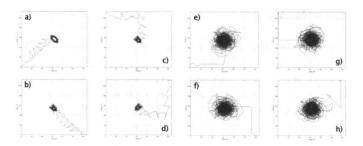

Fig. 22.3: Movement traces of two successful *super-climber* Avidians demonstrating competence at locating the beacon (indicated by red asterisk). In all panels, the x and y axes are the population columns and rows, respectively. Panels a-d: an Avidian with inefficient search but tight orbit; panels e-h: an Avidian with more effective search but a chaotic orbit.

information to evolve highly effective solutions. In the case of *pg-sense-beacon*, the major effect seemed to be to be "fine-tuning" of the Avidian's movement trajectory; in most cases, removal of that instruction either had no visible effect, or caused minor alterations to the movement trajectory but did not abolish beacon-locking ability.

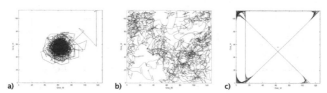

rwqCfiBnjaDwutDADmbfkecGqDDcyxpdgckfxyAwuxDajlDDsjmbsCazaraDcmqhFFpCCjaztnzeh
ulFFwphGdsABwzaCpeFCtlmAyCopzusoezqldDqpaFpDfznBDxlafAtGCjkDCBFkubyBmxoFopGm
wBcrwjjcaoccBvmFmtfyfusDEDrupwrwtiDmmyfvwdsxcvdlkcdEErjBwdvkGjzzuzzcjjolDCzCoBlyq
eimpkkwxbrFukGxctkjlcEcuxFoebdharEzufrCppukxzuvks*ce*GDeDxDgDgz

Fig. 22.4: Example of functional genomics test on Avidian whose movement is shown in Figure 22.3e-h. Axes and beacon as indicated in Figure 22.3. The genome of the organism is represented by the single-letter codes below the three panels as follows: A: *pg-sense-beacon*; B: *pg-sense-distance*; C: *pg-sense-faced-distance-difference*; D: *pg-beacon-move*; E: *rotate-l*; F: *rotate-r*; G: *tumble*. Lower case letters represent standard Avida instructions. Panel a: movement trace when started at (126,112). Panel b: movement when red-highlighted B is replaced by null instruction. Panel c: movement when red-highlighted G is replaced by null instruction.

22.3.1.4 Conclusions Regarding the Movement Subsystem Only

We could identify classes of movement behaviors evolved by Avidians and see them from different perspectives, such as the class of *Climber* movement strategies that approximate gradient ascent. Within the *Climber* class, there were significant differences in the efficiency of individual Avidian runs, particularly between the *suboptimal* and *super* sub-classes, each of which favored different critical instructions. The effectiveness of the solution seemingly depends on the completeness of the information provided: *pg-sense-distance*, which provides an Avidian with the most complete information (the actual squared distance to the beacon), was used in all of the best results, while the incomplete information provided by the other two instructions featured in the suboptimal results. The relatively low fraction of *super-climbers* obtained with both environmental configurations (Table 22.1) indicates that the evolution of near-optimal solutions is not easily achieved under the conditions used here. In contrast to previous work [20], our implicit distance-based gradients are more flexible and more general than an environmentally-explicit, resource-based gradient in which each cell's resource characteristics must be specified individually, effectively providing them with an environmental "signpost" that facilitates evolution of effective strategies. However, the differences between our results and previously obtained ones—particularly the more limited number of CCV outcomes and the occurrence of inconsistent behaviors, which was not previously observed—show that there are important, albeit cryptic differences between environmentally implicit and explicit approaches, and these differences require further investigation. What works well for one approach does not translate easily to the other.

Our results show the potential of evolutionary computation to produce effective movement strategies that lead the evolving agents to lock effectively onto a fixed landmark feature, although their success depends on the completeness of the information provided. Our work shows that such strategies can indeed evolve over an experimental run of reasonable duration. The variety of movement strategies obtained show the interplay between environmental configuration and evolutionary contingency. This interplay is shown by the contrast between the results obtained with PERMISSIVE and STRINGENT settings, with the former more likely to lead to a failed or suboptimal result (Table 22.1). A number of the suboptimal solutions remain sensitive to environmental variables, as revealed by movement assays showing inconsistent outcomes. These suboptimal solutions may provide examples of "satisficing" [20], where imperfect but still "good enough" solutions succeed over ones that are even less effective at addressing the problem. For example, a *suboptimal climber* solution that is inconsistent but works well under at least some conditions is better than one that is ineffective under all conditions (*Cockroach* or *Crown*), although it would be unacceptable in a real-world implementation where global success is needed [20].

Using the implemented features, there are several future directions that can be pursued:

1. discovering the source of inconsistent behavior, and how to mitigate it;

2. improving the evolving agents' frequency of use of the information provided by the beacon-related instructions, and obtaining a greater fraction of effective solutions for static beacon-related movement;
3. introducing a dynamic beacon that can be relocated during either an evolutionary run or even during a movement assay, to examine and improve the generality and "intelligence" of the evolved programs by rewarding the ability for course correction in response to a change in destination. In general, though, these results demonstrate that the features we implemented produce sufficiently good outcomes that they can be included in a more sophisticated scenario using the ecological subsystem.

22.3.2 Ecological Subsystem Results

22.3.2.1 Full-Habitat Experiments

We also used CCV heat maps for initial categorization of results in the full-habitat experiments. These experiments produced a variety of outcomes illustrating differing degrees of habitat fidelity and efficiency of resource use, for which we introduce a new categorization here (Figure 22.5). We also examined sequence data and functional output to determine additional characteristics of each result type (summarized in Figure 22.5). Although a continuum of results with different characteristics was obtained, the density within it was highly non-uniform (Figure 22.5, left column). Few populations evolved high habitat fidelity, while many more produced substantial spatial dispersion (Figure 22.5, right column). We used the variance of average Euclidean distance to habitat center (V_{EDHC}) as the measure of dispersion, which we time-averaged over the last 10,000 updates of the experiment.

22.3.2.2 Post-Hoc Ecological Subsystem-Only Tests

Based on the results obtained from the full-habitat experiments, we performed an additional set of experiments where we disabled the movement subsystem, leaving only the ecological subsystem. We ran these additional experiments to see whether the frequency of strong habitat fidelity would increase. Indeed, populations from these follow-up experiments showed a far greater degree of high habitat fidelity: 98/100 replicates evolved it in the allotted time, producing a *Tower*-type (Figure 22.5) CCV heat map.

22.3.2.3 Conclusions Regarding the Ecological Subsystem

When both of our habitat model's major subsystems were included, we obtained only a small fraction of replicates that evolved moderate (*Obelisk*) to high (*Tower*) habitat fidelity (only 12/100 replicates combined). Instead, most runs produced low-efficiency ecosystems with Avidians more highly dispersed throughout the population grid (*Beveled Mound* and *Smooth Mound*, Figure 22.5). A high occurrence of fidelity obtained when only the ecological subsystem was enabled. Taken in light of the movement-only results, which showed evolution of effective solutions, these results show that the movement and ecological subsystems, as currently implemented, work acceptably well in isolation, but actually interfere with each other when combined. We can identify several possible reasons for this interference:

1. Each subsystem provides conflicting evolutionary pressures: the beacon acts as a magnet, drawing Avidians towards it and rewarding for close proximity, while *smell-mod-resource* makes the Avidians forage in search of high amounts of resource, which may direct them away from the beacon;
2. Having additional movement and rotation instructions present in the genome may offset or override any benefit from *smell-mod-resource* by causing an Avidian to move away from a resource-rich cell before it can eat;
3. The *smell-mod-resource* instruction, as currently implemented, is a pure "action" instruction: That is, it uses environmental information which the Avidian then acts on, but it does not provide any storable information that the Avidian can store and use for calculations, like the *pg-sense-X* instructions do. Also, an Avidian's decisions are based on a greedy evaluation of resource amounts, with no "compromise" between which cell is best for resources and which is best for beacon proximity.
4. As mentioned previously, offspring placement differs between the two setups: random for movement only, and local for the ecological models. Local placement removes a key selective pressure for evolution of navigation ability, since with random placement a newly-placed offspring cannot count on getting high rewards simply from being born near a parent already placed close to the beacon (W. Elsberry, pers. comm.). Hence, the Avidians that evolve in the full-habitat experiments may well be navigationally deficient.

22.3.3 Future Directions Including Habitat Tracking and Improved Resource Sensing

The additions and modifications to Avida that we have presented here make it a platform suitable for investigation of coordinated stasis through habitat tracking, laying the groundwork for tractable experiments. As a first step, we envision an experiment in which a habitat, already containing a diverse community that has been evolving in it for some time, is moved to a new location. How much of the community will follow the habitat, and how much of the community's ecological and phylogenetic integrity will be maintained? The strongest evidence bearing on coordinated stasis would come from simultaneous consideration of phylogeny and ecology [64, 34]. With such an integrated approach, one can in fact make testable

CCV designation and numbers of replicates (/100)	Prominent characteristics
Tower (9) 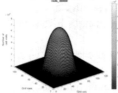	• Effective use of smell by great majority of population ($\geq$95%) • Strong attraction to resource field, high habitat fidelity • High performance of all reactions, low to high ecological generalization (2 – 6 *eat* tasks per Avidian) • Lowest spatial dispersion of Avidians; habitat boundaries clear in CCV (V_{EDHC} = 82.13 ± 7.41)
Obelisk (3)	• Effective use of smell by only some of population ($\leq$50%) • Moderate attraction to resource field, not by all ecotypes • Moderate performance of most reactions, low to high ecological generalization(2 – 6 *eat* tasks per Avidian) • Moderate to high spatial dispersion (V_{EDHC} = 130.79 ± 21.53)
Beveled Mound (50) 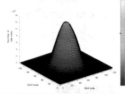	• No effective use of smell • High performance only of basal reactions through non-smell-based detection mechanism; low performance of other reactions, low to moderate ecological generalization (2 – 5 *eat* tasks per Avidian) • Moderate to high spatial dispersion (V_{EDHC} = 121.85 ± 3.16)
Smooth Mound (21) 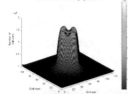	• No effective use of smell • Little-to-no attraction to resource field • Low performance of all reactions, highly variable ecological generalization including pure specialist communities • High spatial dispersion of Avidians (V_{EDHC} = 138.1 ± 5.73)
Irregular (8)	• No effective use of smell • Irregularly-shaped, often asymmetric CCV plot • Low to moderate performance of all reactions; low to moderate ecological generalization (1 – 5 *eat* tasks per Avidian) • Great variability in spatial dispersion (V_{EDHC} = 151.01 ± 20.92)

An additional 9/100 replicates failed to evolve either any resource consumption or any cross-feeding. V_{EDHC} values are means ± 2 s.e.

Fig. 22.5: Summary of outcomes for full-habitat experiments based on CCV heat map

predictions about expected outcomes, only one of which would demonstrate coordinated stasis: namely, the retention of both ecological roles and phylogenetic relationships of the dominant ecotypes before and after the habitat relocation. Any substantial change in either of these outcomes would not indicate the stasis that underlies the habitat-tracking hypothesis.

The results obtained so far already provide a series of possible starting points. The *Tower*, *Obelisk*, and *Beveled Mound* results in particular can be used to represent different evolutionary histories that resulted in different degrees of habitat fidelity, which could then be subjected to a uniform habitat-relocation treatment. Which of these initial states, if any, have major ecotypes with good habitat-tracking abilities that might yield coordinated stasis? We speculate that there is a trade-off between fidelity and tracking: communities that are too tightly bound to the habitat may show poor tracking and therefore fail to exhibit stasis, whereas less tightly bound communities will have ecotypes that can migrate more easily, and might be more likely to show coordinated stasis.

We also plan to improve the performance of resource sensing. As mentioned earlier, it has some shortcomings in its current form. Modifications should rectify these deficiencies and also allow better integration of "smell" and "vision" capabilities, thereby enabling Avidians to choose cells based on the best combination of proximity to the beacon and level of the preferred resource.

22.4 Concluding Remarks

In this contribution, we have taken a proven digital evolution platform and explained how we modified it for experimental examination of the habitat-tracking hypothesis of coordinated stasis. We have presented preliminary results obtained using each of the major new subsystems implemented, showing that they yield at least some positive outcomes for evolution of navigation ability and for resource sensing, but that it is difficult to get them to work well in concert. Despite these difficulties, our results demonstrate the potential for digital evolution to make certain aspects of studying deep-time evolutionary dynamics more explicit than is possible with the fossil record, and for this approach to allow hypothesis testing in a truly experimental manner. Further use and development of digital evolution, especially when combined with rigorous measurement approaches borrowed from community ecology and phylogenetics [61, 34, 64], can yield new insights into problems in paleobiology and macroevolution that cannot be addressed adequately using other approaches.

Acknowledgements The authors thank Dr. Wesley Elsberry (RealPage, Inc., USA) for helpful discussion and comments regarding settings and experimental design, Dr. Laura Grabowski (Dept. of Computer Science, SUNY Potsdam, U.S.A.) for introducing us to the vision model used here, Dr. Josef Voglmeier (Dept. of Food Science, Nanjing Agricultural University, Nanjing, China) for comments that improved the manuscript, and especially to Dr. Richard E. Lenski and Dr. Wolfgang Banzhaf (BEACON Center, Michigan State University, U.S.A) for the invitation to submit this contribution. This research was supported by a grant to GY from the National Natural Science Foundation of China (#31470435). GY gives special thanks to Jing Zhao for assistance with figures, and to Dr. Lenski and Dr. Charles Ofria for additional comments and editorial suggestions.

References

1. Adami, C.: *Digital genetics: unravelling the genetic basis of evolution.* Nature Reviews Genetics **7**, 109–118 (2006)
2. Bailey, S.F., Dettman, J.R., Rainey, P.B., Kassen, R.: *Competition both drives and impedes diversification in a model adaptive radiation.* Proceedings of the Royal Society B: Biological Sciences **280**, 20131,253 (2013)
3. Betancur-R., R., Ortí, G., Stein, A.M., Marceniuk, A.P., Pyron, R.A.: *Apparent signal of competition limiting diversification after ecological transitions from marine to freshwater habitats.* Ecology Letters **15**, 822–830 (2012)
4. Blois, J.L., Zarnetske, P.L., Kassen, R.: *Climate change and the past, present, and future of biotic interactions.* Science **341**, 499–504 (2013)
5. Bonuso, N., Newton, C.R., Brower, J.C., Ivany, L.C.: *Does coordinated stasis yield taxonomic and ecologic stability? Middle Devonian Hamilton Group of central New York.* Geology **30**, 1055–1058 (2002)
6. Brett, C.E.: Coordinated stasis reconsidered: A perspective at fifteen years. In: J. Talent (ed.) Earth and Life: Global biodiversity, extinction intervals and biogeographic perturbations through time, pp. 23–36. Springer Dordrecht (2012)
7. Brett, C.E., Baird, G.C.: Coordinated stasis and evolutionary ecology of Silurian to middle Devonian faunas in the Appalachian basin. In: D. Erwin, R. Anstey (eds.) New approaches to speciation in the fossil record, pp. 285–315. Columbia University Press, New York (1995)
8. Brett, C.E., Bartholomeow, A.J., Baird, G.C.: *Biofacies recurrence in the Middle Devonian of New York state: An example with implications for evolutionary paleoecology.* Palaios **22**, 306–324 (2007)
9. Brett, C.E., Hendy, A.J.W., Bartholomew, A.J., Bonelli, J.R., McLaughlin, P.I.: *Response of shallow marine biotas to sea-level fluctuations: A review of faunal replacement and the process of habitat tracking.* Palaios **22**, 228–244 (2007)
10. Brockhurst, M.A., Colegrave, N., Hodgson, D.J., Buckling, A.: *Niche occupation limits adaptive radiation in experimental microcosms.* PLoS ONE **2**, e193 (2007)
11. Buatois, L., Carmona, N., Curran, H., Netto, R., Mangano, M., Wetzel, A.: The Mesozoic marine revolution. In: M. Mangano, L. Buatois, M.G. Mángano, L.A. Buatois (eds.) The Trace-Fossil Record of Major Evolutionary Events, Topics in Geobiology Vol 40, pp. 19–134. Springer Netherlands (2016)
12. Chow, S.S.: *Adaptive radiation from resource competition in digital organisms.* Science **305**, 84–86 (2004)
13. Cooper, T., Ofria, C.: *Evolution of stable ecosystems in populations of digital organisms.* In: R. Standish, M. Bedau (eds.) Eighth International Conference on Artificial Life, Sydney, Australia, pp. 227–232. MIT Press, Cambridge, MA (2002)
14. DiMichele, W., Behrensmeyer, A., Olszewski, T., Labandeira, C., Pandolfi, J., Wing, S., Bobe, R.: *Long-term stasis in ecological assemblages: Evidence from the fossil record.* Annual Review of Ecology, Evolution, and Systematics **35**, 285–322 (2004)
15. Dineen, A.A., Fraiser, M.L., Sheehan, P.M.: *Quantifying functional diversity in pre- and post-extinction paleocommunities: A test of ecological restructuring after the end-Permian mass extinction.* Earth Science Reviews **136**, 339–349 (2014)
16. Dolson, E., Ofria, C.: *Spatial resource heterogeneity creates local hotspots of evolutionary potential.* In: C. Knibbe, et al. (eds.) Proceedings of the 14th European Conference on Artificial Life, pp. 122–129. MIT Press, Cambridge, MA (2017)

17. Eldredge, N., Gould, S.: Punctuated equilibrium: an alternative to phyletic gradualism. In: T. Schopf (ed.) Models in paleobiology, pp. 82–115. Freeman, Cooper, San Francisco (1972)
18. Eldredge, N., Thompson, J.N., Brakefield, P.M., Gavrilets, S., Jablonski, D., Jackson, J.B.C., Lenski, R.E., Lieberman, B.S., McPeek, M.A., III, W.M.: *The dynamics of evolutionary stasis.* Paleobiology **31**(sp5), 133–145 (2005)
19. Elena, S., Sanjuan, R.: *The effect of genetic robustness on evolvability in digital organisms.* BMC Evolutionary Biology **8**, 284 (2008)
20. Elsberry, W.R., Grabowski, L.M., Ofria, C., Pennock, R.T.: *Cockroaches, drunkards, and climbers: Modeling the evolution of simple movement strategies using digital organisms.* In: 2009 IEEE Symposium on Artificial Life, pp. 92–99. IEEE Press, New York (2009)
21. Grabowski, L., Elsberry, W., Ofria, C., Pennock, R.: *On the evolution of motility and intelligent tactic response.* In: C. Ryan, M. Keijzer, et al. (eds.) Proceedings of the 10th annual conference on Genetic and evolutionary computation - GECCO 2008, pp. 209–216. ACM Press, New York (2008)
22. Hang, D., Torng, E., Ofria, C., Schmidt, T.: *The effect of natural seledtion on the performance of maximum parsimony.* BMC Evolutionary Biology **7**, 94 (2007)
23. Holland, S.: *The stratigraphic distribution of fossils.* Paleobiology **21**, 92–109 (1995)
24. Holland, S., Zaffos, A.: *Niche conservatism along an onshore-offshore gradient.* Paleobiology **37**, 270–286 (2011)
25. III, W.M.: *Ecology of coordinated stasis.* Palaeogeography, Palaeoclimatology, Palaeoecology **127**, 177–190 (1996)
26. Ingalls, B.R., Park, L.E.: *Biotic and taphonomic response to lake level fluctuations in the greater Green River Basin Eocene, Wyoming.* Palaios **25**, 287–298 (2010)
27. Ivany, L.C., Brett, C.E., Wall, H.L.B., Wall, P.D., Handley, J.C.: *Relative taxonomic and ecologic stability in Devonian marine faunas of New York State: A test of coordinated stasis.* Paleobiology **35**, 499–524 (2009)
28. Johansson, J.: *Evolutionary Responses to Environmental Changes: How does Competition affect Adaption?* Evolution **62**, 421–435 (2008)
29. Kerr, B., Riley, M., Feldman, M., Bohannan, B.: *Local dispersal promotes biodiversity in a real-life game of rock-paper-scissors.* Nature **418**, 171–174 (2002)
30. Kidwell, S., Holland, S.: *The quality of the fossil record: Implications for evolutionary analysis.* Annual Review of Ecology and Systematics **33**, 561–588 (2002)
31. Lawrence, D., Fiegna, F., Behrends, V., Bundy, J.G., Phillimore, A.B., Bell, T., Barraclough, T.G.: *Species interactions alter evolutionary responses to a novel environment.* PLoS Biology **10**, e1001,330 (2012)
32. Lenski, R., Ofria, C., Collier, T., Adami, C.: *Genome complexity, robustness and genetic interactions in digital organisms.* Nature **400**, 661–664 (1999)
33. Lenski, R., Ofria, C., Pennock, R., Adami, C.: *The evolutionary origin of complex features.* Nature **423**, 139–144 (2003)
34. Luo, T.T., Heier, L., Khan, Z.A., Hasan, F., Reitan, T., Yasseen III, A.S., Xie, Z.X., Zhu, J.L., Yedid, G.: *Examining community stability in the face of mass extinction in communities of digital organisms.* Artificial Life **24**, 250–276 (2019)
35. de Mazancourt, C., Johnson, E., Barraclough, T.G.: *Biodiversity inhibits species' evolutionary responses to changing environments.* Ecology Letters **11**, 380–388 (2008)
36. McCann, K.S.: *The diversity–stability debate.* Nature **405**, 228–233 (2000)
37. McMeans, B., McCann, K., Humphries, M., Rooney, N., Fisk, A.: *Food web structure in temporally-forced ecosystems.* Trends in Ecology and Evolution **30**, 662–672 (2015)
38. Misevic, D., Ofria, C., Lenski, R.E.: *Experiments with digital organisms on the origin and maintenance of sex in changing environments.* Journal of Heredity **101**, S46–S54 (2010)
39. Monteiro, A.B., Faria, L.D.B.: *Causal relationships between population stability and food-web topology.* Functional Ecology **31**, 1294–1300 (2017)
40. Morris, P., Ivany, L., Schopf, K., Brett, C.: *The challenge of paleoecological stasis: reassessing sources of evolutionary stability.* Proceedings of the National Academy of Sciences **92**, 11269–11273 (1995)

41. Mougi, A., Kondoh, M.: *Diversity of interaction types and ecological community stability.* Science **337**, 349–351 (2012)
42. Mougi, A., Kondoh, M.: *Food-web complexity, meta-community complexity and community stability.* Scientific Reports **6**, 1–5 (2016)
43. Nordbotten, J.M., Stenseth, N.C.: *Asymmetric ecological conditions favor red-queen type of continued evolution over stasis.* Proceedings of the National Academy of Sciences **113**, 1847–1852 (2016)
44. Odling-Smee, F., Laland, K., Feldman, M.: *Niche construction: The neglected process in evolution.* Princeton University Press, Princeton, NJ (2003)
45. Ofria, C., Bryson, D., Wilke, C.: Avida: A software platform for research in computational evolutionary biology. In: A. Adamatzky, M. Komosinski (eds.) Artificial Life Models in Software (2nd. ed.), pp. 3–35. Springer, London (2009)
46. Oliver, T., et al.: *Biodiversity and resilience of ecosystem functions.* Trends in Ecology and Evolution **30**(11), 673–684 (2015)
47. Olszewski, T.D.: *Persistence of high diversity in non-equilibrium ecological communities: Implications for modern and fossil ecosystems.* Proceedings of the Royal Society B: Biological Sciences **279**, 230–236 (2011)
48. Olszewski, T.D., Erwin, D.H.: *Change and stability in Permian brachiopod communities from western Texas.* Palaios **24**, 27–40 (2009)
49. Ostrowski, E.A., Ofria, C., Lenski, R.E.: *Genetically integrated traits and rugged adaptive landscapes in digital organisms.* BMC Evolutionary Biology **15**, 83 (2015)
50. Reichenbach, T., Mobilia, M., Frey, E.: *Mobility promotes and jeopardizes biodiversity in rock–paper–scissors games.* Nature **448**, 1046–1049 (2007)
51. Reymond, C.E., Bode, M., Renema, W., Pandolfi, J.M.: *Ecological incumbency impedes stochastic community assembly in Holocene foraminifera from the Huon Peninsula, Papua New Guinea.* Paleobiology **37**, 670–685 (2011)
52. Rooney, N., McCann, K.S.: *Integrating food web diversity, structure and stability.* Trends in Ecology & Evolution **27**, 40–46 (2012)
53. Sauve, A., Fontaine, C., Thébault, E.: *Structure-stability relationships in networks combining mutualistic and antagonistic interactions.* Oikos **123**, 378–384 (2013)
54. Sepkoski, J.: *A kinetic model of Phanerozoic taxonomic diversity. III. Post-Paleozoic families and mass extinctions.* Paleobiology **10**, 246–267 (1984)
55. Stenseth, N.C., Smith, J.M.: *Coevolution in ecosystems: Red Queen evolution or stasis?* Evolution **38**, 870–880 (1984)
56. Vickerstaff, R., DiPaolo, E.: *Evolving neural models of path integration.* Journal of Experimental Biology **208**, 3349–3366 (2005)
57. Vrba, E.S.: *Environment and evolution: alternative causes of the temporal distribution of evolutionary events.* South African Journal of Science **254**, 752–753 (1985)
58. Wagner, P.J., Kosnik, M.A., Lidgard, S.: *Abundance distributions imply elevated complexity of post-Paleozoic marine ecosystems.* Science **314**, 1289–1292 (2006)
59. Walker, B.L., Ofria, C.: *Evolutionary potential is maximized at intermediate diversity levels.* Artificial Life **13**, 116–120 (2012)
60. Wilke, C., Wang, J., Ofria, C., Lenski, R., Adami, C.: *Evolution of digital organisms at high mutation rates leads to survival of the flattest.* Nature **412**, 331–333 (2001)
61. Yedid, G., Heier, L.: Effects of random and selective mass extinction on community composition in communities of digital organisms. In: P. Pontarotti (ed.) Evolutionary Biology: Mechanisms and Trends, pp. 43–64. Springer, Berlin, Heidelberg (2012)
62. Yedid, G., Ofria, C., Lenski, R.: *Historical and contingent factors affect reevolution of a complex feature lost during mass extinction in communities of digital organisms.* Journal of Evolutionary Biology **21**, 1335–1357 (2008)
63. Yedid, G., Ofria, C.A., Lenski, R.E.: *Selective press extinctions, but not random pulse extinctions, cause delayed ecological recovery in communities of digital organisms.* The American Naturalist **173**, E139–E154 (2009)

64. Yedid, G., Stredwick, J., Ofria, C.A., Agapow, P.M.: *A comparison of the effects of random and selective mass extinctions on erosion of evolutionary history in communities of digital organisms.* PLoS ONE **7**, e37233 (2012)
65. Zaman, L., Meyer, J.R., Devangam, S., Bryson, D.M., Lenski, R.E., Ofria, C.: *Coevolution drives the emergence of complex traits and promotes evolvability.* PLoS Biology **12**, e1002023 (2014)

Chapter 23
Major Transitions in Digital Evolution*

Heather Goldsby, Benjamin Kerr and Charles Ofria

Abstract The astonishing complexity of the world around us is the result of major transitions in evolution where lower-level entities unite to form a higher-level unit: living and reproducing as one. These transitions give rise to many questions within evolutionary biology, but can prove challenging to study due to their rare occurrence in nature. Here, we describe a digital evolution approach where cells, organisms, and worlds all exist within the framework of a computer and as such have rapid generation times and are amenable to experimental control, replaying key events, and precise data tracking. Using this approach we describe our previous experiments exploring fraternal major transitions in evolution — transitions that occur when identical lower-level units (e.g., cells) remain together as one higher-level unit (e.g., a multicellular organism). We have performed experiments to test key hypotheses regarding the formation of higher-level units and the reproductive and task-based division of labor that can evolve once these units are in place. We then describe a series of on-going studies that explore hypotheses regarding the forces that prevent higher-level units from reverting to their lower-level origins, how plasticity may predispose the evolution of division of labor, and how egalitarian transitions (which occur when different lower-level units come together) may occur.

Heather Goldsby
Department of Computer Science and Engineering, and BEACON Center for the Study of Evolution in Action, Michigan State University, East Lansing, MI, 48824, USA e-mail: hjg@msu.edu

Benjamin Kerr
Department of Biology, University of Washington, Seattle, WA 98195, USA, and BEACON Center for the Study of Evolution in Action, Michigan State University, East Lansing, MI, 48824, USA e-mail: kerrb@u.washington.edu

Charles Ofria
Department of Computer Science and Engineering, and BEACON Center for the Study of Evolution in Action, Michigan State University, East Lansing, MI, 48824, USA e-mail: ofria@msu.edu

* This paper was externally peer-reviewed.

© Springer Nature Switzerland AG 2020
W. Banzhaf et al. (eds.), *Evolution in Action: Past, Present and Future*,
Genetic and Evolutionary Computation, https://doi.org/10.1007/978-3-030-39831-6_23

Key words: digital evolution, major transitions in evolution, division of labor, fraternal transition, egalitarian transition, cell-cell signaling, germ-soma differentiation, multicellularity, plasticity

23.1 What Are the Major Transitions in Evolution?

You are walking through the desert at the base of the Chiricahua Mountains when your eyes drop to the ground. As you weave around the shrubs, you count multiple red harvester ant colonies. Each colony contains a throng of ants. Each ant comprises a multitude of cells. Each cell houses multiple chromosomes. And each chromosome possesses scores of genes. As with these ant colonies, living systems are hierarchically organized, with different levels embedded like a series of Russian dolls. However, this hierarchy is not a static feature of life on our planet; rather, it is a product of organic evolution in which higher levels of complexity originate from lower levels. In such a view, individuality is a property that emerges over time as each new level crystallizes [3, 15, 26, 27, 32, 33, 34, 39]. The process of this emergence is known as a major evolutionary transition.

A major transition in evolution occurs when multiple formerly autonomous units at a given level in the hierarchy of life unite to produce a higher-level unit that is able to reproduce [4, 26]. In their seminal book, Maynard Smith and Szathmary cover other types of "major transitions" as well, but here we focus on their subset involving the origins of higher-level of complexity. In his review of their book, Queller ([33]) introduced a useful dichotomy based on the manner by which the lower-level entities end up together. A major transition is fraternal when genetically similar units stay together in the same group. For example, the shift from unicellularity to multicellularity involves genetically identical cells sticking together to form a multicellular organism. This fraternal transition has occurred independently dozens of times throughout the history of life [20]. On the other hand, a major transition is egalitarian when genetically dissimilar units come together to form a group. For example, the origin of the eukaryotic cell traces to an endosymbiotic event where a host cell engulfed an unrelated bacterium (the predecessor to the modern mitochondrion) and the combined organism began to replicate as an individual. This egalitarian transition appears to have occurred only once [42]. Here we will focus on fraternal transitions, but we will return to egalitarian transitions towards the conclusion.

23.2 What Are the Principal Components of Fraternal Transitions?

The first step of a fraternal transition is the formation of the higher-level group from closely-related lower-level units (from Figure 23.1a to Figure 23.1b). This initial step may simply co-opt reproduction of the lower-level unit. For instance, cells that fail to fully separate after reproduction will generate cellular groups, a first step from unicellularity to multicellularity. Importantly, these new kin groups must also reproduce, such that parent-offspring relationships can be established at the group level over time [35]. The subsequent steps in the transition involve various forms of division of labor between the lower-level parts of the higher-level entity (from Figure 1b to Figure 1c). Such division of labor may involve spatial patterning or temporal coordination, or both. The properties defining differentiation between the lower-level parts could be morphological, functional, or even reproductive (e.g., a germ-soma split in multicellular organisms). One advantage of division of labor involves the ability to concurrently perform tasks at the higher level that are impossible for lower-level units to simultaneously achieve. For example, in cyanobacteria filaments, the nitrogenase used in nitrogen fixation is inactivated in the presence of oxygen, which makes nitrogen fixation and oxygenic photosynthesis incompatible pathways; however, both processes can be synchronously performed at the level of the filament as different cells specialize on each activity [16, 21]. Successive rounds of lower-level expansion and division of labor lead to progressively more complex higher-level units (e.g., from Figure 23.1c to Figure 23.1d), which may exhibit functionality at the higher level that was never witnessed at the lower level [14].

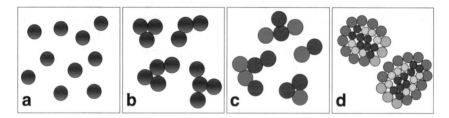

Fig. 23.1: A possible pathway for a fraternal transition in individuality from single cells to more complex multicellular organisms. Individual cells (a) begin to cluster into clonal groups with no phenotypic differentiation (b). Next, groups start exhibiting division of labor such that cells specialize on tasks (c), likely improving the functionality of the group as a whole. Finally, cells physically organize within the organism (d), allowing to more sophisticated forms of coordination, but requiring a non-trivial developmental process for the overall organism to form.

23.3 How Does One Study Major Transitions?

Evolutionary biologists study major transitions in a variety of ways. Some researchers use a comparative approach. For instance, an outgroup taxon that is closely related to a clade whose common ancestor underwent a major transition can be studied for clues to important ingredients favoring the transition. As an example, unicellular choanoflagellates, a sister group to the multicellular animals, may reveal insights into the origins of animal multicellularity [2]. Alternatively, a clade that has recently undergone a variety of steps in a major transition can be explored to gauge the order and repeatability of components of the transition. As an example, the volvocine algae, which transitioned to multicellularity at least 200 million years ago, range from species with just a handful of undifferentiated cells to species with hundreds of cells organized into a spheroid with reproductive differentiation; and phylogenetic analysis has shed light on the evolutionary steps leading to this gradient in complexity [17]. A second approach involves mathematical or simulation-based modeling [8, 37, 41]. For instance, Ispolatov et al. [19] modeled the evolution of cellular aggregation when compartmentalization of different physiological process is advantageous. Such models are intentionally abstract, omitting many details in order to focus our attention on potential factors of critical importance for the major transition. A final approach involves real-time experimental evolution. For example, when single celled budding yeast was propagated under simple gravity selection, multicellular yeast clusters evolved in multiple replicate lines [36].

One alternative that balances tractability with complexity and is thus gaining popularity within the evolutionary community is the use of Artificial life techniques that employ man-made systems (be they computational, robotic, or even biochemical) to emulate natural process. These systems have been able to balance speed and control with "real" evolutionary dynamics. In response to selection pressures, artificial life approaches frequently evolve strategies that were not envisioned by the researchers. However, these extra details make artificial life systems run more slowly than simulations. Experiments frequently take hours and potentially can take days or weeks. In this review, we will focus on one form of experimental evolution and describe how it offers some distinct opportunities in the study of the major evolutionary transitions. The approach in the spotlight is experimental evolution with digital organisms.

23.4 Digital Evolution

Digital organisms exist within a computational environment, where they are subject to mutations, experience natural selection, and exhibit many natural evolutionary dynamics [5, 6, 23, 13]. For digital evolution studies surrounding major transitions in evolution, each experimental replicate consists of a world where digital organisms (which can be either unicellular or multicellular) compete for resources and space (Figure 23.2). Each cell has a genome, which is a list of computational instructions that are executed. The instruction set is Turing complete and thus can theoretically solve any computational problem. By executing instructions within its genome, a cell is able to reproduce forming another cell that can serve as a sister-cell within its organismal body, or as a propagule cell used as a starting point to form another organism. A cell is able to perform the digital equivalent of metabolic work by taking in binary numbers as input and perform logic functions (e.g., NOT, NAND, ORNOT, EQUALS) on them to gain resources from the environment. These resources are necessary for organismal reproduction. As such, organisms race to perform functions, acquire resources, and then reproduce displacing another organism within their world.

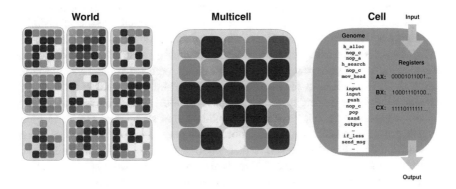

Fig. 23.2: A world consisting of digital multicells, each of which contains up to 25 cells. Each cell has been colored according to its phenotype, where different phenotypes are produced by performing different logic functions. Each cell has a genome, a set of registers it can store values in, and the means to access inputs from the environment and produce outputs to the environment.

Digital evolution fills a unique niche in the study of evolutionary phenomena. Digital organisms can evolve a wide variety of complex strategies in response to evolutionary pressures. Many times organisms develop unexpected and innovative strategies that were not envisioned by the developers of the experiment. The tractable nature of the computational platform – rapid generation times, automated data collection, and many generations – enables large-scale experiments. As a result, digital evolution is a powerful technique for generating new evolutionary hypotheses, testing existing evolutionary hypotheses, and exploring the mechanisms employed by digital organisms in response to evolutionary pressures.

23.5 Case Studies

23.5.1 Formation

An initial step in fraternal transitions is the shift from independence on the part of the lower-level units to inter-dependence within kin groups and, ultimately, replication of the group as a whole at the higher level (e.g., by fragmentation or propagule production). We initially investigate whether limiting the availability of resources to a random subset of cells would be sufficient to encourage a shift of their cellular replication efforts to a higher-level entity, giving up their ability to replicate autonomously [29]. Upon each update of the system, a random set of cells were

notified that a new resource was available. After that cell collected its fill, it would
notify any neighbors that it identified as part of its group to do the same. This process would continue until either the group's limits were met, or the resource was
fully consumed (harming the last round of cells that expended resource in a failed
attempt to collect more). Each simulated cell is a simplified version of a digital
organism consisting of a set of values that described its behaviors, including a probability that offspring would be labeled as part of their same group, a preference for
offspring replacing non-group members over group members (potentially not repli-
cating at all if a cell is in the middle of a group), and a preference for hoarding
resources locally or sharing them in a group-level pool. Over time, cells evolved to
adjust these values so that groups were maintained at an optimal size for a given
level of provided resources, group members were never over-written, and resources
were all shared at the group level. In fact, when cells were given multiple levels of
groups that they could be a part of (that is, groups of groups were possible and had
access to more abundant resources), populations sometimes progressed through the
second transition, storing resources with the high-level group and regulating its size
appropriately (Figure 23.3).

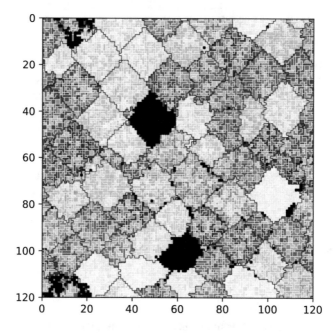

Fig. 23.3: A typical evolved population where organisms underwent multiple transitions in
individuality. Cells first linked into to small groups (single solid colors bounded by a
white border) and then to larger groups (collections of small groups with the same hue,
bounded by a black border). The largest groups are the maximum size that can be
sustained by a single resource.

23.5.2 Task-based Division of Labor

Division of labor, where individuals within groups specialize to improve group function, is a key part of fraternal transitions in evolution [26]. For example, within eusocial insect colonies, some insects are reproductive (e.g., the queen), while others are
not (workers) [18]. Workers can further specialize on particular types of work, such
as soldier, undertaker, and nurse. Within multicellular organisms, cellular specialization gives rise to germ cells (used to form new organisms), and cells comprising
the somatic tissue. These latter cells can have many different specialized roles – skin
cell, myocyte, neuron. Because of the central role of division of labor in fraternal
transitions, it is critical to understand the forces that give rise to it and the pathways
of its evolution. Within our work, we have explored several different forces that
promote division of labor [9, 10, 11, 12, 13, 14].

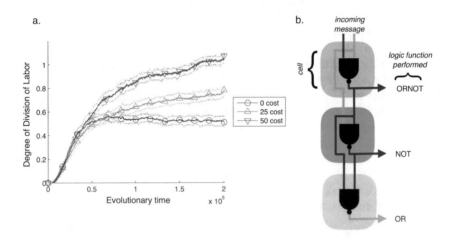

Fig. 23.4: The effect of task-switching costs on the evolution of division of labor. In (a) we see
that as task switching costs increase from 0 to 50 units, over evolutionary time,
multicells evolve to exhibit more division of labor (measured using Shannon mutual
information among tasks and cells). Figure (b) depicts a problem decomposition
strategy evolved by the multicells. Cell color again represents phenotype, indicating the
type of task performed by the cell. Each cell takes in a message (consisting of two
numbers), NANDs the numbers together and exports the result. The first cell performs
the ORNOT task and sends a message to the second cell consisting of the ORNOT and
OR result. The second cell NANDs these numbers together to export the NOT task. It
then sends a message consisting of the ORNOT and NOT results. The third cell uses
these results to perform the OR task.

Adam Smith, the famous economist, declared the avoidance of costs associated
with switching tasks to be one of the greatest benefits of division of labor [38]. In one
study, we demonstrated that by applying task-switching costs, digital cells placed

into a multicellular organism framework would evolve division of labor. Moreover, cells within organisms that exhibited division of labor became so dependent upon their sister cells that they lost the ability to independently perform tasks or replicate. This loss of functionality at the lower level indicates a loss of individuality at the cellular level, thus indicating task-switching costs may not only drive the evolution of division of labor, but also lead to a shift in autonomy from the lower-level to the higher-level (Figure 23.4).

23.5.3 Reproductive Division of Labor

One type of specialization that can occur in groups is reproductive division of labor [1], where most lower-level units focus on life-processes needed to sustain the higher level unit, while some others are responsible for producing a higher-level offspring. Superficially, it seems that any lower-level unit that gives up its ability to propagate outside of the higher-level unit will be at an evolutionary disadvantage; as such, reproductive division of labor is a key hallmark that an evolutionary transition has been completed.

A possible explanation for the benefit of reproductive division of labor comes from the "dirty work hypothesis" [13]. This hypothesis states that if some metabolic functions that would otherwise be beneficial also cause mutagenic damage, such functions should be performed by non-reproductive "somatic cells" while more pristine genetic material is maintained inside of reproductive "germ cells" . As such, the higher-level organism is able to still have most of its cells contributing to its metabolic success, while also ensuring that its offspring has as few mutations as possible.

In our earlier work [13], we tested the dirty work hypothesis in digital organisms. In this study, each multicell was composed of 25 cells, which replicated as a group after metabolizing a threshold amount of resource (typically 500 units). At the early stages of evolution, lower-level cells shared the burden of performing a small number of tasks each until the threshold was reached. New higher-level offspring were seeded with a single, randomly chosen propagule-eligible cell from the parent multicell that replicated internally to fill the new multicell. Over time, cells evolved to divide labor, where some cells performed many tasks and seriously damaged their genomes, while others performed none. This strategy markedly increased overall resource metabolism and created a substantial speedup in the frequency with which multicells produced offspring; a large fraction of those offspring weren't able to grow (since they were initialized using cells with damaged genomes), but enough were healthy, resulting in a net competitive advantage. Finally, after additional evolutionary time, task-performing cells recused themselves from being chosen as a seed for a new multicell (marking themselves propagule-ineligible) in favor of their less mutated kin, thus becoming proper soma for a protected germline (Figure 23.5).

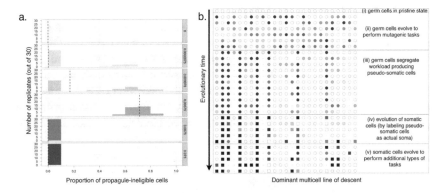

Fig. 23.5: The evolution of reproductive division of labor. (a.) Each histogram depicts the proportion of propagule-ineligible cells within a multicell for a given mutagen level. At an intermediary mutagen level, a non-zero proportion of propagule-ineligible cells evolve and perform a large amount of the dirty work – as a result they are considered true somatic cells and the multicells exhibit reproductive division of labor. (b.) The line of descent of one replicate that evolved reproductive division of labor. Here, each row represents a multicell along the line of descent. Circles represent propagule-eligible cells. Squares represent propagule-ineligible cells. The depth of shading (blue for propagule eligible; red for propagule ineligible) represent the amount of dirty work performed by the cell.

23.6 Future Big Questions

23.6.1 Mechanisms Underlying Formation/Capitalization

The digital evolution framework allows experimental exploration of not only the conditions favoring major transitions, but also the mechanisms underlying components of these transitions. For example, how is division of labor actually achieved inside digital multicells? In order for different digital cells to display different phenotypes, one of a few possible situations must hold [14]. Cells could break symmetry by using stochastic information such as information taken in from the environment or received in communications from other cells. Alternatively, cells could embrace phenotypic plasticity, conditioning their functionality on their environment or information provided by other cells. In previous work, we have found that cells within multicells will often use either environment-based or communication-based plasticity to achieve division of labor [13, 14]. This observation raises an interesting open question. Does pre-existent phenotypic plasticity expedite the evolution of division of labor in digital multicells? It is possible to select for phenotypic plasticity in in response to environmental states [7, 22]. Even if such plastic machinery evolves in populations of single digital cells, does it get co-opted to facilitate division of labor within a multicell context? Effective cell-cell communication relies on more than just plasticity – cells need to both respond to textitand generate states. Within a digital platform, we can allow cells to have impact on their local environment (e.g., by sending messages to neighbors); such impact can be described as a kind of "niche construction" [30]. Again we can ask whether the niche construction and plasticity available in a single cell context are reformulated into communication systems within a multicell context, thereby accelerating the evolution of communication-based division of labor (Figure 23.6). In this way, the digital evolution framework opens exciting possibilities for the mechanistic underpinnings of major transitions.

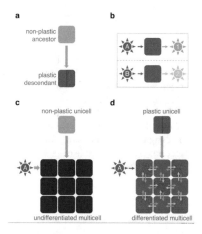

Fig. 23.6: Can plasticity and niche construction facilitate the evolution of division of labor? (a) Consider the evolution of phenotypic plasticity in a single cell from a non-plastic ancestor. Here the plastic cell responds differently to distinct environments and is represented by a half-blue and half-red individual. (b) More specifically, let us assume that environment A favors a certain task denoted by a red cell, while environment B favors a different task denoted by a blue cell. Furthermore, a cell that performs the "red task" emits signal 1 (denoted in orange) and a cell that performs the "blue task" emits signal 2 (denoted in green). The evolution of a population of single cells under variable conditions (sometimes environment A, sometimes environment B) may result in the evolution of phenotypic plasticity (and niche construction)– a single cell that conditions its task performance (and resulting emitted signal) on the environment it experiences. (c) Suppose the non-plastic/non-constructing single cell ancestor is evolved under circumstances favoring multicellularity. Because there is no machinery for breaking symmetry among cells and dividing labor, it may take many evolutionary steps for the multicell to differentiate, even if an environmental stimulus is provided to part of the multicell. (d) The plastic/constructing single cell is pre-equipped with machinery that can be co-opted to achieve division of labor once multicellularity evolves. For instance, suppose the upper left cell of the multicell experiences environmental stimulus A. This cell performs the red task in response and emits a signal to its cardinal neighbors. Now if cells evolve to read signals 1 and 2 as stimuli B and A (or if cells emit signals B and A instead of 1 and 2), then the checkerboard pattern of task division results. Here, phenotypic plasticity (conditionality of phenotypic state on stimulus input) and niche construction (impact on local environment based on phenotypic state) set the stage for division of labor.

23.6.2 Avoiding Reversion

Once a fraternal transition occurs, it is not guaranteed that the new higher-level entity (e.g., multicellular organism, eusocial insect colony) will persist. Forces such as the competition between the lower-level units and higher-level unit can lead to competition and cheating, which can result in disruptive lower-level elements, such as reproductive workers and cancerous cells, that destroy the higher-level unit [28, 40]. Less attention has been given to the possibility that developmental variants can revert to the primordial state thus undoing the advances made by major transitions, especially if the conditions that favored the major transition have shifted. This gives rise to the question: How do major transitions avoid reversion? Libby et. al have suggested that two types of ratcheting mutations could maintain the higher-level unit [24]. First, mutations may increase the fitness of the newly formed multicellular organism at the expense of the cell. Second mutations may decrease the probability a mutation will cause a reversion to the unicellular state. Within the first category an intriguing idea presents itself: Could the many ways in which the higher-level unit capitalizes on the transition (e.g., increased efficiency, division of labor [25]) serve to entrench the higher-level unit within its current state (Figure 23.7)? Digital evolution approaches are a tractable option for studying this question by observing a major transition, where organisms begin as unicellular, transition to multicellular, and then exhibit to division of labor. We can then control environmental conditions to make multicellularity less desirable and assess when organisms remain entrenched in their multicellular state and when they revert to unicellularity.

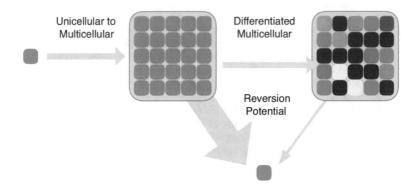

Fig. 23.7: Can division of labor lower the potential for reversion to the unicellular state? Undifferentiated multicells all exhibit the phenotypic traits necessary for survival and thus are expected to have a high probability to revert to unicells if such a reversion becomes beneficial. Cells in a differentiated multicells, on the other hand, exhibit only a subset of phenotypic traits and as such would often require more substantial changes for a reversion to be successful.

23.6.3 Egalitarian Transitions

In most of our discussion above we have focused on fraternal major transitions, but the exploration of egalitarian transitions using digital evolution is an area that is rife with possibilities. Because genetically unrelated lower-level units comprise the higher-level unit in such transitions, some attention must be given to how different lower-level units stay associated across generations. This means we must specify the manner by which higher-level entities reproduce.

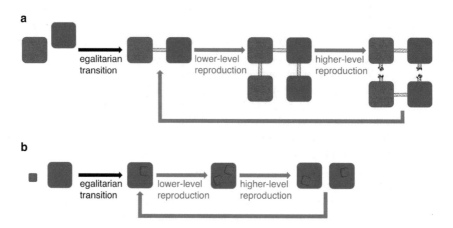

Fig. 23.8: llustrations of life cycles after an egalitarian major transition. (a) Suppose cells are endowed with the ability to attach to and detach from other cells. The first step towards an egalitarian transition could be the attachment of two different cell types (red and blue). However, reproduction of this higher-level unit is not trivial. Here, we outline one possibility. Suppose that both cells reproduce while offspring attach to their parents. Then the offspring attach to one another and afterward both detach from their parents. This would complete the higher-level life cycle (in gray arrows). Such a life cycle requires coordination of multiple lower-level events (reproduction, attachment, detachment). (b) Suppose cells are asymmetric and one cell type (blue) can engulf another (red). The first step towards an egalitarian transition could be the engulfing event. However, reproduction of this higher-level unit also requires further consideration. Suppose the engulfed red cell reproduces faster, such that two red cells are present before the engulfing blue cell reproduces. If both offspring inherit one of the red cells, then a higher-level life cycle is complete (in gray arrows). Here a difference in reproductive rate between cell types is critical, but such a life cycle might be easier to accomplish than the coordinated attachment/detachment in part a.

The simplest possibility is that an offspring aggregate is formed by fragmentation or random sampling of the parent aggregate (see [13]). A more nuanced possibility is to enable lower-level units to evolve instructions for attaching to, detaching from, engulfing and expelling other lower-level units (Figure 23.8). In this way, the lower-level gains control over the very formation of the higher-level units, and their reproduction. Because an egalitarian transition involves the unification of genetically textitdissimilar units, the higher-level unit can start with a pre-packaged division of labor. Nonetheless, how such division of labor evolves in the context of the higher-level unit remains an interesting question. For instance, some division of labor might be necessary to ensure that different lower-level units are transmitted reliably across higher-level generations. Although strict reproductive division of labor is not expected in egalitarian transitions [34], reproductive restraint at the lower-level may play a fundamental role in higher-level heritability [13]. Both the conditions favoring the fusion of lower-level units and the mechanisms ensuring higher-level stability are worth exploration. Thus, we see egalitarian transitions as an exciting new target for digital evolution studies.

23.7 Conclusions

Major transitions are the focus of many of the most important and most interesting questions throughout evolutionary biology. Unfortunately, these evolutionary events are rare, thus making them difficult to study either in nature or even in a traditional laboratory setting. Digital evolution, however, provides us with an evolving system more amenable to these sorts of investigations. Digital organisms have already been used as a study system to address hypotheses on group formation during fraternal transitions, division of labor, and fundamental shifts in individuality. Ongoing studies are continuing to look of other aspects of major evolutionary transitions, such as the causes of entrenchment in a multicellular state, the role of plasticity in evolving coordination of division of labor, and how any of these factors play out during egalitarian transitions. These studies are also providing insights into computer science where processors are shifting from getting individually faster each year to having more and more cores that must be coordinated in a sophisticated manner to harness their processing power, and natural organisms can provide insight into the algorithms that should be used. Just like major transitions have a profound effect on biology, so too may their insights profoundly effect computer science.

Acknowledgements This work was supported by the BEACON Center for the Study of Evolution in Action (NSF Cooperative Agreement DBI-0939454), NSF grant DEB-1655715, and NSF CAREER award to B.K. (DEB0952825). We thank Erik Goodman for his constant support of interdisciplinary exploration, including our work, and leadership of the BEACON center.

References

1. Beshers, S.N., Fewell, J.H.: *Models of division of labor in social insects.* Annual Review of Entomology **46**(1), 413–440 (2001)
2. Brunet, T., King, N.: *The origin of animal multicellularity and cell differentiation.* Developmental Cell **43**, 124–140 (2017)
3. Buss, L.W.: *The Evolution of Individuality.* Princeton University Press, Princeton, NJ (1987)

4. Calcott, B., Sterelny, K.: *The Major Transitions in Evolution Revisited*. MIT Press, Cambridge, MA (2011)

5. Chow, S.S., Wilke, C.O., Ofria, C., Lenski, R.E., Adami, C.: *Adaptive radiation from resource competition in digital organisms*. Science **305**, 84–86 (2004)

6. Clune, J., Goldsby, H.J., Ofria, C., Pennock, R.T.: *Selective pressures for accurate altruism targeting: Evidence from digital evolution for difficult-to-test aspects of inclusive fitness theory*. Proc Biol Sci **278**, 666–674 (2011)

7. Clune, J., Ofria, C., Pennock, R.T.: *Investigating the emergence of phenotypic plasticity in evolving digital organisms*. In: European Conference on Artificial Life 2007, pp. 74–83. Springer, Berlin (2007)

8. Gavrilets, S.: *Rapid transition towards the division of labor via evolution of developmental plasticity*. PLoS Comput Biol **6**, e1000805 (2010)

9. Goldsby, H.J., Dornhaus, A., Kerr, B., Ofria, C.: *Task-switching costs promote the evolution of division of labor and shifts in individuality*. Proceedings of the National Academy of Sciences **109**, 13686–13691 (2012)

10. Goldsby, H.J., Knoester, D.B., Clune, J., McKinley, P.K., Ofria, C.: *The evolution of division of labor*. In: G. Kampis, I. Karsai, E. Szathmáry (eds.) Proceedings of the European Conference on Artificial Life (ECAL-2009), pp. 10–18. Springer, Berlin (2009)

11. Goldsby, H.J., Knoester, D.B., Kerr, B., Ofria, C.: *The effect of conflicting pressures on the evolution of division of labor*. PloS One **9**, e102713 (2014)

12. Goldsby, H.J., Knoester, D.B., Ofria, C.: *Evolution of division of labor in genetically homogeneous groups*. In: Proceedings of the Genetic and Evolutionary Computation Conference (GECCO-2010), pp. 135–142. ACM Press (2010)

13. Goldsby, H.J., Knoester, D.B., Ofria, C., Kerr, B.: *The evolutionary origin of somatic cells under the dirty work hypothesis*. PLoS Biology **12**, e1001858 (2014)

14. Goldsby, H.J., Serra, N., Dyer, F., Kerr, B., Ofria, C.: *The evolution of temporal polyethism*. Artificial Life **13**, 178–185 (2012)

15. Griesemer, J.: *The units of evolutionary transition*. Selection **1**, 67–80 (2000)

16. Herrero, A., Stavans, J., Flores, E.: *The multicellular nature of filamentous heterocyst-forming cyanobacteria*. FEMS Microbiology Reviews **40**, 831–854 (2016)

17. Herron, M.D., Hackett, J.D., Aylward, F.O., Michod, R.E.: *Triassic origin and early radiation of multicellular volvocine algae*. PNAS **106**, 3254–3258 (2009)

18. Hoelldobler, B., Wilson, E.O.: *The Superorganism: The Beauty, Elegance, and Strangeness of Insect Societies*. WW Norton & Company, New York, NY (2009)

19. Ispolatov, I., Ackermann, M., Doebeli, M.: *Division of labour and the evolution of multicellularity*. Proc R Soc Lond B **279**, 1768–1776 (2012)

20. Knoll, A.H.: *The multiple origins of complex multicellularity*. Annual Review of Earth and Planetary Sciences **39**, 217–239 (2011)

21. Kumar, K., Mella-Herrera, R.A., Golden, J.W.: *Cyanobacterial heterocysts*. Cold Spring Harbor Perspectives in Biology **2**, 1–19 (2010)

22. Lalejini, A., Ofria, C.: The evolutionary origins of phenotypic plasticity. In: Proceedings of the Artificial Life Conference 2016, pp. 372–379. MIT Press, Cambridge (2016)

23. Lenski, R.E., Ofria, C., Pennock, R.T., Adami, C.: *The evolutionary origin of complex features*. Nature **423**, 139–144 (2003)

24. Libby, E., Conlin, P.L., Kerr, B., Ratcliff, W.C.: *Stabilizing multicellularity through ratcheting*. Phil. Trans. R. Soc. B **371**(1701), 20150444 (2016)

25. Libby, E., Ratcliff, W.C.: *Ratcheting the evolution of multicellularity*. Science **346**(6208), 426–427 (2014)

26. Maynard Smith, J., Szathmary, E.: *The Major Transitions in Evolution*. Oxford University Press, Oxford, UK (1995)

27. Michod, R.E.: *Darwinian Dynamics: Evolutionary Transitions in Fitness and Individuality*. Princeton University Press, Princeton, NJ (1999)

28. Michod, R.E., Roze, D.: *Cooperation and conflict in the evolution of multicellularity*. Heredity **86**, 1-7 (2001)

29. Moreno, M.A., Ofria, C.: *Toward open-ended fraternal transitions in individuality*. Tech. rep., PeerJ Preprints 6:e27275v1 (2018)
30. Odling-Smee, F.J., Laland, K.N., Feldman, M.W.: *Niche construction: the neglected process in evolution*. Princeton University Press (2003)
31. Ofria, C., Wilke, C.O.: *Avida: A software platform for research in computational evolutionary biology*. Artificial Life **10**, 191–229 (2004)
32. Okasha, S.: *Evolution and the Levels of Selection*. Oxford University Press, Oxford, UK (2006)
33. Queller, D.C.: *Cooperators since life began*. The Quarterly Review of Biology **72**, 184–188 (1997)
34. Queller, D.C.: *Relatedness and the fraternal major transitions*. Philosophical Transactions of the Royal Society of London, Series B **355**, 1647–1655 (2000)
35. Rainey, P., Kerr, B.: *Cheats as first propagules: a new hypothesis for the evolution of individuality during the transition from single cells to multicellularity*. Bioessays **32**, 872–880 (2010)
36. Ratcliff, W.C., Denison, R.F., Borrello, M., Travisano, M.: *Experimental evolution of multicellularity*. Proceedings of the National Academy of Sciences **109**(5), 1595–1600 (2012)
37. Rueffler, C., Hermisson, J., Wagner, G.P.: *Evolution of functional specialization and division of labor*. PNAS **109**, E326–E335 (2012)
38. Smith, A.: *The Wealth of Nations*. W. Strahan and T. Cadell, London (1776)
39. Strassmann, J.E., Queller, D.C.: *The social organism: Congresses, parties and committees*. Evolution **64**, 605–616 (2010)
40. Strassmann, J.E., Queller, D.C.: *Evolution of cooperation and control of cheating in a social microbe*. Proceedings of the National Academy of Sciences **108**(Supplement 2), 10855–10862 (2011)
41. Willensdorfer, M.: *On the evolution of differentiated multicellularity*. Evolution **63**, 306–323 (2009)
42. Zimorski, V., Ku, C., Martin, W.F., Gould, S.V.: *Endosymbiotic theory for organelle origins*. Current Opinion in Microbiology **22**, 38–48 (2014)

Part V
Evolutionary Applications

Chapter 24
Rise of Evolutionary Multi-Criterion Optimization: A Brief History of Time with Key Contributions

Kalyanmoy Deb

Abstract One of the main success stories in the evolutionary computation (EC) field is the use of EC framework to solve multi-criterion optimization problems. These problems give rise to a set of trade-off Pareto-optimal solutions, instead of a single optimal solution; hence a population-based EC framework is a natural choice for solving them. Starting in the early nineties with a few parameter-dependent algorithms, the research and application of evolutionary multi-criterion optimization (EMO) algorithms has become a field of its own, by attracting mathematicians, computer scientists, engineers, economists, managers, and entrepreneurs. In this chapter, we provide a chronological account of key contributions which kept the field alive, useful, and vibrant. A bibliometric study of published materials on the topic is also provided to paint a picture of the rise and the popularity of the field.

Key words: Evolutionary multi-criterion optimization, multi-criterion decision making, evolutionary computation.

24.1 Introduction

Traditionally, optimization refers to a task in which a single pre-defined objective is either minimized or maximized to find a single optimal or a near-optimal feasible solution represented by a variable vector that satisfies a set of constraint functions. While the above-described single-objective optimization makes up most of the theoretical, algorithmic, and practical studies in the optimization literature, multi-

Koenig Endowed Chair Professor
Department of Electrical and Computer Engineering
Michigan State University, East Lansing, MI 48824, USA
e-mail: kdeb@egr.msu.edu
http://www.coin-laboratory.com/

© Springer Nature Switzerland AG 2020   351
W. Banzhaf et al. (eds.), *Evolution in Action: Past, Present and Future*,
Genetic and Evolutionary Computation, https://doi.org/10.1007/978-3-030-39831-6_24

objective optimization is more pragmatic and has certainly become a major focus in academics and industries.

Multi-objective optimization refers to a task of minimizing or maximizing not one, but multiple conflicting objectives simultaneously. In an engineering design context, minimizing fabrication cost and simultaneously maximizing design quality, are two such conflicting objectives. Theoretically, such conflicting objectives give rise to a set of trade-off optimal solutions (known as Pareto-optimal solutions [2, 34]), instead of a single solution, of which individual optimal solutions to each objective are members. Other optimal solutions are called compromise solutions to the two objectives. Since both objectives are of importance, usually one of the compromise solutions is the target, but clearly the choice of a single preferred solution from the optimal set becomes an important decision-making task which is subjective and involves a number of decision-makers (DMs). Thus, when multiple conflicting objectives are present, a solution process should be comprised of at least two tasks – a multi-objective optimization task in which a number of trade-off solutions are to be found and a multi-criterion decision-making (MCDM) task which allows DM's preference information to be analyzed to select a single Pareto-optimal solution.

Since the first task of multi-objective optimization attempts to find multiple optimal solutions, evolutionary computation (EC) methods become potential candidates, mainly due to their population approach. Started in early nineties, barring a single study in 1984, the research and application of Evolutionary Multi-objective Optimization (EMO) has become a complete field of its own. Over the years, EMO has encroached into various other EC fields and, importantly, have taken the EC field outside engineering and computing fields. In this chapter, we provide a brief history of key events and contributions describing how the EMO field has grown in the past 25 years.

All bibliographic analysis done in June 2018 and presented in this chapter are done with 72,575 documents identified from the SCOPUS database with several keywords from evolutionary multi-objective optimization related phrases. Figure 24.1 shows two spurts of exponential growth in the publication of EMO papers – one until about 2010 and another which is presently underway. We believe the first burst was caused by the publication of efficient EMO algorithms for solving two and three-objective problems. The second is mostly due the publication of new algorithms for solving four or more objective problems. Publications of six to seven thousand papers a year definitely makes an EMO as one of the emerging fields within EC.

Figure 24.2 shows that while about 50% of the 72,575 documents are in engineering and computer science area, the remaining half is spread over many other science, practice, and business areas. Particularly, current applications to medicine, genetics, social sciences, and astronomy, are beyond the wildest dreams of early EMO researchers.

In the remainder of this chapter, we describe the events which caused the EMO surge to begin in Section 24.2. Section 24.3 provides a systematic account of events responsible for the rise of the EMO field. Section 24.4 presents briefly a list of current research topics of interest to EMO researchers and Section 24.5 highlights

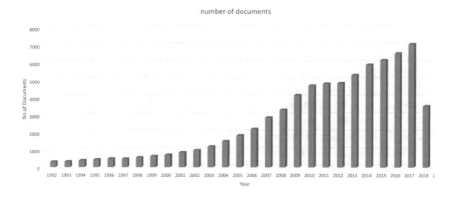

Fig. 24.1: Growth of number of EMO papers over the years. Survey was done on 30 June 2018.

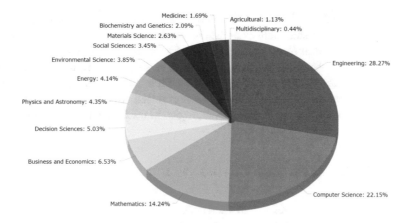

Fig. 24.2: Discipline-wise EMO papers, indicating about 50 % papers come from
non-engineering and non-computer science fields.

a list of topics for future research and application. Finally, Section 27.5 makes a
summary of the chapter and presents a broad picture of the countries and researchers
who have significantly contributed to the advancement of the EMO field.

24.2 The Beginning

Early EC researchers highlighted the use of multiple conflicting objectives to find
pragmatic solutions, but no real EC algorithm was suggested until 1984 by David
Schaffer [39]. However, his VEGA approach failed to maintain a well-distributed

set of trade-off solutions. David E. Goldberg, in his seminal GA book [25] in 1989, suggested the inclusion of a *niche-preserving* operator to maintain multiple diverse solutions. Three groups of researchers from three continents followed Goldberg's suggestion and developed three different successful EMO algorithms during 1993-95: Fonseca and Fleming's MOGA [23], Horn et al.'s NPGA [27], and Srinivas and Deb's NSGA [41]. Each of these algorithms modified the selection operator of an evolutionary algorithm to emphasize non-dominated and less-crowded solutions using a mathematical partial ordering concept and a niche-preserving method, respectively. While each of these methods required a crucial parameter to be tuned for a problem, the publication of these algorithms marked the beginning of the EMO revolution.

24.3 The Rise

The ability to find multiple tradeoff solutions by considering two conflicting objectives, such as cost and quality or cost and emission, through a population-based evolutionary algorithm made sense, and as if the scientific and industrial community were simply waiting for such algorithms to be suggested. EMO research dramatically rose starting around1995 with many different algorithms suggested, many different problems solved, and many new issues raised. We attempt to account for some key events responsible for EMO's rapid rise in the following subsections.

24.3.1 EMO Test Problem Construction

Following the initial suggestion of fundamental EMO methods in 1995, the lead EMO researchers realized that there was a need of effective and tunable test problems which would enable them to develop efficient EMO algorithms. Before 1999, only a few two or three-variable problems involving two or at most three objectives were available [45] and it was not even clear what aspects of a multi-objective optimization algorithm these problems were trying to test. In 1999, Deb [6] suggested a multi-objective test-problem construction procedure by bringing in well-understood single-objective problem difficulty concepts. The idea was to use three separate functionals to construct two objectives which could exhibit different kinds of convergence and diversity-preserving difficulties. The idea was eventually realized by the suggestion of a six-problem test suite, now largely known as ZDT (Zitzler-Deb-Thiele) test suite [47] suggested in 2000. While ZDT problems were all two-objective problems, later the same group of EMO researchers, came up with a scalable test problem construction method for utilizing any number of objectives and a eight-problem scalable test suite DTLZ (Deb-Thiele-Laumanns-Zitzler) in 2001 [19]. Both ZDT and DTLZ problems have helped EMO researchers develop their own algorithms and compare with other existing algorithms in a manner which computer science researchers were used to. In some sense, these problems made EMO research legitimate and useful in the computer science discipline. Figure 24.3 demonstrates how these two test-problem suites were cited over the years. A few

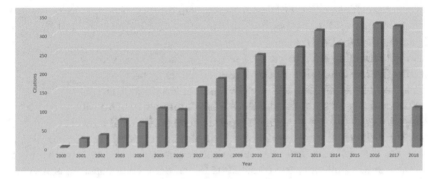

Fig. 24.3: Growth of papers citing ZDT and DTLZ test suites.

other test-problems were also suggested and used extensively [28, 36].

24.3.2 EMO Performance Metrics

Test-problems allowed EMO researchers to sharpen their algorithms against various problem complexities. However, to evaluate their results, there was a need for performance metrics to quantify the success of an algorithm. Compared to single-objective studies, in which there is a clear and obvious single performance measure directly related to the objective function value, in multi-objective problems, multiple performance criteria were needed for an accurate evaluation of two sets of trade-off solutions [49]. Deb [7] argued that two functional performance metrics – one for convergence and another for diversity achievement – would be practically enough for multi-objective optimization studies. While different metrics were suggested for each aspect, EMO researchers always felt the need to have a single performance metric to compare two algorithms. Two metrics – hypervolume [48] and inverse generational distance (IGD) [46] – are now popularly used as an aggregate measure of convergence and diversity, despite valid criticisms against their use [30]. Hypervolume does not need any prior knowledge of the Pareto-optimal front but becomes computationally expensive with an increase in the number of objectives. IGD requires the knowledge of Pareto-optimal front and is therefore useful for testing an algorithm on test-problems with a known Pareto-optimal front.

24.3.3 Efficient EMO Algorithms

Having different test-problems for an algorithm and having metrics to evaluate
that algorithm's performance, the next big task for EMO researchers was to design
and develop efficient algorithms to improve from the first-generation parameterized
EMO algorithms. A few efficient algorithms [31, 48] were suggested, but the one
that catapulted the EMO research beyond engineering and computing paradigms and
is still contributing after 16 years of its publication was the Elitist Non-dominated
Sorting GA (NSGA-II) [9]. NSGA-II used a modular approach in implementing emphasis for non-dominated, less-crowded, and elite solutions in a manner that does
not demand any additional parameter - a break-through which EMO algorithm developers were looking for. A simple working code of NSGA-II is available at devel-
oper's website (http://www.coin-laboratory.com) and commercial versions are also available from various software companies. A steady growth of citations – about 2,000 a year – of the 2002 IEEE Transaction paper on NSGA-II is
shown Figure 24.4.

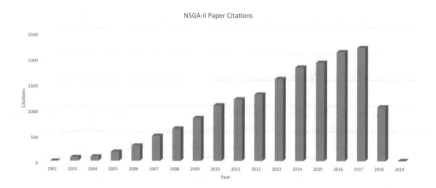

Fig. 24.4: Growth of citations of the NSGA-II paper published in IEEE Transactions on EC [9].

A chart of the main disciplines in which NSGA-II has been used so far is shown
in Figure 24.5. Besides engineering and computer science, NSGA-II has been applied to a wide variety of non-computing specific fields, such as environmental science, social sciences, medicine, etc.

24.3.4 EMO Conference Series

With an all-round interest from the academic and industrial communities, the leading researchers of EMO planned an international conference on EMO to facilitate a
platform for EMO researchers to share their current research and to engage in col-

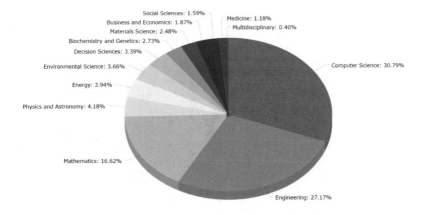

Fig. 24.5: Sources of citing the NSGA-II paper.

laborations. The first conference was arranged in Zurich, Switzerland in 2001 with a single-track format. The conference was a big success, as about 30% additional participants showed up on the day of the conference. Three groups of EMO researchers also proposed to host the next one. An EMO steering committee was then formed and a biannual format for the EMO conference series was decided by the committee. After nine successful EMO conferences, the author and Erik D. Goodman will host the 10th EMO conference in March 2019 at Michigan State University.

Without a dedicated journal, EMO conference proceedings stayed as a reliable and main source of early EMO papers. Figure 24.6 shows how EMO conference papers have been cited over the years, averaging around 700 citations per year. Particularly, the citation pattern of the first three EMO conference proceedings (2001, 2003 and 2005) indicate that despite being old, some of the EMO conference papers are still well-cited. Of them, three papers of 2005 EMO conference (MO-PSO [40], DEMO [44] and S-MOEA [21] are each cited more than 100 times so far.

24.3.5 EMO Journal Publications

EMO papers have also dominated the two mainstream EC journals since early 2000. Figure 7(a) shows an excerpt taken on 30 June 2018 from IEEE Transactions on Evolutionary Computation (TEC) website indicating the top five popular articles among all papers published since its inception in 1997. Of them, three papers – 2002 NSGA-II paper, 2007 MOEA/D paper, and 2014 NSGA-III paper – are from the EMO field. Also, for MIT Press's Evolutionary Computation Journal (ECJ), four out of five most-cited papers – 1995 NSGA paper, 2000 ZDT paper, 2008 approximate hypervolume paper, and 2000 PAES paper – are from the EMO field (Figure 7(b)). Besides the top five papers, 24.6% of all TEC published papers and

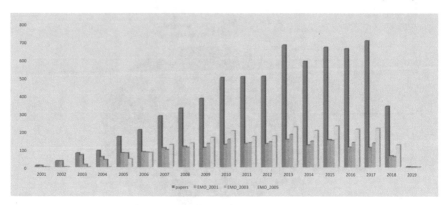

Fig. 24.6: Growth of EMO conference series papers.

15.2% of all ECJ published papers are in the field of EMO. These are remarkable numbers and demonstrate clearly the impact EMO has made in shaping the research landscape of EC.

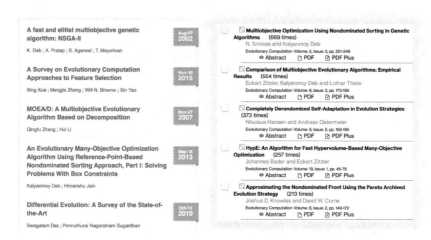

(a) IEEE TEVC on 30 June 2018.   (b) MIT Press's Evolutionary Computation journal
                                          on 30 June 2018.

Fig. 24.7: Top five most popular papers of different journals.

24.3.6 EMO-MCDM Fusion

With efficient and well-tested EMO algorithms being available, the next question was 'How do we choose a single preferred solution from the obtained trade-off set?'. This forced some of the EMO researchers to look into the literature for an answer. In their search, they discovered the multi-criterion decision-making (MCDM) field. MCDM has been actively developing multi-objective optimization algorithms for single preferred solution discovery using a systematic decision-making processes since early seventies. The author and Juergen Branke from EMO joined hands with Kaisa Miettinen and Ralph Steuer from the MCDM field to co-organize the Dagstuhl seminar series in Saarbrucken, Germany entitled 'Practical Methods of Multi-objective Optimization' starting in 2004. The week-long seminar allowed 30 EMO and 30 MCDM researchers to come to a single room and discuss ways of: solving multi-objective problems starting from problem formulation, developing efficient hybrid optimization methods, and finally decision-making. The seminar has been repeated a number of times and one major outcome was the publication of an edited book [1] in which most chapters are jointly authored by researchers from both EMO and MCDM fields.

Such a collaboration between two contemporary fields of similar interests helped both researchers to better understand the practices of the other and together develop hybrid and interactive methods. Besides individual collaborations, every EMO conference now organizes an MCDM track and every MCDM conference organizes an EMO track, making a more substantial impact of one field on to the other.

24.3.7 EMO Commercialization

It is important also to note the contribution of the optimization software companies in the field's development. Siemens's HEEDS software, embedded with a multi-objective SHERPA procedure, Estico's ModeFrontier with a number of EMO algorithms, VR&D's VisualDoc with NSGA-II and AMGA procedures, and Dassault Systemes's SIMULA and ISIGHT softwares have made multi-objective optimization popular among industries. In addition, a number of public domain multi-objective softwares (MOEA Framework, PISA, jMetal, NIMBUS, to name a few) have facilitated EMO's wide-spread use.

24.4 Current Status

Currently, research in EMO field is focused in multiple directions. Below, we provide a list of few popular topics and then subsequently provide a more detailed discussion on some of the most popular ones.

- Evolutionary Many-Objective Optimization (EMaO) for handling more then three conflicting objectives.
- Meta-modeling based EMO for handling problems involving computationally expensive evaluation models.
- Interactive EMO-MCDM methodologies for performing multi-objective or many-objective optimization and decision-making.
- Other meta-heuristics-based multi-objective optimization. methods.
- Multi-objectivization for aiding solution of single-objective search and optimization methods using additional helper objectives.
- 'Innovization': Discovery of innovative knowledge from multi-objective optimization.
- Reduction of correlated objectives for efficient EMO application.

NSGA-II, SPEA, and similar algorithms which attempted to find a well-distributed and well-converged set of solutions without any additional information did not scale well beyond three-objective problems. A new generation of algorithms based on the concept of decomposition emerged in 2007, with the introduction of MOEA/D [46]. A pre-defined set of reference directions were provided for EMO population members to *follow* in order to reach to the Pareto-optimal front. Since a different philosophy of algorithms is needed for handling more than three objectives, these problems are known as many-objective optimization problems in the EMO literature. NSGA-III [14] is an extension of NSGA-II to solve these many-objective problems. The difference between a clustering based NSGA-II and NSGA-III is evident from results on three-objective DTLZ2 problem in Figures 8(a) and 8(b). NSGA-III like algorithms have shown to be effective in solving up to about 20 objective problems.

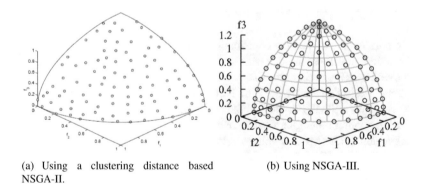

(a) Using a clustering distance based NSGA-II.   (b) Using NSGA-III.

Fig. 24.8: Three-objective problem solved by different methods.

Many practical problems involve expensive evaluation routines, requiring an hour or sometimes days to evaluate a single solution. To optimize such problems, every solution during the optimization process cannot be evaluated using the high-fidelity and expensive evaluation procedures. Meta-modeling methods are used to model an approximate model of objective and constraint functions from a few high-fidelity evaluated solutions. The derived meta-models are then used to run the optimization process until there is a need for a new meta-model. In multi-objective optimization, a recent study [13] suggested a taxonomy of various frameworks for meta-modeling all objective functions and constraints separately to meta-modeling only a single aggregate selection function derived from objective and constraint functions. Further studies are needed to adaptively choose a suitable meta-modeling framework for a problem class.

EMO algorithms can find multiple trade-off solutions and a subsequent multi-criterion decision-making (MCDM) approach can then choose a single preferred solution. But this serial a-posteriori EMO-MCDM approach may not be the most efficient approach, particularly for many-objective optimization problems. An interactive EMO-MCDM method in which a suitable MCDM approach is executed and used to alter the EMO procedure after a few generations may be a better approach. One such approach – progressively interactive EMO or PI-EMO [16] algorithm – was proposed in 2010. Further such studies are needed to combine EMO and MCDM tasks together to develop pragmatic approaches, which can eventually coded in commercial softwares.

While EMO algorithms are developed mostly on the genetic algorithm framework, the concepts are also being applied to other evolutionary and meta-heuristics frameworks, such as on differential evolution [32], particle swarm optimization [4, 35], evolution strategy [29], etc.

EMO methods maintain a diversity of solutions within a population due to simultaneous emphasis of more than one conflicting objectives. If diversity is desired, a *helper* objective providing a diversity estimate of the population, can be added to a single-objective optimization formulation to construct a bi-objective optimization problem. In problems involving bloating in GP [5], design space exploration [42], constraint handling [10], and others, such a multi-objectivization method has been successfully used. These ideas are indicative to more future innovations which can be tried in other different problem-solving scenarios.

When many objectives are present, in certain problems, some of them may become correlated as the solutions progress towards the Pareto-optimal front. If the redundant objectives can be identified by principal component analysis or other more sophisticated machine learning methods, the redundant objectives can be eliminated or combined together to reduce the dimensionality of resulting problem. Although some studies [38] exist, they need to be integrated with the mainstream EMO algorithms for handling real-world problems.

Multiple Pareto-optimal solutions, once found, are all high-performing solutions in the context of the problem. Although one of them will be eventually chosen after a decision-making task, all of them can be analyzed to decipher any common properties involving variables that are shared among them. Such common properties, found by a feature selection method applied to EMO-obtained trade-off solutions, can stay as derived *knowledge* pertaining to the optimal solutions and can be used in subsequent application tasks. This knowledge discovery approach is highly pragmatic and is termed as 'Innovization' – a task of deriving innovative solution principles from multi-objective optimization [17]. 'Innovized' principles, if learned during the optimization process, can also be used in subsequent generations to improve the convergence property of an EMO algorithm. Current research studies are being pursued in this direction [24].

24.5 The Future

The above developments and applications of EMO research brings out a number of salient future research directions, which we briefly mention here.

- Visualization of higher than three-dimensional Pareto-optimal solutions.
- Evolutionary Massive-Objective Optimization (EMsO) for handling more then 20 conflicting objectives.
- EMO with other practicalities, such as uncertainties, combinatorial problems, dynamically changing problems, etc.
- Multi-level, multi-objective optimization algorithms.
- Theory based EMO for guaranteed convergence.
- EMO implementation on parallel hardware.
- Large-scale industrial, societal or business application problems.

For problems having more than three objectives, a suitable visualization of the obtained trade-off solutions becomes an important matter. This is essential for DMs to make a visual picture of trade-offs and other properties of solutions. Existing methods, such as parallel coordinate plots, scatter plots, radviz [26] are far from being practical and useful.

Academic research on EMO must produce efficient algorithms for handling 20 or more objectives, although very few problems from practice have demanded more than 15 objectives so far. Are the existing EMaO algorithms efficient for evolutionary massive optimization (EMsO) problems or different types of algorithms must be developed for such cases [33]?

With passing time, EMO researchers will be forced to answer questions related to the theoretical optimality of EMO-obtained solutions. A recent method using KKT optimality conditions [8] must be corroborated with further ideas for the EMO methods to be recognized by mathematical optimization community.

EMO algorithms are excellent for parallel and distributed computing platforms. Some ideas and implementations have already been proposed [20, 43], but more efficient implementations are needed. Further, EMO's GPU-based implementation must be given more priority.

EMO methodologies must be extended to handle different types of practicalities that industry or other practical problems will offer. Uncertainty in variables or parameters, dynamic nature of certain variables or parameters of the problem, hierarchical problems with multiple levels, etc. require changes in existing methods. Although some EMO algorithms exist [11, 12, 15, 22], more studies are definitely needed.

24.5.1 Large-Scale Applications

EMO and MCDM methods are ready to be applied to large-scale industrial, societal, and business related problems. We present here one such application that the author has conducted along with Prof. Erik Goodman and Dr. Oliver Chikumbo, which earned them the 'Wiley Practice Prize' from International Society on Multi-Criterion Decision Making in 2013.

It is a land use management problem in which local tribes of New Zealand along with the local government wanted to change the type of land uses (about 100 options involving different vegetable pastures, animal gazing, shrubs, etc.) in each of 315 paddocks. It required 14 different objectives to be either minimized or maximized. To make the changes practical, we, in consultation with Governmental agencies, decided to make all necessary changes over 10 years from the start of the planning horizon. The objectives were categorized into a triple bottom-line scenario in which the changes are related to environmental effects, economics of converting and maintaining the land, and productivity from the land. Two types of decision-makers (tribes and government) and their preferences (more productivity with less cost for tribes, and less environmental effect for government) were clear from the onset of the description of the problem [3].

A calculation with the above numbers will dictate that there are $\left((100)^{315}\right)^{10}$ different land use alterations possible. Moreover, the presence of 14 different conflicting objectives makes several solutions Pareto-optimal, which requires a multi-criterion decision task to finally settle on a single preferred solution. Our proposal has been to use an EMO algorithm to tackle the original astronomically large search space problem and find a few (to the tune of about 50 to 100) non-dominated trade-off solutions. In this study, reference point based NSGA-II [18] was applied and a set of 100 trade-off solutions are obtained. These solutions are presented in a parallel coordinate plot (PCP) in Figure 24.9. The trade-off among the solutions are clear from the criss-crossing of the lines between different objective axis, shown in 14 vertical lines.

Relevant solutions from these 100 solutions are then shown to two decision-making groups from their apparently un-preferred solutions towards their preferred solutions. This intermediate study ended up with four trade-off solutions (shown as Scenarios 1 to 4 in the figure) for further analysis. Since both groups could not settle with a single solution, a specific MCDM method – analytical hierarchy process (AHP) [37] – involves decision-makers to make pairwise comparisons to arrive at a single winning solution. The procedure involves both groups of decision-makers to compare objectives and solutions in pairs and comes up with a preference number in a 1 to 9 scale. Thereafter, a mathematical procedure identifies the winner by the highest idealized priority. In our case, Scenario 3 turned out to be winner.

We believe that the principle of using an EMO method to do most of the bullwork in reducing the original large number of options to a few, followed by a systematic decision-making approach to reduce the number of preferred options to a handful, and then ending up with an AHP-like DM-involved, analytical approach

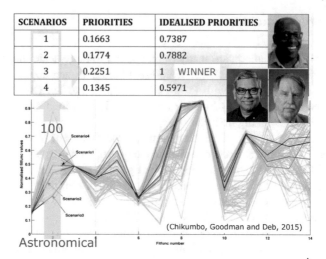

SCENARIOS	PRIORITIES	IDEALISED PRIORITIES	
1	0.1663	0.7387	
2	0.1774	0.7882	
3	0.2251	1 WINNER	
4	0.1345	0.5971	

Fig. 24.9: EMO-MCDM Principle for a large-scale application problem: (i) EMO finds a few (around a few hundreds) solutions from an astronomically large search space, (ii) MCDM is performed involving relevant decision-makers to analyze EMO-found solutions and select a handful (usually 3-5) solutions, and (iii) if needed, another MCDM approach (AHP, for example) can be used to pick a single winning solution.

to select a single solution is generic and can be applied to complex practical multi-criterion optimization problems.

24.6 Conclusions

This chapter has attempted to make an account of the rise of an emerging field – Evolutionary Multi-Criterion Optimization – within the realm of evolutionary computation. It is difficult to put all details of more than 25 years of activities of a field in a chapter, but we have divided the chapter into three main sections describing rise, current and future of the field. The chapter has also described a large-scale, real-world societal application using the cutting-edge developments of the EMO field.

From a 25-year old thought of using a parameterized EC framework to find a few trade-off solutions in a two-objective problem to solving a representative 14-objective real-world problem involving an astronomical search space with two groups of real decision-makers protecting their own interests and embracing ideas from another contemporary MCDM field in its solution process definitely seems like a significant progress, which all EMO and EC researchers and applicationists should be proud of.

Analyzing over 75,000 research publications from SCOPUS database, we have observed that EMO has taken the evolutionary computation concept outside the computer science and engineering communities, and proliferated into medicine, basic sciences, and business areas. Figure 24.10 shows a country-wise use of EMO, indicating that EMO is popularly used worldwide, and most importantly, in a collaborative manner between countries, which may be rare in most emerging fields.

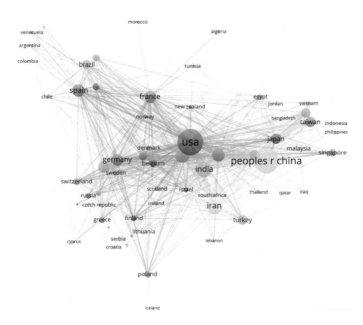

Fig. 24.10: Country-wise EMO papers.

We conclude the chapter by showing a collaborative network of various EMO researchers in terms of their citation network in Figure 24.11. It is abundantly clear that EMO is not restricted to work of a few, but a large number of researchers from around the world.

Acknowledgements Having Prof. Erik D. Goodman as a colleague and collaborator, my primarily algorithm-centric research on EMO has taken a well-grounded practical shape. My specific discussions with Prof. Goodman during the land use management study had been most enjoyable. It has been extremely rewarding to work with him on a daily basis on different EMO and other problems. I dedicate this study to him, knowing how much he will enjoy the numbers and facts describing the rise of a field in which he has contributed so much.

The author acknowledges help from Dr. Aparna Basu and Shivam Mani Tripathi from South Asian University, New Delhi, India and Yashesh Dhebar and Zhichao Lu of Michigan State University in preparing the bibliometric charts.

This material is based in part upon work supported by the National Science Foundation under Cooperative Agreement No. DBI-0939454. Any opinions, findings, and conclusions or recommendations expressed in this material are those of the author(s) and do not necessarily reflect the views of the National Science Foundation.

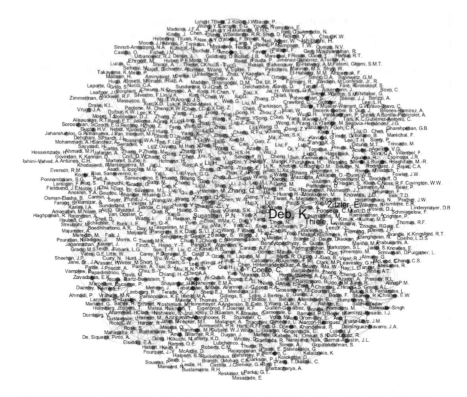

Fig. 24.11: Network of EMO authors in terms number of collaborations and number of citations.

References

1. Branke, J., Deb, K., Miettinen, K., Slowinski, R.: *Multiobjective optimization: Interactive and evolutionary approaches.* Springer-Verlag, Berlin, Germany (2008)
2. Chankong, V., Haimes, Y.Y.: *Multiobjective Decision Making Theory and Methodology.* New York: North-Holland (1983)
3. Chikumbo, O., Goodman, E., Deb, K.: *Triple bottomline many-objective-based decision making for a land use management problem.* Journal of Multi-Criterion Decision Analysis: Optimization, Learning and Decision Support **22**(3-4), 133–159 (2015)
4. Coello, C.A.C., Lechuga, M.S.: *MOPSO: A Proposal for Multiple Objective Particle Swarm Optimization.* In: Congress on Evolutionary Computation (CEC'2002), vol. 2, pp. 1051–1056. IEEE Service Center, Piscataway, New Jersey (2002)
5. De Jong, E.D., Watson, R.A., Pollack, J.B.: *Reducing bloat and promoting diversity using multi-objective methods.* In: Proceedings of the Genetic and Evolutionary Computation Conference (GECCO-2001), pp. 11–18 (2001)
6. Deb, K.: *Multi-objective genetic algorithms: Problem difficulties and construction of test problems.* Evolutionary Computation Journal **7**(3), 205–230 (1999)
7. Deb, K.: *Multi-objective optimization using evolutionary algorithms.* Wiley, Chichester, UK (2001)
8. Deb, K., Abouhawwash, M.: *A optimality theory based proximity measure for set based multi-objective optimization.* IEEE Transactions on Evolutionary Computation **20**(4), 515–528 (2016)

9. Deb, K., Agrawal, S., Pratap, A., Meyarivan, T.: *A fast and elitist multi-objective genetic algorithm: NSGA-II*. IEEE Transactions on Evolutionary Computation **6**(2), 182–197 (2002)
10. Deb, K., Datta, R.: *A bi-objective constrained optimization algorithm using a hybrid evolutionary and penalty function approach*. Engineering Optimization **45**(5), 503–527 (2013)
11. Deb, K., Gupta, H.: *Introducing robustness in multi-objective optimization*. Evolutionary Computation **14**(4), 463–494 (2006)
12. Deb, K., Gupta, S., Daum, D., Branke, J., Mall, A., Padmanabhan, D.: *Reliability-based optimization using evolutionary algorithms*. IEEE Trans. on Evolutionary Computation **13**(5), 1054–1074 (2009)
13. Deb, K., Hussein, R., Roy, P., Toscano, G.: *A taxonomy for metamodeling frameworks for evolutionary multi-objective optimization*. IEEE Transactions on Evolutionary Computation **23**, 104–116 (2018)
14. Deb, K., Jain, H.: *An evolutionary many-objective optimization algorithm using reference-point based non-dominated sorting approach, Part I: Solving problems with box constraints*. IEEE Transactions on Evolutionary Computation **18**(4), 577–601 (2014)
15. Deb, K., Sinha, A.: *An efficient and accurate solution methodology for bilevel multi-objective programming problems using a hybrid evolutionary-local-search algorithm*. Evolutionary Compuation **18**(3), 403–449 (2010)
16. Deb, K., Sinha, A., Korhonen, P., Wallenius, J.: *An interactive evolutionary multi-objective optimization method based on progressively approximated value functions*. IEEE Transactions on Evolutionary Computation **14**(5), 723–739 (2010)
17. Deb, K., Srinivasan, A.: *Innovization: Innovating design principles through optimization*. In: Proceedings of the Genetic and Evolutionary Computation Conference (GECCO-2006), pp. 1629–1636. New York: ACM (2006)
18. Deb, K., Sundar, J., Uday, N., Chaudhuri, S.: *Reference point based multi-objective optimization using evolutionary algorithms*. International Journal of Computational Intelligence Research **2**(6), 273–286 (2006)
19. Deb, K., Thiele, L., Laumanns, M., Zitzler, E.: *Scalable test problems for evolutionary multi-objective optimization*. In: A. Abraham, L. Jain, R. Goldberg (eds.) Evolutionary Multiobjective Optimization, pp. 105–145. London: Springer-Verlag (2005)
20. Deb, K., Zope, P., Jain, A.: *Distributed computing of Pareto-optimal solutions using multi-objective evolutionary algorithms*. In: Proceedings of the Second Evolutionary Multi-Criterion Optimization (EMO-03) Conference (LNCS 2632), pp. 535–549 (2003)
21. Emmerich, M., Beume, N., Noujoks, B.: *An EMO algorithm using the hypervolume measure as selection criterion*. In: Proceedings of Evolutionary Multi-Criterion Optimization Conference (EMO-2005), pp. 62–76. Berlin: Springer-Verlag (2005)
22. Farina, M., Deb, K., Amato, P.: *Dynamic multiobjective optimization problems: Test cases, approximations, and applications*. IEEE Transactions on Evolutionary Computation **8**(5), 425–442 (2000)
23. Fonseca, C.M., Fleming, P.J.: *Genetic algorithms for multiobjective optimization: Formulation, discussion, and generalization*. In: Proceedings of the Fifth International Conference on Genetic Algorithms, pp. 416–423. San Mateo, CA: Morgan Kaufmann (1993)
24. Gaur, A., Deb, K.: *Effect of size and order of variables in rules for multi- objective repair-based innovization procedure*. In: Proceedings of Congress on Evolutionary Computation (CEC-2017) Conference. Piscatway, NJ: IEEE Press (2017)
25. Goldberg, D.E.: *Genetic Algorithms for Search, Optimization, and Machine Learning*. Reading, MA: Addison-Wesley (1989)
26. Hoffman, P.E., Grinstein, G., Pinkney, D.: *Dimensional anchors: A graphic primitive for multidimensional multivariate information visualizations*. In: Proceedings of the 1999 Workshop on New Paradigms in Information Visualization and Manipulation in Conjunction with the Eighth ACM Internation Conference on Information and Knowledge Management, pp. 9–16 (1999)
27. Horn, J., Nafploitis, N., Goldberg, D.E.: *A niched Pareto genetic algorithm for multi-objective optimization*. In: Proceedings of the First IEEE Conference on Evolutionary Computation, pp. 82–87 (1994)

28. Huband, S., Barone, L., While, L., Hingston, P.: *A scalable multi-objective test problem toolkit.* In: Proceedings of the Evolutionary Multi-Criterion Optimization (EMO-2005). Berlin, Germany: Springer (2005)
29. Igel, C., Hansen, N., Roth, S.: *Covariance matrix adaptation for multi-objective optimization evolutionary computation.* Evolutionary Computation Journal **15**(1), 1–28 (2007)
30. Ishibuchi, H., Imada, R., Setoguchi, Y., Nojima, Y.: *How to specify a reference point in hypervolume calculation for fair performance comparison?* Evolutionary Computation **26**(3), 411–440 (2018)
31. Knowles, J.D., Corne, D.: *M-PAES: A memetic algorithm for multiobjective optimization.* In: Proceedings of Congress on Evolutionary Computation (CEC-2000), pp. 325–332 (2000)
32. Kukkonen, S., Lampinen, J.: *GDE3: The third evolution step of Generalized Differential Evolution.* In: Proceedings of the 2005 Congress on Evolutionary Computation (CEC 2005), pp. 443–450. Edinburgh, Scotland (2005)
33. Li, K., Deb, K., Altinoz, T., Yao, X.: *Empirical investigations of reference point based methods when facing a massively large number of objectives: First results.* In: 9th International Conference on Evolutionary Multi-Criterion Optimization (EMO-2017), pp. 390–405 (2017)
34. Miettinen, K.: *Nonlinear Multiobjective Optimization.* Kluwer, Boston (1999)
35. Mostaghim, S., Teich, J.: *Strategies for finding good local guides in multi-objective particle swarm optimization (MOPSO).* In: 2003 IEEE Swarm Intelligence Symposium Proceedings, pp. 26–33. IEEE Service Center, Indianapolis, Indiana, USA (2003)
36. Okabe, T., Jin, Y., Olhofer, M., Sendhoff, B.: *On test functions for evolutionary multi-objective optimization.* In: Parallel Problem Solving from Nature (PPSN VIII), pp. 792–802 (2004)
37. Saaty, T.L.: *Decision Making for Leaders: The Analytic Hierarchy Process for Decisions in a Complex World.* Pittsburgh, Pennsylvania: RWS Publications (2008)
38. Saxena, D., Duro, J.A., Tiwari, A., Deb, K., Zhang, Q.: *Objective reduction in many-objective optimization: Linear and nonlinear algorithms.* IEEE Transactions on Evolutionary Computation **77**(1), 77–99 (2013)
39. Schaffer, J.D.: Some experiments in machine learning using vector evaluated genetic algorithms. Ph.D. thesis, Nashville, TN: Vanderbilt University (1984)
40. Sierra, M.R., Coello, C.A.C.: *Improving PSO-based multi-objective optimization using crowding, mutation and epsilon-dominance.* In: Proceedings of Evolutionary Multi-Criterion Optimization Conference (EMO-2005), pp. 505–519. Berlin: Springer-Verlag (2005)
41. Srinivas, N., Deb, K.: *Multi-objective function optimization using non-dominated sorting genetic algorithms.* Evolutionary Computation Journal **2**(3), 221–248 (1994)
42. Tamara, U.: *Exploring Structural Diversity in Evolutionary Algorithms.* CreateSpace Independent Publishing Platform (2012)
43. Toro, F., Ortega, J., Paechter, B.: *Parallel single front genetic algorithm: Performance analysis in a cluster system.* In: Proc. Intl. Parallel and Distributed Processing Symposium, pp. 1–8. IEEE Press (2003)
44. Tusar, T., Fillipic, B.: *DEMO: Differential evolution for multiobjective optimization.* In: Proceedings of Evolutionary Multi-Criterion Optimization Conference (EMO-2005), pp. 520–533. Berlin: Springer-Verlag (2005)
45. Veldhuizen, D.V.: Multiobjective evolutionary algorithms: Classifications, analyses, and new innovations. Ph.D. thesis, Dayton, OH: Air Force Institute of Technology (1999). Technical Report No. AFIT/DS/ENG/99-01
46. Zhang, Q., Li, H.: *MOEA/D: A multiobjective evolutionary algorithm based on decomposition.* Evolutionary Computation, IEEE Transactions on **11**(6), 712–731 (2007)
47. Zitzler, E., Deb, K., Thiele, L.: *Comparison of multiobjective evolutionary algorithms: Empirical results.* Evolutionary Computation **8**(2), 125–148 (2000)
48. Zitzler, E., Thiele, L.: *Multiobjective evolutionary algorithms: A comparative case study and the strength Pareto approach.* IEEE Transactions on Evolutionary Computation **3**(4), 257–271 (1999)
49. Zitzler, E., Thiele, L., Laumanns, M., Fonseca, C.M., Fonseca, V.G.: *Performance assessment of multiobjective optimizers: An analysis and review.* IEEE Transactions on Evolutionary Computation **7**(2), 117–132 (2003)

Chapter 25
Doing Research at the Intersection of Arts and Science

Francisco Fernández de Vega

Abstract In this chapter we reflect about evolutionary art, past and present, and how best to evaluate it for positioning in the art world. A personal journey that went through Michigan State University's Garage Lab, led to a series of projects at the intersection of arts and science. This has enabled me to reflect on the synergies between both areas from the point of view of evolutionary computation.

So far, researchers have discussed how to design appropriate fitness functions for aesthetic evaluations, and have resorted to interaction to improve quality evaluation. We suggest that both the public and art critics, represented by experts, scholars, museums, art galleries and international art contests, be included in the evaluation process of evolutionary art. Only then, will art created by evolutionary processes have a chance to become a new trend that competes on an equal footing with other types of art.

25.1 A Personal Journey

In 1999, when I was a young doctoral student, under the supervision of professor Marco Tomassini, I was eager to learn as many aspects of evolutionary computation as I could. Previously, I had obtained a position at a regional government administration that guaranteed me the opportunity to develop my professional career in my home town with a good salary. Although that certainly allowed me to get married, start a my family, and earn my life, that was not entirely satisfying.

After studying artificial intelligence in college, I discovered artificial life, cellular automata and evolutionary algorithms in science books, and decided to make an attempt in the area, even though a professor of control theory had told me during my master's studies in 1995 that these algorithms, although fashionable in earlier

Francisco Fernández de Vega
University of Extremadura, Sta Teresa de Jornet 38, 06200 Mérida, Spain, e-mail: `fcofdez@unex.es`

W. Banzhaf et al. (eds.), *Evolution in Action: Past, Present and Future*,
Genetic and Evolutionary Computation, https://doi.org/10.1007/978-3-030-39831-6_25

years, were no longer of interest. I do not want to blame anybody in the area of control theory of lack of vision, but I decided to not trust him and to continue my particular quest for an exciting research topic: artificial evolution.

So, with my wife's support and after three years working in the same government office, I decided to quit my job, enter academia and start pursuing a new goal, that of becoming a researcher in evolutionary algorithms. My willingness to move forward pushed me to try my particular version of the American dream: to go to the United States to meet some of the leaders in the field and work in their laboratories, so that I could benefit from their knowledge and enthusiasm while sharing experiences with other international doctoral students.

This is the main reason I traveled to East Lansing and stayed three months at the Garage laboratory, under the guidance of Professors William F. Punch and Erik Goodman during the summer of 1999. It was the year the first GECCO was held in Orlando -my first conference in the EA area- and I had the opportunity to hear from pioneers like John Holland. But the trip took me from this first Gecco conference in Florida to Michigan and the Garage lab with my family. We all have unforgettable memories of that summer, like visiting Niagara Falls, Toronto, New York, etc. I also experienced some difficulties I had to overcome: my first presentation to a group of leading researchers and students in an international laboratory. My difficulty with English had me quite worried, and I must thank Professor Goodman and his team for patience with my ineffective English at the time. I also demonstrated impressive ability to freeze the parallel computer that was being used by all members of the lab at the time, with massive runs of my parallel genetic programming experiments that took up all resources and preventing others from using the computer during night.

That summer was a start. Although I had previous experience in European research laboratories, such as the one run by my PhD advisor in Lausanne, this was my first meeting with non-European researchers. I encountered leading researchers, pioneers in evolutionary computing, sharing their enthusiasm with students with very different life experiences from around the world. New lifelong friendships were struck in the Garage lab, and I still keep in touch with some of the colleagues that I met there as students for the first time.

25.1.1 Being Inspired by the Leaders

When landing at Michigan State University, I was given grant to use the parallel computer available there. At that time I was interested in parallel and distributed models of Evolutionary Algorithms, particularly Parallel Genetic Programming, and had never before had the opportunity to launch my PVM based simulations on a computer with 6-cpus. The Garage lab was well known because of some of the most widely available tools for EA researchers were created there. In my case, LilGp, developed under the supervision of Bill Punch, was one of my preferred tools, because of its simplicity and the language employed to program it: C. I must say that over the years, I have followed the path towards functional programming, that I discovered in a hidden book about LISP in the library of my home town. How this book reached that library is a mystery to me, but it opened up a world of possibilities that I've been able to explore more deeply after completing my PhD. Today it allows me to deploy a strategy to successfully teach functional programming and recursiveness to young students without any previous programming experience. But at the time, when I was a visiting PhD student at Garage, C was my preferred programming language.

In such an environment I got to know other topics of research and interests displayed not only by other PhD students, but also by the leading researchers. Through the years I have come to know the wide range of interests that Erik Goodman displays, and I guess those interests are fundamental to a productive research life. They open the mind, provide inspiration and, make life more enjoyable. I have found that I share some of Erik's hobbies related to the arts, particularly music performance: he is well known because of his blue grass group, and I am involved in the promotion of Jazz music in my region. I guess music, and arts in general, allows researchers to be more balanced, and in the end, to find easier paths to interesting research topics. In my case, I have tried to connect these other activities with my main line of research over the years, so that I can enjoy research, and discover new areas while enjoying these other activities.

25.1.2 The Art of Evolutionary Computation

In 2015, I had the opportunity to invite Erik Goodman to give the Keynote at the 10th Spanish Conference on Evolutionary Algorithms. The story of this national conference is also connected with my initial steps in this area. The first two GECCO conferences allowed me to know the main players, and most of the colleagues with whom I work today. When I visited Garage, I learned that Gecco had been created from the International Conference on Genetic Algorithms (ICGA), that was organized a couple of years earlier at Michigan State University, led by Prof. Goodman's team. In the following years a number of Spanish colleagues played with the idea of organizing a Spanish conference on EAs, and finally decided to do it in the city where I work, Mérida, in 2002. That was the first EA conference. But then, after a successful series of conferences in subsequent years, I had the privilege to organize the 10th EA conference again in Mérida, and decided to invite Erik Goodman as our plenary speaker. At that occasion he took the opportunity to describe the area and the challenges we will face in the coming years. And I cannot forget how much we enjoyed together one of the social events, held at the Wine Science Museum in my hometown of Almendralejo, with Jazz music while food was being served. Music, and arts in general, speak directly to the soul of human beings, yet it is also a perfect area for research.

Thus, during these twenty years since my visit to Garage Lab, I have tried to connect arts and science through evolutionary algorithms, so that we can expand the frontiers of research while enjoying art and music. In the next section, I describe some findings and reflections on the conjunction of the arts and EAs reached during the last decade. Although I have enjoyed facing a number of problems related to music by means of EAs, such as audio analysis, music transcription, 4-part harmonization and music genre classification [14, 19, 20, 19], I will focus on evolutionary art and address the issue of legitimate evaluation of quality of results.

25.2 The Fitness Dilemma in Evolutionary Art

The application of EAs to art and design is no longer a new topic. For more than three decades researchers have confronted the question of how to adapt the algorithm to this inherently creative process that requires proper evaluation of aesthetic components. This art movement could be included within the broader concept of *Generative Art*, and is part of a well-known trend that has developed through the 20th century, the aim of which is to generate art by means of physical, chemical, mechanical or computer processes [6]. Regarding computer based art, an eye-catching exhibition was organized in London by the end of the sixties, where graphic art, poetry, music, etc, were generated by means of autonomous computer programs [13].

The concept of *art* is not something immovable. On the contrary, it is evolving and changing throughout history. As an example of its inherent capability of change and adaptation we can refer to the *Saloon des Refusés* (1863), an alternative to the official *Paris Salon*, that showed paintings that were rejected by the selection committee of the official annual showcase of French art [2]. Since then, the impressionist movement was first accepted by the public and then by the critics. Impressionism was followed in the early 20th century by other art movements, including expressionism, cubism, futurism, dadaism, surrealism, etc.

Similarly, if we focus on music, *purists* did not considered jazz in the early Twenties to be as serious as traditional classical music. This new music was born and grew in America as a result of the crossbreeding of cultures and traditions from various continents. Initially despised as race music, jazz conquered the public's preferences and was finally accepted by critics who understood its importance. Some of its main colored performers and composers are considered today to belong to the most important musicians in history, such as Louis Armstrong and Duke Ellington [12]. But it is important to note that the public's acceptance of this music was soon followed by scholars who also discovered, studied and accepted its importance in the history of music. Therefore, a process of convergence in the recognition of jazz -starting with the audience and then gripping the experts- has led this music to be recognized as a category of art. But this is not always the case, and we frequently see how the public surrenders to *pop music*, which lacks any attribute accepted by scholars.

This lack of convergence can also be noticed from the other side: new art movements considered as *academic art* have not been accepted by the public, despite the effort of experts. A well-known example is *serialism* and *dodecaphonism*, promoted by the Second Vienna School led by A. Schoenberg. Although it proved productive in association with other art forms, such as cinema, it arouses little interest among the audiences of classical music [3].

Similarly, if we look to plastic art, public opinion was very shocked when Duchamp's "Fountain" was considered art by scholars. Yet, as stated by Young and Priest, "Fountain is not art simply because it is called art. It is art because Duchamp deemed it so with his signature and exhibition, and the message of this deeming was recognized and, over time, accepted by members of the art world" [25]. Nevertheless, public opinion has not changed significantly despite academic appreciation.

These examples allow us to see that the process of convergence between public appreciation and expert opinion does not always happen.

Maybe a deeper reflection is necessary about what kind of art is more aligned with human nature, given that we are the final consignee, about the very meaning of *art*. We should thus consider, for instance, whether it has been misinterpreted during the 20th century, embodying components that should be removed. As in other social areas, stretching the meaning of a concept too much can lead to the loss of semantic information.

We can therefore summarize that the change and evolution of the concept of *art*, through history has been influenced not only by scholars and *critics*; museums that exhibit what they consider as genuine *art*; but also by the public, whose favor drives some artists but not others. Nor can we forget the importance of auction rooms to launch some emergent artists. In our opinion, both, public and critic, are the two sides of the same coin: the question of what art is. This is the perspective we adopt from here on.

Unfortunately, if we focus on evolutionary art -and those results obtained by means of evolutionary processes, especially considering plastic art [15]- not all of the roles played by those involved in the art world (museums, galleries, critics, audiences) are usually taken into account. As we said before, if some of them are missing in an art project, the real meaning of art may be lost.

We more deeply analyze in this chapter what we have already stated before [21]: the path toward a proper evaluation of evolutionary art. We describe some recent experiences that we believe improve the quality of results obtained, and how to involve both the public and critics in the evaluation process, so that evolutionary art is accepted in the art world as it deserves.

The rest of the paper is organized as follows: Section 25.3 describes the area, and some of the proposals for evaluating evolutionary art. Section 25.4 describes our approach to improving evolutionary art. Section 25.5 presents some experiments we have developed in the last few years, and finally Section 25.6 summarizes our conclusions.

25.3 State-of-the-Art

When researchers first decided to apply EAs to art projects they encountered a problem, which is at the core of any EC algorithm: how to correctly evaluate the aesthetic value of an evolved work. As we will see below, this problem, associated with the intrinsic value of a work of art has been crucial through the history of art, and is today of paramount importance in evolutionary art [11].

What researchers have traditionally done is to use human beings to assess fitness, eschewing a computational evaluation, thus producing a fundamental change in the evolutionary process, giving rise to an interactive version of the algorithm which is now widely applied [17].

As a result, many software tools available for evolutionary art, such as PicBreeder, rely on user criteria for deciding what is aesthetically of interest [16].

Many attempts have been made to understand and develop the best way of analyzing aesthetic quality in evolutionary art [10, 7]. Although some attempts have tried to capture an adequate measure of artistic quality in a mathematical equation, such as Birkhoff and the Aesthetic Measure, The Golden Ratio, Zipfs Law, Fractal Dimension, Gestalt Principles, The Rule of Thirds, etc, to date the results have been discouraging [11]. In addition, public opinion has been analyzed with survey-based procedures, but results were not satisfactory mainly due to user fatigue [23].

Methods based on utilizing human offers a key insight: the participation of a large number of different users giving their opinion on what is most appreciated in a work of art, do not guarantee adequate results. But if that is not enough, what other elements should be included?

Recently, some researchers have raised the possibility of generating a Turing Test for the arts [1], which can be applied to evolutionary art. Unfortunately it is not easy to see how some of the main components of the test should be designed:

- Who should play the part of the jury?
- Should we only consider art generated by professional artists, or art produced by anyone when comparing results with human art?
- When an evolutionary work of art cannot be distinguished from that produced by a human, is it enough to give it a seal of artistic quality?

These are three basic questions that we address in the next section, describing some of the proposals that we have developed in the last decade, and some of the results obtained when trying to propose new models more adequate for evolutionary art.

25.4 Methodology: Ways to Assess Evolutionary Art

To date, good intentions dominate evolutionary art. Typically, an IEA, which is traditionally employed in this context, works as follows: Every algorithmic operation, except fitness evaluation, is carried out as usual. In the case of fitness measurement, users decide on the quality of the product. But user fatigue has shown to progressively affect the evaluation process given the large number of generations and individuals required by the algorithm [5]. Some proposals have certainly been described which avoid this problem -mainly based on the use of non-intrusive devices responsible for analyzing the behavior and feelings of users when observing a work of art [23]- but these are usually exceptions to the standard configuration of the algorithm.

Once the execution of the algorithm is complete, the best individual is considered the art product. But in evolutionary art that should not be enough: the evolved work of art should be shown to the public, for feedback, in the same way as with art produced by human artists. However, this is not what we see in most cases: results are shown to the evolutionary computing community in their own congresses and journals, but art exhibits are seldom organized to show the works to a non-specialist audience.

This way of proceeding is quite far from how the art world actually works. We shall examine in detail the three points highlighted above, and we shall propose how we can get closer to the international art circuit, so that evolutionary art will someday occupy the place it deserves.

25.4.1 The Turing Test for Art

As described above, the difficulty for accurately measuring aesthetic quality of a computer-generated work of art led some researchers to propose a Turing test for art: if a human jury cannot distinguish between a work produced by a computer and another produced by a human artist, we could say that the first one has passed the Turing Test for the arts.

Yet, there are some reasons for questioning the usefulness of a test like this. First, the undecidability about the origin of a work -human or computer-created- does not have to automatically give it a seal of quality. In traditional art, the importance of the recognition of the artist behind the work is mainly due to historical issues, academic considerations, public favor, and finally, the art market which evaluates the quotation of the signing artist [18]. In [4], a similar discussion about the Turing test for the arts comes to the conclusion that asking whether a work was painted by a human being or by a computer is the wrong question. Here, Colton states that the right question would be what artwork people would buy. This question is aligned with our previous discussion of audience preferences on the kind of music they like to listen to. Therefore, regardless who produced a work, a human being or a computer, this is not guarantee of acceptance by the public.

On the other hand, we doubt whether such a proof (the Turing test) makes sense: some artistic trends that were born in the 20th century produce results that can easily be described as computer-generated, even when a human artist is in charge of producing them. Consider for instance *serialism* and *atonality* in music, which are somehow -though not only- based in stochastic processes. Although they are carried out by human beings, they are easily emulated by computers, in comparison to more standard tonal classical music which requires a large number of rules that must be embodied (or learned) [19]. Similarly, the use of digital tools by artists could also mislead jurors.

In short, the Turing test does not seem to be the right tool to judge the results produced by evolution.

25.4.2 About Human Artists

Among the best-known evolutionary art projects (see Picbreeder [16]), the evolutionary process takes place in a kind of *crowd-painting* approach. Many users are invited to participate and interact using a web tool that allows the collective evolution of images.

A question arises: Is this the right procedure if we are looking for a quality works of art? We do not think so. Instead, we need to also include within the algorithm itself necessary ingredients to achieve such a goal. In our opinion, we need the collaboration of experienced human artists, not just users, who can provide an aesthetic evaluation of the partial results, allowing the algorithm to be based on an adequate evaluation of fitness, and thus to progress toward the goal, given that to date no good automatic evaluation of fitness has been devised. Only when machine learning processes can appropriately model the behavior of artists in the future can human artists be replaced. However, at this moment the participation of the artist is essential, and the EA should play the role of a Computer Assisted Art tool, until better automatic evaluation techniques have been devised.

We have followed this approach in some of our recent work, allowing the artist to be part of the evolutionary algorithm [22]. However, we think that although necessary, the participation of artists in the algorithm may not be sufficient. To certify the quality of a work, we think a final evaluation is required, in the same way that after an evolutionary algorithm completes execution, the best solution is checked to see if the quality is sufficient or an additional run is required. The main difference is that in the latter case, the standard fitness function is typically applied for final evaluation, whereas in evolutionary art this should not be so, as described below.

In the art world, there are a number of players in charge of the *final* evaluation. This includes museums, art galleries, international competitions and the public when they participate in art exhibitions and auctions and finally decide the quality of a given work.

In the field of EAs, some relevant international congresses have frequently organized international art competitions. We may mention ACM GECCO Evolutionary Art, Design and Creativity Competition [9], which has allowed an international art jury to evaluate the best works of evolutionary art.

These competitions provides answer to some of the controversial questions highlighted in the previous section: the constitution of a jury and an international competition provides the means for a specialized group to deliver a verdict. We can thus refer, for instance, to our experience with some of our collective evolutionary works of art, such as *"XY"*, which we sent to the ACM GECCO 2013 art competition (see figure 25.1) [9] and was selected as the winner. Although this work somehow departed from traditional IEA, using instead the *Unplugged Evolutionary Algorithm* [24], the work was favorably considered by the jury.

Fig. 25.1: "XY": ACM Gecco 2013 Evolutionary Art, Design and Creativity competition winner. Four out of the sixty works produced are displayed.

However, this competition takes place at a conference devoted to EC, and the jury is composed mainly of renowned researchers in the field who are also artists. This means that both the scope of the competition and the professional jury imposed a *bias* and established a perfect *sandbox* for evolutionary works of art. We believe that a further step is required if the goal is to position evolutionary art where it deserves to be.

25.4.3 The Jury

The composition of the jury is crucial for a proper evaluation of the quality of a work. Usually in the world of art, curators are in charge of deciding the works to be exhibited. Similarly, in art competitions, the selection of a competent jury is crucial to ensure the quality of winning works. Thus, both international art competitions as well as curatorial art exhibitions in art museums and galleries provide the necessary filter to select what deserves to be shown as work of art.

We therefore think that evolutionary art must go to standard art contests to compete face-to-face with man-made works of art in the context of traditional art. Similarly, trying to show the work in museums and art galleries will provide the information needed to know the quality of a work. This should be thus the final evaluation step required for this special EA product.

Below we describe the experience we gathered with work developed over the last five years, that has allowed us to face this final evaluation process, with a tour through art galleries and international competitions, places that are far away from the territory of EAs. Through this journey, we have also tried to involve the public: we know that favor of the public is another ingredient that allows works of art to survive.

25.5 Experiments and Results

We have developed the Unplugged Evolutionary Algorithm, which goes one step further than the IEA by allowing users to perform every step of the evolutionary algorithm and not just the evaluation step, since 2012 [22].

During these years we have tried to apply recommendations described above in a series of stages, which has allowed us to improve the evaluation process for artistic results, in a way that can be useful as a sample for other works of Evolutionary Art, and that we believe will finally allow to position this new type of art in the world of art.

As we will see below, we have tried to approach the quality assessment process as close to the art world as possible, which in some ways takes us away from the EA field.

25.5.1 Art Galleries and Their Audiences

In the last five years three collective works of art have been produced: *"XY"*, *"XYZ"* and finally *"The horizon project"*. From the very beginning we realized the need to organize art exhibitions in order to get feedback from the audience.

As for the first work, XY, we decided to show it at EC conferences. Therefore, we had exhibitions in Cancún (IEEE CEC 2013), Madrid and Mérida (National Spanish Conferences on AEs 2013 and 2015). These events allowed us to learn the opinion of participants about the work, and we detected some problems associated with the use of surveys to obtain information, namely *fatigue*, which also occurs in IEAs when a user must provide fitness values (see figure 25.2). Nonetheless, the information obtained came from people involved in EAs, given the conferences where the art exhibitions were organized, although we would have preferred a more skeptic audience. Moreover, data obtained by means of surveys had to be post-processed, preventing feedback from being used within the algorithm kernel. In any case, the information is useful for other experiments to be launched.

Fig. 25.2: Using surveys to obtain information from the audience.

Recently, we have set up an interactive art exhibition that allows automatic analysis of user perception by means of *kinnect* based computer systems [23]. The results obtained show two interesting things. First, they correlate and are consistent with survey-based data, and second, a much larger amount of data can be collected, given the non-intrusiveness of the approach, which allows for better analysis of audience perception. Figure 25.3 shows the interface that displays the information captured by one of the kinnect devices located along with each work of art exhibited.

Although it is still a work in progress, we believe that this type of automatic data retrieval system would allow automatically providing an EC algorithm with audience perception data while the evolutionary process is still producing the work of art. Automatic audience analysis has already been used in the context computer generated art (check for instance S. Colton's work on emotional modeling in [4]). Yet, there is still room for improvement and IEAs can greatly benefit from this kind of installation, where user participation is facilitated.

Given our thesis that a proper location for an artist's work is the art gallery, we tested the idea directly. We thus organized art exhibitions of *XYZ* between 2015 and 2017 in several art galleries. The first exhibition took place in Vancouver at the "Back Gallery Project". We scheduled this event on the same dates of the IEEE

Fig. 25.3: Capturing information from the audience without disturbing the experience: a kinnect
device is in charge of obtaining users mood (happy, engaged ...).

CEC in the city, although we organized it without the support of the conference.
Therefore, the event was exclusively devoted to art in an art gallery. Similarly, a
second exhibition was presented at the Gallery Louchard in Paris in October 2017,
during the Artificial Evolution conference. Surveys were provided in both cases to
the audience attending the art exhibitions that has allowed us to confirm the problem
of fatigue, but also to see the interest of the public and art galleries with these new
kind of art movement.

A general positive feedback was retrieved from these experiences. Firstly, we
confirmed the interest of art galleries to display evolutionary art. Secondly, it is
important to know the public opinion on the work displayed, and the way they understand the evolutionary approach. Both are useful evaluation tools for the results
produced by means of evolutionary algorithms, and we are confident this information retrieved will allow to improve evolutionary art in the future.

Our conclusion is that art galleries can be a good place to show evolutionary
art; moreover, art galleries can actually be interested in this kind of work if the
quality is appropriate. Therefore, evolutionary art exhibitions in art galleries provide
a guarantee of the quality of the works on display, as they are primarily interested
in doing business, and are only interested in exhibiting works that they believe can
be sold.

25.5.2 The Definitive Evaluation: International Art Competition

As stated before, the best evaluation process regardless of how a work of art was
produced is to submit it to an international art competition where the work competes
with many others produced by different artists and techniques. This is what we did
with our third collective work, *The Horizon Project* [8] (see figure 25.4).

We thus submitted it to the *Show Your World 2017 competition*[1] that was held in New York at the *MC Gallery* in Manhattan. We were fortunate that our work was selected, the only one produced by means of an evolutionary approach.

To the best of our knowledge, this is probably one of the first times that a work of art of this type is evaluated and selected as finalist in an open international art competition, a pure competition in the art world, with a specialized jury, and exhibited in an international exhibition of curated art. However, we think this should be the general case for evolutionary art to be recognized as a new trend of interest in the art world. Otherwise, it will remain a picturesque research area of a minority for EA experts.

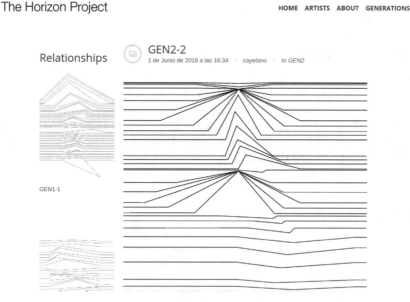

Fig. 25.4: The Horizon Project, available at: http://merida.herokuapp.com/

25.6 Conclusions

Although in traditional evolutionary algorithms a fitness function is employed to measure the quality of candidate solutions every generation, we believe that a different approach is required in evolutionary art. Firstly, artists must be somehow involved in the evolutionary process, so that the algorithm can be considered a computer assisted creativity tool. The artists can help in the evaluation step, but they can

---

[1] http://www.reartiste.com/juried-exhibition-show-your-world-2017/

also be involved in other genetic operators, being their knowledge and experience the main principles that guide the way the algorithm works. Secondly, the participation of artists may also allow the application of machine learning techniques so that the algorithm can learn from human artists and thus be improved in the future on the basis of that learning. Finally, information and feedback must be collected from the public at art exhibitions held in museums, art galleries, etc.; the opinion of critics, professionals, scholars and curators of the art world, responsible for assessing their quality, make reviews in the media and be part of the juries of competitions is crucial if we want evolutionary art to compete on equal terms with other works produced by recognized human artists. Both the public and the critics are the two sides of the same coin that provides the required seal of quality and will allow the recognition of this new artistic movement.

Acknowledgments

This work has been supported by Spanish Ministry of Economy and Competitiveness (MINECO) projects: EphemeCH (TIN2014-56494-C4-1...4-P), and DeepBio (TIN2107-85727-C4- 1...4-P), and Consejería de Comercio y Competitividad, Junta de Extremadura, under the European Regional Development Fund FEDER, projects IB16035, GR15068 and GR15130.

References

1. Boden, M.A.: *The Turing test and artistic creativity*. Kybernetes **39**(3), 409–413 (2010)
2. Boime, A.: *The Salon des Refusés and the evolution of modern art*. Art Quarterly **32**, 411–426 (1969)
3. Botstein, L.: *Schoenberg and the audience: Modernism, music, and politics in the twentieth century*. Schoenberg and his World pp. 19–54 (1999)
4. Colton, S.: *Creativity versus the perception of creativity in computational systems*. In: AAAI spring symposium: Creative intelligent systems, vol. 8 (2008)
5. Frade, M., Fernández de Vega, F., Cotta, C.: *Evolution of artificial terrains for video games based on accessibility*. In: European Conference on the Applications of Evolutionary Computation, pp. 90–99. Springer (2010)
6. Galanter, P.: *What is generative art? Complexity theory as a context for art theory*. In: 6th Generative Art Conference (GA 2003) (2003)
7. den Heijer, E., Eiben, A.E.: *Comparing aesthetic measures for evolutionary art*. In: European Conference on the Applications of Evolutionary Computation, pp. 311–320. Springer (2010)
8. Hernández, P., Fernández de Vega, F., Cruz, C., Albarrán, V., García, M., Navarro, L., Gallego, T., García, I.: *The Horizon project: Emotion in lines*. Journal Art and Science **1**(1), 1–9 (2017)
9. Loiacono, D.: *GECCO-2013 competitions*. ACM SIGEVOlution **6**(2), 27–28 (2014)
10. Machado, P., Cardoso, A.: *NEvAr–the assessment of an evolutionary art tool*. In: Proceedings of the AISB00 Symposium on Creative & Cultural Aspects and Applications of AI & Cognitive Science, Birmingham, UK, vol. 456 (2000)
11. McCormack, J.: *Open problems in evolutionary music and art*. In: Workshops on Applications of Evolutionary Computation, pp. 428–436. Springer (2005)

12. Peress, M.: *Dvořák to Duke Ellington: A Conductor Explores America's Music and Its African American Roots*. Oxford University Press on Demand (2004)
13. Reichardt, J.: *Cybernetic Serendipity: The Computer and the Arts*. Praeger (1969)
14. Reis, G., Fernández de Vega, F., Ferreira, A.: *Automatic transcription of polyphonic piano music using genetic algorithms, adaptive spectral envelope modeling, and dynamic noise level estimation*. IEEE Trans. Audio, Speech & Language Processing **20**(8), 2313–2328 (2012)
15. Romero, J.J.: *The art of artificial evolution: A handbook on evolutionary art and music*. Springer Science & Business Media (2008)
16. Secretan, J., Beato, N., D Ambrosio, D.B., Rodriguez, A., Campbell, A., Stanley, K.O.: *Picbreeder: Evolving pictures collaboratively online*. In: Proceedings of the SIGCHI Conference on Human Factors in Computing Systems, pp. 1759–1768. ACM (2008)
17. Takagi, H.: *Interactive evolutionary computation: Fusion of the capabilities of ec optimization and human evaluation*. Proceedings of the IEEE **89**(9), 1275–1296 (2001)
18. Thompson, D.: *The $12 million stuffed shark: The curious economics of contemporary art*. Macmillan (2010)
19. Fernández de Vega, F.: *Revisiting the 4-part harmonization problem with gas: A critical review and proposals for improving*. In: Evolutionary Computation (CEC), 2017 IEEE Congress on, pp. 1271–1278. IEEE (2017)
20. Fernández de Vega, F., Cotta, C., Reck Miranda, E.: *Special Issue on Evolutionary Music*. Soft Comput. **16**(12), 1995–1996 (2012)
21. Fernández de Vega, F., Cruz, C., Hernández, P.: *Propuestas de mejora para la evaluación del arte evolutivo*. In: MAEB 2018, Granada, Spain. In press (2018)
22. Fernández de Vega, F., Cruz, C., Navarro, L., Hernández, P., Gallego, T., Espada, L.: *Unplugging evolutionary algorithms: An experiment on human-algorithmic creativity*. Genetic Programming and Evolvable Machines **15**(4), 379–402 (2014)
23. Fernández de Vega, F., García, M., Merelo, J., Aguilar, G., Cruz, C., Hernández, P.: Analyzing evolutionary art audience interaction by means of a Kinect based non-intrusive method. In: Studies in Computational Intelligence series, vol. 785, pp. 321–347 (2018)
24. Fernández de Vega, F., Navarro, L., Cruz, C., Chavez, F., Espada, L., Hernandez, P., Gallego, T.: *Unplugging evolutionary algorithms: On the sources of novelty and creativity*. In: Evolutionary Computation (CEC), 2013 IEEE Congress on, pp. 2856–2863. IEEE (2013)
25. Young, D., Priest, G.: *It is and it isn't*. Aeon (Sept. 2016)

Chapter 26
Making Better Use of Repair Templates in Automated Program Repair: A Multi-Objective Approach

Yuan Yuan and Wolfgang Banzhaf

Abstract The automation of program repair can be coached in terms of search algorithms. Repair templates derived from common bug-fix patterns can be used to determine a promising search space with potentially many correct patches, a space that can be effectively explored by GP methods. Here we propose a new repair system, ARJA-p, extended from our earlier ARJA system of bug repair for JAVA, which integrates and enhances the performance of the first approach that combines repair templates and EC, PAR. Empirical results on 224 real bugs in Defects4J show that ARJA-p outperforms state-of-the-art repair approaches by a large margin, both in terms of the number of bugs fixed and of their correctness. Specifically, ARJA-p can increase the number of fixed bugs in Defects4J by 29.2% (from 65 to 84) and the number of correctly fixed bugs by 42.3% (from 26 to 37).

Key words: Program repair, evolutionary multi-objective optimization, genetic programming, repair templates

26.1 Introduction

Automated program repair [9, 32] aims to fix bugs in software automatically and has shown promise recently. Such techniques generate a patch for a bug that can satisfy a specification. Our study focuses on the test-suite based program repair in JAVA where the specification is given by a test suite.

Yuan Yuan
BEACON Center for the Study of Evolution in Action and Department of Computer Science and Engineering, Michigan State University, East Lansing, MI, USA e-mail: yyuan@cse.msu.edu

Wolfgang Banzhaf
BEACON Center for the Study of Evolution in Action and Department of Computer Science and and Engineering, Michigan State University, East Lansing, MI, USA e-mail: banzhaf@msu.edu

© Springer Nature Switzerland AG 2020
W. Banzhaf et al. (eds.), *Evolution in Action: Past, Present and Future*,
Genetic and Evolutionary Computation, https://doi.org/10.1007/978-3-030-39831-6_26

A test suite should contain at least one initially failing, negative test case which triggers the bug to be repaired plus any number of initially passing, positive test cases that define the expected functionality of the program. In terms of test-suite driven repair, a bug is said to be *fixed* or *repaired*, if a patch can allow the modified program to pass the entire test suite. Such a patch is called *test-adequate* [26].

GenProg [20, 21] is among the most well-known approaches for test-suite based repair. This approach uses three types of statement-level mutations/edits (i.e., replace a destination statement with another, insert another statement before a destination or delete a destination statement) to rearrange the extant code of the buggy program. To explore the search space, GenProg uses genetic programming [3, 16] to search for potential patches that fulfill the test suite. However, due to the randomness of mutation operations, GenProg often generates patches which simply overfit the test suite [36]. To relieve this problem, Kim et al. [15] proposed PAR, an approach using common *fix patterns* (e.g., adding a null checker) manually learned from human-written patches. In PAR, a *repair template* is the central concept, a kind of program transformation schema derived from fix patterns. Unlike GenProg, PAR generates new program variants by using such predefined repair templates. To find a patch, PAR also employs an EC technique like GenProg, but focuses the search on more meaningful program transformations compared to GenProg, resulting in better chances to produce test-adequate or correct patches. PAR was the first repair approach that combined repair templates with EC techniques.

However, the performance of PAR is far from satisfactory. As reported by Le et al. [19], a reimplementation of PAR for JAVA can only fix very few bugs correctly in the Defects4J [13] dataset. Note that there are generally two key elements in a successful repair approach: the search space and the search algorithm [36]. The search space should contain as many correct patches as possible while the search algorithm should be powerful enough to explore such a large search space. PAR has limitations in both regards. Regarding the search space, although the templates used in PAR define potentially useful program transformations, they are often not sufficiently general and cover only a very limited number of fix patterns. Regarding the search algorithm, PAR uses an EC framework similar to that of GenProg, which recombines and mutates high-granularity edits via crossover and mutation operators. Recent studies [34, 45] have shown, however, that evolving such high-level units strongly limits the ability to effectively traverse a search space, which may be a reasons why GenProg usually generates patches that are equivalent to a single functionality deletion [36].

Our work is motivated by recent progress on both issues. Several repair approaches such as ELIXIR [38], SPR [24] or Cardumen [28] use a richer set of repair templates than PAR to generate program variants, achieving state-of-the-art performance on well-known datasets of bugs. Also, very recent studies [34, 45] suggest that evolving patches with lower-granularity patch representations via advanced EC techniques can lead to a substantially improved search ability.

Given the above, we want to exploit the benefits of both, recent template based approaches (they work over a richer and more promising search space) and EC approaches, more powerful than previous ones, in order to improve over the PAR algorithm. We hence develop a new repair approach for Java, extended from ARJA [45], called ARJA-p. In ARJA-p, the repair templates in PAR are made more general to cover a larger set of fix patterns for potential use in patches. To bridge the gap between template based edits (usually in the expression-level) and the lower-granularity patch representation of ARJA (only applies to statement-level edits), we execute the templates offline, thereby abstracting various template based edits into two types of statement-level edits (i.e., replacement and insertion). We can then introduce a lower-granularity patch representation for template based repair. Lastly, we formulate program repair as a multi-objective search problem and use a classical multi-objective evolutionary algorithm (i.e., NSGA-II [8]) to search for simpler patches, following the paradigm of search-based software engineering [2, 11, 12].

ARJA-p is evaluated on 224 real bugs in Defects4J [13] and compared to other state-of-the-art approaches. We can show that ARJA-p outperforms all the other approaches with a significant margin. Overall, ARJA-p is able to increase the number of bugs fixed in Defects4J by 29.2% (from 65 to 84) and the number of correctly fixed bugs even more, by 42.3% (from 26 to 37). Notably, ARJA-p can correctly fix several multi-location bugs, impossible for most of the existing repair approaches.

The rest of this paper is structured as follows. Section 26.2 provides background knowledge and a motivating example. Section 26.3 describes our repair approach in detail. Section 26.4 presents the experimental design and Section 26.5 reports our empirical results.

26.2 Background and Motivation

Search-based repair approaches determine a search space potentially containing correct patches and employ metaheuristic search techniques or random search to find test-adequate patches.

GenProg [20, 21] is a representative approach that uses genetic programming (GP) to search for test-adequate patches. This approach is based on the redundancy assumption [4, 29] (i.e., the ingredients for a fix exist elsewhere in the current program) and performs three kinds of statement-level edits, replacement, insertion and deletion. [22] studied the influence of different solution representations and genetic operators in GenProg; [35] presents RSRepair that replaces GP in GenProg with random search; [39] suggested a set of anti-patterns to inhibit GenProg from generating nonsensical patches; [34] introduce a lower-granularity patch representation and several related crossover operators. PAR [15] is a pioneering approach that exploits repair templates to generate bug fixes. ARJA-p extends repair templates iof PAR to accommodate more useful fix patterns and conducts more effective search via evolutionary multi-objective optimization. ARJA [45] is a GP based approach for JAVA, characterized by a novel patch representation, multi-objective search and several auxiliary techniques for speeding up fitness evaluation and reducing the search space. Unlike ARJA, which follows the redundancy assumption, ARJA-p uses templates to generate fix ingredients either for replacement or for insertion and puts them into two separate evolvable segments. Other typical search-based approaches include AE [40], SPR [24], HDRepair [19], ACS [42], ssFix [41] and Cardumen [28].

Besides search-based approaches, semantics-based approaches [7, 14, 17, 30, 31, 33, 43] are other techniques that have been extensively studied. There, semantic constraints are inferred from the given test-suite which are then used to generate test-adequate patches by solving the resulting constraint satisfaction problem. Very recently, some emerging techniques (e.g., deep learning) have been introduced into program repair, leading to several novel repair systems [6, 10, 25, 23, 38].

Another line of research focuses on the empirical aspects of program repair, including the problem of patch overfitting [36, 44], performance evaluation of different repair systems [18, 26] and analysis of real-world bug fixes [27, 48].

26.2.1 Brief Introduction to PAR

PAR [15] is an automatic program repair technique based on repair templates. Like GenProg, PAR takes a buggy program and a test suite with at least one negative test as input, with the goal to find a patch that allows all test cases to pass. Unlike GenProg which uses random statement replacement, as well as insertion and deletion to edit a program, PAR exploits repair templates to generate new program variants. Each repair template represents a common way to fix a specific kind of bug. For example, a specific bug is the access to a `null` object reference, and a common fix is to add `if` statement to check whether the object is `null` (this template is called "Null Pointer Checker" in PAR). PAR collects 10 repair templates by manually inspecting human-written patches and adopts an evolutionary process to use these templates.

The overall procedure of PAR is summarized as follows: First, PAR uses a simple fault localization strategy to find a number of suspicious statements. Its evolutionary process starts with an initial population of program variants iterating through two tasks: reproduction and selection. In reproduction, each program variant derives a new one by applying templates to the selected suspicious statements. In the selection, a tournament selection scheme chooses better (in terms of passing tests) program variants for the next generation. Iterations are stopped when a program variant passes all the tests.

From an EC perspective, PAR uses an evolutionary process similar to that in GenProg, but PAR only relies on template-based mutation and does not use any crossover at all. In other words, individual programs in the population of PAR do not exchange information with each other, so good genetic material cannot be propagated from one individual to another.

26.2.2 A Motivating Example

In this subsection, we take a bug as an example to highlight the key insights underlying ARJA-p which motivates our algorithm design.

Figure 26.1 (a) shows a correct patch for the real bug Math98 in Defects4J [13]. The two methods `operate(BigDecimal[] v)` and `operate(double[] v)` implement similar functionality: multiply the current matrix $\mathbf{A}_{m \times n}$ by a n-dimensional vector $\mathbf{x}$ (stored in the array `v`) and return the product $\mathbf{Ax}$ that is a m-dimensional vector. However, in the buggy program, a vector with size n (i.e., `v.length`) is used to store the m-dimensional vector $\mathbf{Ax}$ by mistake. To correctly fix the bug, `v.length` in lines 4 and 11 should both be changed to m (represented by the variable `nRows` in the code).

Based on two modifications in Figure 26.1 (a), if we make an additional modification as shown in Figure 26.1 (b), the synthesized patch (containing three edits) can still pass the whole test suite but is indeed incorrect. To understand this, we have to look at the meaning of line 3 in Figure 26.1 (b): This line checks whether the current matrix $\mathbf{A}$ can be multiplied by vector $\mathbf{x}$. Obviously, the only requirement is that the column dimension of $\mathbf{A}$ is equal to the vector dimension of $\mathbf{x}$. The original program checked this correctly, but the test-adequate (but incorrect) patch adds a further condition `isSingular()` (lines 4–5) that judges whether matrix $\mathbf{A}$ is singular. The patched program introduces some unexpected program behavior: if $\mathbf{A}$ is not singular, it can be multiplied by vector $\mathbf{x}$ with any dimension. However, this behavior cannot be detected by the test suite associated with Math98.

```
1   // BigMatrixImpl.java
2   public BigDecimal[] operate(BigDecimal[] v) throws ... {
3       ...
4   -     final BigDecimal[] out = new BigDecimal[v.length];
5   +     final BigDecimal[] out = new BigDecimal[nRows];
6       ...
7   }
8   // RealMatrixImpl.java
9   public double[] operate(double[] v) throws ... {
10      ...
11  -     final double[] out = new double[v.length];
12  +     final double[] out = new double[nRows];
13      ...
14  }
```

```
1   // BigMatrixImpl.java
2   public BigDecimal[] operate(BigDecimal[] v) throws ... {
3   -     if (v.length != this.getColumnDimension()) {
4   +     if (v.length != this.getColumnDimension() &&
5   +         isSingular()) {
6           throw new IllegalArgumentException("...");
7       }
8       ...
9   }
```

(a) Correct patch   (b) An additional modification

Fig. 26.1: Correct patch and test-adequate patch for Math98.

We can make three observations using this example:

(1) The bug fixing Math98 needs to replace a qualified name (i.e., `v.length`) with a variable `nRows`. However, this fix pattern cannot be handled by any of the 10 repair templates defined in PAR [15], because such a replacement in PAR can only be applied to a method parameter whereas `v.length` is an array index.

(2) The correct patch shown in Figure 26.1(a) requires multi-line changes that fix multiple buggy locations. A single modification in either line 4 or 11 cannot produce

a functionally correct program. To repair such multi-location bugs is hard or even impossible for almost all existing repair approaches. Since PAR uses an evolutionary framework similar to GenProg it can change multiple locations of a program in principle. However, [36] showed that seemingly complex patches generated by such search mechanisms are equivalent to single line modifications in the overwhelming majority of cases. PAR can only fix 4 bugs correctly in Defects4J, none of which was claimed to be a multi-location bug [19].

(3) As shown in Figure 26.1(b), an additional modification will turn the correct patch into a test-adequate but incorrect one. This implies that simpler or smaller patches should be preferred. With a weak test suite, looking for a smaller patches avoids making undesirable modifications. However, most of existing repair approaches including PAR do not explicitly take simplicity of a patch into account.

Here we aim to improve template-based repair approaches, particularly PAR, in the following way. First, we extend the repair templates used in PAR so as to make them more general and accommodate more fix patterns. Second, we propose a new evolutionary framework with a lower-granularity patch representation for template-based repair, which allows to fix multi-location bugs by leveraging its stronger search ability. Third, we formulate program repair as a multi-objective search problem and use evolutionary multi-objective optimization techniques to discover smaller patches.

26.3 Approach

The input of ARJA-p is a buggy program associated with a number of JUnit tests. Among these tests, there is at least one negative test. The others are positive tests defining the expected program functionality. The basic goal of ARJA-p is to modify a buggy program so as to allow it to pass all tests cases. ARJA-p consists of the following four main steps.

(1) **Fault Localization:** Given an input, ARJA-p first applies a spectrum-based fault localization technique called Ochiai [1] to locate a list of likely-buggy statements (LBSs). Each LBS is given a suspiciousness value $susp \in [0,1]$ by Ochiai, which indicates the likelihood of this LBS to contain the bug. In ARJA-p, we only consider some of the LBSs returned by fault localization in order to reduce the search space. This is determined by two parameters denoted by γ_{min}, the minimal suspiciousness value, and n_{max}, the maximal number of LBSs.

(2) **Generating Potential Fix Ingredients:** After fault localization, we apply the predefined repair templates to each LBS in the reduced set. By executing the transformations specified by repair templates each LBS can derive a number of new statements, which we call *ingredient statements* to be either inserted before or replacing the corresponding LBS.

(3) **Test Filtering:** Before entering into the evolutionary search, we conduct coverage analysis to filter out those positive test cases that are not influenced by the manipulation of selected LBSs. Specifically, we run the positive tests one by one,

and record the statements visited by this test. If none of the LBSs is touched, this positive test can be ignored. Thus, modified program variants can be validated on a reduced test suite, which speeds up fitness evaluation in our MOEAs.

(4) **Searching Test-Adequate Patches:** Now that we have selected a number of LBSs, each of which with a number of ingredient statements for replacement or insertion we try to find test-adequate patches consisting of replacement/insertion edits, with smaller patches preferred. In ARJA-p, we formulate this problem as a multi-objective combinatorial optimization/search problem and use MOEAs to explore the search space.

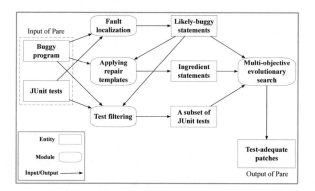

Fig. 26.2: Overview of our repair approach (i.e.,ARJA-p).

Figure 26.2 shows an overview of the proposed repair approach. In the remainder of this section, we detail the second and fourth steps (i.e., how to apply the repair templates to generate potential fix ingredients and how to evolve the patches via MOEAs), which are two characteristic procedures in ARJA-p.

26.3.1 Repair Templates

ARJA-p uses 7 repair templates that are mainly derived from templates in PAR. Each template specifies a type of transformation of code.

(1) **Element Replacer:** This template replaces an abstract syntax tree (AST) node element in a LBS with another compatible one. Table 26.1 lists the elements that can be replaced and illustrates alternative replacers for each kind of element. Note that templates "Parameter Replacer" and "Method Replacer" used in PAR are just a subset of a template here. Some replacement rules shown in Table 26.1 with a widened type follow, e.g. rules from ELIXIR [38] or REFAZER [37].

(2) **Method Parameter Adder or Remover:** This template is applicable for the method invocation that has overloaded methods. To restrict the complexity, ARJA-p only considers to add or remove a single element from the parameter list of the

Table 26.1: List of Replacement Rules for Different Elements

Element	Format	Replacer
Variable	`x`	(i) The visible fields or local variables with compatible type (ii) A compatible method invocation in the form of `f()` or `f(x)`
Field access	`this.a` or `super.a`	The same as above
Qualified name	`a.b`	The same as above
Method name	`f(...)`	The name of another visible method with compatible parameter and return types
Primitive type	e.g., `int` or `double`	A widened type, e.g., `float` to `double`
Boolean literal	`true` or `false`	The opposite boolean value
Number literal	e.g., `1` or `0.5`	Another number literal located in the same method
Infix operators	e.g., `+` or `>`	A compatible infix operator e.g., `+` to `-`, `>` to `>=`
Prefix/Postfix operators	e.g., `++` or `--`	The opposite prefix/postfix operator e.g., `++` to `--`
Assignment operators	e.g., `+=` or `*=`	The opposite assignment operator e.g., `+=` to `-=`, `*=` to `\=`
Conditional expression	`a ? b : c`	`b` or `c`

current method invocation. The order of current parameters can be rearranged provided that their types are compatible to the corresponding types declared in the overloaded method. When adding a parameter, ARJA-p collects all fields and local variables within the scope of an LBS's location, and candidates to be added must be type-compatible with the corresponding parameter type in the method declaration.

(3) **Boolean Expression Adder or Remover:** This template is applicable for an LBS that has a conditional branch. Take the `if` statement as example and suppose the LBS is like `if (c1 && c2){...}`. When adding a boolean expression in the predicate, this template collects all boolean expressions that are in the same file with the LBS, and those within the scope of the LBS's location can be alternatives to be added. Suppose `c3` is chosen, ARJA-p uses the following four transformation schemas to edit the predicate in the LBS: (i) `c1 && c2 && c3` (ii) `c1 && c2 || c3` (iii) `c1 && c2 && !c3` (iv) `c1 && c2 || !c3`. When removing a term in the predicate, ARJA-p can select any one (e.g., `c1` or `c2`) to remove.

Note that ARJA-p extends this template from PAR in two ways: The added boolean expression can be any one from the file that meets the scope, not just the one in the predicate. Second, ARJA-p adopts more transformation schemata.

(4) **Null Pointer Checker:** For a LBS, this template first extracts all objects in the statement that have object references (e.g., `o1` and `o2`). Then it creates a predicate (e.g., `o1!= null && o2 != null`) to assure that all these objects cannot be `null` when executing the LBS. With this predicate, ARJA-p uses the following 6 transformation schemata to manipulate the LBS.

(i) `if (o1 != null && o2 != null)buggyStatement;`

(ii) `if (!(o1 != null && o2 != null))return sth;`

(iii) `if (!(o1 != null && o2 != null))throw exception;`

(iv) `if (!(o1 != null && o2 != null))break;`

(v) `if (!(o1 != null && o2 != null))continue;`

(vi) `if (o1 == null)o1 = new Obj1();`

`if (o2 == null)o2 = new Obj2();`

The first schema makes the LBS a part of the `if` statement whereas each of the others inserts the entire `if` statement before the LBS. The second and third schemata need to be instantiated since they require another `return` and `throw` statement, respectively. For `return` statements, the return type of the method containing the LBS is first checked, then the corresponding values according to this type are returned: (i) `boolean`: return true or false; (ii) `void`: return nothing; (iii) the other primitives: return 0 or 1; (iv) an object: return null. ARJA-p always uses the last `return` statement in the method. As for `throw` statements, ARJA-p collects alternative thrown exceptions in three ways: (i) find the method declaration where the LBS located and use its declared thrown exception types; (ii) if the considered objects are in method parameters, consider the `IllegalArgumentException`; (iii) consider the exception types defined in the buggy program that are extended from the `NullPointerException`. The last schema uses the basic constructor to initialize `null` objects having references.

To avoid compile errors, sometimes not all 6 schemata are applicable. Specifically, the first schema can only be applied to an LBS which is not a variable declaration statement and the fourth and fifth schema can only be used if the LBS is in a `for` or `while` loop.

We introduce three further repair templates that are similar to "Null Pointer Checker". They use the same 6 schemata mentioned above to edit a buggy program, but their predicates in the `if` statement check different contexts.

(5) **Range Checker:** This template mainly checks whether all array or list element accesses are valid in an LBS (i.e., indices cannot exceed the upper and lower bounds of the size of an array or list). Different from PAR, it also considers to check the validity of `char` access (in the form of `charAt` or `substring`) for `String` objects since `String` is a list of characters and is frequently used in Java.

(6) **Cast Checker:** This template checks whether, in each class-casting expression, the variable or expression to be converted is an instance of casting type (using `instanceof` operator).

(7) **Divide-by-Zero Checker:** This template checks whether all the divisors are not equal to 0. It is not used in PAR.

26.3.2 Offline Execution of Templates

As described in Section 26.3.1, repair templates in ARJA-p enrich those used in
PAR, so that ARJA-p can potentially handle more fix patterns. Our system exploits repair templates in a way different from PAR and other related approaches
[19, 37, 41]. PAR executes templates *on-the-fly* (i.e., during the evolutionary process), whereas our system uses them *offline*. Specifically, ARJA-p executes all possible transformations defined by templates for all considered LBSs *before* searching
for patches. Each LBS can derive a number of new statements and each of these
statements can either replace the LBS or be inserted before it. These template based
repair actions (some occur in the AST node level) are abstracted into two kinds
of statement-level edits (i.e., replacement and insertion). Figure 26.3 illustrates this
abstraction for a supposed LBS of `a.add(x,y)`. In order to avoid combinatorial ex-
plosion, ARJA-p only applies a template to a single node at a time. For example,
ARJA-p does not use the template "Element Replacer" to simultaneously change `a`
and `add` in `a.add(x,y)`.

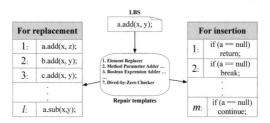

Fig. 26.3: Illustration of the offline template execution.

After offline template execution, repair templates are invisible to the evolutionary
search in ARJA-p. Only two kinds of statement-level edits are visible to the search.

26.3.3 Evolving Patches

Suppose *n* likely-buggy statements (LBSs) are selected by fault localization, with
each of them having two sets of ingredient statements (one for replacement and another for insertion) that are produced through the offline execution of templates. We
can then view template based patch generation as a three-level decision process: (1)
Choose to edit some statements among *n* LBSs; (2) Select which operation ("replace" or "insert before") to apply for each LBS to be edited; (3) Choose statements
from the corresponding ingredient statements for replacement/insertion. ARJA-p
uses a lower-granularity patch representation that properly decouples the search
subspaces of potentially buggy locations, operation types and replacement/insertion
statements. Based on this representation, we can evolve patches via multi-objective

evolutionary algorithms (MOEAs) that effectively explore this search space. The details are described as follows.

Patch Representation

To encode a patch as a MOEA individual, we first sequence the n LBSs in a random order. For the j-th LBS, $j = 1, 2, \ldots, n$, its associated two sets of ingredient statements are denoted as R_j (for replacement) and I_j (for insertion) respectively, and the statements in R_j/I_j are numbered from 1 to $|R_j|/|I_j|$ in any order. These ID numbers are fixed throughout the evolutionary process.

A solution (i.e., a patch) to the program repair problem is represented as a collated vector $\mathbf{x} = (\mathbf{b}, \mathbf{c}, \mathbf{u}, \mathbf{v})$, with four parts each being a vector of size n itself. $\mathbf{b} = (b_1, b_2, \ldots, b_n)$ is a binary vector, where $b_j \in \{0, 1\}$, $j = 1, 2, \ldots, n$ indicates whether the j-th LBS is to be edited or not. $\mathbf{c} = (c_1, c_2, \ldots, c_n)$ is also a binary vector, where $c_j = 0$ ($c_j = 1$) means the "replace" ("insert before") operator is chosen for the j-th LBS. $\mathbf{u} = (u_1, u_2, \ldots, u_n)$ is an integer vector and u_j indicates that patch $\mathbf{x}$ selects the u_j-th ingredient statement in the set R_j. Similar to $\mathbf{u}$, $\mathbf{v} = (v_1, v_2, \ldots, v_n)$ is also an integer vector in which v_j means the v_j-th statement in the set I_j is selected by patch $\mathbf{x}$. Suppose the j-th LBS is a.add(x,y), Figure 26.4 illustrates the patch representation in ARJA-p. Figure 26.5 describes how to apply a patch $\mathbf{x}$ to the buggy program (i.e., decoding procedure) so as to obtain a modified program.

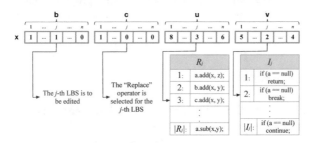

Fig. 26.4: Illustration of the patch representation in ARJA-p.

Fitness Function

The automated program repair is formuated as a multi-objective search problem. To evaluate the fitness of each individual $\mathbf{x}$, we employ a multi-objective function to simultaneously minimize the *patch size* (denoted by $f_1(\mathbf{x})$) and a *weighted failure rate* (denoted by $f_2(\mathbf{x})$). Patch size is defined as $f_1(\mathbf{x}) = \sum_{i=1}^{n} b_i$, refering to the number of edits contained in the patch. Weighted failure rate is defined as

```
Input: A patch x = (b, c, u, v); the buggy program; n LBSs; .
Output: A modified program.
1  for j = 1 to n do
2      if bⱼ = 1 then
3          st ← the j-th LBS;
4          if cⱼ = 0 then
5              st* ← the uⱼ-th statement in the set Rⱼ;
6              Replace st with st*;
7          else
8              st* ← the vⱼ-th statement in the set Iⱼ;
9              Insert st* before st;
```

Fig. 26.5: The procedure to apply a patch.

$$f_2(\mathbf{x}) = \frac{|\{t \in T_f \mid \mathbf{x} \text{ fails } t\}|}{|T_f|} + w \times \frac{|\{t \in T_c \mid \mathbf{x} \text{ fails } t\}|}{|T_c|} \qquad (26.1)$$

where T_f is the set of negative test cases, T_c is the reduced set of positive test cases, and $w \in (0, 1]$ is a global parameter used to emphasize the passing of negative test cases. $f_2(\mathbf{x})$ measures how well the modified program behaves in terms of passing the given test cases. $f_2(\mathbf{x}) = 0$ means $\mathbf{x}$ does not fail any test case and is thus a test-adequate patch.

By simultaneously minimizing these two objectives, we introduce search bias toward smaller patches. Note that if the modified program fails to compile or runs out of time, $f_1(\mathbf{x})$ and $f_2(\mathbf{x})$ are both set to $+\infty$. Moreover, $f_1(\mathbf{x}) = 0$ is meaningless since no modifications would be made to the buggy program. Should $f_1(\mathbf{x})$ reach 0, we reset both $f_1(\mathbf{x})$ and $f_2(\mathbf{x})$ to $+\infty$, in order to eliminated this solution by selection.

Population Initialization

We initialize the population by combining prior knowledge and randomness: For each individual $\mathbf{x}$, $\mathbf{b}$ is initialized using fault localization information. Suppose $susp_j$ is the suspiciousness of the j-th LBS, then b_j is initialized to 1 with probability $susp_j \times \mu$ and to 0 with $1 - susp_j \times \mu$, where $\mu \in (0, 1)$ is a predefined parameter. The remaining three parts are randomly initialized.

Genetic Operators

Genetic operators crossover and mutation are used to produce offspring in MOEAs. In ARJA-p, crossover and mutation are applied to each part of the patch representation separately, in order to inherit good traits from parents. For $\mathbf{b}$ and $\mathbf{c}$ we employ half uniform crossover (HUX) and bit-flip mutation. For $\mathbf{u}$ and $\mathbf{v}$), we use single-point crossover and uniform mutation due to their integer nature. Figure 26.6 demonstrates how crossover and mutation are performed on two parents.

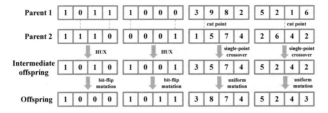

Fig. 26.6: Illustration of the crossover and mutation in ARJA-p. Only one offspring shown.

Multi-Objective Search

As a selection (including mating and environmental selection) framework, we use NSGA-II [8], a classical Pareto dominance-based MOEA. Note that recently proposed MOEAs like θ-DEA [46] and MOEA/D-DU [47] can also be used here.

The NSGA-II search procedure works as follows: First, an initial population with N (the population size) individuals is generated using the initialization strategy introduced in Section 26.3.3. Then the algorithm iterates over generations until a termination criterion is met. In the g-th generation, binary tournament selection [8] and the genetic operators of Section 26.3.3 are applied to the current population P_g so as to create an offspring population Q_g of size N. Then the fast non-dominated sorting and crowding distance comparison [8] (based on two objectives formulated in Section 26.3.3) are used to select the best N individuals from the union population $P_g \cup Q_g$, arriving at the next population P_{g+1}.

When the search is terminated, the obtained non-dominated solutions with $f_2 = 0$ are output as test-adequate patches found by ARJA-p. If there is no such solution, ARJA-p has failed to repair the bug.

26.4 Experimental Design

Our evaluation is conducted on four open-source Java projects (i.e., Chart, Time, Lang and Math) from Defects4J [13], a database widely used for evaluating Java repair systems [6, 19, 26, 28, 38, 41, 42, 45]. Table 26.2 shows the the basic information of the four projects. There are 224 real-world bugs in total: 26 bugs from Chart (C1–C26), 27 bugs (T1–T27) from Time, 65 bugs (L1–L65) from Lang and 106 bugs (M1–M106) from Math. Note that Defects4J indeed contains another project, namely Closure. We do not use Closure because its customized testing format is incompatible with GZoltar [5], a third-party fault localization tool used in our repair system.

Table 26.3 shows the basic parameter setting for ARJA-p in our empirical study, where n is the number of LBSs determined by γ_{min} and n_{max} together (see Section 26.3). We run 5 random trials in parallel for each bug. Each trial is terminated after

Table 26.2: The descriptive statistics of Defects4J dataset

Project	ID	#Bugs	#JUnit Tests	Source KLoC	Test KLoC
JFreeChart	C	26	2,205	96	50
Joda-Time	T	27	4,043	28	53
Commons Lang	L	65	2,295	22	6
Commons Math	M	106	5,246	85	19
Total		224	13,789	231	128

3 hours following the practice in refs. [26, 28]. Our experiments are performed on HPC machines with 2.4 GHz Intel Xeon E5 Processors and 20 GB memory.

Table 26.3: The parameter setting for ARJA-p in the experiments

Parameter	Description	Value
N	Population size	40
γ_{min}	Threshold for the suspiciousness	0.1
n_{max}	Maximum number of LBSs considered	60
w	Refer to Section 26.3.3	0.5
μ	Refer to Section 26.3.3	0.06
p_c	Crossover probability	1.0
p_m	Mutation probability	$1/n$

26.5 Results and Discussions

Table 26.4 shows the bugs in Defects4J that can be fixed (i.e., test-adequate patches are found) and correctly fixed by ARJA-p. Note that only non-dominated solutions with $f_2 = 0$ are meaningful for program repair. Among all 224 bugs considered, ARJA-p can generate test-adequate patches for 84 bugs.

A major concern raised recently [36] was whether test-adequate patches are correct beyond passing the test suite. Following previous work [26, 38, 41, 42, 45], we manually checked the correctness of these patches found by ARJA-p. A test-adequate patch is deemed as *correct* if it is exactly the same as or semantically equivalent to a human-written patch. After a careful manual study, we confirmed that ARJA-p found correct patches for 37 bugs.

We compare ARJA-p with 11 state-of-the-art repair tools in terms of the number of bugs fixed (i.e., test-adequate) and correctly fixed. The 11 tools for comparison are jGenProg [26] (an implementation of GenProg for Java), jKali [26] (an implementation of Kali for Java), xPAR (a reimplementation of PAR by Le et al. [19]), Nopol [26, 43], HDRepair [19], ACS [42], ssFix [41], JAID [6], ELIXIR [38],

Table 26.4: The bugs for which the test-adequate patches and the correct patches are synthesized by ARJA-p

Project	Test-Adequate	Correct
Chart	C1, C3, C4, C5, C7, C10, C11, C13, C14, C15, C17, C19 ,C24, C25, C26	C1, C4, C10, C11, C14, C17,C19, C24
	$\Sigma = 15$	$\Sigma = 8$
Time	T4, T11, T14, T17, T20	T4
	$\Sigma = 5$	$\Sigma = 1$
Lang	L7, L16, L20, L21, L22, L24, L27, L33, L34, L39, L41, L44, L45, L47, L50, L51, L57, L58, L59, L60, L61, L63	L20, L24, L33, L34, L39, L47, L57, L59, L61
	$\Sigma = 22$	$\Sigma = 9$
Math	M2, M3, M5, M6, M7, M22, M24, M25, M28, M30, M32, M34, M40, M42, M49, M50, M56, M57, M58, M62, M63, M65, M70, M71, M73, M75, M77, M78, M79, M80, M81, M82, M84, M85, M88, M89, M94, M95, M96, M98, M104, M105	M5, M22, M25, M30, M34, M56, M57, M58, M70, M75, M79, M80, M82, M89, M94, M98, M105
	$\Sigma = 42$	$\Sigma = 19$
Total	84 (37.5%)	37 (16.5%)

ARJA [45] and Cardumen [28], which cover almost all the tools that have ever been tested on Defects4J. Table 26.5 shows our comparison results. Note that for xPAR, HDRepair and Cardumen, some results were not reported by the original authors. ARJA-p performs best and indeed outperforms all 11 other techniques by a large margin, both in terms of the number of bugs fixed and correctly fixed. Specifically, ARJA-p is able to increase the highest number of fixed bugs in Defects4J by 29.2%, from 65 (achieved by Cardumen) to 84; it increases the highest number of correctly fixed bugs in Defects4J by an even larger margin, 42.3%, from 26 (achieved by ELIXIR) to 37. Moreover, xPAR can only correctly fix 3 bugs, which is much less than the number achieved by ARJA-p. Since our ARJA-p is based on PAR, the much improved performance of ARJA-p strongly demonstrates the improvements over PAR.

Figure 26.7(a) presents a Venn diagram showing the intersections of fixed bugs among ARJA-p, ARJA and Cardumen. We select ARJA and Cardumen since they are the best-performing tools among the 11 existing ones in terms of producing test-adequate patches. ARJA-p is able to fix 22 bugs that neither ARJA nor Cardumen could fix. But although ARJA-p fixes more bugs, ARJA and Cardumen can also

Table 26.5: Comparison with 11 techniques in terms of the number of bugs fixed and
correctly fixed (Test-Adequate/Correct).

Project	ARJA-p	jGenProg	jKali	xPAR	Nopol	HDRepair
Chart	**15/8**	7/0	6/0	NA/0	6/1	NA/2
Time	**5/1**	2/0	2/0	NA/0	1/0	NA/1
Lang	**22/9**	0/0	0/0	NA/1	7/3	NA/7
Math	**42/19**	18/5	14/1	NA/2	21/1	NA/6
Total	**84/37**	27/5	22/1	NA/3	35/5	NA/16

Project	ACS	ssFix	JAID	ELIXIR	ARJA	Cardumen
Chart	2/2	7/2	4/4	7/4	9/3	15/NA
Time	1/1	4/0	0/0	3/**2**	4/1	6/NA
Lang	4/3	12/5	8/5	12/8	17/4	7/NA
Math	16/12	26/7	8/7	19/12	29/10	37/NA
Total	23/18	49/14	20/16	41/26	59/18	65/NA

"NA" means the data is not available.

fix 16 and 19 bugs, respectively, that cannot be fixed by ARJA-p. This indicates
that state-of-the-art repair approaches are complementary to each other, and further
attempts at combinations should be envisioned.

Figure 26.7(b) shows a Venn diagram to compare ARJA-p with two other state-
of-the-art approaches, ACS and JAID, in terms of correct bug fixing. Note that the
ELIXIR paper reported the highest number of bugs correctly fixed in the literature
but did not show which bugs were correctly fixed, so we cannot compare with it here.
The Figure shows that ARJA-p, ACS and JAID can uniquely generate a correct patch
for 25, 10 and 9 bugs, respectively, compared to their counterparts. Again, these
systems exhibit good complementarity in terms of producing correct bug fixes.

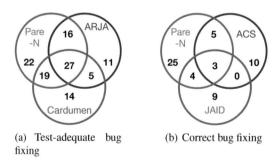

(a) Test-adequate   bug   (b) Correct bug fixing
fixing

Fig. 26.7: Venn diagram of bugs for which test-adequate patches (ARJA-p, ARJA and Cardumen)
and correct patches (ARJA-p, ACS and JAID) are found.

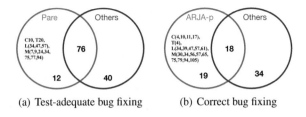

(a) Test-adequate bug fixing (b) Correct bug fixing

Fig. 26.8: Venn diagram of bugs for which test-adequate patches and correct patches are found. Our results are compared with the combined results of all the other techniques mentioned in Table 26.5.

Finally, we compare the results of ARJA-p with the combined results of all 11 existing techniques. Figure 26.8 shows that our technique can uniquely find test-adequate patches for 12 bugs and uniquely generate correct patches for 20 bugs. To our knowledge, each of these 12 bugs (listed in Figure 26.8(a)) is fixed for the first time, and each of 20 bugs (listed in Figure 26.8(b)) is repaired correctly for the first time.

26.5.1 Results on Multi-Location Bugs

Among the 37 bugs correctly fixed by ARJA-p, C14, C19, L20, L34, L47, L61, M22 and M98 are deemed multi-location bugs, since the correct patch by ARJA-p for these 8 bugs contains at least two edits at multiple buggy locations.

```
// ToStringStyle.java
static Map<Object, Object> getRegistry() {
-   return REGISTRY.get() != null ? REGISTRY.get() :
-       Collections.<Object, Object>emptyMap();
+   return REGISTRY.get();
}
static boolean isRegistered(Object value) {
    Map<Object, Object> m = getRegistry();
+   if (!(m != null)) return false;
    return m.containsKey(value);
}
```

Fig. 26.9: Correct patch generated by ARJA-p for the bug L34.

For M22 and M98, ARJA-p can synthesize a correct patch that is exactly the same as a human-written patch. As for the remaining 6 bugs, ARJA-p generates a correct repair semantically equivalent to a human-provided one. It is quite challenging for correctly fixing these bugs. For example, for C14, ARJA-p generates a correct patch that executes the "Null Pointer Checker" template at 4 different LBSs (located in two different Java files). Another interesting example is fixing bug L34, which is shown in Figure 26.9. To correctly fix this bug, ARJA-p uses two kinds of templates for two LBSs: "Element Replacer" for lines 3–4 and "Null Pointer Checker" for line 10. A human-written patch for L34 differs in that it replaces line 10 with `return m !=` `null && m.containsKey(value);`. Obviously, this modification is functionally equivalent to the null pointer check done by ARJA-p. Note that L61 can also be seen as a single-location bug in terms of a human-written patch that modifies only a single LBS. However, that patch is not in the search space of ARJA-p.

To our knowledge, only two existing approaches (i.e., ACS [42] and ARJA [45]) reported correct fixes for multi-location bugs on Defects4J. For the 8 multi-location bugs correctly fixed by ARJA-p, ACS can only generate correct patches for two of them (i.e., C14 and C19), while ARJA can generate three (i.e., L20, M22 and M98). Bugs L34, L47, L61 are correctly fixed by ARJA-p for the first time. The strength of ARJA-p in fixing multi-location bugs demonstrates the power of evolutionary multi-objective search.

26.5.2 Contribution of Repair Templates

ARJA-p uses 7 repair templates. It is interesting to understand the contribution of each template on the number of bugs fixed (test-adequate or correctly), which should provide insight into the strength and weakness of ARJA-p. If multiple patches are obtained for a bug, we just randomly choose one for analysis. Note that a bug fix may involve more than one template (e.g., L34 in Figure 26.9).

Table 26.6 summarizes the contribution of templates. It is clear that "ER" is the most useful template for the performance of ARJA-p on Defects4J. This template contributes to the generation of a test-adequate patch for 54 bugs and a correct one for 27 bugs. "BEAR" helps to synthesize a test-adequate patch for 17 bugs (just behind "ER"), however only 1 of them is identified as correct. "MPAR" and "NPC" make moderate contribution for both test-adequate and correct bug fixing. "DC", "CC" and "RC" do not contribute much to ARJA-p's performance on Defects4J, and "RC" even contributes nothing.

Table 26.6: Contribution of each repair template

Template	Test-Adequate	Correct
Element Replacer (ER)	54	27
Method Parameter Adder or Remover (MPAR)	6	3
Boolean Expression Adder or Remover (BEAR)	17	1
Null Pointer Checker (NPC)	10	7
Range Checker (RC)	0	0
Cast Checker (CC)	2	1
Divide-By-Zero Checker (DC)	1	1

It is not surprising that "ER" contributes most since it can manipulate many more types of AST nodes than other templates. "BEAR" tries to synthesize a condition by exploiting the intrinsic redundancy of a program, so the synthesized condition may not be so accurate. This could be the reason that most of the test-adequate patches contributed by this template are indeed incorrect. A promising way to enhance the strength of "BEAR" is to incorporate a precise condition synthesis technique like that in ACS [42]. "NPC" contributes much more than three similar templates (i.e., "RC", "CC" and "DC"), implying that the null pointer bug could be quite common in Java. Moreover, the patch produced by "NPC" has a relatively high probability to be correct, may be because the manipulations performed by "NPC" are usually harmless. Note that although "RC", "CC" and "DC' have limited contribution here, they could be helpful for ARJA-p on other bug datasets.

26.5.3 Value of Test Filtering

We select the latest buggy versions of the four projects considered in Defects4J (i.e., C1, T1, L1 and M1) as the subject programs to examine the effect of test filtering. For each buggy program, we vary γ_{min} (i.e., the threshold of suspiciousness) from 0 to 0.2. For each γ_{min} value chosen, we use our test filtering procedure to obtain a subset of original JUnit tests and record two metrics associated with this reduced test suite: the number of tests and the execution time. Fig. 26.10 shows the influence of γ_{min} on the two metrics for each subject program. For comparison purposes, the two metrics have been normalized by dividing by the original number of JUnit tests and the CPU time consumed by the original test suite respectively. Note that the fluctuations of CPU time in Fig. 26.10 are due to the dynamic behavior of the computer system.

Fig. 26.10 shows that test filtering can achieve substantial benefits in terms of reduction of computational costs. As we increase γ_{min} the number of tests that needs to be considered and the corresponding CPU time consumed both decrease significantly and quickly. Even if we set γ_{min} to a very small value, test filtering can result in a considerable reduction of CPU time. Suppose as an example, we set $\gamma_{min} = 0.01$, then CPU time reduction is about 59% for C1, 31% for T1, 97% for L1 and 37% for M1. Generally, if we set γ_{min} to a larger value, we can consider a smaller test suite for fitness evaluation. However, it is not desirable to use too large a γ_{min} (e.g., 0.5), because it could result in the repair approaches missing the actual faulty location. In practice, we choose a moderate value for γ_{min} (e.g., 0.1) to strike a compromise. Normally, test filtering can significantly speed up the fitness evaluation in such a case. For example, if we set γ_{min} to 0.1 for M1, the number of tests considered is reduced from 5,246 to 118, and the CPU time for one fitness evaluation is reduced from 210 seconds to 3.4 seconds. Suppose the termination criterion of ARJA-p is 2,000 evaluations, then we can save up to 115 hours for just a single repair trial.

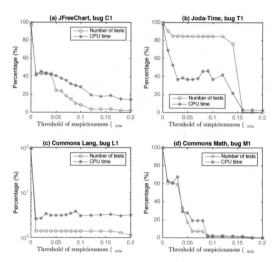

Fig. 26.10: Illustration of the value of the test filtering procedure. The base 10 logarithmic scale is used for the y axis in (c).

By conducting a post-run validation of the obtained patches on the original test suite we confirm that if a patch can pass the reduced test suite, it can also pass the original one, without any exception.

26.6 Conclusion

In this chapter we have examined a new program repair approach for Java, ARJA-p. It combines very recent techniques from template based and EC based repair approaches to enhance performance of PAR. Our technique is compared with almost all the existing techniques (11 tools) that have ever been evaluated on Defects4J dataset. The empirical results on 224 real bugs in Defects4J show that ARJA-p can generate test-adequate patches for 84 bugs and correct patches for 37 bugs outperforming state-of-the-art approaches with a considerable margin. ARJA-p can correctly fix several multi-location bugs that cannot be handled by almost any existing repair approaches. We have also found that there exists good The complementarity in performance shown between ARJA-p and other state-of-the-art techniques like ACS and ARJA imply that a combination of principles of existing approaches is a promising avenue to further improve repair effectiveness of evolutionary program repair methods.

Acknowledgements W.B. gratefully acknowledges support from the Koza endowment at MSU, made possible with a generous gift by John R. Koza. This would not have been possible without the BEACON Center for the Study of Evolution in Action under the capable leadership of Erik Goodman.

References

1. Abreu, R., Zoeteweij, P., Van Gemund, A.J.: *An evaluation of similarity coefficients for software fault localization.* In: Dependable Computing, 2006. PRDC'06. 12th Pacific Rim International Symposium on, pp. 39–46. IEEE (2006)
2. Banzhaf, W.: Some remarks on code evolution with genetic programming. In: A. Adamatzky, S. Stepney (eds.) Inspired by Nature, pp. 145–156. Springer (2018)
3. Banzhaf, W., Nordin, P., Keller, R.E., Francone, F.D.: *Genetic Programming: An Introduction.* Morgan Kaufmann, San Francisco (1998)
4. Barr, E.T., Brun, Y., Devanbu, P., Harman, M., Sarro, F.: *The plastic surgery hypothesis.* In: Proceedings of the 22nd ACM SIGSOFT International Symposium on Foundations of Software Engineering, pp. 306–317. ACM (2014)
5. Campos, J., Riboira, A., Perez, A., Abreu, R.: *Gzoltar: An Eclipse plug-in for testing and debugging.* In: Proceedings of the 27th IEEE/ACM International Conference on Automated Software Engineering, pp. 378–381. ACM (2012)
6. Chen, L., Pei, Y., Furia, C.A.: *Contract-based program repair without the contracts.* In: Proceedings of the 32nd IEEE/ACM International Conference on Automated Software Engineering, pp. 637–647. IEEE Press (2017)
7. D'Antoni, L., Samanta, R., Singh, R.: *QLOSE: Program repair with quantitative objectives.* In: International Conference on Computer Aided Verification, pp. 383–401. Springer (2016)
8. Deb, K., Pratap, A., Agarwal, S., Meyarivan, T.: *A fast and elitist multiobjective genetic algorithm: NSGA-II.* IEEE transactions on evolutionary computation **6**(2), 182–197 (2002)
9. Gazzola, L., Micucci, D., Mariani, L.: *Automatic software repair: A survey.* IEEE Transactions on Software Engineering **45**, 34–67 (2017)
10. Gupta, R., Pal, S., Kanade, A., Shevade, S.: *DeepFix: Fixing Common C Language Errors by Deep Learning.* In: AAAI, pp. 1345–1351 (2017)
11. Harman, M., Jones, B.F.: *Search-based software engineering.* Information and software Technology **43**(14), 833–839 (2001)
12. Harman, M., Mansouri, S.A., Zhang, Y.: *Search-based software engineering: Trends, techniques and applications.* ACM Computing Surveys (CSUR) **45**(1), 11 (2012)
13. Just, R., Jalali, D., Ernst, M.D.: *Defects4j: A database of existing faults to enable controlled testing studies for java programs.* In: Proceedings of the 2014 International Symposium on Software Testing and Analysis, pp. 437–440. ACM (2014)
14. Ke, Y., Stolee, K.T., Le Goues, C., Brun, Y.: *Repairing programs with semantic code search (t).* In: Automated Software Engineering (ASE), 2015 30th IEEE/ACM International Conference on, pp. 295–306. IEEE (2015)
15. Kim, D., Nam, J., Song, J., Kim, S.: *Automatic patch generation learned from human-written patches.* In: Proceedings of the 2013 International Conference on Software Engineering, pp. 802–811. IEEE Press (2013)
16. Koza, J.R.: *Genetic Programming: On the programming of computers by means of natural selection.* MIT Press, Cambridge, MA (1992)
17. Le, X.B.D., Chu, D.H., Lo, D., Le Goues, C., Visser, W.: *JFIX: Semantics-based repair of Java programs via symbolic PathFinder.* In: Proceedings of the 26th ACM SIGSOFT International Symposium on Software Testing and Analysis, pp. 376–379. ACM (2017)
18. Le, X.B.D., Lo, D., Le Goues, C.: *Empirical study on synthesis engines for semantics-based program repair.* In: Software Maintenance and Evolution (ICSME), 2016 IEEE International Conference on, pp. 423–427. IEEE (2016)
19. Le, X.B.D., Lo, D., Le Goues, C.: *History driven program repair.* In: Software Analysis, Evolution, and Reengineering (SANER), 2016 IEEE 23rd International Conference on, vol. 1, pp. 213–224. IEEE (2016)
20. Le Goues, C., Dewey-Vogt, M., Forrest, S., Weimer, W.: *A systematic study of automated program repair: Fixing 55 out of 105 bugs for $8 each.* In: Software Engineering (ICSE), 2012 34th International Conference on, pp. 3–13. IEEE (2012)

21. Le Goues, C., Nguyen, T., Forrest, S., Weimer, W.: *GenProg: A generic method for automatic software repair*. IEEE Transactions on Software Engineering **38**(1), 54–72 (2012)
22. Le Goues, C., Weimer, W., Forrest, S.: *Representations and operators for improving evolutionary software repair*. In: Proceedings of the 14th annual conference on Genetic and evolutionary computation, pp. 959–966. ACM (2012)
23. Long, F., Amidon, P., Rinard, M.: *Automatic inference of code transforms for patch generation*. In: Proceedings of the 2017 11th Joint Meeting on Foundations of Software Engineering, pp. 727–739. ACM (2017)
24. Long, F., Rinard, M.: *Staged program repair with condition synthesis*. In: Proceedings of the 2015 10th Joint Meeting on Foundations of Software Engineering, pp. 166–178. ACM (2015)
25. Long, F., Rinard, M.: *Automatic patch generation by learning correct code*. In: ACM SIGPLAN Notices, vol. 51, pp. 298–312. ACM (2016)
26. Martinez, M., Durieux, T., Sommerard, R., Xuan, J., Monperrus, M.: *Automatic repair of real bugs in Java: A large-scale experiment on the Defects4j dataset*. Empirical Software Engineering pp. 1–29 (2016)
27. Martinez, M., Monperrus, M.: *Mining software repair models for reasoning on the search space of automated program fixing*. Empirical Software Engineering **20**(1), 176–205 (2015)
28. Martinez, M., Monperrus, M.: *Open-ended Exploration of the Program Repair Search Space with Mined Templates: the Next 8935 Patches for Defects4J*. arXiv preprint arXiv:1712.03854 (2017)
29. Martinez, M., Weimer, W., Monperrus, M.: *Do the fix ingredients already exist? An empirical inquiry into the redundancy assumptions of program repair approaches*. In: Companion Proceedings of the 36th International Conference on Software Engineering, pp. 492–495. ACM (2014)
30. Mechtaev, S., Yi, J., Roychoudhury, A.: *Directfix: Looking for simple program repairs*. In: Proceedings of the 37th International Conference on Software Engineering-Volume 1, pp. 448–458. IEEE Press (2015)
31. Mechtaev, S., Yi, J., Roychoudhury, A.: *Angelix: Scalable multiline program patch synthesis via symbolic analysis*. In: Software Engineering (ICSE), 2016 IEEE/ACM 38th International Conference on, pp. 691–701. IEEE (2016)
32. Monperrus, M.: *Automatic software repair: A bibliography*. ACM Computing Surveys (2017)
33. Nguyen, H.D.T., Qi, D., Roychoudhury, A., Chandra, S.: *Semfix: Program repair via semantic analysis*. In: Proceedings of the 2013 International Conference on Software Engineering, pp. 772–781. IEEE Press (2013)
34. Oliveira, V.P.L., de Souza, E.F., Le Goues, C., Camilo-Junior, C.G.: *Improved representation and genetic operators for linear genetic programming for automated program repair*. Empirical Software Engineering pp. 1–27 (2018)
35. Qi, Y., Mao, X., Lei, Y., Dai, Z., Wang, C.: *The strength of random search on automated program repair*. In: Proceedings of the 36th International Conference on Software Engineering, pp. 254–265. ACM (2014)
36. Qi, Z., Long, F., Achour, S., Rinard, M.: *An analysis of patch plausibility and correctness for generate-and-validate patch generation systems*. In: Proceedings of the 2015 International Symposium on Software Testing and Analysis, pp. 24–36. ACM (2015)
37. Rolim, R., Soares, G., D'Antoni, L., Polozov, O., Gulwani, S., Gheyi, R., Suzuki, R., Hartmann, B.: *Learning syntactic program transformations from examples*. In: Proceedings of the 39th International Conference on Software Engineering, pp. 404–415. IEEE Press (2017)
38. Saha, R.K., Lyu, Y., Yoshida, H., Prasad, M.R.: *Elixir: Effective object oriented program repair*. In: Proceedings of the 32nd IEEE/ACM International Conference on Automated Software Engineering, pp. 648–659. IEEE Press (2017)
39. Tan, S.H., Yoshida, H., Prasad, M.R., Roychoudhury, A.: *Anti-patterns in search-based program repair*. In: Proceedings of the 2016 24th ACM SIGSOFT International Symposium on Foundations of Software Engineering, pp. 727–738. ACM (2016)
40. Weimer, W., Fry, Z.P., Forrest, S.: *Leveraging program equivalence for adaptive program repair: Models and first results*. In: Proceedings of the 28th IEEE/ACM International Conference on Automated Software Engineering, pp. 356–366. IEEE Press (2013)

41. Xin, Q., Reiss, S.P.: *Leveraging syntax-related code for automated program repair.* In: Proceedings of the 32nd IEEE/ACM International Conference on Automated Software Engineering, pp. 660–670. IEEE Press (2017)
42. Xiong, Y., Wang, J., Yan, R., Zhang, J., Han, S., Huang, G., Zhang, L.: *Precise condition synthesis for program repair.* In: Proceedings of the 39th International Conference on Software Engineering, pp. 416–426. IEEE Press (2017)
43. Xuan, J., Martinez, M., DeMarco, F., Clement, M., Marcote, S.L., Durieux, T., Le Berre, D., Monperrus, M.: *Nopol: Automatic repair of conditional statement bugs in Java programs.* IEEE Transactions on Software Engineering **43**(1), 34–55 (2017)
44. Yang, J., Zhikhartsev, A., Liu, Y., Tan, L.: *Better test cases for better automated program repair.* In: Proceedings of the 2017 11th Joint Meeting on Foundations of Software Engineering, pp. 831–841. ACM (2017)
45. Yuan, Y., Banzhaf, W.: *ARJA: Automated Repair of Java Programs via Multi-Objective Genetic Programming.* IEEE Transactions on Software Engineering (2020, in press)
46. Yuan, Y., Xu, H., Wang, B., Yao, X.: *A new dominance relation-based evolutionary algorithm for many-objective optimization.* IEEE Transactions on Evolutionary Computation **20**(1), 16–37 (2016)
47. Yuan, Y., Xu, H., Wang, B., Zhang, B., Yao, X.: *Balancing convergence and diversity in decomposition-based many-objective optimizers.* IEEE Transactions on Evolutionary Computation **20**(2), 180–198 (2016)
48. Zhong, H., Su, Z.: *An empirical study on real bug fixes.* In: Proceedings of the 37th International Conference on Software Engineering-Volume 1, pp. 913–923. IEEE Press (2015)

Chapter 27
From Biological to Computational Arms Races – Studying Cyber Security Dynamics

Una-May O'Reilly and Erik Hemberg

Abstract The design of computational arms races can draw upon the compelling inspiration of biological arms races. To study cyber security attack-defense dynamics, we have abstracted a description of biological adversarial ecosystems to design an adversarial computational system. The system has elements and processes with abstracted biological analogs. It centers on engagements. Engagements feature adversarial actors (predator/attacker, prey/defender) competing with conflicting objectives culminating in a measurable performance-based outcome. Adversarial dynamics are controlled by coevolution, which selects for better adversaries over multiple engagements using aggregate engagement performance as fitness. Altogether, this system abstracted from nature is capable of population-based, arms race dynamics arising from interacting, evolving adversaries.

Key words: arms race, dynamics, cyber security, adversarial, coevolution, Distributed Denial of Service, coevolutionary algorithm, network, peer-2-peer, decision support, grammar

27.1 Introduction

Adversarial advantages have coevolved in many natural contexts – between plants and animals, between species, and within a species, [3, 5, 10, 22]. They emerge from conditions where competition is necessary for survival. Some astonishing examples are predation competitions where one side, as predator, seeks an upper hand to feed and the other side, as prey, aims to evade being eaten. The speed of cheetahs and the yellow-colored crab spider (*Misumena vatia*), Figure 1(a), masking itself

Una-May O'Reilly
Massachusetts Institute of Technology, e-mail: unamay@csail.mit.edu

Erik Hemberg
Massachusetts Institute of Technology, e-mail: hembergerik@csail.mit.edu

© Springer Nature Switzerland AG 2020   409
W. Banzhaf et al. (eds.), *Evolution in Action: Past, Present and Future*,
Genetic and Evolutionary Computation, https://doi.org/10.1007/978-3-030-39831-6_27

on goldenrod sprays, waiting to entrap a wasp, are only two examples of evolved predatory advantages among many, see [16, 35] for more. In addition, traits such as horns or sharp, long canines have evolved from sexual selection, itself a competitive context. Other physical advantages have evolved for defensive purposes. Spiders (Figure 1(a)) and chameleons (Figure 1(b)) respectevly make use of passive and active camouflage [27]. Cacti (Figure 1(c)) have evolved spines [18]. In an example of what is called "*aposematism*", a skunk (Figure 1(d)) has an offensive squirting order and definitive coloring that preempts attack [26]. Snakes and various bugs have evolved to be poisonous. The Viceroy butterfly (Figure 1(e)) has evolved mimicry of the Monarch (Figure 1(f)). Defensive behavior can also manifest as enzymatic adaptation rather than overt physical adaptation. For example, the African variegated grasshopper (*Zonocerus variegatus*) and the Asian cinnabar moth (*Tyria jacobaeae*), feed on toxic plants in order to protect themselves from predators, see [6]. In an example of dual defenses, wood tiger moths produce two different fluids that target different predators and defensive contexts. One fluid results in a wing odor that repels birds before even tasting. The other fluid is for ground level protection when the insect can not fly due to colder temperatures or is coming out of its pupa [17].

Our interest in the evolution of biological adversarial adaptation is motivated by cyber security. We observe that, in cyberspace, there is a plethora of "attack surfaces", e.g. network layers, humans, and applications – environments that have contentious resources and thus are ripe for becoming a ground or host to adversarial engagements. These surfaces bear high level resemblance to the complex environments within which biological adversarial adaptation plays out. They have a myriad of resources for which adversaries compete. Networks offer communication resources for legitimate users and these fall prey to malicious consumption by denial of service attackers. Data, another cyber resource, polarizes actors. One side maintains and protects (and also uses) it while properties of data – including privacy or confidentiality spur non-legitimate attacks by means of ransomware to deny access to it, or ATPs - advanced persistent threats, for exfiltrating it.

We also observe each and all attack surfaces being frequently attacked by multiple actors of malicious intent. Attacking actors (or simply attackers) come in tactical varieties (e.g. one variety mounts a denial of service attack by leveraging a network protocol and another by application layer vulnerability) and within a variety, each are different in terms of specific tools and techniques. Demarcating them as "species" is arguable as is terming multiple sets of actors "populations". Most importantly, in cyberspace, there are arms races. As a result of success or failure, both adversaries - attackers and defenders, adapt in attempts to improve for future engagements. Actors undergo adaptation to improve upon or maintain their performance on their objective. We see defensive measures being mounted, such as patches, code updates, new access schemes, new authentication protocols, etc. Continually adapting defensive strategies emerge to protect, while analogous attack strategies emerge to compromise, given the new protections.

(a) A spider(Misumena vatia) [29]   (b) A chameleon [34]

(c) A cactus with spines [32]   (d) A skunk [30]

(e) A Viceroy Butterfly [31]   (f) A Monarch butterfly [33]

Fig. 27.1: Adversarially defense and attack adapted plants and animals.

Obviously these predatory interactions of a human nature are, in details, different from those of biology[1]. Adaptations in nature take place on a much longer time scale, through biological mechanisms whereas, cyber attacks can adapt in days and humans or machine learning direct the adaptation. However, it is our position that, at a helpful and intuitive abstracted level, nature and cyberspace are analogous.

In this contribution, we present a design abstraction for a computational system capable of stylized cyber security arms races. Our goal is to study the dynamics of cyber systems under attack by computationally modeling and simulating them. Ul-

---

[1] Another consideration is the importance of distinguishing whether an attacker is human or the software being used.

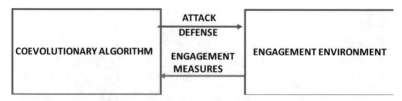

Fig. 27.2: A computational system capable of stylized cyber security arms races. The left hand
        side component uses a competitive coevolutionary algorithm that evolves two
        competing populations: attacks and defenses. The right hand side component from a
        computational perspective is a modeling or simulation environment. The environment
        is initialized with a mission and can be reset each time it is passed an attack-defense
        pairing to run an engagement. It first installs the defense, then it starts the mission and
        the attack. It evaluates the performance of the adversaries relative to their objectives
        and returns appropriate measurements.

timately we aim to provide defensive designers with information that allows them
to anticipate attacks and design for resilience and resistance before their attack surfaces are attacked.

Summarized as three elements, our abstraction (which is realized computationally), consists of :

- *Adversaries*: Two adversarial populations – *attackers* and *defenders*.
- *Engagements*: An engagement is an attack on a defense. Engagements take
  place in an *engagement environment*.
- *A competitive coevolutionary algorithm* controlling the adversarial engagements and population-based evolution.

These fit together as two connected modules - one executing the engagements and
the other executing the evolutionary algorithm, shown figuratively in Figure 27.2,
with the competitive coevolutionary algorithm selecting attack and defense pairings
to engage.

Solely emulating or simulating cyber arms races is not sufficient to practically
inform the design of better, anticipatory defenses. To this end, our system has an additional decision support component, named ESTABLO [23, 25], see Figure 27.3.
The system supports a number of use cases using simulation and emulation of varying model granularity. These include: *A)* Defending a peer-2-peer network against
Distributed Denial of Service (DDOS) attacks [8] *B)* Defenses against spreading
device compromise in a segmented enterprise network [11], and *C)* Deceptive defense against the internal reconnaissance of an adversary within a software defined
network [20]

We proceed as follows: Section 27.2 presents a brief context on coevolutionary
search algorithm precedents and describes them in more detail. In Section 27.3 we
expand the description of our system. In Section 27.4 we provides an example applying to cyber security and network attacks. Section 27.5 summarizes and addresses
future work.

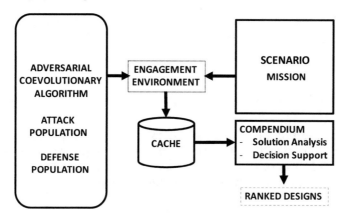

Fig. 27.3: The system including the compendium and cache that assists the decision support component named ESTABLO.

27.2 Related Work and Background

Competitive coevolutionary algorithms are well established and not of our invention. We draw from the theoretical and empirical efforts of research that precedes us. Competitive coevolutionary algorithms are typically introduced at a very general abstraction level. The two competing populations are generally called *solutions* and *tests*. Tests compete to stump solutions and solutions compete to correctly pass tests. This general abstraction supports the algorithms' use in software engineering problems, theoretical social games and toy problems, as well as cyber security.

The strategy of testing the security of a system by trying to successfully attack it is somewhat analogous to software fuzzing [15]. Fuzzing tests software adaptively to search for bugs while adaptive attacks test defenses. In contrast to software where a bug is fixed by humans, we are interested in approaches that also adapt a defense. This forms a novel *counter attack*. Fuzzing is driven by genetic algorithms (GA) whereas, to drive cyber arms races in which both adversaries adapt, our approach uses the coupled GAs we call competitive coevolutionary algorithms. Cyber security investigative systems vary in complexity, level of abstraction and resolution. Examples address computer security [13, 28, 36] and network dynamics in particular, e.g., in CANDLES – the Coevolutionary, Agent-based, Network Defense Lightweight Event System of [24], attacker and defender strategies are coevolved in the context of a single, custom, abstract computer network defense simulation.

Coevolutionary algorithms can encounter problematic dynamics where tests are unable improve solutions, or drive toward a solution that is the *a priori* intended goal. There are accepted *remedies* to specific coevolutionary pathologies [4, 7, 21]. They generally include maintaining population diversity so that a search gradient is always present and using more explicit memory, e.g. a *Hall of Fame* or an archive, to prevent regress [14]. The pathologies of coevolutionary algorithms are similar to those encountered by generative adversarial networks (GANs) [2, 9].

27.3 System Description

We now provide further details on each element of the system: Adversaries (27.3.1), Engagements and their environments (27.3.2), and Competitive coevolutionary algorithms (27.3.3). Last, we detail the decision support module that supports selection of a superior anticipatory defensive configuration (27.3.4).

27.3.1 Adversaries

The system consists of two adversarial populations – *attacks* and *defenses*. The primary objective of an *attack* is to maximize disruption to a resource, thus impairing a mission. The primary objective of a *defense* is to complete a mission by minimizing disruptions to the resource. Both attack and defense can have secondary objectives related to the cost of their behavior. We design attack and defense behavior by defining grammars that support the expression of different variations of attacks and defenses. An attack or defense is a "sentence" in the grammar's language. We implement a generator to form sentences from the grammar by rewriting, see Figure 27.4 and [19]. Very specific to the machine learning techniques that we use, i.e. genetic programming and grammatical evolution [12, 19], sentences (i.e. attacks and defenses) are computational functions that are executable in the layers of the engagement environment where the defense is installed and where the simulation interfaces with the attack.

The system's grammars currently express a low level abstraction of network attack and defensive behavior. For example, attacks target a variable number of network links, by specifying, for a link, the attack starting time, its duration and denial strength. It is plausible that an adversary could, in the future, instead be represented by a plan and the space of all possible plans would be defined by a grammar. The plan would represent fulfillment of a goal the adversary can be assumed to hold, e.g target a specific enterprise. It would be derived by a different generator. The generator would derive a plan that draws upon a set of tactics, (e.g. a network protocol or application layer vulnerability) and techniques (e.g. different bot compromise exploits) of the adversary that are described by the grammar. As sentences, plans would be executable in the engagement environment's initialization and simulation layers.

In our implementation, a grammar is input in Backus Naur Form (BNF). The BNF input is parsed to construct a context free grammar representation. The (rewrite) rules of the generator express how a sentence, i.e. attack or defense, can be composed by rewriting a start symbol such as "attack" or "defense". Because the system implements a technique known as grammatical evolution (GE) [19], the adversaries are fixed length integer vectors. The generator decodes these vectors to control its rewriting. As a result, different vectors generate different attacks or defenses. For solving different problems, it is only necessary to change the BNF grammar, engagement environment and fitness function of the adversaries. This modularity, as well as the reusability of the parser and rewriter are efficient software engineering and problem solving advantages. In addition, the grammar helps informing the designer how the system works and its assumptions, i.e. threat model. This information enables conversations and validation at the domain level and, overall, contributes to confidence in solutions from the system.

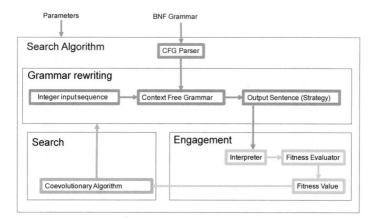

Fig. 27.4: Overview of the Grammatical Evolution search heuristic. A BNF grammar and search parameters are used as input in the search for solutions. The generator rewrites the start symbol under the control of the integer input to form a sentence (attack or defense). Fitness is calculated by executing the sentence's plan or defense and measuring its performance on its objective(s). The search component, a coevolutionary algorithm, modifies the population using two central mechanisms: fitness based selection and random variation of the integer vectors.

27.3.2 Engagements and the Engagement Environment

An engagement is an attack on a defense. Engagements take place in an *engagement environment* that is initialized with a mission to complete and a set of resources such as network services that support the mission. The defense is first installed in the environment and then, while the mission runs, the attack is launched. The scenario (mission and resources) and attacks are simulated, emulated or actually computed. Engagements have outcomes that match up to objectives; they are phrased in terms of mission completion (primary objective) and resource usage (second objective). Note that, unlike nature, the system does not replicate predator-prey competition in the sense that the prey can lose its life or the predator could starve. Its abstracts away these details as well as where an engagement takes place. See the next section 27.3.3 for details of the algorithm that controls fitness-based selection and replication via inheritance and random variation. Implementation-wise, the engagement environment component can support a problem-specific network testbed, simulator or model. The abstraction level of the use case determines the choice of a simple or more detailed mod-sim or even the actual engagement environment. Mod-sim is appropriate when testbeds incur long experimental cycle times or do not abstract away irrelevant detail.

27.3.3 Coevolutionary Algorithms

Our system makes direct use of competitive coevolutionary algorithms and one of their standard designs - the two population model [21]. This model maintains two *populations* of competing types, in our case, attack and defense. It calculates the fitness of each population member (in both attack and defense populations) by summarizing its ability to successfully engage one or more members from the adversarial population, given its objective(s). It also directs selection and variation per standard evolutionary algorithm logic, see Figure 27.5 for a flowchart.

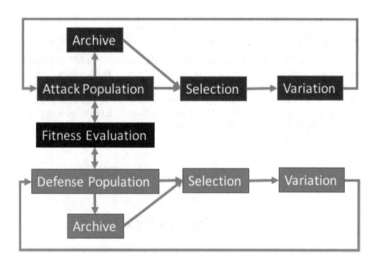

Fig. 27.5: Flowchart of a coevolutionary algorithm. Each population performs selection, variation and evaluation. An archive can provide a "memory" and improve the search efficiency by storing high performing solutions.

Our system exploits different coevolutionary algorithms to help it generate diverse behavior. The algorithms, for further diversity, use different "solution concepts", i.e. measures of adversarial success. Because engagements are frequently computationally expensive and have to be pairwise sampled from two populations each generation, the system has a number of enhancements that enable efficient use of a fixed budget of computation or time. The different competitive coevolutionary algorithms vary in synchronization of the two populations and solution concepts [20, 23].

27.3.4 ESTABLO System Decision Support

Competitive coevolution poses general challenges when used for design optimization. The following ones in our system make it difficult to present a designer with clear information derived solely from multiple simulation runs [23, 25]:

1. Attacks (and defenses) of different generations are not comparable because fitness is based solely on the composition of the defense (attack) population at each generation. So no precisely identifiable champion emerges from running the algorithm.
2. From multiple runs, with one or more algorithms, it is unclear how to automatically select a "best" attack or design.

The system's decision support module, ESTABLO, addresses these challenges. At the implementation level, the engagements and results of every run of any of the system's coevolutionary algorithms are cached. Later, offline, ESTABLO filters these results and moves a subset this to its *compendium*. To prepare for the decision support analysis, it then competes all the attacks in the compendium against all defenses and ranks them according to multiple criteria. ESTABLO also provides visualizations and comparisons of adversarial behaviors. This information informs the decision process of a defensive manager.

27.4 System Use Case

In this section, we describe one use case of the system. It is motivated by two principal factors. 1. Distributed denial of service (DDOS) attacks have high, critical impact. Using compromised servers that are maliciously controlled, a DDOS attack simultaneously targets and consumes a network's resources at critical choke points. 2. DDOS attacks, over time, adapt. For example, see Figure 27.6 which shows a typical quarterly shift in targets. Figure 27.7 shows a time series of botnet size and how it locally escalates prior to an attack, peaks at the attack, and then declines after defenses have reacted. The overall decline in the botnet size is due to "intraspecies" competition among botnets; more varieties have arisen and are competing for servers on the Internet to compromise. These figures are published by a large content delivery network and cloud service provider that provides cloud security solutions [1].

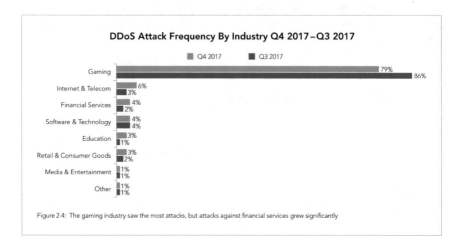

Fig. 27.6: Target types for DDOS, from [1].

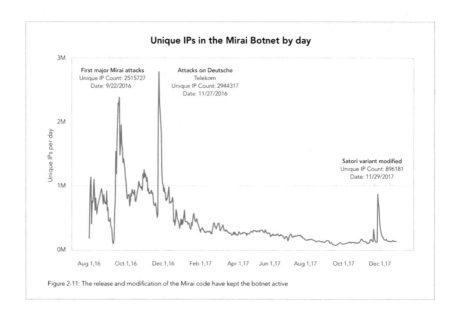

Fig. 27.7: Devices in the MIRAI botnet, from [1].

This use case assumes that defenses are mounted from a peer-to-peer (P2P) network because a P2P network, even without adversarial hardening, offers robust and resilient means of securing mission reliability. A P2P network's intrinsic defense is due mainly to its distributed properties that allow self-repair after nodes drop out and new nodes appear and that disperse data and compute with redundancy eliminating a single point of failure. We assume a structured P2P network overlaying a private network. The project, named RIVALS [8], assists in developing P2P network defense strategies against DDOS attacks. It models adversarial DDOS attack and defense dynamics to help identify robust network design and deployment configurations that support mission completion despite an ongoing attack.

RIVALS models DDOS attack strategies using a variety of behavioral languages ranging from simple to complex. A simple language, e.g. allows a strategy to select one or more network servers to disable for some duration. Defenders can choose one of three different network routing protocols: shortest path, flooding and a peer-to-peer ring overlay to try to maintain their performance. A more complex language allows a varying number of steps over which the attack is modulated in duration, strength and targets. It can even support an attack learning a parameterized condition that controls how it adapts during an attack, i.e. "online" based on feedback it collects from the network on its impact. Defenders have simple languages related to parameterizations of P2P networks that influence the degree to which resilience is traded off with service costs. A more complex language allows the P2P network to adapt during an attack based on local or global observations of network conditions. Attack completion and resource cost minimization serve as attacker objectives. Mission completion and resource cost minimization are the reciprocal defender objectives. See [8] for more details.

One scenario that we investigate involves the defense optimizing the placement of its mobile assets, given a node-level DDOS attack threat model. It does this by fielding feasible paths between the nodes that host the assets which support the tasks. A mobile asset is, for example, mobile personnel or a software application that can be served by any number of nodes. For example, a task can be the connection that allows personnel to use a software application.

An example of visualization with ESTABLO is shown in Figure 27.8. ESTABLO filtered high performing attacks and defenses that were cached from a large number of runs of different coevolutionary algorithms executed under the same Mobile Asset Placement scenario. After comparing the filtered attacks (and defenses) against each other to yield multiple scores on different metrics, in this visualization an attack is scored by score aggregation. All the attacks are then sorted on the aggregation and plotted in sorted order. Scores indicating on how the same attack also performed in the arms race from which it evolved and against a hand-derived set of defense are shown in different plot lines. We color the attacks by the algorithm variant that evolved them.

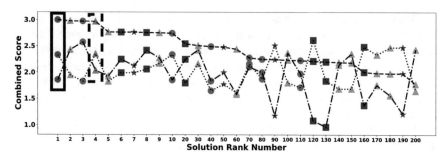

Fig. 27.8: Mobile Asset Placement scenario ESTABLO visualization. The x-axis shows a sorted
subsample of attackers (note, the top 10 are shown and then every tenth) and the y-axis
shows the ranking score. The ranking is done on the scores from the compendium and
shown in the top line. Scores for the same attack on the run on which it evolved and
versus unseen test sets are shown on separate lines. The algorithm used to evolve the
attacker is shown by the marker and the color. The attacker in the box with the solid
line is the top ranked solution from the Combined Score ranking schemes. The solution
in the dashed box is the top ranked solution from another ranking scheme called
Minimum Fitness.

27.5 Conclusion

We have described one modest way in which adversarial phenomena in biology ab-
stractly translate to cyber security. Given that there are typically imperfect matches
in analogy, analogies are nonetheless helpful when designing a computational adversarial system. For a system consisting of adversaries, engagements and a competitive
coevolutionary process, we presented a use case and showed how defensive solutions can be harvested from its arms races. Future work includes extending the system's use cases to support other network cyber security applications, improving the
sophistication of attack and defense behavior, and investigating "intra-species", i,e.
within population, competition and cooperation. The latter direction reflects the observation that different DDOS botnets have emerged to compete for Internet servers
to compromise.

Acknowledgements This chapter celebrates the outstanding academic career of Dr. Erik Goodman. We are inspired by his technical work and his important contributions that have helped grow
the field of genetic programming from its inception. This material is based upon work supported
by DARPA. The views and conclusions contained herein are those of the authors and should not
be interpreted as necessarily representing the official policies or endorsements. Either expressed or
implied of Applied Communication Services, or the US Government.

References

1. Akamai: Akamai Security Report (2017)
2. Arora, S., Ge, R., Liang, Y., Ma, T., Zhang, Y.: *Generalization and Equilibrium in Generative Adversarial Nets (GANs)*. In: International Conference on Machine Learning, pp. 224–232 (2017)
3. Beddard, F.E.: *Animal coloration: An account of the principal facts and theories relating to the colours and markings of animals*. London, S. Sonnenschein & Company (1892)
4. Bongard, J.C., Lipson, H.: *Nonlinear system identification using coevolution of models and tests*. IEEE Transactions on Evolutionary Computation **9**(4), 361–384 (2005)
5. Cott, H.B.: *Adaptive coloration in animals*. Methuen; London (1940)
6. Eat and let die: Insect feeds on toxic plants for protection from predators. www.sciencedaily.com/releases/2012/02/120221090240.htm (2012). Christian-Albrechts-Universitaet zu Kiel.
7. Ficici, S.G.: Solution concepts in coevolutionary algorithms. Ph.D. thesis, Brandeis University (2004)
8. Garcia, D., Lugo, A.E., Hemberg, E., O'Reilly, U.M.: *Investigating coevolutionary archive based genetic algorithms on cyber defense networks*. In: Proceedings of the Genetic and Evolutionary Computation Conference Companion, GECCO 17, pp. 1455–1462. ACM, New York, NY, USA (2017)
9. Goodfellow, I., Pouget-Abadie, J., Mirza, M., Xu, B., Warde-Farley, D., Ozair, S., Courville, A., Bengio, Y.: *Generative adversarial nets*. In: Advances in Neural Information Processing Systems, pp. 2672–2680 (2014)
10. Harborne, J.B.: *Biochemical aspects of plant and animal coevolution*. Academic Press (1978)
11. Hemberg, E., Zipkin, J.R., Skowyra, R.W., Wagner, N., O'Reilly, U.M.: *Adversarial co-evolution of attack and defense in a segmented computer network environment*. In: Proceedings of the Genetic and Evolutionary Computation Conference Companion, pp. 1648–1655. ACM (2018)
12. Koza, J.R.: *Genetic Programming: On the Programming of Computers by Means of Natural Selection*. MIT Press, Cambridge, MA (1992)
13. Lange, M., Kott, A., Ben-Asher, N., Mees, W., Baykal, N., Vidu, C.M., Merialdo, M., Malowidzki, M., Madahar, B.K.: *Recommendations for model driven paradigms for integrated approaches to cyber defense*. Tech. rep., US Army Research Laboratory Computational and Information Sciences (2017)
14. Miconi, T.: *Why coevolution doesn't "work": Superiority and progress in coevolution*. In: European Conference on Genetic Programming, pp. 49–60. Springer Berlin Heidelberg (2009)
15. Miller, B.P., Fredriksen, L., So, B.: *An empirical study of the reliability of unix utilities*. Communications of the ACM **33**(12), 32–44 (1990)
16. Morse, D.H., Fritz, R.S.: *Experimental and observational studies of patch choice at different scales by the crab spider Misumena vatia*. Ecology **63**(1), 172–182 (1982)
17. Moths: A different weapon against each enemy. www.sciencedaily.com/releases/2017/09/170927095639.htm (2017). University of Jyväskylä
18. Niklas, K.J.: *The evolutionary biology of plants*. University of Chicago Press (1997)
19. O'Neill, M., Ryan, C.: *Grammatical evolution: Evolutionary automatic programming in an arbitrary language, Genetic Programming Series*, vol. 4. Springer (2003)
20. Pertierra, M.: Investigating coevolutionary algorithms for expensive fitness evaluations in cybersecurity. Master's thesis, Massachusetts Institute of Technology (2018)
21. Popovici, E., Bucci, A., Wiegand, R.P., De Jong, E.D.: Coevolutionary principles. In: Handbook of Natural Computing, pp. 987–1033. Springer (2012)
22. Poulton, E.B.: *The colours of animals: Their meaning and use, especially considered in the case of insects*. D. Appleton (1890)
23. Prado Sanchez, D.: Visualizing adversaries - transparent pooling approaches for decision support in cybersecurity. Master's thesis, Massachusetts Institute of Technology (2018)

24. Rush, G., Tauritz, D.R., Kent, A.D.: *Coevolutionary agent-based network defense lightweight event system (candles)*. In: Proceedings of the Companion Publication of the 2015 on Genetic and Evolutionary Computation Conference, pp. 859–866. ACM (2015)

25. Sanchez, D.P., Pertierra, M.A., Hemberg, E., O'Reilly, U.M.: *Competitive coevolutionary algorithm decision support*. In: Proceedings of the Genetic and Evolutionary Computation Conference Companion, pp. 300–301. ACM (2018)

26. Sherratt, T.N.: *The coevolution of warning signals*. Proceedings of the Royal Society of London B: Biological Sciences **269**(1492), 741–746 (2002)

27. Stevens, M., Merilaita, S.: *Animal camouflage: Current issues and new perspectives*. Philosophical Transactions of the Royal Society of London B: Biological Sciences **364**(1516), 423–427 (2009)

28. Thompson, B., Morris-King, J., Cam, H.: *Controlling risk of data exfiltration in cyber networks due to stealthy propagating malware*. In: Military Communications Conference, MILCOM 2016-2016 IEEE, pp. 479–484. IEEE (2016)

29. Wikimedia-Commons: Misumena vatia with wasp (1998). Picture taken by Olaf Leillinger on 1998-04-30, License: CC-BY-SA-2.0/DE and GNU FDL

30. Wikimedia-Commons: Striped skunk (5710971209) (2005). This file is licensed under the Creative Commons Attribution 2.0 Generic license.

31. Wikimedia-Commons: Viceroy butterfly (2005). This file is licensed under the Creative Commons Attribution-Share Alike 3.0 Unported license. Subject to disclaimers.

32. Wikimedia-Commons: Fishook Barrel Ferocactus wislizeni (2007). By Susan Lynn Peterson Sue in az - Own work, CC BY 3.0

33. Wikimedia-Commons: Monarch in May (2007). Created: 29 May 2007 By Kenneth Dwain Harrelson, CC BY-SA 3.0

34. Wikimedia-Commons: An Indian chameleon wildlife in Andhra Pradesh India 2016 (2016). This file is licensed under the Creative Commons Attribution 2.0 Generic license.

35. Wilson, A.M., Lowe, J., Roskilly, K., Hudson, P.E., Golabek, K., McNutt, J.: *Locomotion dynamics of hunting in wild cheetahs*. Nature **498**(7453), 185 (2013)

36. Winterrose, M.L., Carter, K.M.: *Strategic evolution of adversaries against temporal platform diversity active cyber defenses*. In: Proceedings of the 2014 Symposium on Agent Directed Simulation, p. 9. Society for Computer Simulation International (2014)

Chapter 28
Small Implementation Differences Can Have Large Effects on Evolvability

"Even the largest avalanche is triggered by small things"
— Vernor Vinge

Jory Schossau and Arend Hintze

Abstract A key challenge in Evolutionary Computation is fitness landscape design. While the objective of the optimization process is often predefined, the actual fitness function, computational substrate, mutation scheme, and selection function must often be chosen by the experimenter. These choices are recognized to possibly have a large impact on experimental results. Here we investigate implementation differences of the encoding method and elaborate on its potential impact.

Key words: Evolutionary computation, genetic algorithm, markov brain, mutation scheme, logic gates

28.1 Introduction

In the field of Evolutionary Computation the principles of inheritance, variation, and selection are used to optimize a population of agents (sometimes called candidate solutions). Agents are evaluated according to their performance on some problem defined by the experimenter. After evaluation and selection, mutations are applied to those individuals who are allowed to transmit to the next generation. After populations have gone through many generations, the final population ideally contains some agents that perform well given the experimenter's original problem.

Jory Schossau
Department of Integrative Biology
Michigan State University
e-mail: jory@msu.edu

Arend Hintze
Department of Computer Science
Department of Integrative Biology
Michigan State University
e-mail: hintze@msu.edu

© Springer Nature Switzerland AG 2020   423
W. Banzhaf et al. (eds.), *Evolution in Action: Past, Present and Future*,
Genetic and Evolutionary Computation, https://doi.org/10.1007/978-3-030-39831-6_28

There are many proposed methods for performing evolutionary computation to discover one or more evolved agents that solve particular problems or classes of problems. Each method relies on representing the problem such that variations of potential solutions may be explored. Biologically-inspired methods rely on a genome as heritable material necessitating a translation step, whereas non-biologically-inspired methods use a modifiable algorithm directly representing a solution, thereby avoiding a translation step.

Older systems like Avida [1] use a straightforward encoding where each site in the genome represents a command of the Avida program, so the genome directly encodes the resulting program. Because of this encoding, mutations in Avida occur at the scale of program instructions. Other systems like NEAT [17] or Markov Brains [8] have a more complex encoding. These systems form complex networks of logic in the form of interconnected artificial neurons or logic gates. For example, these Markov Brain networks receive input, perform computations, and create output that in turn can be used to control agents in a virtual environment [10, 11, 3] or be used as outputs for classifiers [5, 12]. Each computational component (artificial neuron or logic gate) in these systems is encoded by multiple sites where one site specifies the connectivity, another the logic, and yet another the single weight or probability the logic may rely on, such as in an artificial neuron. Here we will be focusing exclusively on Markov Brains. For a detailed description, please see Hintze 2017 [8].

Previous work in the field of Evolutionary Game Theory (a form of evolutionary ecology) showed that differences in encoding can have a large effect in both evolutionary history and final population, assuming non-infinite bounds on population size and evolution time [2]. Therefore, it seems possible such a difference could be observable in more complex, indirect-encoded and behavior-driven methods, such as Markov Brains. To investigate difference requires several implementations of encodings to test.

Historically, the encoding used for Markov Brains has changed every occasion a new implementation was proposed, especially for deterministic logic gates [6, 11]. We will systematically investigate how these changes and other variations affect the evolvability of the Markov Brain system in two different fitness landscapes, and show that the implementation can have an effect on the evolutionary trajectory and outcome. Lastly, we applied a recently proposed opportunistic method called the *Buffet Method* as an automatic machine learning (AutoML) [18, 15] method to allow the use of a subset of encodings, and we show this performs decently on average while overcoming the problem that we are unable to predict how a chosen encoding will affect results [9].

28.2 Experiment

The original encoding of *deterministic logic* Markov Brain gates was Marstaller 2010 [10]. In this implementation, each logic gate receives one of a set of binary input patterns and deterministically produces one of a set of binary output patterns. The original encoding of *probabilistic logic* gates is accomplished using a probability matrix which contains a probability for each input-output pattern pairing (see Figure 28.1B). Each element in this table is encoded by a single site in the genome. The input pattern that a gate receives determines which row of the matrix will be used to produce the output pattern. However, the row of the matrix has been populated by arbitrary values from the genome, which in turn should be used as probabilities that define the likelihood for outputs to be generated. Therefore, the values in each row of the table are normalized (the row sums to 1.0), such that the row represents a probability mass function and the values represent individual probabilities of producing each complete output pattern.

The same code that is used to implement a probabilistic logic gate was later used to implement a second method of encoding for deterministic logic gates, except that the probability table needed to be adapted to reflect the deterministic nature of the new gate. In lieu of normalizing the row of values read from the genome, the maximum value in the row was identified, then the highest value was set to 1.0, and all other values in the same row were set to 0.0 (see Figure 28.1). Effectively this encoding turns a probabilistic system into a deterministic one. Multiple mutations were often necessary to change the values in the table sufficiently to affect the logic of the gate (data not shown). One could argue that the epistasis [19, 14, 13] between these mutations is high. This became the canonical, or *Traditional*, encoding for deterministic gates of Markov Brains.

The second time deterministic logic gates were implemented was 2013 in order to model the evolution of swarming in the presence of predation [11]. It was recognized that the earlier implementation might impose too much epistasis to allow for a smooth adaptation of logic gates. Instead of the logic table encoding previously described, each input was encoded by a single site and was used to encode the corresponding output. Specifically, the bit pattern encoded by each site was directly used to encode the bit pattern that each logic gate returned as an output (see Figure 28.2A). This leads to a new relation between mutations and their effects, such that now a single mutation determines the new function of a logic gate for a specific input. While this encoding still had epistasis, now each mutation was potentially much more powerful in its effects. We call this encoding the *Single Site* encoding.

The last change to this encoding was done when the software framework MABE was introduced [4]. In MABE the output of a deterministic logic gate for each input is now encoded in multiple sites. Each site determines the output of each of the output wires independently. This encoding has the least epistasis (similar to decomposable gates [16]) but also the mutational effects of each mutation is the smallest. This encoding is called the *Independent Bit* encoding, shown in Figure 28.2B.

Obviously there are endless other possible methods to encode the logic of deterministic logic gates. The question is one of sufficiency: Are there observable differences in evolvability with respect to the encoding of these three gates? This cannot be answered for all possible problem domains since different fitness landscapes imply different evolutionary trajectories and thus different optimal fitness landscape traversals. Therefore, we test the evolvability of populations of agents

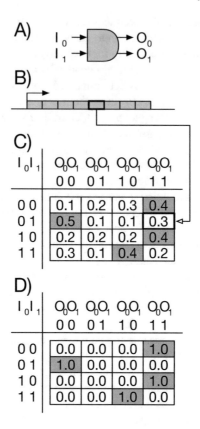

Fig. 28.1: Illustration of the *Traditional* encoding for deterministic logic gates. We assume a binary logic gate that receives two inputs I_0 and I_1, performs a computation, and returns the output of the computation to O_0 and O_1 (panel A). The logic and connection are encoded genetically. For that a sequence of number is read until a specific pair of numbers (42 followed by a 213) is found, which indicates the beginning of a gene, and with that the definition of a logic gate (panel B). Each site now defines a single probability, and these together define the probability matrix of the logic gate (panel C). This matrix is now translated into a deterministic matrix, by finding the highest value in each row (gray) and turning that to 1.0 while all other values become 0.0 (panel D).

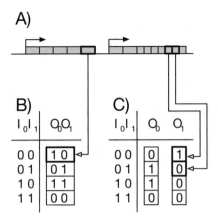

Fig. 28.2: Illustration of the *Single Site* (B) and the *Independent Bit* (C) encoding. Each site of each gene (A) can be interpreted to either define the entire output pattern (B) or each bit individually (C). In the first case, a mutation can potentially change all outputs of a gate at the same time (B), while in the case of the independent bit encoding multiple mutations are required, one for each output individually.

utilizing deterministic gates with one of the three encodings presented above on the two following tasks.

28.2.1 Foraging Task

The Foraging task is as described in [4], originally intended for the study of speciation and foraging behavior. Used here, the environment is a 2-dimensional grid wherein a single agent may move by discrete grid cell increments. The grid is regularly shaped (6×6) in size, and is surrounded by an impenetrable wall. The grid may be populated with two types of collectible resources representing food. The agent is evaluated for 400 time steps. The agent can either turn 45 deg left or right, move forward, or consume the immediate resource sharing its location. If a resource is consumed and the agent later moves away, then the vacant spot is replenished with a new random resource from the available types. The agent can only perceive resources at its immediate location, allowing for only three sensations: resource type 1, resource type 2, or empty. Agents explore their environments separately, and will not interact with each other directly.

In this task, resource acquisition is used to determine reproductive fitness. Fitness is calculated by tallying 1 point of fitness for each resource collected. However, if a type of resource consumed differs from the previous one, then a task-switching cost (1.4 points) is subtracted in addition to the 1 point fitness gain resulting in a net loss of 0.4 points for that transition in resource type. This penalty discourages random foraging and encourages strategies that minimize switching over time. Agents receive five binary inputs every update. The first two encode the state of the current grid location ($1, 0$ = resource one, $0, 1$ = resource two, $0, 0$ = empty). The next three inputs encode the state of the grid location in front of the agent in the same way, including the "wall" state. the agent provides 2 binary outputs ($0, 0$ = no action, $0, 1$ = turn right, $1, 0$ = turn left, $1, 1$ = move forward).

28.2.2 Associative Memory Task

The associative learning task uses the number of unique locations visited along a predefined path to determine reproductive success (fitness). Similar to the Foraging task, this task defines a 2-dimensional grid wherein a single agent may move about one cell at a time. On every update the agent is given three inputs: one to encode if the agent is on a straight section of the predefined path, one input to encode if the path is progresses to the Left at the current location, and one input if the path progresses to the Right at the current location. The fitness tally is increased for each unique location visited along the path, and decreased for every time step the agent is not on the path. In this way agents that spend time on non-path regions are treated as less fit than those that spend time exploring the path. Between each fitness evaluation the set of inputs provided for when the path "turns left" and "turns right" are randomized. No special information is provided indicating the end of the path. At the beginning an agent's lifetime the agent should first discover the mapping between signs and turn directions. This behavior, if performed, is similar to exploration. After this exploration phase, the agent then may utilize that information to properly follow the path, which appears like an exploitation phase [7] (visual not shown). The original work for this task was performed in AVIDA [1] and this extension of the task to associative learning was proposed by Anselmo Pontes in yet unpublished dissertation work.

28.3 Results and Discussion

We performed 400 independent evolutionary experiments ("biological" replicates in the classical science sense) for all three types of gate encodings in both fitness landscapes of the Foraging task and Associative Memory task. Populations were allowed to evolve for $30,000$ generations, with genomes initialized with 18 gates for the individual encoding conditions, and 6 of each gate encoding for the

All condition. We used a circular genomic encoding, per site mutation probability of 0.005, per site copy insertion probability of 0.00002, and per site deletion probability of 0.0002, which are all standard default MABE framework parameters. The rest of the settings are easily readable in the configuration files stored with the experimental data in the repository for this publication `https://gitlab.msu.edu/jory/mbencoding`.

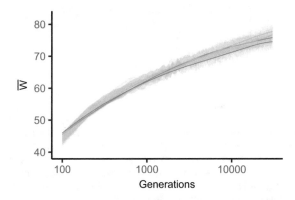

Fig. 28.3: Evolution of populations with one or all gate encoding methods available for the Foraging task. Blue line represents the *Single Site*, green the *Independent Bit* method, orange *All* three methods, and purple the *Traditional* method. Vertical axis is raw fitness evaluated for the task. Horizontal axis is time in generations. Error bars are 95% confidence intervals of the mean. Lines are averages of lines of descent from 400 replicates. The differentiation of conditions toward the end of evolution shows that differences in encoding can make a difference in results, especially for comparisons.

We find that in the Foraging task the *All* encoded gate populations tend to evolve higher final fitness than the populations with *Traditional* encoded gates (Figure 28.4). The *Independent Bit* encoded gate populations achieved an average amount of final fitness relative to the *Single Site* and *All* encoded gate populations, however the *Traditional* encoded gate populations performed worse than the *Independent Bit* encoded gate populations. Evolutionary trajectories during the earlier stage of evolutionary share very similar fitness gains and only begin to differentiate as fitness gains slow down: standard deviation of a cross-section of all data in early evolution (generation 5,000) reveals less diversity than in the late observations (27,000 $\leq$ generation $\leq$ 29,000) with standard deviations of 20 and 23 respectively (Figure 28.3).

In the Associative Memory task evolutionary trajectory variation is reversed: standard deviation of a cross-section of all data in early evolution (generation 5,000) reveals more diversity than in the late observations (27,000 $\leq$ generation $\leq$ 29,000) with standard deviations of 365 and 140 respectively (Figure 28.5). Another reversal is the relative ordering of average fitness between conditions. *Single Site* encoded gate populations performed worse or equivalent to the other conditions (Fig-

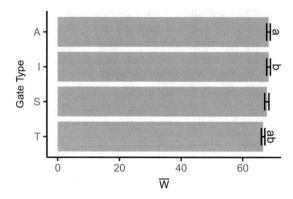

Fig. 28.4: Final raw fitness for populations evolved with one or all gate encoding methods available for the Foraging task. 'A' represents *All* encoding methods, 'I' the *Independent Bit* method, 'S' the *Single Site* method, and 'T' the *Traditional* method. Values are averages of lines of descent from 400 replicates near the end of evolution ($27,000 \leq$ generations $\leq 29,000$) of 30,000 generations. Error bars are 95% confidence intervals of the mean. Values with matching symbols are statistically significantly different determined through Mann-Whitney U comparison with Bonferroni correction, significance level of 0.05.

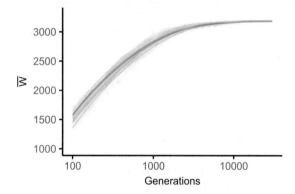

Fig. 28.5: Evolution of populations with one or all gate encoding methods available for the Associative Memory task. Blue line represents the *Single Site*, green the *Independent Bit* method, orange *All* three methods, and purple the *Traditional* method. Vertical axis is raw fitness evaluated for the task. Horizontal axis is time in generations. Error bars are 95% confidence intervals of the mean. Lines are averages of lines of descent from 400 replicates. The differentiation of conditions toward the end of evolution shows that differences in encoding can make a difference in results, especially for comparisons.

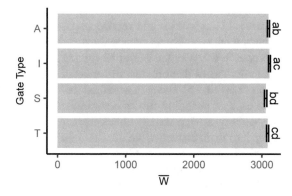

Fig. 28.6: Final raw fitness for populations evolved with one or all gate encoding methods available for the Associative Memory task. 'A' represents *All* encoding methods, 'I' the *Independent Bit* method, 'S' the *Single Site* method, and 'T' the *Traditional* method. Values are. Values are averages of lines of descent from 400 replicates near the end of evolution ($27,000 \leq$ generations $\leq 29,000$) of $30,000$ generations. Error bars are 95% confidence intervals of the mean. Values with matching symbols are statistically significantly different determined through Mann-Whitney U comparison with Bonferroni correction, significance level of 0.05.

ure 28.6). Note that this reversal and reordering is particular to the comparison of these two tasks and the selected encoding methods.

This shows that indeed the different effects of the encodings depend on the fitness landscape. The obvious question is if it is possible to predict evolutionary behavior from the encoding? If it is not easy to predict the evolutionary differences from the encodings, then it becomes interesting to find a method by which no *a priori* selection of encoding must be made to yield decent results. Imagine a new fitness landscape or task for the Markov Brain. Instead of a cumbersome investigation into which of the encodings is optimal for the new environment, it would be great to have some indicator for which encoding to choose, or one encoding that works optimally across all problems. At the same time, this might be impossible. Given the idea of the "no free lunch theorem," which "state[s] that any two optimization algorithms are equivalent when their performance is averaged across all possible problems" [20]. We conjecture that the same should be true for the implications of the various types of encodings: The evolvability of systems with different encodings might be the same when averaged across all possible fitness landscapes. This implies that while specific encodings accelerate adaptation in specific environments, we would not be able to find an optimal one for all environments. Additionally, there may be more than one quantifiable property of the fitness landscape (for example, the degree of epistasis) that we can measure in order to predict the optimal method for encoding. This creates an interesting puzzle: we know that some encodings will perform better than others, but we do not know in what situations.

We previously proposed the Buffet Method that can remedy this enigma. The idea of the buffet method is to provide an evolving population with many options, of which at least one of them is highly relevant to the fitness function, or exploration of the fitness landscape. In this way the Buffet Method is similar to Auto Machine Learning (AutoML) approaches that provide a meta-level optimization of a lower-level optimization process [18, 15]. Using this method we not only evolve a Markov Brain that uses one type of encoding for its deterministic logic gates, but allow for all three at the same time. Specifically, this means that during the translation phase of the genome into logic elements, we use three different identifiers, or start codons, indicating which gate encoding method to use for that logic element. This way evolution-through-selection might explore which encoding methods yield faster adaptation. We already showed that this approach works for different types of gates (neuronal, logic, genetic programming type) but not for different encodings of the same gate.

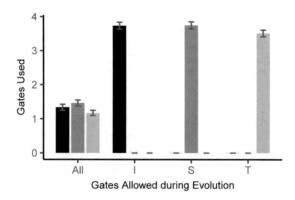

Fig. 28.7: Average prevalence of evolved encoding method usage by experimental condition. 'A' represents *All* encoding methods, 'I' the *Independent Bit* method, 'S' the *Single Site* method, and 'T' the *Traditional* method. Black represents the *Independent Bit* method, medium gray the *Single Site* method, and light gray the *Traditional* method. Values represent average prevalence of encoding methods employed by agents toward the end of the line of descent, averaged over 400 replicates along the line of descent and only timepoints near the end of the line of descent ($27,000 \leq$ generations $\leq 29,000$) of $30,000$ generations. Error bars are the 95% confidence interval of the mean. Note that the non-*All* experimental conditions only allowed one type of encoding to evolve, so their distributions are singular. The *All* encoding method shown here results in agents that have evolved to use all three types of encodings, possibly proportional to their prevalence in the single encoding conditions.

In order to test if the Buffet Method allows faster adaptation when applied to different types of encodings, we ran 400 replicate experiments in both tasks while allowing the evolving populations to make use of any of the three types of gate encodings. This experimental condition we label as the *All* method. We find that in the Foraging task the *All* method produced average population fitness by the end of evolution, relative to the other methods, but notably is not the worst. In the Associative Memory task the *All* method produced similar final fitness to the other methods, but showed faster adaptation than the worst method (*Single Site*) early in evolution. There may be historical contingency due to early changes in gate encoding composition that might limit or enhance evolutionary potential. Indeed, encoding method distribution in the *All* method is proportional to prevalence among single-method populations, possibly reflecting their evolutionary utility.

Our results suggest that the method of encoding in an evolutionary computation can produce measurable differences in experimental outcomes. One consequence of this conclusion is that experimenters need to take care which encoding method they use when they choose one, if they choose only one. Additionally, the Buffet Method can be used when *a priori* knowledge of encoding method and fitness landscape pairing is insufficient to make a confident decision. Lastly, this conclusion hints that the job of benchmarking various evolutionary computation methods for comparison is made even more difficult because encoding method is yet another dimension of freedom that either should be acknowledged when making comparisons, or a normalization discovered. One interesting extension to these results may include an investigation of the fidelity in how the *All* method distribution of methods reflects evolutionary prevalence of each of the individual methods in solo-method populations. This distribution might allow further understanding and optimization for particular tasks. A similar insight from a different perspective was made in [9].

28.4 Conclusion

Historical use of Markov Brains produced several implementations. Here we showed that under certain circumstances, these encodings can have significant but weak effects, and that different encodings are optimal in different contexts, shown here using only two fitness landscapes. Exploring more encodings and fitness landscapes will likely help to build a more detailed understanding of the relation between encoding and fitness landscape, and might even yield general insights into what kind of problem domains benefit from certain types of encodings and their associated mutational operators.

Acknowledgements The changes in encodings we tested here, occurred during the first 10 years of the BEACON Center for the Study of Evolution in Action. As we showed here, sometimes it is not only the big model decisions that matter, but the small ones too. While we all recognize the importance of leadership that Erik Goodman provided as head of BEACON, we also want to thank him for all the small details that we think made a difference: the metaphorical pat on the back, the excitement Erik shared, and the empathy Erik Goodman showed when we were ourselves in difficult situations. Thank you Erik, you made a small and a big differences!

References

1. Adami, C., Brown, C.T.: *Evolutionary learning in the 2D artificial life system Avida.* In: Artificial Life IV, vol. 1194, pp. 377–381. Cambridge, MA: MIT Press (1994)
2. Adami, C., Schossau, J., Hintze, A.: *Evolutionary game theory using agent-based methods.* Physics of Life Reviews **19**, 1–26 (2016)
3. Albantakis, L., Hintze, A., Koch, C., Adami, C., Tononi, G.: *Evolution of integrated causal structures in animats exposed to environments of increasing complexity.* PLoS Computational Biology **10**(12), e1003,966 (2014)
4. Bohm, C., CG, N., Hintze, A.: *MABE (Modular Agent Based Evolver): A Framework for Digital Evolution Research.* Proceedings of the European Conference of Artificial Life, 14 pp. 76–93 (2017)
5. Chapman, S., Knoester, D.B., Hintze, A., Adami, C., et al.: *Evolution of an artificial visual cortex for image recognition.* In: Proc European Conference on Artificial Life (ECAL-2013), pp. 1067–1074. MIT Press, Cambridge, MA (2013)
6. Edlund, J.A., Chaumont, N., Hintze, A., Koch, C., Tononi, G., Adami, C.: *Integrated information increases with fitness in the evolution of animats.* PLoS Comput Biol **7**(10), e1002,236 (2011)
7. Grabowski, L.M., Bryson, D.M., Dyer, F.C., Ofria, C., Pennock, R.T.: *Early evolution of memory usage in digital organisms.* In: Proceedings ALIFE-2010, pp. 224–231. MIT Press, Cambridge, MA (2010)
8. Hintze, A., Edlund, J.A., Olson, R.S., Knoester, D.B., Schossau, J., Albantakis, L., Tehrani-Saleh, A., Kvam, P., Sheneman, L., Goldsby, H., et al.: *Markov Brains: A technical introduction.* arXiv preprint arXiv:1709.05601 (2017)
9. Hintze, A., Schossau, J., Bohm, C.: The evolutionary buffet method. In: W. Banzhaf, L. Spector, L. Sheneman (eds.) Genetic Programming Theory and Practice XVI, Genetic and Evolutionary Computation, pp. 17–36. Ann Arbor, USA (2019)
10. Marstaller, L., Hintze, A., Adami, C.: *The evolution of representation in simple cognitive networks.* Neural computation **25**(8), 2079–2107 (2013)
11. Olson, R.S., Hintze, A., Dyer, F.C., Knoester, D.B., Adami, C.: *Predator confusion is sufficient to evolve swarming behaviour.* Journal of the Royal Society Interface **10**(85), 20130,305 (2013)
12. Ortiz, A., Bradler, K., Hintze, A.: *Episode forecasting in bipolar disorder: Is energy better than mood?* Bipolar Disorders (2018)
13. Østman, B., Hintze, A., Adami, C.: *Impact of epistasis and pleiotropy on evolutionary adaptation.* Proceedings of the Royal Society of London B: Biological Sciences **279**(1727), 247–256 (2011)
14. Phillips, P.C.: *Epistasis: The essential role of gene interactions in the structure and evolution of genetic systems.* Nature Reviews Genetics **9**(11), 855 (2008)
15. Real, E., Moore, S., Selle, A., Saxena, S., Suematsu, Y.L., Tan, J., Le, Q., Kurakin, A.: *Large-scale evolution of image classifiers.* In: Proceedings of the 34th International Conference on Machine Learning-Volume 70, pp. 2902–2911 (2017)
16. Schossau, J., Albantakis, L., Hintze, A.: *The role of conditional independence in the evolution of intelligent systems.* In: Proceedings of the Genetic and Evolutionary Computation Conference Companion, pp. 139–140. ACM (2017)
17. Stanley, K.O., Miikkulainen, R.: *Evolving neural networks through augmenting topologies.* Evolutionary Computation **10**(2), 99–127 (2002)
18. Thornton, C., Hutter, F., Hoos, H.H., Leyton-Brown, K.: *Auto-WEKA: Combined Selection and Hyperparameter Optimization of Classification Algorithms.* In: Proceedings of the 19th ACM SIGKDD International Conference on Knowledge Discovery and Data Mining, KDD '13, pp. 847–855. ACM, New York, NY, USA (2013)
19. Whitlock, M.C., Phillips, P.C., Moore, F., Tonsor, S.J.: *Multiple fitness peaks and epistasis.* Annual Review of Ecology and Systematics **26**(1), 601–629 (1995)
20. Wolpert, D.H., Macready, W.G.: *Coevolutionary free lunches.* IEEE Transactions on Evolutionary Computation **9**(6), 721–735 (2005)

Chapter 29
Surrogate Model-Driven Evolutionary Algorithms: Theory and Applications

Subhrajit Dutta and Amir H. Gandomi

Abstract Engineering optimization problems are challenging to solve mainly due to their numerical modeling and analysis complexities. This chapter deals with the efficient use of surrogate model-driven evolutionary algorithms, built hierarchically for the solution of large-scale computation intensive optimization problems. In most optimization problems, the majority of computation is involved in repetitive function calls to evaluate the system response/bahaviour under consideration. The quality solutions depends on the system response estimation, and in most cases high fidelity models are used to get accurate results. Conventional evolutionary algorithms require a great number of such high fidelity function calls. Here, we use low cost surrogate models or metamodels, which approximate the original model mathematically, but significantly reduce the computation cost for a desired accuracy level. The surrogate model training requires a small amount of evaluations of the original model at support points. The hierarchical surrogate model-based PSO algorithms we propose are tested on a range of large-scale design optimization problems and compared with other well-known surrogate modeling techniques.

Key words: Surrogate models, polynomial chaos expansion, kriging, evolutionary algorithm, particle swarm

Subhrajit Dutta
Department of Civil Engineering, National Institute of Technology Silchar, Cachar, Assam 788010, INDIA e-mail: subhrajit.nits@gmail.com

Amir H. Gandomi
Faculty of Engineering & Information Technology, University of Technology Sydney, Ultimo, NSW 2007, AUSTRALIA e-mail: gandomi@uts.edu.au

© Springer Nature Switzerland AG 2020
W. Banzhaf et al. (eds.), *Evolution in Action: Past, Present and Future*,
Genetic and Evolutionary Computation, https://doi.org/10.1007/978-3-030-39831-6_29

29.1 Introduction

Evolutionary algorithms (EAs) comprise a set of nature-inspired algorithms, inspired by Charles Darwin's classical evolution theory and have been successfully applied as solution schemes for complex optimization problems. In many science and engineering problems, high-fidelity complex analysis is needed in order to simulate the real behaviour of the systems. Moreover, the nature of real problems demands the effect of uncertainty – aleatory and epistemic – to be considered in this analysis for more realistic responses and predictions. With the advancement of technology and computational resources, the optimization community has given due consideration to complex high-fidelity models to obtain more realistic practical optimization solutions. For example, in the aerospace and automobile industries, efficient computer simulations are performed to avoid expensive physical experiments, with the aim of improving the quality and performance of engineered products and devices, but using a limited computation effort. Recent developments in EAs make use of surrogate models to solve computation-intensive optimization problems [11, 13, 36].

Surrogate models are computationally inexpensive mathematical models that mimic the original high-fidelity model in the domain of interest. Such a pseudo-model is constructed using the model responses at certain special points and adopting a learning algorithm to obtain its parameters/coefficients. In general, a surrogate model is created to approximate the responses over a range of input (design) variables. As the dimensionality of input variable(s) increases, adaptation in the surrogate model is required to match the solution accuracy. Thus, there is a trade-off between the surrogate cost and its desired accuracy level. Surrogate models are extensively used for solutions involving optimal designs [6, 25], and reliability-based optimization [22, 30], but little emphasis has been laid on exploring their behavior when coupled to EAs. There is a definite need for comparison of various surrogate models coupled with EAs in solving real optimization problems. This will provide sufficient insight on the surrogate selection process. Hence, it is of paramount importance to do a comprehensive comparison, considering essential aspects such as, robustness, accuracy, simplicity, computation cost etc.

On the other hand, the selection of a proper optimization algorithm plays a vital role for solution accuracy, particularly when complicated objective/constraint functions with multiple local minima are present. Also, the optimizer should able to manage the intensive simulation with a limited amount of available computational resources. Recent developments in heuristic/evolutionary optimization algorithms have provided researchers with a choice to formulate computationally attractive algorithms to solve complex design optimization problems. Some of the well-known EAs include genetic algorithms (GAs) [21], simulated annealing (SA) [28], ant colony optimization (ACO) [8], artificial bee colony (ABC) algorithm [26], bat algorithm (BA) [39], cuckoo search (CS) [17], particle swarm optimization (PSO) [27], harmony search (HS) [29], firefly algorithm (FA) [38], and krill herd (KH) algorithm [16] among others. Practical implementation of optimization problems using EAs, are documented in references [2, 10, 11, 32].

In this chapter, some well-known surrogate models are reviewed in the context of optimization problems. The performance of these surrogate models coupled with an EA – particle swarm optimizer – in solving structural optimization under uncertainty problems of high dimensionality and complexity will be illustrated. This chapter is organized as follows. In Section 29.2, surrogate models – polynomial chaos expansion (PCE) and kriging (KRG) are introduced. In Section 29.3, the hierarchical surrogate-driven PSO algorithm is discussed. In Section 29.4, a comparative study of the PCE and KRG-based PSO is performed on two practical optimization problems with high computational complexity.

29.2 Surrogate Models

29.2.1 Polynomial Chaos Expansion

Let us consider a physical system which can be idealized by a computational model $\mathcal{M}$. Consider a vector of random variables ξ with support $\mathcal{D}_\xi$ and described by the marginal/joint probability density function (PDF) f_ξ. Considering a finite variance of the physical model $Y = \mathcal{M}(\xi)$, such that:

$$\mathbb{E}[Y^2] = \int_{\mathcal{D}_\xi} \mathcal{M}^2(\xi) f_\xi \, d\xi \tag{29.1}$$

Then the polynomial chaos expansion (PCE) of $\mathcal{M}(\xi)$ is defined as [20]

$$Y = \mathcal{M}(\xi) = \sum_\alpha y_\alpha \Psi_\alpha(\xi) \tag{29.2}$$

where Ψ_α are the multivariate orthonormal polynomials, α are the multi-indices that map the multivariate Ψ_α to their corresponding base coordinates, which are denoted as the deterministic PCE coefficients y_α. In PCE, the model response is characterized by random basis function and deterministic coefficients as given by the expansion in Eq. 29.2. A square-integrable random variable, random vector, or random process can be written in a mean-square convergent series using random orthonormal polynomial bases known as PCE for Hermite bases, and generalized PCE for other bases [20]. For Hermite bases, the random variables should follow normal/Gaussian distributions. Whereas non-normal random variables must be transformed to standard normal random space using isoprobabilistic transformations. For a practical implementation, the expansion of Y must be truncated by retaining only significant terms. Extending to M random variables, the complete N-order PCE is defined as the set of all multidimensional Hermite polynomials Ψ whose degree does not exceed N [35]:

$$Y(\xi_1, \xi_2 \ldots \xi_M) = \sum_{\alpha=0}^{P-1} y_\alpha \Psi_\alpha(\xi) \qquad (29.3)$$

where P is the number of terms with degree $\leq N$, such that:

$$P = \sum_{k=0}^{N} C_k^{M+k+1} = \frac{(M+N)!}{M!N!} \qquad (29.4)$$

The key point in constructing surrogates by PCE lies in the determination of its deterministic coefficients. In general, the PCE coefficient y_α are computed using two approaches: intrusive approaches (e.g., Galerkin scheme) or non-intrusive approaches (e.g., projection, least-square regression). Here, a non-intrusive regression based approach is discussed. The term non-intrusive indicates that the chaos coefficients are evaluated over a set of input realizations $\xi = \{\xi_1, \xi_2 \ldots \xi_M\}$, referred to as the design of experiments (DoE). Considering $Y = \{\mathcal{M}(\xi_1), \mathcal{M}(\xi_2) \ldots \mathcal{M}(\xi_M)\}$ as the vector of outputs of the computational model for each sample point, the set of coefficients y_α is then calculated by minimizing the least-square residual of the surrogate (PCE) approximation of the "true" computational model:

$$\hat{y} = \underset{y_\alpha}{\mathrm{argmin}} \frac{1}{M} \sum_{i=1}^{M} \left(\left(\mathcal{M}(\xi_i) - \sum_{\alpha=0}^{P-1} y_\alpha \Psi_\alpha(\xi_i) \right) \right)^2 \qquad (29.5)$$

29.2.2 Kriging

Kriging (KRG), also known as Gaussian process modeling, is a popular metamodeling technique to mathematically approximate the numerically simulated data which are deterministic/probabilistic in nature. In optimization problems, kriging can act as an efficient surrogate by substituting the original model in the simulation, and also as an interpolator, to obtain the value of model response for any intermediate design point within the 'design variable' domain of interest. The construction of a kriging model is a two-stage process. First, a regression (or, trend) function R is constructed based on the simulated input-output data vector(s). Then, a Gaussian process (GP) Z is built for the residuals with zero mean and autocovariance, $\sigma^2 R(\mathbf{x}, \mathbf{x}')$. Here, σ^2 is the variance of the GP and $R(\mathbf{x}, \mathbf{x}')$ is the autocorrelation function of the input vector. A kriging model of response (output) Y, which depends on a vector of input variables $\mathbf{x}$, is formulated as

$$Y(\mathbf{x}) = R(\mathbf{x}) + Z(\mathbf{x}) \qquad (29.6)$$

The creation of a kriging surrogate focuses primarily on obtaining the parameters (mean and covariance) of the Gaussian process Z. A maximum likelihood estimator (MLE) is used for obtaining these parameters [15], in which Z is modeled with a zero mean, a finite variance, and a correlation matrix $\mathbb{C}$. Depending on the form of regression, there are variants of kriging surrogates, of which the most widely used are the 'simple', 'ordinary' and 'universal' types. *Simple kriging* assumes the regression function to be a known constant, while in *ordinary kriging* the regression function is an unknown constant. *Universal kriging* (UK) treats the regression function to be a generalized polynomial expression. In UK, the trend function is represented as a linear combination of simpler regression functions

$$Y(\mathbf{x}) = \sum_i \beta_i f_i(\mathbf{x}) + Z(\mathbf{x}) \tag{29.7}$$

where β_i is the weighted coefficient vector and f is a family of regression functions, such as a mixed polynomial of multiple order. The main aim of the construction of the KRG model is to estimate the hyperparameters, β_i and σ^2. Values of these hyperparameters can be estimated by maximizing the likelihood function, which is defined considering the noise, $Z(\mathbf{x}) = Y(\mathbf{x}) - \sum \beta f(\mathbf{x})$ as a correlated Gaussian vector. Further details on the mathematical formulation of kriging surrogates can be found in books, such as [15] and [33].

Selection of the trend function is difficult and depends on the problem at hand. However, for optimization problems, the use of 'ordinary kriging' is recommended by [14]. Kriging surrogates have been used efficiently as a surrogate model in various optimization problems recently: for optimal design of aeroengine turbine disc [24], optimal designs of a short column, trusses, bracket structures, shells and wind turbines under uncertainty [9, 22, 23].

29.2.3 Surrogate Model Accuracy

Once the surrogate model is built, it is important to verify its accuracy in comparison to the "true" computational model. Several statistical error estimators exist, out of which an effective measure is given by the mean square error (MSE)

$$\text{MSE} = \mathbb{E}\left[\mathcal{M}(\mathbf{X}) - \hat{\mathcal{M}}(\mathbf{X})\right]^2 \tag{29.8}$$

where the validation set $\mathbf{X} = \{x_1, x_2 \ldots x_{\text{val}}\}$ consists of samples of the input variables that do not belong to the experimental design (ED) used to construct the surrogate model, $\mathcal{M}$. A lower value of MSE corresponds to a more accurate surrogate. A variant estimator of MSE is the relative mean square error (RMSE)

$$\text{RMSE} = \frac{\text{MSE}}{\sigma_Y^2} \tag{29.9}$$

with σ_y^2 being the variance of the computational model response.

In practical problems, a sufficiently large validation set (typically in the range of 10^4 to 10^5) is required to obtain a stable estimate of the MSE. Mostly MSE/RMSE are computationally demanding in nature for high-fidelity numerical models since it requires computing model responses for a large sample set. To overcome this issue, an error estimator is developed based on the ED used for surrogate model construction. In statistical learning, the leave-one-out error (Err_{LOO}), or the relative leave-one-out error (ε_{LOO}) estimate is computed, which performs very well in terms of the estimation bias and the mean-square error [1]. Statistically, ε_{LOO} gives a measure of the coefficient of determination ($R^2 \approx 1 - \varepsilon_{LOO}$), while Err_{LOO} is related to the well-known error estimator called predicted residual sum of squares (PRESS).

29.3 Surrogate-Model Driven Optimization Algorithms

The solution scheme of a surrogate-driven optimization problem has two major parts: the optimization part, and the computation of optimization objective and constraints. In general, a double loop approach is adopted for solving this optimization problem, wherein a surrogate model is used in the inner loop to compute the optimization parameters of interest, and an optimizer in the outer loop to obtain the global optimum design variable values of interest. The level of complexity in large-scale optimization problems [32] demands innovative approaches to solve such problems. To this end, researchers started adopting ideas from nature and translated them to solve problems in multi/interdisciplinary problems. This led to the application of heuristic or evolutionary algorithms (EAs) to solve challenging optimization problems, particularly given uncertainties in design. Optimization problems, that are traditionally solved by the classical gradient-based approaches, are constantly encroached by EAs due to their (i) robustness and flexibility, (ii) reasonable computational effort, albeit not compromising the solution accuracy, (iii) intuitive mathematical formulation, and (iv) ability to handle stochastic and dynamic information [18, 19]. Also, such techniques provide solutions in a heuristic way as compared to gradient-based approaches, which suffer from drawbacks such as curse of dimensionality and accounts for continuity and differentiability of the objective/constraint functions. Development of EAs, such as genetic algorithm (GA), evolutionary strategies (ES), particle swarm optimization (PSO) expands the scope to develop computationally efficient simulation schemes to tackle large-scale optimization problems. In this chapter, we will focus on a widely used population-based EA – particle swarm optimization.

Particle swarm optimization (PSO) was first introduced by [27] inspired by the concept of the social behaviour of animals, such as bird flocking and fish schooling. PSO gained popularity due to its robust stochastic search algorithm, and today, researchers across the globe work extensively with PSO to tackle optimization problems encountered in science and engineering. PSO uses the "flying" movement of "particles" in a swarm and has a mechanism to search for global and local optimum positions [34]. Each particle in the swarm is considered as a point in a D-dimensional design variable vector space. The kth particle is characterized by its current (mth iteration) position vector, $\mathbf{x}_{k,m}$ and velocity vector, $\mathbf{v}_{k,m}$. For each iteration of the algorithm, the current particle's position is considered as an optimal solution. The objective function (also known as the 'fitness function') is then computed based on the current position and the *personal best* fitness value is stored in a variable, *pbest*. If the particle's position corresponding to *pbest* is "better" than that in the previous iterations, then that position of particles is stored as $\mathbf{p}_{k,m}$. Another best solution is obtained based on the global fitness value (*gbest*) when considering the entire particle population. The global best position of the particle related to *gbest* is stored as $\mathbf{p}_{g,m}$. The primary optimal search objective is to explore a better position and fitness value by updating $\mathbf{x}_{k,m}$, *pbest* and *gbest*. The mechanism of updating the position for each particle, from iteration m to $m+1$, so as to obtain the global best fitness value is formulated as [27]

$$v_{k,m+1} = wv_{k,m} + c_1 r_1 (p_{k,m} - x_{k,m}) + c_2 r_2 (p_{g,m} - x_{k,m}) \tag{29.10}$$

$$x_{k,m+1} = x_{k,m} + v_{k,m+1} \tag{29.11}$$

where r_1 and r_2 are uniform random numbers between 0 and 1, c_1 and c_2 are positive constants that randomly pull/push each particle towards the optimum region. 'w' is the inertia weight that controls the global and local search ability of the swarm by balancing between exploration and exploitation [12]. The parameters used in the algorithm are discussed in detail in the literature along with their selection criteria for better convergence [34, 37]. A particle population size used in PSO typically varies in the range of 20 to 50 for most practical problems [12], however, the exact population size depends on the complexity of the problem at hand. In general, the initially generated swarm in the PSO algorithm is not stable and it is of practical importance to define the stability/convergence of the particles over PSO iterations. [37] et al. states that the time behaviour of particles in PSO depends on the eigenvalues of the dynamic matrix, $\mathbb{A}$ with its element being a function of the parameters: 'w', 'c_1' and 'c_2'. The necessary and sufficient condition for convergence of the algorithm is that both the eigenvalues (real or complex) of matrix $\mathbb{A}$ must have an absolute value less than 1 [37]. In the present work, this eigenvalue criterion is adopted for convergence.

The proposed surrogate model-driven PSO algorithm for optimization is performed following these steps, also illustrated in the flowchart in Figure 29.1:

1. A total of K_{part} PSO particles are randomly generated. The kth particle is characterized by the design variables.
2. During the first PSO iteration, for the kth design variable (particle) the response parameters required for the objective function and the constraints are obtained using the computationally lighter surrogate model. It must be noted that a surrogate model can be constructed separately as discussed in Section 29.2. The deterministic/probabilistic parameters are estimated by simulation using a sampling technique in the surrogate modeling framework.
3. The kth particle is accepted if the constraints are satisfied, and it is rejected otherwise.

4. Successive PSO iterations with K_{part} particles are performed until the objective
   function is minimized/maximized and the optimum design variables are obtained. As mentioned before, the convergence is based on the eigenvalue-based
   criterion used by [37].

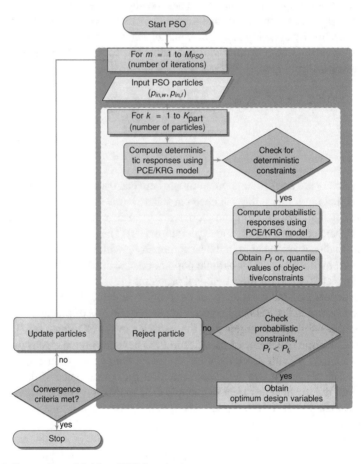

Fig. 29.1: Surrogate model-driven PSO flowchart

29.4 Applications

29.4.1 A Steel Transmission Tower

A three-dimensional steel transmission tower (Figure 29.2) is considered for testing the efficiency of the surrogate models in optimization, following the work of [32]. The height of the tower is 16 m, with a triangular base of width 6.93 m as shown in Figure 29.2. A vertical load, $V = 2$ kN is applied to all the joints, while an uncertain horizontal load, H with mean value 8 kN is applied to the top nodes in the global X direction.

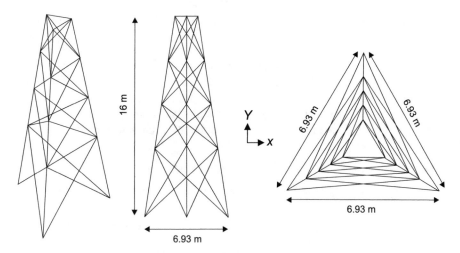

Fig. 29.2: A steel transmission tower with its elevation and plan

The optimization problem solved here is a minimization of the tower weight (W), with the design variables being the dimensions of the tower elements. Circular hollow sections (CHS) are used to design this tower following the load and resistance factor design practice. For the design variable vector, the dimension variables are the outer diameter, d_0 and the thickness, t of the CHS. Three types of optimization constraints are considered – axial stress constraint to avoid yielding, compressive force for buckling, and nodal displacement constraints. The allowable values for these constraints are obtained from Indian Standard (IS) code of practice [5]. A geometric nonlinear finite element (FE) analysis is carried out using the stiffness method to obtain model responses [4]. Fixed boundary conditions are applied at the base of the tower, with zero nodal displacement and rotation. It must be noted that each such FE analysis takes roughly 1 hour on a four-core 3.2 GHz processor with 4.0 gigabytes RAM to compute. Hence, the computational demand shoots up significantly making optimization simulation prohibitive. The probabilistic modeling of input/design variables is given in Table 29.1, with the design variables (d_0 and

t) modeled as random variables so as to perform a robust design optimization. The stress and buckling constraints are taken to be reliability-based, while the displacement constraint is deterministic. The present reliability-based optimization problem is formulated as

$$\underset{\mu_{d_0}, \mu_t}{\text{minimize}} \quad F = \mu_W$$

subject to: $\quad \mathbb{P}\left(p_{1_{\max}} \geq p_{\text{alw}}\right) \leq 10^{-4} \qquad$ probabilistic constraint $\qquad$ (29.12)

$$\mathbb{P}\left(P_{c_{\max}} \geq P_e\right) \leq 10^{-4} \qquad \text{probabilistic constraint}$$

$$d_{\max} \leq d_{\text{alw}} \qquad\qquad \text{deterministic constraint}$$

where μ_W is the mean weight of the tower, which depends on the mean values of the design variables, μ_{d_0} and μ_t. $p_{1_{\max}}$ is the absolute maximum principal member axial stress for all loading cases and p_{alw} is the allowable axial stress. $P_{c_{\max}}$ is the absolute maximum compressive force for all loading cases and P_e is the critical Euler buckling force in compression. The maximum value of nodal displacement, $d_{\max}$ is restricted to an allowable deterministic value, $d_{\text{alw}} = 200$ mm. It must be noted that the optimization function parameters ($p_{1_{\max}}$, $P_{c_{\max}}$ and $d_{\max}$) are obtained using a computational expensive geometric nonlinear FE analysis. A target failure probability of 10^{-4}, that is typical for civil structures, is chosen here [10].

Table 29.1: Probabilistic characterization of input/design random variables [32]

Random variables (X)	Probability distribution	Mean (μ_X)	CoV* (V_X)
Modulus of elasticity, E	Normal	2×10^8 kN/m^2	7%
Yield stress, f_y	Normal	3.5×10^5 kN/m^2	10%
Horizontal load, H	Normal	8 kN	37.5%
CHS outer diameter, d_0	Normal	150 mm	10%
CHS thickness, t	Normal	5 mm	5%

*Coefficient of variation

Here, the optimization objective function is the mean weight of the tower. The surrogate models (PCE & KRG) contain input/design variables as listed in Table 29.1. For construction of both PCE and kriging model, 1200 Latin hypercube samples of input/design variables are used. A third order PCE ($N = 3$) is constructed for the response parameters. The chaos coefficients/hyperparameters are generated using the non-intrusive regression method. The probability density function (PDF) of W estimated using the PCE and kriging is plotted in Figure 29.3(a), and is compared with the "true" FE model PDF estimated using 10^4 crude Monte Carlo simulation (MCS) [31]. Since the target failure probability for the optimization problem was set to 10^{-4}, a total of 10^4 realizations are used for MCS [3]. The comparison shows a qualitatively good match, depicting that PCE/KRG can provide a better approximation of the "true" FE model for the chosen problem. To verify the goodness-of-fit quantitatively, leave-one-out error (ε_{LOO}) for W is obtained as 2.635% and 4.735% for PCE and kriging respectively.

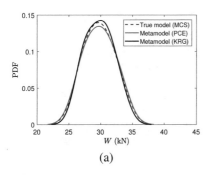 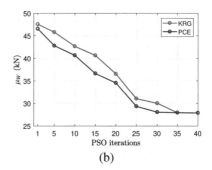

(a)   (b)

Fig. 29.3: (a) Comparison of surrogate model response PDFs. (b) Convergence of objective
function

The reliability-based design optimization (RBDO) problem is next solved using
the formulation given in Eq. 29.12 in order to verify the accuracy of a surrogate-
driven optimization algorithm. The convergence history of the objective function,
μ_W, is shown in Figure 29.3(b). Both the PCE and kriging model based results are
close to those obtained using the MCS scheme. Based on the PCE-PSO algorithm,
the optimum CHS dimension values for the RBDO are: $d_0 = 164.3$ mm and $t = 4.47$
mm (the objective function for these optimum values is, $\mu_W = 27.88$). However, it
must be noted that the normalized CPU time required by MCS is almost 129 times
larger than that by PCE and 113 times than by kriging. Such time saving makes a
high dimensional optimization under uncertainty problem tractable. In this case, the
PCE model is found to be more efficient than kriging. These results are reported in
Table 29.2.

Table 29.2: Comparison of optimization results with normalized CPU time

	PCE	KRG	MCS
No. of analyses	1200	1200	10,000
CPU time	1	1.15	129
Minimized μ_W (kN)	27.88	27.89	27.83

29.4.2 A Tensile Fabric Composite Structure

The next practical application considered here is a real conic tensile fabric composite structure. These structures are nowadays extensively used as permanent rooftops
to cover large areas, for example in stadiums, shopping malls, vehicle parking etc.
Tensile membrane structures (TMS) are typically designed with wind as the gov-

erning load, however, snow and rain loads may also become dominant based on the location under consideration. The TMS considered here is adopted from a recent studies on TMS design optimization under uncertainty [10, 11]. The details of this structure are shown in Figure 29.4. Additional details for analysis are:

- The fabric composite (membrane) yarn directions are: warp along radial and fill along circumferential directions (see Figure 29.4a).
- Material properties of the membrane: Modulus of elasticity, $E = 600.0$ kN/m; Poisson's ratio, $v = 0.4$; Design/nominal yield stress, $f_y = 40.0$ kN/m.
- Nominal wind load intensity, $W_n = 1.0$ kN/m^2.

The membrane is discretized with constant-strain triangular (CST) elements that have 3 translational degrees of freedom per node $\{U_X, U_Y, U_Z\}$ along the global $\{X, Y, Z\}$ directions. Proper symmetric boundary conditions are applied for the quarter part of the membrane, in addition to the support boundary conditions. A finite difference based dynamic relaxation technique [10] is implemented in MATLAB to obtain the nodal deformations of the TMS in global directions. The initial prestress is applied along the element warp (radial) and fill (circumferential) directions as shown in Figure 29.4a. The quarter part of TMS meshed with triangular CST elements is shown in Figure 29.4d. The flexible membrane remains in stable condition due to the existing initial prestress applied along yarn (warp and fill) directions. Membrane structures are primarily designed to resist the action of gusty winds. A TMS subjected to a design wind load, needs to remain in a stable shape and avoid wrinkling and tearing failures. Such failures are governed by the membrane principal stress values. For an optimal performance of these structures, membrane deformation must be minimized. Hence, the structural response parameters of interest for the optimum design of TMS are: the absolute maximum principal stress ($p_{1_{max}}$) to ensure no tearing, the absolute minimum principal stress ($p_{2_{min}}$) to ensure no wrinkling/slackness and the total nodal deformation (f_δ). These are defined as

$$f_\delta = \sum_j \delta_j = \sum_j \sqrt{U_{X_j}^2 + U_{Y_j}^2 + U_{Z_j}^2}$$

$$\text{for} \quad j = 1, 2, \ldots, J_{\text{node}}$$

(29.13)

$$p_{1_{max}} = \max_l p_{1_l} \quad \text{for} \quad l = 1, 2, \ldots, L_{\text{elem}}$$

(29.14)

$$p_{2_{min}} = \min_l p_{2_l} \quad \text{for} \quad l = 1, 2, \ldots, L_{\text{elem}}$$

(29.15)

where L_{elem} and J_{node} are the total number of elements and nodes, respectively, used in the finite element model of the TMS. U_{X_j} is the jth nodal displacement along X-direction.

Due to the inherent uncertainty in wind loads, there is a definite need to quantify the uncertainty that propagates to the above structural responses. For this uncer-

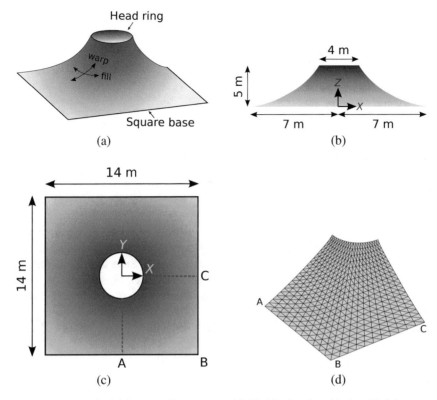

Fig. 29.4: (a) A tensile fabric composite structure with (b) side elevation, (c) plan, (d) finite
element model

tainty quantification, the wind load intensity (W) – which needs to be characterized
probabilistically – is the only random input variable considered here. The cumulative distribution function (CDF) for W is obtained based on past statistical studies.
W follows an Extreme Type I/Gumbel distribution with a coefficient of variation,
$V_W = 0.37$, and a bias factor, $\lambda_W = 0.78$ [11].

The optimization under uncertainty (OUU) problem for an optimal performance
of TMS is discussed here. The optimization problem formulation is adopted from
[10]. The final objective is to obtain optimum initial prestress values ($p_{in,w}$ and $p_{in,f}$)
to be applied along the membrane warp and fill directions, respectively, that is supposed to maximize the TMS stability under the uncertain wind loads. The stability
of the membrane is guaranteed by minimizing the overall membrane deformation.
To this end, the objective is set to minimize the total nodal deformation (f_δ), which
can also be interpreted as the maximization of TMS stability. Considering the random nature of wind intensity (W), the mean of the total nodal deformation (μ_{f_δ}) is
minimized here, which gives a robust estimate of the objective function. In addition,
the constraints are also formulated based on the TMS stability with no wrinkling,
slacking and tearing of the fabric both in the presence and the absence of wind loads.

The design requirement on tearing failure is considered to be probabilistic. Based on typically accepted failure levels of civil engineering structures, the probability of tearing failure is limited to 10^{-4}: $P_f = \mathbb{P}(p_{1_{max}} \geq f_y) \leq P_{f_t} = 10^{-4}$, where f_y is the yield stress of the membrane material under tension [11]. The overall OUU problem adopted from [11] is given as

$$\underset{p_{in,w}, p_{in,f}}{\text{minimize}} \quad F = \mu_{f_\delta}$$

subject to:

$$
\begin{array}{lll}
p_{1_{max}} \leq f_y & \text{without wind load} & (29.16) \\
p_{2_{min}} \geq 0 & \text{without wind load} & \\
\mathbb{P}(p_{1_{max}} \geq f_y) \leq 10^{-4} & \text{with wind load} &
\end{array}
$$

First, the probability density functions (PDFs) are plotted just to compare the system responses obtained from PCE and kriging models with the "true" computational model counterpart. Figure 29.5a and 29.5b show comparisons of the PCE/kriging-based PDF with the "true" PDF. The "true" PDF is obtained from validation set size of 200 crude MCS for f_δ and $p_{1_{max}}$. In a quantitative sense, both the PCE/KRG-based metamodel provides better accuracy with a satisfactory leave-one-out error (ε_{LOO}). The best fit is obtained using a PCE model with the ε_{LOO} values of 1.164% for f_δ and 2.904% for $p_{1_{max}}$.

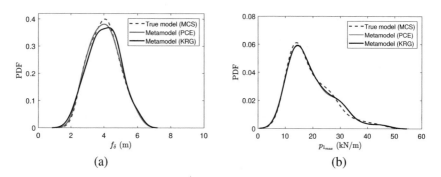

(a)   (b)

Fig. 29.5: Comparison of surrogate models with the "true" model PDFs

Based on the surrogate model-driven optimization algorithms, the best optimum initial prestress values for this optimization problem are: $p_{in,w} = 4.923$ kN/m and $p_{in,f} = 3.320$ kN/m with the objective function for this optimum point is, $\mu_{f_\delta} = 1.622$ m. These results are close to those obtained using 10^4 MCS: $p_{in,w} = 4.960$ kN/m and $p_{in,f} = 3.255$ kN/m with $\mu_{f_\delta} = 1.621$ m. However, the number of dynamic relaxation (DR)-based true model analyses (each analysis takes roughly 1.5 h on a four-core 3.2 GHz processor with 4.0 gigabytes RAM) in the PCE/kriging based optimization is 1/10th of that in the MCS based optimization. In this case, too, PCE is found to perform better than kriging surrogate. These results are reported in Table 29.3.

Table 29.3: Comparison of optimization results with computation cost

	PCE	KRG	MCS
No. of DR analyses	1000	1000	10,000
Optimum $p_{in,w}$, $p_{in,w}$ (kN/m)	(4.923, 3.320)	(5.003, 3.265)	(4.960, 3.255)
minimized μ_{f_δ} (m)	1.622	1.626	1.621

Some features, that make the evolutionary optimizer i.e. PSO efficient (compared to other gradient-based approaches) in the context of complex large-scale optimization under uncertainty problems, are: (a) the PSO algorithm is simple and uses less parameters in a logical sense; (b) the stability of the algorithm is less affected by the selection of the initial population or candidate solutions; (c) the algorithm is less sensitive to the "dimensionality" of the problem at hand; and (d) PSO can provide faster convergence for large-scale structures with many degrees of freedom. Hence, PSO, being a robust stochastic optimizer here, is found to be one of the strongest candidates for the determination of practical optimization solutions within manageable computation time.

29.5 Concluding Remarks

This chapter introduces surrogate models in evolutionary optimization of physical systems. State-of-the-art surrogate models were introduced and their computer implementations along with features were discussed in the context of optimization problems. A double-loop optimization formulation is discussed with an evolutionary optimizer – particle swarm optimization. Data-driven non-intrusive approaches based on least-square technique are used for the determination of hyperparameters/coefficients of the surrogates. For verification of the metamodels, error estimate measures are determined. Comparative studies of the well-known surrogates are performed on numerical models with high dimensionality and computational complexity.

As evident from other recent works on the application of surrogate models in optimization, better parameterization of the standard algorithms such as the use of sparse data representations and more efficient designs of experiments, are expected to improve the surrogate models in general. Further work is in progress to compare the RBDO solutions obtained by PCE-PSO/KRG-PSO methods with the other well-known and efficient RBDO methods using evolutionary algorithms, such as the ones proposed by [7]. Also, a more comprehensive comparison considering aspects like accuracy, robustness, computation cost, simplicity etc. for the surrogate models can provide more insight on the optimization problem at hand. Such studies can provide the optimization community with information on the selection of a robust surrogate model for a larger class of complex engineering problems.

Acknowledgements The authors gratefully acknowledge the opportunity to contribute this chapter in honor of Professor Erik Goodman. Professor Goodman is one of the pioneers in the evolutionary computation research area, in particular it is worthwhile mentioning his PhD work on the use of Genetic Algorithms to solve real problems for the first time in research. The authors also acknowledge the reviewers and editors for their constructive comments in enhancing the quality of this chapter.

References

1. Allen, D.: *The prediction sum of squares as a criterion for selecting prediction variables.* Tech. rep., Department of Statistics, University of Kentucky (1971)
2. Aoues, Y., Chateauneuf, A.: *Benchmark study of numerical methods for reliability-based design optimization.* Structural and Multidisciplinary Optimization **41**(2), 277–294 (2009)
3. Au, S.K., Beck, J.L.: *Estimation of small failure probabilities in high dimensions by subset simulation.* Probabilistic Engineering Mechanics **16**(4), 263–277 (2001)
4. Bathe, K.J.: *Finite Element Procedures.* PHI Learning Private Limited, India (2010)
5. BIS: *Indian standard for general construction in steel: Code of practice. IS:800.* Bureau of Indian Standards (2007)
6. Chaudhuri, A., Haftka, R.T.: *Efficient global optimization with adaptive target setting.* AIAA Journal **52**(7), 1573–1577 (2014)
7. Deb, K., Gupta, S., Daum, D., Branke, J., Mall, A.K., Padmanabhan, D.: *Reliability-based optimization using evolutionary algorithms.* IEEE Transactions on Evolutionary Computation **13**(5), 1054–1074 (2009)
8. Dorigo, M., Stützle, T.: *Ant Colony Optimization.* MIT Press (2004)
9. Dubourg, V., Sudret, B., Bourinet, J.M.: *Reliability-based design optimization using kriging surrogates and subset simulation.* Structural and Multidisciplinary Optimization **44**(5), 673–690 (2011)
10. Dutta, S., Ghosh, S., Inamdar, M.M.: *Reliability-based design optimisation of frame-supported tensile membrane structures.* ASCE-ASME Journal of Risk and Uncertainty in Engineering Systems, Part A: Civil Engineering **3**(2), G4016001 (2017)
11. Dutta, S., Ghosh, S., Inamdar, M.M.: *Optimisation of tensile membrane structures under uncertain wind loads using PCE and kriging based metamodels.* Structural and Multidisciplinary Optimization **57**(3), 1149–1161 (2018)
12. Eberhart, R.C., Shi, Y.: *Particle swarm optimization: Developments, applications and resources.* In: Proceedings of the IEEE Conference on Evolutionary Computation, ICEC, vol. 1, pp. 81–86 (2001)
13. Filomeno Coelho, R., Lebon, J., Bouillard, P.: *Hierarchical stochastic metamodels based on moving least squares and polynomial chaos expansion: Application to the multiobjective reliability-based optimization of space truss structures.* Structural and Multidisciplinary Optimization **43**(5), 707–729 (2011)
14. Forrester, A.I.J., Keane, A.J.: *Recent advances in surrogate-based optimization.* Progress in Aerospace Sciences **45**(1-3), 50–79 (2009)
15. Forrester, A.I.J., Sobester, A., Keane, A.J.: *Engineering Design via Surrogate Modelling: A Practical Guide.* Wiley, Chichester, UK (2008)
16. Gandomi, A.H., Alavi, A.H.: *Krill herd: A new bio-inspired optimization algorithm.* Communications in Nonlinear Science and Numerical Simulation **17**(12), 4831–4845 (2012)
17. Gandomi, A.H., Yang, X., Alavi, A.H.: *Cuckoo search algorithm: A metaheuristic approach to solve structural optimization problems.* Engineering with Computers **29**(1), 17–35 (2013)
18. Gandomi, A.H., Yang, X.S., Talatahiri, S., Alava, A.H.: *Metaheuristic Applications in Structures and Infrastructures.* Elsevier, Waltham, UK (2013)
19. Gandomi, A.H., Yang, X.S., Talatahiri, S., Alava, A.H.: *Metaheuristis in Water, Geotechnical and Transportation Engineering.* Elsevier, Waltham, UK (2013)

20. Ghanem, R., Spanos, P.D.: *Stochastic Finite Elements: A Spectral Approach.* Springer-Verlag, Berlin, Germany (1991)

21. Goldberg, D.E.: *Genetic Algorithm in Search, Optimization and Machine Learning.* Addison-Wesley, Boston, MA (1989)

22. Hao, P., Wang, B., Tian, K., Li, G., Du, K., Niu, F.: *Efficient optimization of cylindrical stiffened shells with reinforced cutouts by curvilinear stiffeners.* AIAA Journal **54**(4), 1350–1363 (2016)

23. Hu, W., Choi, K.K., Cho, H.: *Reliability-based design optimization of wind turbine blades for fatigue life under dynamic wind load uncertainty.* Structural and Multidisciplinary Optimization **54**(4), 953–970 (2016)

24. Huang, Z., Wang, C., Chen, J., Tian, H.: *Optimal design of aeroengine turbine disc based on kriging surrogate models.* Computers and Structures **89**(1-2), 27–37 (2011)

25. Jin, R., Du, X., Chen, W.: *The use of metamodeling techniques for optimization under uncertainty.* Structural and Multidisciplinary Optimization **25**(2), 99–116 (2003)

26. Karaboga, D., Basturk, B.: *A powerful and efficient algorithm for numerical function optimization: Artificial Bee Colony (ABC) algorithm.* Journal of Global Optimization **39**(3), 459–471 (2007)

27. Kennedy, J., Eberhart, R.: *Particle swarm optimization.* In: IEEE International Conference on Neural Networks - Conference Proceedings, vol. 4, pp. 1942–1948 (1995)

28. Kirkpatrick, S., Gelatt Jr., C.D., Vecchi, M.P.: *Optimization by simulated annealing.* Science **220**(4598), 671–680 (1983)

29. Lee, K.S., Geem, Z.W.: *A new meta-heuristic algorithm for continuous engineering optimization: Harmony search theory and practice.* Computer Methods in Applied Mechanics and Engineering **194**(36-38), 3902–3933 (2005)

30. Moustapha, M., Sudret, B., Bourinet, J.., Guillaume, B.: *Quantile-based optimization under uncertainties using adaptive kriging surrogate models.* Structural and Multidisciplinary Optimization **54**(6), 1403–1421 (2016)

31. Nowak, A.S., Collins, K.R.: *Reliability of Structures, 2nd Ed.* CRC Press, Boca Raton, USA (2013)

32. Papadrakakis, M., Lagaros, N.D., Plevris, V.: *Design optimization of steel structures considering uncertainties.* Engineering Structures **27**(9), 1408–1418 (2005)

33. Santner, T.J., Williams, B.J., Notz, W.I.: *The Design and Analysis of Computer Experiments.* Springer series in Statistics, Springer-Verlag, Berlin, Germany (2003)

34. Shi, Y., Eberhart, R.C.: *Parameter selection in particle swarm optimization.* In: Proceedings of the 7th International Conference on Evolutionary Programming VII, vol. 1447, pp. 591–600 (1998)

35. Soize, C., Ghanem, R.: *Physical systems with random uncertainties: Chaos representations with arbitrary probability measure.* SIAM Journal on Scientific Computing **26**(2), 395–410 (2005)

36. Sun, C., Jin, Y., Cheng, R., Ding, J., Zeng, J.: *Surrogate-assisted cooperative swarm optimization of high-dimensional expensive problems.* IEEE Transactions on Evolutionary Computation **21**(4), 644–660 (2017)

37. Trelea, I.C.: *The particle swarm optimization algorithm: Convergence analysis and parameter selection.* Information Processing Letters **85**(6), 317–325 (2003)

38. Yang, X..: *Engineering Optimization: An Introduction with Metaheuristic Applications.* John Wiley & Sons (2010)

39. Yang, X., Gandomi, A.H.: *Bat algorithm: A novel approach for global engineering optimization.* Engineering Computations (Swansea, Wales) **29**(5), 464–483 (2012)

Chapter 30
Mechatronic Design Automation: A Short Review

Zhun Fan, Guijie Zhu and Wenji Li

Abstract This paper gives a short review on mechatronic design automation (MDA) whose optimization method is mainly based on evolutionary computation techniques. The recent progress and research results of MDA are summarized systematically, and the challenges and future research directions in MDA are also discussed. The concept of MDA is introduced first, research results and potential challenges of MDA are analyzed. Then future research directions, focusing on constrained multi-objective optimization, surrogate-assisted constrained multi-objective optimization, and design automation by integrating constrained multi-objective evolutionary computation and knowledge extraction, are discussed. Finally, we suggest that MDA has great potential, and may be the next big technology wave after electronic design automation (EDA).

Key words: Mechatronic Systems, Design Automation, Evolutionary Design, Bond Graph (BG)/ Genetic Programming (GP), Evolutionary Optimization.

30.1 Introduction

Mechatronics is a type of hybrid system that consists of mechanical, electrical, pneumatic, hydraulic and control subsystems. Therefore, the design of mechatronics is different from the design of traditional mechanical, electronic and hydraulic systems.

In the design process of mechatronic systems, several types of energy conversion need to be fused [10]. In addition, the design of continuous and/or discrete controllers may also need to be considered in mechatronic systems. As a result, MDA

Zhun Fan, Guijie Zhu, and Wenji Li
Shantou University, Shantou, Guangdong, 515063, China
e-mail: zfan@stu.edu.cn

© Springer Nature Switzerland AG 2020
W. Banzhaf et al. (eds.), *Evolution in Action: Past, Present and Future*,
Genetic and Evolutionary Computation, https://doi.org/10.1007/978-3-030-39831-6_30

needs to consider the automatic concurrent design of controllers and controlled objects. Since a mechatronic system is usually very complicated, it is so far difficult to achieve a good strategy for automatically designing and optimizing such a complex system.

The design of mechatronic systems usually relies on the long-term experience of engineers, which entails long design cycles and frequent modifications. In addition, the result of the design is not guaranteed to be optimal. Thus, the research on MDA is important to help designers improve design performance and efficiency.

The remainder of the paper is organized as follows. Section 30.2 overviews recent work on MDA, including a discussion on issues and challenges in MDA. Section 30.3 gives future research directions on MDA. Finally, conclusions are drawn in Section 30.4.

30.2 Mechatronic Design Automation (MDA)

Mechatronic systems with the properties of intelligence, flexibility and multi-functionalities are becoming important and have received broad attention in recent years. As a special type of mechanical system, mechatronics is a full-featured and powerful system composed of electronic devices and mechanical components. Fig. 30.1 shows the different characteristics between electronic systems and pro-mechanical systems (including micro-electro-mechanical system, mechatronic system and pure mechanical system). The properties of coupling and modularity of the above-mentioned systems are also illustrated in Fig. 30.1.

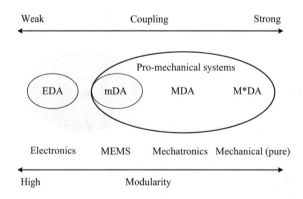

Fig. 30.1: The relationships of design automation of different systems

In Fig. 30.1, EDA, mDA, MDA and M*DA represent the design automation of electronic system, micro-electro-mechanical system(MEMS), mechatronic system and pure mechanical system, respectively. EDA is easy to be realized due to the high modularity and low coupling of digital electronic systems. The pure mechanical system has the lowest modularity and the strongest coupling. Thus, the implementation of M*DA is much more difficult than that of EDA, mDA and MDA. At present, EDA and mDA have already made great progress. The methods and experiences from EDA and mDA can be transferred to help to optimize the design of mechatronic systems.

The most significant difference between MDA and EDA is that the former contains a multi-domain physical system integrated with control systems [51]. Mechatronics is an essential stage for the evolution of modern products, which contains many components from different engineering fields, such as mechanical, electrical, hydraulic and control engineering. Chakrabarti [8] proposed a kind of MDA framework which can generate a series of conceptual designs that meet pre-set requirements. However, the dynamic behavior of the designed mechatronic system has not been studied. Campbell [7] studied and developed an agent-based MDA framework, which has the capability to adapt to dynamic environments. However, it lacks a detailed analysis of the dynamic behavior of the designed system. Behbahani [5] proposed a concept of mechatronic design quotient (MDQ), which can integrate multiple design objectives into one single objective. Then, the formulated optimization problem is solved by using single objective optimization algorithms. However, when the multiple design objectives are conflicting with each other, the performance of this method can not be guaranteed. In fact, when the geometry of the Pareto front of the formulated problem is convex, this method can only find two endpoints, a fact that can be proved theoretically [33]. Thus, multi-objective optimization algorithms are more commonly used methods to solve the mechatronic design optimization problems with more than one objective.

Bond graphs (BGs) are an unified modeling method for multi-domain systems [55]. BGs have already been widely used in modeling various of real-world physical systems such as robots [32], hybrid electric vehicles [21] and wind turbine systems [35], etc. Fig. 30.2 shows an example of a single BG model that can uniformly represent resonator units in three different fields, including mechanical, electrical, and micro-electro-mechanical systems. Since BGs can clearly represent topologies of a system, it becomes an excellent candidate tool in searching open-ended design spaces. Tay et al. [42] utilized BG to automatically generate the design of a mechatronic system that meets the pre-defined design specifications, in which a genetic algorithm (GA) is used to search in the design space. Finger and Rinderle [22] proposed to apply the BG method to conduct the generation process from pre-defined design specifications to physical implementations that meet these design specifications. Seo et al. [39] proposed an automatic design methodology called BG/GP for mechatronic systems, which combines BG and genetic programming (GP). Compared with other methods, the proposed BG/GP method has obvious advantages which are shown in Table 30.1.

From Table 30.1, it can be observed that BG, GA and GP have different properties. BG can be used for the modeling and effective evaluation of multi-domain systems. GP and GA can both search the design space for optimizing the design. However, compared with GA, GP has more advantages due to its strong capability of searching open-ended design spaces. Therefore, GP can optimize the topology

Table 30.1: Comparisons of various design methods [14]

Properties	Design Methods				
	BG	GA	GP	BG/GA	BG/GP
Multi-domain	✓			✓	✓
Topological Variation		✓			✓
Developmental Process		✓	✓	✓	✓
Automated Synthesis		✓	✓	✓	✓
Design Optimization		✓	✓	✓	✓
Efficient Evaluation	✓			✓	✓

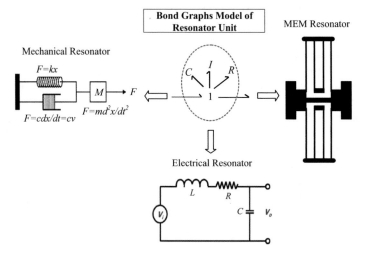

Fig. 30.2: One bond graph represents resonators in different application domains

and parameters of a mechatronic system simultaneously. BG/GA integrates BG and GA, and BG/GP integrates BG and GP, which has the capability to deal with topological variation that is not included in BG/GA.

In BG/GP, the BG is used for the modeling of multi-domain systems, while GP is used for the automatic exploration of the open-ended design spaces. Fig. 30.3 shows the mapping from genotype to phenotype in BG/GP method. The BG serves as an intermediate medium from the GP tree to the final physical realization, which is analogy to the mapping from genotype to phenotype. BG/GP not only can automatically perform open-ended topological search, but also optimize parameters of a system at the same time.

The BG/GP proposed by Fan et al. has already been applied to the design of electrical and mechatronic systems, such as analog filters [39], electric filters [19] and the driver system of a printer [20]. At the same time, Wang et al. [45] pro-

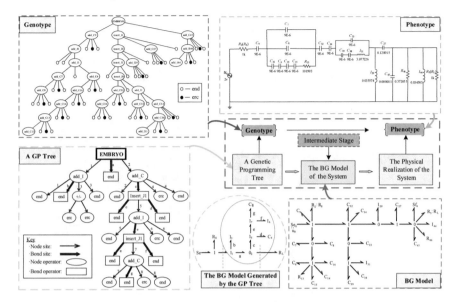

Fig. 30.3: An example of genotype-phenotype mapping

posed a knowledge-based evolutionary design framework for mechatronic systems by combining the BG/GP method with human knowledge, as shown in Fig. 30.4. This framework is demonstrated by a quarter-car suspension control system synthesis and a MEMS bandpass filter design application. Through the above-mentioned cases, the automated design method based on BG/GP provides new ideas for designing of mechatronic systems, which has potential to improve the performance of existing design schemes.

The traditional design methods for mechatronic systems often have long design cycles. Therefore, improving the search efficiency of automated design algorithms is important. Hu et al. [26] proposed a hierarchical hair competition (HFC) model which has the ability to avoid local premature convergence. Oduguwa et al. [36] proposed an intelligent design framework which can integrate the human knowledge and judgement. Zhang et al. [54] proposed a competitive mechanism based on a multi-objective particle swarm optimizer. Wang [43] proposed a hierarchical surrogate-assisted evolutionary algorithm for optimizing the airfoil of a flying wing configuration whose fitness evaluation is time-consuming and computationally expensive. The proposed hierarchical surrogate model is embedded in the covariance matrix adaptation evolution strategy (CMA-ES) to solve the RAE2822 airfoil optimization problem. The search efficiency of the algorithm can be improved by using these mechanisms.

As we all know, mechatronic systems usually consist of multiple sub-systems which come from different domains. Inspired by the study of symbiosis in nature, Potter et al. [38] proposed a general coevolution framework. They designed a rule-based control system for autonomous robots by using this co-evolution framework.

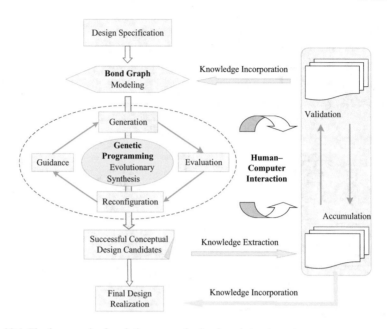

Fig. 30.4: The framework of evolutionary synthesis of mechatronic systems

Although a general architecture of co-evolution has been proposed, the problem of modeling dynamic systems and the selection of different specific species was not well solved. Wiegand et al. [47] provided experimental validation analysis of various collaboration mechanisms and presented some basic recommendations on how to choose a mechanism for a particular problem.

Wang et al. [46] proposed a unified mechatronics modeling and brain-limb collaborative evolution design method. An automobile suspension system was designed as shown in Fig. 30.5. Compared with the traditional method, this method integrated the features of unified modeling of multi-domain systems and an open topological search. It can help designers get a set of more advanced and optimal designs. Burmester et al. [6] presented a model driven development approach called MechatronicUML [4, 23] for the design of self-optimizing mechatronic syetems. They used the proposed MechatronicUML approach to design discrete and continuous control syetems. Fan and Dupuis studied the evolutionary design of hybrid mechatronic systems with continuous and discrete properties [12, 13]. They combined the lookahead controller, hybrid BG, and GP to design the DC-DC converter [13] and used a Finite State Machine (FSM), hybrid BG, and GP to design multiple-tank system [12] automatically.

Robot systems, as a sort of complex mechatronic systems, have received great attention from the industry and academy. For example, Asea Brown Boveri Ltd (ABB), a well-known robot company, has established a long-term cooperative relationship with the team of Professor Peter Krus from Linköping University.

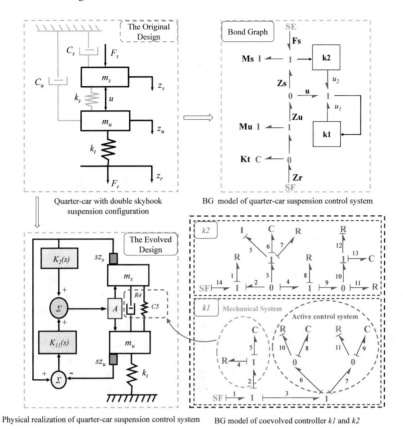

Fig. 30.5: Cooperative coevolutionary synthesis of quarter-car suspension control systems

They have conducted in-depth research on design automation for robot systems. Tarkian [41] clearly defined the concept of "design automation" and used a multi-objective evolutionary algorithm to optimize the design of robotic manipulators. In 2000, Lipson [31] designed computer-generated robotic systems by using evolutionary computation and employing a 3D-printer to prototype them. In [25], a design automation method for a soft robot was proposed, which combined the evolutionary algorithm and the Gaussian mixture model to perform an open topological search. Jamwal et al. [27] proposed a fuzzy sorting selection method based on multi-objective evolutionary algorithms to optimize a three-degree-of-freedom wearable ankle rehabilitation robot. In terms of lightweight design of manipulators, Zhou and Bai [56] designed a service manipulator by using an integrated design optimization approach, where manipulator kinematics, dynamics, drive-train design and strength analysis by means of finite element analysis (FEA) were generally considered. Yin et al. [50] proposed a method for designing a lightweight manipulator, in which the quadratic Lagrange algorithm was used to optimize the structure and the drive trains of the manipulator.

30.3 Future Research Directions

In the 21st century, mechatronic systems have received growing attention as an emerging discipline. The above-mentioned research is devoted to solving some challenging problems in MDA, which stimulates the future development of this discipline. Some issues and directions that deserve further study are listed as follows.

30.3.1 Constrained Multi-Objective Optimization

Generally, in real-world design problems, a designer considers not only one single design objective, but multiple conflicting objectives with a set of constraints, simultaneously. Therefore, the design of mechatronic systems can be formulated as a constrained multi-objective optimization problem (CMOP).

Constrained multi-objective evolutionary algorithms (CMOEAs) are commonly used methods to solve CMOPs, because they can achieve a set of feasible and non-dominated solutions in a single run. Currently, CMOEAs can be generally classified into two categories according to their selection mechanisms. One is the dominance-based CMOEAs, and the other is the decomposition-based CMOEAs.

In dominance-based CMOEAs, solutions are selected the next generation based on non-dominated ranks. Typical examples include NSGA-II-CDP [11] and SP [48]. In decomposition-based CMOEAs, a CMOP is decomposed into a set of constrained single objective optimization subproblems, and each subproblem is solved in a collaborative way. Representative examples include MOEA/D-CDP [28], C-MOEA/D [1], MOEA/D-Epsilon [49], and MOEA/D-SR [28].

Recently, two decomposition-based CMOEAs, MOEA/D-IEpsilon [16] and MO-EA/D-ACDP [15], have been proposed for solving CMOPs with large infeasible regions. In MOEA/D-IEpsilon [16], the epsilon level, which is used to relax constraints, is set dynamically according to the ratio of feasible to total solutions in the current population. Experimental results indicate that MOEA/D-IEpsilon is significantly better than four other decomposition-based CMOEAs, including MOEA/D-Epsilon [49], MOEA/D-SR [28], MOEA/D-CDP [28] and C-MOEA/D [1]. In MOEA/D-ACDP [15], the proposed angle-based constrained dominance principle (ACDP) is embedded in MOEA/D to solve CMOPs. Experimental results demonstrate that the proposed MOEA/D-ACDP [15] is significantly better than the state-of-the-art CMOEAs, including C-MOEA/D [1], MOEA/D-CDP [28], MOEA/D-Epsilon [49], MOEA/D-SR [28], NSGA-II [11] and SP [48].

To get across infeasible regions more efficiently, Zhun et al. [17] proposed a push and pull search (PPS) framework. In the push stage, a multi-objective evolutionary algorithm (MOEA) is used to explore the search space without considering any constraints, which can help to get across infeasible regions very quickly. In the pull stage, a CMOEA with improved epsilon constraint-handling is applied to pull the population to the feasible and non-dominated regions. Experimental results indicate that the proposed PPS method is very effective and efficient in solving CMOPs.

To promote research on constrained multi-objective optimization, Zhun et al. [18] proposed a difficulty-adjustable and scalable (DAS) test suite with three primary types of difficulty, which reflect various types of challenges presented by real-world optimization problems, in order to characterize the constraint functions in CMOPs. Nine CMOPs and nine constrained many-objective optimization problems (CMaOPs), called DAS-CMOP1-9 and DAS-CMaOP1-9, were proposed to evaluate the performance of two popular CMOEAs, MOEA/D-CDP and NSGA-II-CDP and two popular constrained many-objective evolutionary algorithms (CMaOEAs), CMOEA/DD and CNSGA-III, respectively. Experimental results indicate that these methods can not solve these problems efficiently, which stimulates researchers to continue to develop new CMOEAs and CMaOEAs to solve the suggested DAS-CMOPs and DAS-CMaOPs.

In general, to solve CMOPs efficiently, a single constraint-handling mechanism is not enough. A future research direction is to dynamically invoke appropriate constraint-handling mechanisms to search according to the state of the evolving population of a CMOEA.

30.3.2 Surrogate-Assisted Constrained Multi-objective Optimization

In most engineering design problems, the evaluation of objectives and constraints is expensive. Some objectives and/or constraints can only be calculated by doing physical experiments or calling simulation software, such as aerodynamic shape design, structural design, large scale circuit design, pharmaceutical design, etc. Thus, the evaluation process is time- and money-consuming. Optimization problems with the above-mentioned characteristics are also called expensive optimization problems. At present, the most representative work for solving expensive optimization problems is the surrogate-assisted evolutionary algorithm. By using surrogate models in the evolutionary process, the number of fitness evaluation can be reduced significantly.

In recent years, research on surrogate-assisted evolutionary algorithms has attracted increasing attention [9, 29, 34, 37]. For example, Jin et al. [30] proposed an evolutionary algorithm that can effectively solve nonlinear constrained optimization problems. An approximate model was built for each constraint function with increasing accuracy. Experimental results suggest that the proposed method is competitive compared to state-of-the-art methods for solving nonlinear constrained optimization problems. Chugh et al. [9] proposed a surrogate-assisted reference vector guided evolutionary algorithm for computationally expensive many-objective optimization problems. It adopted the Kriging model to approximate each objective function to reduce computational costs. Sun et al. [40] proposed a surrogate-assisted cooperative swarm optimization method for solving high-dimensional expensive problems. In the proposed method, a surrogate-assisted particle swarm optimization (PSO) algorithm and a surrogate-assisted social learning-based algorithm cooperatively search for the global optimum.

Mechatronic systems are a kind of complex and multi-energy domain coupled system which consists of components from different engineering fields. The design process of mechatronic systems is also complex, time-consuming and computationally expensive. Therefore, the research on surrogate-assisted CMOEAs to solve optimization problems of mechatronic systems efficiently is a direction worthy of further study.

30.3.3 Design Automation by Integrating Constrained Multi-Objective Evolutionary Computation and Knowledge Extraction

In the evolutionary process of CMOEAs, a large amount of data is generated, which contains a lot of knowledge related to the optimization problem. However, in traditional CMOEAs, these data are not mined, which results in a huge waste of resources.

As we all know, machine learning methods can assist MOEAs to improve search efficiency in selection and recombination processes [44, 52, 53]. Moreover, machine learning methods [2, 3] have the capability to acquire knowledge automatically and to refine knowledge bases, such as discovering new concepts and new models, finding errors in the knowledge bases, and optimizing and simplifying knowledge, etc. The fusion of machine learning methods with evolutionary algorithms not only improves the performance of the algorithms, but also acquires some design knowledge. The obtained knowledge can be transformed to other related scenarios, and generate some innovative designs [24]. Therefore, knowledge-driven optimization, by fusing CMOEAs and machine learning methods, is a very promising research direction for design automation of mechatronic systems.

30.4 Conclusion

We provide a preliminary overview of research work in MDA. With the growing amount and size of mechatronic systems being developed, the need for design automation for mechatronic systems is paramount. In MDA, evolutionary optimization, such as the surrogate-assisted CMOEA, has been shown to be successful at exploring large search spaces of optimization problems of mechatronic systems with expensive fitness evaluation, which has great potential for solving the design optimization problems of mechatronic systems. In the future, MDA integrating knowledge extraction and surrogate-assisted CMOEAs will further improve the performance of design automation algorithms and generate more innovative designs.

Acknowledgements This research work was supported by the Key Lab of Digital Signal and Image Processing of Guangdong Province, the National Natural Science Foundation of China under Grant 61175073, as well as the Hong Kong, Macao & Taiwan Science and Technology Cooperation Innovation Platform in Universities in Guangdong Province (2015KGJH2014).

References

1. Asafuddoula, M., Ray, T., Sarker, R., Alam, K.: *An adaptive constraint handling approach embedded MOEA/D*. In: Evolutionary Computation (CEC), 2012 IEEE Congress on, pp. 1–8. IEEE (2012)
2. Bandaru, S., Ng, A.H., Deb, K.: *Data mining methods for knowledge discovery in multiobjective optimization: Part A-Survey*. Expert Systems with Applications **70**, 139–159 (2017)
3. Bandaru, S., Ng, A.H., Deb, K.: *Data mining methods for knowledge discovery in multiobjective optimization: Part B-New developments and applications*. Expert Systems with Applications **70**, 119–138 (2017)
4. Becker, S., Dziwok, S., Gerking, C., Heinzemann, C., Schäfer, W., Meyer, M., Pohlmann, U.: *The Mechatronic UML method: model-driven software engineering of self-adaptive mechatronic systems*. In: Companion Proceedings of the 36th International Conference on Software Engineering, pp. 614–615. ACM (2014)
5. Behbahani, S., Silva, C.W.D.: *System-based and concurrent design of a smart mechatronic system using the concept of mechatronic design quotient (MDQ)*. IEEE/ASME Transactions on Mechatronics **13**(1), 14–21 (2008)
6. Burmester, S., Giese, H., Münch, E., Oberschelp, O., Klein, F., Scheideler, P.: *Tool support for the design of self-optimizing mechatronic multi-agent systems*. International Journal on Software Tools for Technology Transfer **10**(3), 207–222 (2008)
7. Campbell, M.I.: *The A-Design invention machine: A means of automating and investigating conceptual design*. Ph.D. thesis, Carnegie Mellon University (2000)
8. Chakrabarti, A., Bligh, T.P.: *An approach to functional synthesis of solutions in mechanical conceptual design. Part I: Introduction and knowledge representation*. Research in Engineering Design **6**(3), 127–141 (1994)
9. Chugh, T., Jin, Y., Miettinen, K., Hakanen, J., Sindhya, K.: *A surrogate-assisted reference vector guided evolutionary algorithm for computationally expensive many-objective optimization*. IEEE Transactions on Evolutionary Computation **22**(1), 129–142 (2018)
10. Coelingh, E., de Vries, T., Amerongen, J.: *Automated performance assessment of mechatronic motion systems during the conceptual design stage*. In: Proc. 3rd Intl Conf. on Adv. Mechatronics, pp. 472–477 (1998)
11. Deb, K., Pratap, A., Agarwal, S., Meyarivan, T.: *A fast and elitist multiobjective genetic algorithm: NSGA-II*. IEEE Transactions on Evolutionary Computation **6**(2), 182–197 (2002)
12. Dupuis, J.F., Fan, Z., Goodman, E.: *Evolutionary design of discrete controllers for hybrid mechatronic systems*. International Journal of Systems Science **46**(2), 303–316 (2015)
13. Dupuis, J.F., Fan, Z., Goodman, E.D.: *Evolutionary design of both topologies and parameters of a hybrid dynamical system*. IEEE Transactions on Evolutionary Computation **16**(3), 391–405 (2012)
14. Fan, Z.: *Mechatronic Design Automation: An Emerging Research and Recent Advances*. Nova Science Publishers, Incorporated (2010)

15. Fan, Z., Fang, Y., Li, W., Cai, X., Wei, C., Goodman, E.: *MOEA/D with angle-based constrained dominance principle for constrained multi-objective optimization problems.* arXiv preprint arXiv:1802.03608 (2018)
16. Fan, Z., Li, W., Cai, X., Huang, H., Fang, Y., You, Y., Mo, J., Wei, C., Goodman, E.: *An improved epsilon constraint-handling method in MOEA/D for CMOPS with large infeasible regions.* arXiv preprint arXiv:1707.08767 (2017)
17. Fan, Z., Li, W., Cai, X., Li, H., Wei, C., Zhang, Q., Deb, K., Goodman, E.: *Push and pull search for solving constrained multi-objective optimization problems.* Swarm and Evolutionary Computation **44**, 665–679 (2019)
18. Fan, Z., Li, W., Cai, X., Li, H., Wei, C., Zhang, Q., Deb, K., Goodman, E.D.: *Difficulty adjustable and scalable constrained multi-objective test problem toolkit.* arXiv preprint arXiv:1612.07603 (2016)
19. Fan, Z., Seo, K., Hu, J., Goodman, E.D., Rosenberg, R.C.: *A novel evolutionary engineering design approach for mixed-domain systems.* Engineering Optimization **36**(2), 127–147 (2004)
20. Fan, Z., Wang, J., Goodman, E.: *Exploring open-ended design space of mechatronic systems.* International Journal of Advanced Robotic Systems **1**(4), 295–302 (2004)
21. Filippa, M., Mi, C., Shen, J., Stevenson, R.C.: *Modeling of a hybrid electric vehicle powertrain test cell using bond graphs.* IEEE Transactions on Vehicular Technology **54**(3), 837–845 (2005)
22. Finger, S., Rinderle, J.R.: *A transformational approach to mechanical design using a bond graph grammar.* In: Proceedings of the First ASME Design Theory and Methodology Conference. American Society of Mechanical Engineers, Montreal (1989)
23. Giese, H., Schäfer, W.: Model-driven development of safe self-optimizing mechatronic systems with mechatronic uml. In: Assurances for Self-Adaptive Systems, pp. 152–186. Springer (2013)
24. Gupta, A., Ong, Y.S., Feng, L.: *Insights on transfer optimization: Because experience is the best teacher.* IEEE Transactions on Emerging Topics in Computational Intelligence **2**(1), 51–64 (2018)
25. Hiller, J., Lipson, H.: *Automatic design and manufacture of soft robots.* IEEE Transactions on Robotics **28**(2), 457–466 (2012)
26. Hu, J., Goodman, E., Seo, K., Fan, Z., Rosenberg, R.: *The hierarchical fair competition (HFC) framework for sustainable evolutionary algorithms.* Evolutionary Computation **13**(2), 241–277 (2005)
27. Jamwal, P.K., Hussain, S., Xie, S.Q.: *Three-stage design analysis and multicriteria optimization of a parallel ankle rehabilitation robot using genetic algorithm.* IEEE Transactions on Automation Science and Engineering **12**(4), 1433–1446 (2015)
28. Jan, M.A., Khanum, R.A.: *A study of two penalty-parameterless constraint handling techniques in the framework of MOEA/D.* Applied Soft Computing **13**(1), 128–148 (2013)
29. Jin, Y.: *Surrogate-assisted evolutionary computation: Recent advances and future challenges.* Swarm and Evolutionary Computation **1**(2), 61–70 (2011)
30. Jin, Y., Oh, S., Jeon, M.: *Incremental approximation of nonlinear constraint functions for evolutionary constrained optimization.* In: Evolutionary Computation (CEC), 2010 IEEE Congress on, pp. 1–8. IEEE (2010)
31. Lipson, H., Pollack, J.B.: *Automatic design and manufacture of robotic lifeforms.* Nature **406**(6799), 974 (2000)
32. Margolis, D., Karnopp, D.C.: *Bond graphs for flexible multibody systems.* Journal of Dynamic Systems, Measurement, and Control **101**(1), 50–57 (1979)
33. Miettinen, K.: *Nonlinear multiobjective optimization*, vol. 12. Springer Science & Business Media (2012)
34. Miranda-Varela, M.E., Mezura-Montes, E.: *Constraint-handling techniques in surrogate-assisted evolutionary optimization. an empirical study.* Applied Soft Computing **73**, 215–229 (2018)
35. Mojallal, A., Lotfifard, S.: *Multi-physics graphical model based fault detection and isolation in wind turbines.* IEEE Transactions on Smart Grid **9**, 5599–5612 (2017)

36. Oduguwa, V., Roy, R., Farrugia, D.: *Development of a soft computing-based framework for engineering design optimisation with quantitative and qualitative search spaces.* Applied Soft Computing **7**(1), 166–188 (2007)
37. Pan, L., He, C., Tian, Y., Wang, H., Zhang, X., Jin, Y.: *A classification based surrogate-assisted evolutionary algorithm for expensive many-objective optimization.* IEEE Transactions on Evolutionary Computation **23**, 74–88 (2018)
38. Potter, M.A., Jong, K.A.D.: *Cooperative coevolution: An architecture for evolving coadapted subcomponents.* Evolutionary computation **8**(1), 1–29 (2000)
39. Seo, K., Fan, Z., Hu, J., Goodman, E.D., Rosenberg, R.C.: *Toward a unified and automated design methodology for multi-domain dynamic systems using bond graphs and genetic programming.* Mechatronics **13**(8-9), 851–885 (2003)
40. Sun, C., Jin, Y., Cheng, R., Ding, J., Zeng, J.: *Surrogate-assisted cooperative swarm optimization of high-dimensional expensive problems.* IEEE Transactions on Evolutionary Computation **21**(4), 644–660 (2017)
41. Tarkian, M.: Design automation for multidisciplinary optimization: A high level CAD template approach. Ph.D. thesis, Linköping University Electronic Press (2012)
42. Tay, E.H., Flowers, W., Barrus, J.: *Automated generation and analysis of dynamic system designs.* Research in Engineering Design **10**(1), 15–29 (1998)
43. Wang, H., Doherty, J., Jin, Y.: *Hierarchical surrogate-assisted evolutionary multi-scenario airfoil shape optimization.* In: 2018 IEEE Congress on Evolutionary Computation (CEC), pp. 1–8. IEEE (2018)
44. Wang, H., Jin, Y., Jansen, J.O.: *Data-driven surrogate-assisted multiobjective evolutionary optimization of a trauma system.* IEEE Transactions on Evolutionary Computation **20**(6), 939–952 (2016)
45. Wang, J., Fan, Z., Terpenny, J.P., Goodman, E.D.: *Knowledge interaction with genetic programming in mechatronic systems design using bond graphs.* IEEE Transactions on Systems Man & Cybernetics Part C **35**(2), 172–182 (2005)
46. Wang, J., Fan, Z., Terpenny, J.P., Goodman, E.D.: *Cooperative body–brain coevolutionary synthesis of mechatronic systems.* AI EDAM **22**(3), 219–234 (2008)
47. Wiegand, R.P., Liles, W.C., Jong, K.A.D.: *An empirical analysis of collaboration methods in cooperative coevolutionary algorithms.* In: Proceedings of the 3rd annual conference on genetic and evolutionary computation, pp. 1235–1242. Morgan Kaufmann Publishers Inc. (2001)
48. Woldesenbet, Y.G., Yen, G.G., Tessema, B.G.: *Constraint handling in multiobjective evolutionary optimization.* IEEE Transactions on Evolutionary Computation **13**(3), 514–525 (2009)
49. Yang, Z., Cai, X., Fan, Z.: *Epsilon constrained method for constrained multiobjective optimization problems: some preliminary results.* In: Proceedings of the Companion Publication of the 2014 Annual Conference on Genetic and Evolutionary Computation, pp. 1181–1186. ACM (2014)
50. Yin, H., Huang, S., He, M., Li, J.: *A unified design for lightweight robotic arms based on unified description of structure and drive trains.* International Journal of Advanced Robotic Systems **14**(4), 1–14 (2017)
51. Youcef-Toumi, K.: *Modeling, design, and control integration: a necessary step in mechatronics.* IEEE/ASME Transactions on Mechatronics **1**(1), 29–38 (1996)
52. Zhang, J., Zhou, A., Zhang, G.: *A classification and Pareto domination based multiobjective evolutionary algorithm.* In: Evolutionary Computation (CEC), 2015 IEEE Congress on, pp. 2883–2890. IEEE (2015)
53. Zhang, X., Tian, Y., Cheng, R., Jin, Y.: *A decision variable clustering-based evolutionary algorithm for large-scale many-objective optimization.* IEEE Transactions on Evolutionary Computation **22**(1), 97–112 (2018)
54. Zhang, X., Zheng, X., Cheng, R., Qiu, J., Jin, Y.: *A competitive mechanism based multi-objective particle swarm optimizer with fast convergence.* Information Sciences **427**, 63–76 (2018)
55. Zhong, J.: *Coupling design theory and method of complex electromechanical systems.* China Machine Press (2007)

56. Zhou, L., Bai, S.: *A new approach to design of a lightweight anthropomorphic arm for service applications*. Journal of Mechanisms and Robotics **7**(3), 031,001 (2015)

Chapter 31
Evolving SNP Panels for Genomic Prediction

Ian Whalen, Wolfgang Banzhaf, Hawlader A. Al Mamun and Cedric Gondro

Abstract The use of genetic variation (DNA markers) has become widespread for prediction of genetic merit in animal and plant breeding and it is gaining momentum as a prognostic tool for propensity to disease in human medicine. Although conceptually straightforward, genomic prediction is a very challenging problem. Genotyping organisms and recording phenotypic traits are time consuming and expensive. Resultant datasets often have many more features (markers) than samples (organisms). Therefore, models attempting to estimate the effects of markers often suffer from overfitting due to the curse of dimensionality. Feature selection is desirable in this setting to remove markers that do not appreciably affect the trait being predicted and amount to statistical noise. We present a differential evolution system for feature selection in genomic prediction problems and demonstrate its performance on simulated data. Code is available at: `https://github.com/ianwhale/tblup`.

Key words: differential evolution, evolutionary computation, genomic prediction, feature selection

Ian Whalen
Beacon Center for the Study of Evolution in Action and Department of Computer Science and Engineering, Michigan State University, East Lansing, MI, USA e-mail: `whalenia@msu.edu`

Wolfgang Banzhaf
Beacon Center for the Study of Evolution in Action and Department of Computer Science and Engineering, Michigan State University, East Lansing, MI, USA e-mail: `banzhafw@msu.edu`

Hawlader A. Al Mamun
CSIRO Data61, Commonwealth Scientific and Industrial Research Organisation, Canberra, ACT, Australia e-mail: `hawlader.almamun@data61.csiro.au`

Cedric Gondro
Beacon Center for the Study of Evolution in Action and Department of Animal Science, Michigan State University, East Lansing, MI, USA e-mail: `gondroce@msu.edu`

© Springer Nature Switzerland AG 2020
W. Banzhaf et al. (eds.), *Evolution in Action: Past, Present and Future*,
Genetic and Evolutionary Computation, https://doi.org/10.1007/978-3-030-39831-6_31

31.1 Introduction

The use of DNA markers for prediction of genetic merit has become widespread in plant and animal breeding and is gaining momentum as a prognostic tool for susceptibility to disease in human medicine. Meuwissen, Hayes, and Goddard [40] introduced the idea of using a very large number of genotypic markers to predict phenotypes. This process is known as genomic prediction and tasks a system with estimating the joint effects of thousands of markers, usually single nucleotide polymorphisms (SNP) on a trait. For agricultural applications, these estimated SNP effects are then used to predict phenotypes or breeding values for new individuals that do not have trait information but do have genotype (marker) information. Over the past ten years, genomic prediction has been widely adopted in genomic selection [21] in agriculture [29, 37] and in human studies [1]. Hayes, Bowman, Chamberlain, and Goddard [27] emphasize its value, touting genomic selection as the most significant advancement for the dairy industry in the last two decades.

Although conceptually straightforward, genomic prediction is a very challenging problem. Genotyping and trait recording are costly and time demanding exercises; the result is that most genomic datasets will have hundreds of thousands or even millions of markers for which effects need to be simultaneously estimated from usually only a few thousand phenotyped individuals. This means that the datasets are underdetermined (also known as the $p \gg n$ problem) and suffer from overfitting due to the curse of dimensionality. In effect, genomic prediction can be treated as a high dimensionality, sparse data problem and, consequently, suffers from the same issues as other problems in this domain. Most notably being that the prediction models derived by statistical inference are sub-optimal since the accuracy of the parameter estimates (marker effects) rapidly decays as the number of features that needs to be estimated increases. The accuracy of prediction is also conditional on the genetic architecture of the traits – the interplay between genotypes and phenotypes is complex and varies widely from trait to trait; e.g. highly heritable traits regulated by a few genes of large effect are easier to predict than traits regulated by many genes with small effects and with a low heritability [28]. Moreover, there are still various other factors that will also influence the accuracy of prediction such as marker density (if the data is not at full sequence resolution), the effective population size (N_e), measures of linkage disequilibrium and family relationships [9, 20, 55], population stratification [41], sample size, reliability of phenotypes [19], and the methodology used to estimate marker effects [9].

For these reasons, genomic datasets are prime candidates for feature selection techniques. However, popular methods for genomic feature selection are often statistically-based; e.g. genome wide association studies (GWAS) which aim to identify, in the case of sequence data, the causal variants of a given trait or, when SNP arrays are used, the markers that are in high linkage disequilibrium with the causal variants [30, 53]. These approaches are limited to local searches of the feature space since they are conditioned on the supporting statistical evidence. On the other end of the spectrum, all markers are simultaneously used for prediction irrespective of them having or not a functional role on the trait – this is the main method currently adopted for genomic prediction.

Even though quantitative traits are largely polygenic with hundreds or thousands of variants influencing a trait, it still stands to reason that not every single genetic variant across the genome will have a real effect on every single trait. This suggests that current methods lead to sub-optimal accuracy of genomic prediction, especially with sequence data, due to background noise introduced by the large number of spurious non-causative variants included in the prediction models. Under this ratio- nale, we suggest that better prediction models are attainable by using only subsets of markers that are truly informative of a given trait. In this paper we suggest that genomic prediction should be treated as a feature selection problem and that it is amenable to non-statistical methods since they are potentially better at performing global searches of the feature space. Herein we discuss a non-statistical approach for genomic prediction through the use of an evolutionary computation (EC) technique called differential evolution (DE) [50] and compare its performance to mainstream methods.

31.2 Previous Work

31.2.1 Genomic Prediction

Genomic prediction is the process of using a large number of genetic markers to pre- dict phenotypic traits [40]. There are two main approaches used to estimate marker effects. The first approximates a traditional infinitesimal model that assumes all markers—usually single nucleotide polymorphisms (SNP)—contribute a non-zero value to the genetic variance and that SNP effects are normally distributed. The sec- ond approach is based on nonlinear methods that emphasize certain genomic regions and allow marker effects to come from distributions other than a Gaussian.

Linear methods like ridge regression best linear unbiased prediction (RRBLUP) [54] and genomic best linear unbiased prediction (GBLUP) [25, 52] follow the as- sumption that all markers have some nonzero, normally distributed effect[1]. Such methods are well-studied and have been applied across many domains [51, 57]. We point out that all linear methods like the ones mentioned share a common flaw. As- suming that all markers contribute a nonzero effect leads to loci that do not affect the output being assigned an effect value. This amounts to the linear model fitting to statistical noise, which will be detrimental to performance. Thus, feature selection is desirable in order to remove these non-informative marker sites.

The non-linear methods for genomic prediction include Bayes A, Bayes B [40], Bayes C [26], Bayesian Lasso [11], and Bayes R [15]. These methods mainly differ in their assumptions about what distributions the marker effects should follow. Even though a large proportion of the variants might be allocated to a distribution with

[1] These methods are equivalent [24].

very small to zero effects, these methods will still assign a nonzero posterior density to most variants. Hence the number of variants to be used for prediction is still very large and, in the same manner as the linear methods, the Bayesian methods also have limited discrimination between markers with and without an actual effect.

Which of these methods performs better depends on the underlying genetic architecture of the trait, e.g. Bayesian methods tend to outperform BLUP approaches when the trait is less polygenic. In practice, differences in prediction accuracy between methods have been generally very small. While these methods have well characterized statistical properties they are constrained by the underlying model assumptions and, given the dimensionality of the solution space, even very small estimates of effects in non-informative markers will, collectively, reduce prediction accuracy. This is an increasing problem with the increasing number of genetic variants to predict from.

31.2.2 Feature Selection

Feature selection is a subdiscipline of a larger class of techniques known as dimensionality reduction, a well studied problem in machine learning. Feature selection seeks to retain a subset of the original set of features, rather than transform them in some way. This can be preferable to feature transformation—which constructs some function of all input features—since it is interpretable and can lead to deeper understanding about what markers affect a trait most significantly. Genomic data often has the pervasive quality of orders of magnitude more features than samples ($p \gg n$ problem), making it a natural candidate for dimensionality reduction. However, work has been relatively limited to statistical filtering [48]. Here, we present relevant filter and wrapper methods.

Filter methods are often used as a preprocessing step in machine learning problems, functioning independently of any actual modeling. Features are selected based on some statistical relationship with the predicted output and possibly other features. Such examples are univariate correlation significance values or redundancy [7, 44]. Filter methods work by suppressing the least interesting variables, leaving the more promising variables to be used to learn a predictive model. Filter methods tend to be quite computationally efficient and are robust to overfitting, making them a popular choice for genomic classification [3, 8, 12]. To a large extent a genome-wide association study can be viewed as a filtering method – in its simplest form, a GWAS is just a univariate correlation on each SNP that calls attention to significant markers in the genome and is a standard technique to identify genomic regions of interest for a trait [30, 53].

Wrapper methods are of particular interest here because they include EC. These methods iteratively update a feature subset over time, preferring those that perform better according to some measure. Classically, EC is an effective population-based heuristic search inspired by biological principles [14]. Evolutionary computation includes a diverse catalog of methods like genetic algorithms (GAs) [32, 22], genetic programming [36, 4], evolutionary strategies [47, 49], and ant colony optimization [13]. Storn and Price [50] introduced DE as a greedier alternative to genetic algorithms and evolutionary strategies. There are common themes throughout all EC methods. A group of *individuals* that each represent a solution to some problem have a *fitness* assigned to them based on an objective function. In general, EC has the luxury of being able to use almost any objective function—non-differentiable or otherwise—due to its gradient-free nature. Individuals then combine—or share information—with each other. Those with more desirable fitness values are *selected* to remain in the population and guide a search toward the global optimum.

Feature selection with EC has been applied to a variety of domains problems. Raymer et al. demonstrated the efficacy of dimensionality reduction with a GA on a protein water-binding site identification problem, showing better performance than sequential feature selection methods [46]. Firpi and Goodman showed in [17] that particle swarm optimization can also produce similar results to a GA in multiple applications. Luque-Baena, et al. [38] showed that a simple GA could outperform the state of the art on a cancer pathway identification and classification task. Furthermore, it was shown in [39] that a GA outperforms simple sequential feature selection methods. Feature subsets discovered in [39] were also deemed more biologically relevant, potentially furthering the understanding about the traits being predicted.

Feature selection using DE is relatively new, with first successes being shown in 2008 [34]. In that work, DE was shown to outperform other wrapper methods like particle swarm optimization and genetic algorithms in an electroencephalogram classification task. The method presented here is the technique in [2, 16]. Both works deal with using DE to do feature selection in cattle applications. The original applications performed well compared with random search in [16] and GBLUP in [2]. We further contribute here by the addition of methods to control overfitting.

31.3 Methods

31.3.1 Differential Evolution

Differential evolution is a real-valued optimization technique originally introduced by Storn and Price [50]. The strategy for DE introduced here is the original implementation. See Algorithm 1 for an overview which is abstracted into four main parts: evaluation, mutation, crossover, and selection. We present the relevant descriptions of each of these operations here.

The foundation of DE is a population of n candidate solutions, all of which are vectors in $\mathbb{R}^d$, where d is the dimensionality of the given problem. Each solution, $\mathbf{X}_i = [x_{1,i}, \ldots, x_{d,i}]$ has an associated fitness, which is some performance metric to be maximized. At the beginning of the search, each vector is randomly initialized according to a uniform random distribution, so that $0 \leq x_{j,i} < 1, \forall j$.

Evaluation

The least complex operation in DE is evaluation. Evaluation simply assigns each vector in the population a fitness based on an objective function. See Section 31.3.2 for specifics on how a feature subset is extracted from a real vector and assigned a fitness.

Mutation

As noted, we use the original mutation rule presented in [50]. Known as DE/rand/1, the rule uses a *mutation factor* hyperparameter $F > 0$. A *donor vector*, $\mathbf{V}_i$, is created for each vector in the population, $\mathbf{X}_i$, according to the following equation

$$\mathbf{V}_i = \mathbf{X}_a + F \cdot (\mathbf{X}_b - \mathbf{X}_c). \tag{31.1}$$

Where a, b, c are unique random integers from 1 to population size n^2.

Crossover

Also known as parameter mixing, crossover combines information across solutions. Again, we use the method in [50] known as uniform—or binomial—crossover. For each index j in $\mathbf{X}_i$ and $\mathbf{V}_i$, the following is applied to create *trial vector* $\mathbf{U}_i$

$$u_{j,i} = \begin{cases} v_{j,i} & \text{if } rand[0,1) < C_r \text{ or } j = j_{rand} \\ x_{j,i} & otherwise. \end{cases} \tag{31.2}$$

Here, $C_r \in [0,1]$ is the crossover rate hyperparameter, $rand[0,1)$ is a uniform random number in the half-open range $[0,1)$, and j_{rand} is a uniform random integer from $[1,d]$ that is generated once per generation for each solution in the population. The purpose of j_{rand} is to ensure at least one index is crossed-over with the donor vector for each vector in the population.

Selection

The standard selection operator at each generation is a simple tournament selection between $\mathbf{X}_i$ and $\mathbf{U}_i$. If the fitness of $\mathbf{U}_i$ is greater than the fitness of $\mathbf{X}_i$, it replaces $\mathbf{X}_i$ and continues on to the next generation in the ith index of the population.

Note that more recent methods combine DE with other heuristics [6, 35] or use more sophisticated update rules [33, 45]. However, we prefer the original imple-

---

[2] Storn and Price originally named this update rule as a mutation [50]. For algorithms like GAs and genetic programming, a mutation carries out changes on a single individual in the population—rather than the multiple shown in Equation 31.1.

Algorithm 1 Differential Evolution

Input: Population size n, dimensionality d, generations g
Output: Solution P_{best}
  1: $P = \{\mathbf{X}_i | \mathbf{X}_i \in [0,1)^d, 1 \leq i \leq n\}$
  2: evaluate(P)
  3: **for** 1 to g **do**
  4:     $P' = \{\}$
  5:     **for** $i = 1$ to n **do**
  6:         $\mathbf{V}_i = $ mutate($\mathbf{X}_i$)
  7:         $\mathbf{U}_i = $ crossover($\mathbf{V}_i, \mathbf{X}_i$)
  8:         $P' = P' \cup \{\mathbf{U}_i\}$
  9:     **end for**
10:     evaluate(P')
11:     $P = $ selection(P, P')
12: **end for**
13: **return** P_{best}

---

mentation for its generality and success across a variety of domains as noted in [10].

31.3.2 Random Keys

Differential evolution in its standard form is a real valued optimization technique. Therefore, some accommodation must be made to obtain indices of a feature subset from a vector of real numbers. Here, we use a technique called *random keys* [5]. The random key technique is well known in EC due to its use in combinatorial optimization tasks such as scheduling [42]. Random keys represent solutions to a combinatorial problem as a real valued vector that is somehow decoded to produce a solution that is always valid in the objective space. This is in contrast to traditional binary encoding that—when acted on by operators like mutation and crossover— may no longer be a valid solution (e.g., a solution selects more than the desired number of features after crossover).

We demonstrate the random key decoding process used for feature selection with an example. Consider a scenario with a five dimensional data set. Each solution in the DE population is then a vector in $\mathbb{R}^5$. The vector

$$[0.08, 0.53, 0.91, 0.34, 0.18].$$

decodes to the feature ordering

$$[3, 2, 4, 5, 1].$$

The feature ordering is obtained by finding indices of the original solution vector in sorted order. More explicitly, observe that 0.91 is at index 3 in the example. Hence, 3 is the first value of the decoding since 0.91 is the largest value in the vector, and so on. If the task was to select two features from the original five, the features at indices 3 and 2 would be selected. Therefore, over time, the DE search will tend to increase the values in the solution vector at indices that tend to increase fitness. A fitness function is then applied to the obtained feature subset. In our case, this is the absolute value of Pearson's correlation between predicted and true phenotypes using RRBLUP [54].

31.3.3 Self-adaptive Differential Evolution

The above description of DE leaves out discussion on tuning the associated hyperparameters F, C_r, and N_p. These values often have a dramatic influence on the convergence of DE [33, 45]. As a result, some effort has gone into alleviating the choice of F and C_r through *self-adaptive* methods that "learn" these values throughout the course of a DE experiment. In the method presented below, this is done by observing which particular settings create trial vectors that successfully enter the next population.

Qin and Suganthan present Self-adaptive Differential Evolution (SaDE) in [45] as a way to learn not only the F and C_r parameters, but which mutation method to use as well. For each individual, a donor vector has probability p of being created with DE/rand/1 and probability $1 - p$ of being created with DE/current-to-best/1. Where p is initialized to be 0.5 and updated by calculating

$$p = \frac{ns_1 \cdot (ns_2 + nf_2)}{ns_1 \cdot (ns_2 + nf_2) + ns_2 \cdot (ns_1 + nf_1)}. \tag{31.3}$$

Where ns_1 and nf_1 are the number of trial vectors that were produced with donor vectors from DE/rand/1 that entered the next population (a *success*) and the number that did not (a *failure*), respectively. The values ns_2 and nf_2 have the same definition, but count the number of successes and failures for DE/current-to-best/1. Finally, the first 50 generations of the search do not update p to allow some time for the algorithm to stabilize and learn meaningful success and failure rates [45].

Values of F and C_r are newly generated for each individual solution vector at each generation. F is not learned using any particular scheme in SaDE, and is simply randomly sampled from the normal distribution $\mathcal{N}(0.5, 0.3^2)$, then clipped to fall in the range $(0, 2]$. The authors state that C_r is much more important to the performance of DE and chose to adjust it based on the trajectory of the search [45]. To do so, a C_r is sampled for each index in the population every 5 generations from $\mathcal{N}(C_{rm}, 0.1^2)$. Then, similarly to the mutation strategy, every 25 generations, C_{rm} is recalculated based on the values of C_r that successfully produced trial vectors that entered the next population. This method has proved successful on many test problems, so it will be applied here as well.

31.3.4 Seeded Initial Population

In order to incorporate domain knowledge, the results of a GWAS can be included in the initial DE population. Through *seeding*, the indices corresponding to the s most significant SNPs are marked with a value of 1 in some vector in the population. Since all solution vectors are initialized in the range $[0, 1)$, these indices will form the subset for that particular vector. Due to the greedy nature of DE, the search will never reach a fitness value that is worse than the seeded initial vector.

31.3.5 Heritability Thresholding

The concept of heritability is well known to geneticists studying genomic prediction. At a high level, heritability[3] is the proportion of variance in a phenotype that can be explained only by effects in the genotype. Heritability, h^2 is quite useful since it can be used to define h (i.e., $\sqrt{h^2}$), which is the theoretical limit on prediction for a genomic prediction task. In practice, heritability can be estimated for a trait by analyzing the behavior of a trait through familial lines using a restricted maximum likelihood technique [18, 43].

Reaching values equal to or greater than h means the search has begun to overfit to the validation set since it has gone above the highest possible accuracy. The peculiarity of h should be emphasized. It is quite uncommon for any predictive modeling problem to have a hard threshold for performance. Usually a practitioner shoots for some value that is as high as possible. The method explored here is an initial look into using this value for preventing overfitting in search based feature subset selection.

A rudimentary method is to simply stop the search when some statistic of the population reaches $h(1 + \alpha)$ for some small $\alpha \in (-1, 1)$. Where these possible measures could be the maximum, minimum, median, or mean fitness of the population. This method is based on the fact that we can treat h as a hard threshold and the idea that it could be better to simply stop searching rather than continue a search that is already known to be overfit to the validation set.

---

[3] For this discussion, heritability is limited to "narrow-sense" heritability which is captured by additive affects of alleles in a genotype [23].

31.4 Data

31.4.1 Simulation

To demonstrate the performance of DE, simulated data is favorable to control the complexities introduced in genomic data. Specifically, the number of QTL, desired trait heritability, h_{in}^2, and absence of epistatic effects are controlled for through the following process. For this study, $h_{in}^2 = 0.4$ Using the genotype matrix $\mathscr{X} \in \{0, 1, 2\}^{n \times d}$, q columns are uniformly chosen the QTL. Let $\beta^* = [\beta_1, \dots, \beta_q]^T$ be the true, simulated QTL effects. For this study $q = 100$. First, $\beta_i \sim \mathscr{N}(0, 1)$, $\forall i$. Then, β^* is adjusted by calculating the variance of each allele. For diploid organisms, there are three genotypes: heterozygous (AB) and homozygous (AA, BB). To determine the genetic variance, we calculate the rate that each genotype occurs (i.e. total occurrences of a particular genotype divided by total number of sample in $\mathscr{X}$). Then the genotypic variance is simply the variance of these three values, namely V_G. The true genetic values are then calculated by

$$
\mathbf{t}_g = \frac{\mathscr{X}_q \beta^*}{V_G h_{in}^2 t_v}, \tag{31.4}
$$

where $\mathscr{X}_q \in \{0, 1, 2\}^{n \times q}$ is the matrix consisting of the q columns that were chosen to be QTL, t_v is the desired output trait variance, and the division corresponds to an element-wise division of $\mathscr{X}_q \beta^* \in \mathbb{R}^n$. Here, $t_v = 40$. The vector $\mathbf{t}_g$ is then centered at zero by subtracting its mean. Finally, to calculate the actual phenotypes, $\mathbf{y}$, "environmental" noise is added to $\mathbf{t}_g$. This is done by calculating $\mathbf{y} = \mathbf{t}_g + \varepsilon$. To calculate $\varepsilon, \mathbf{e} = [e_1, \dots, e_q]$, $e_i \sim \mathscr{N}(0, [t_v(1 - h_{in}^2)]^2)$ is first sampled, then

$$
\varepsilon = \mathbf{e} \cdot \sqrt{\frac{(1 - h_{in}^2) t_v}{var(\mathbf{e})}}. \tag{31.5}
$$

When estimated, the heritability of the simulated trait will be approximately equal to the desired heritability, h_{in}^2. See Figure 31.1 for a Manhattan plot describing the results of the simulation using the genotypes of 7539 sheep with 48,588 SNPs.

31.4.2 Splitting

Data splitting is an important part of wrapper based feature selection as it can control some overfitting a method displays. Here, the data is split into three groups. First, a testing set is removed from the data that will not be used in any way during the search. This will serve as external validation for the DE search process to report results. Then, with the remaining data a few choices can be made. The simplest method uses no cross-validation and splits the data again into training and valida-

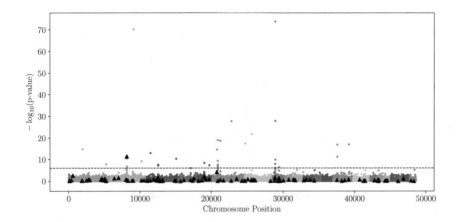

Fig. 31.1: A Manhattan plot describing the results of a GWAS on simulated data. Higher values indicate SNPs with more significant effects. Black triangles mark the randomly distributed true QTL in the simulation. The dotted line shows the Bonferroni corrected significance threshold $0.05/48588$.

tion. The training set is then used to train the BLUP model in the fitness evaluation and the validation set is used to obtain the prediction accuracy. However, since the DE search assigned fitness based on the same validation set for the entire experiment, it is likely that the search will overfit to the validation set. In practice, we used a 64%/16%/20% train/validation/test split.

To combat this, three cross-validation schemes are proposed: (1) *intergenerational* cross-validation which does k-fold cross-validation over the course of the search, changing the validation set at each generation. More concretely, at generation g_i, validation set $k \bmod g_i$ will be used to calculate prediction accuracy, and the data is used as training. (2) Intragenerational cross-validation performs k-fold cross-validation at every fitness evaluation. This method increases the computation time required by a factor of k. But, intuitively, may provide a more reliable fitness value. (3) Monte Carlo cross-validation uniformly samples a random subset of the data to be used as validation. The intuition behind this method is to drive the search through obtaining solutions that better generalize across the entire validation set. For the k-fold methods used in the following experiments, $k = 5$.

31.5 Results

The evolutionary parameters used for the results presented below are presented in Table 31.1. A fixed subset size of 1000 was used in these experiments, though more sophisticated methods can be used to choose (or even search) this value. Differential

Table 31.1: A table of evolutionary parameters used for the DE search.

Parameter (symbol)	Value
Generations (g)	5000
Population Size (N_p)	50
Crossover Rate (C_r)	0.8
Mutation Factor (F)	0.5
Replicates	10
Subset Size	1000

evolution without any additional features to improve performance will be referred to as *vanilla* DE where there is ambiguity.

31.5.1 Baseline

As a baseline, vanilla DE will be compared to two common genomic prediction methods using the entire genome: (1) GBLUP; a functionally equivalent method to SNPBLUP. (2) BayesR; this is considered the state-of-the-art method for these experiments. The BayesR experiments use 50,000 iterations with a burn-in of 20,000. In addition to these methods, random search is also presented using 500 uniformly sampled subsets of size 1000 that are evaluated with SNPBLUP.

See Figure 31.2 for the results of this baseline study. It is clear that DE alone on this problem was not enough to be competitive with the state of the art method. However, there is something to be said for the comparison against GBLUP. GBLUP is essentially the fitness function for DE, which shows that the same accuracy was obtained using 50× less markers. As expected, random search provides much worse solutions than any other baseline methods, showing that DE is accomplishing something. Now that this baseline has been established, the goal is to reduce the gap between DE and BayesR.

31.5.2 Controlling Overfitting

As evidence of overfitting, consider Figure 31.2(b), the fitness convergence plot for ten replicates of vanilla DE. As was discussed in Section 31.3.5, there is a theoretical maximum on performance for genomic prediction tasks. The convergence plot shows that the validation accuracy in the baseline experiment exceeded $h \approx 0.63$. This suggests that the feature subsets selected after this point were overfit to the fixed validation set—which may have led to the underperforming testing results in

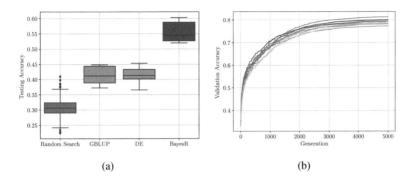

(a)   (b)

Fig. 31.2: (a) Boxplots comparing the vanilla DE experiment to the various baseline methods. The random search results were obtained by evaluating 500 uniformly sampled feature subsets. (b) Convergence plot for the maximum fitness at each generation for all 10 replicates in the vanilla DE experiment.

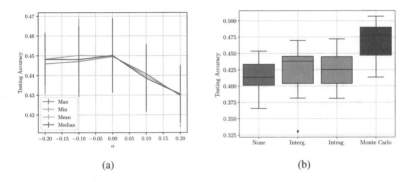

(a)   (b)

Fig. 31.3: (a) Plots of testing accuracy for each statistic used with a given α in the heritability threshold formula $h(1 + \alpha)$. Error bars show standard deviation. (b) Boxplots comparing the different cross-validation strategies.

Figure 31.2(a). This is a well known problem in wrapper method feature selection [48].

To remedy this, two preventative measures are proposed. First, the heritability thresholding discussed in Section 31.3.5 will be carried out for varying values of α in the formula $h(1 + \alpha)$. More concretely, when some statistic of the population reaches $h(1 + \alpha)$ the search will be stopped. Second, the cross-validation schemes discussed in Section 31.4.2 will be carried out as well.

The results of the thresholding experiments are presented in Figure 31.3(a). The plot shows the thresholding experiment for $\alpha \in \{-0.2, -0.1, 0, 0.1, 0.2\}$. The statistics of the population used to compare to $h(1 + \alpha)$ were the minimum, maximum, mean, and median validation accuracy. It is clear that there was no significant difference between any one statistic used. However, nonnegative α values showed a

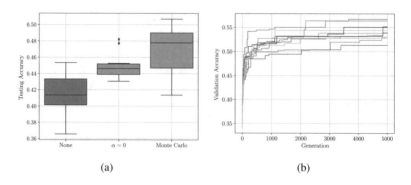

(a)   (b)

Fig. 31.4: (a) Boxplots comparing the DE with no overfitting control, the $\alpha = 0$ using minimum fitness heritability threshold experiment, and DE with Monte Carlo cross-validation. (b) The convergence plot of the 10 DE with Monte Carlo cross-validation experiments.

decline in testing accuracy as α increased. This means that as validation accuracy increased past the value of h, testing accuracy decreased. There was no real trend in negative values of α.

Figure 31.3(b) shows the results of the cross-validation experiments. There was no significant difference between intergenerational and intragenerational cross-validation. However, Monte Carlo showed a significant improvement over both strategies. Some indication as to why this may have occurred is shown in Figure 31.4(b). The convergence graph for the Monte Carlo DE search shows it never reaching $h = 0.63$. Figure 31.4(a) shows the comparison of this method to the thresholding method with $\alpha = 0$. It is clear that some method to control overfitting increases testing performance over vanilla DE, however, Monte Carlo search likely has an advantage since it controls overfitting while continuing to search.

31.5.3 Improving Performance

The two methods proposed to improve performance are seeding the initial population (see Section 31.3.4) and self-adaptive DE (see Section 31.3.3). The first is intended to start the search out at a good performance using the results of a GWAS. The self-adaptive method is intended to simply boost performance through better convergence properties and to askewing the need for a stringent parameter sweep on C_r and F. In addition, these methods were combined and tested for efficacy.

Figure 31.5 shows the results of the combination experiments. Using seeding does not provide a significant boost in performance compared with vanilla DE. Intuitively, using seeding without an overfitting control mechanism simply leads to a search that converges quickly to an overfit solution. Hence, the results shown in Figure 31.5 do not show a significant difference from the unseeded experiment.

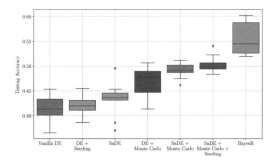

Fig. 31.5: Boxplots comparing the combination experiments and the BayesR results.

Similarly, the self-adaptive method shows a nominal improvement over vanilla DE. Once Monte Carlo cross-validation is applied to the more sophisticated self-adaptive method, the variance displayed in the SaDE experiments with no cross-validation was greatly reduced along with increasing over performance. When the self-adaptive method was combined with seeding, a slight—but not significant—increase in performance was observed. In all cases, DE underperformed when compared to the state-of-the-art method, BayesR.

31.5.4 Validation

The results presented above were a hand tuning of the DE system by adding in many components intended to increase performance. To ensure our resulting "best" configuration was not only well suited to the dataset used in the tuning experiments, a new phenotype is considered.

For this validation study, a real phenotype is used with the same sheep genotypes. In other words, our genotype is still 7,539 rows with 48,588. The heritability of the trait is estimated to be $h^2 \approx 0.16$. This value was obtained using the restricted maximum likelihood approach implemented in the NAM R package [56]. Therefore, the theoretical bound on prediction accuracy is $h \approx 0.4$.

Similarly to the baseline study, we compared DE against random search, GBLUP, and BayesR. To verify that the SaDE method with seeding and Monte Carlo does perform well, we also included it in the baseline. The results of this study are presented in Figure 31.6. The maximum obtained accuracy by the SaDE search method was 0.3993, which is on par with the estimated theoretical maximum of 0.4. However, this does not suggest DE is always better than the common methods in a non-simulated environment. As was pointed out previously, the trait being predicted is not very heritable. Because of this, it is likely that the performance of BayesR suffered greatly in comparison to the simulated environment where the effects of the QTL—and markers in linkage disequilibrium with the QTL—were large and well

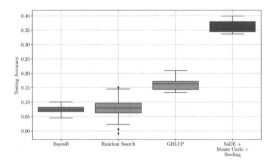

Fig. 31.6: Boxplots comparing the baseline methods to the best obtained DE method. Random search was again done by evaluating 500 uniformly sampled feature subsets.

defined. That said, these results do suggest that DE based feature selection may be better at identifying marker subsets that better capture relationships between animals, while removing noisy markers.

31.6 Discussion

With 1,000 SNP, feature selection with DE performed on par with BayesR and outperformed RRBLUP, even though the latter is mostly due to the structure of the simulation – with real data, the differences are generally negligible. In practice, at this point, we expect the three methods to be largely comparable with each other. But it is noteworthy that by using the DE, the same results are achievable with 1,000 SNP instead of almost 50,000. The DE captured a reasonable proportion of the *real* underlying causal variants and other features in the data that approximated the genetic architecture. This does bring us closer to prediction models based on real causative variants which has several advantages in relation to whole-genome models: 1) there is an immediate benefit for industry to have smaller panels since the production costs are lower and it will enable wider adoption of the technology; 2) whole-genome methods rely on relationships between individuals, the accuracy of prediction in distantly related ones is very low – functional panels can be expected to hold accuracy irrespective of the genetic distances between the discovery and validation populations; 3) can provide new biological insights and novel candidate targets for clinical intervention.

However, performance is still lacking. We saw good results with 1,000 SNP but of course, for this particular data set the objective was to evolve a panel of high accuracy with only 100 SNP. Even the panel with 1,000 SNP only included 15 QTL, with a sizable proportion of the accuracy still coming from the DE using the SNP to optimize the relationship structure between individuals in a manner that maximized the accuracy. For comparison purposes it is interesting to SaDE that simply using the top 100 SNP from the GWAS does a better job than DE, RRBLUP or BayesR. This is, by design, just an artifact of the simulation since many QTL have very high LOD scores (Figure 31.1) which are much higher than usually observed with complex polygenic traits; but it does suggest that there is scope to use GWAS results to *seed* the DE runs in the future.

We believe that the underperfomance of the DE with 100 SNP is largely attributable to overfitting of the data. During the search, there is only a finite amount of data to evaluate and assign fitness with and the search is guided by the performance on the validation set. Therefore, it is more likely to select SNPs that perform well on the validation data set alone, leading to overfitting. More evidence of this comes from observations on the heritability of our trait. Because $h^2 = 0.3855$, we know that about 39% of the variation in the phenotype is accounted for by the effects from the QTL in the genotypes. We can expect our best possible accuracy to be $h^2/\sqrt{h^2}$. Figure 31.7 shows the convergence of the 1,000 feature subset DE experiment. This shows our search obtaining values much higher than this expected maximum accuracy of 0.6208, which of course should not be possible. Hence, our search overfits to our validation set, which is a well known problem of wrapper methods in feature selection [48].

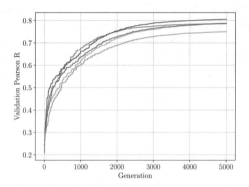

Fig. 31.7: Convergence plot for the 1000 feature subset DE experiment. The separate lines show the five replicates.

The smaller subset experiments likely converge into local optima quickly and their searches stall before finding good solutions. This may be due to the comparative dimensionalities of the encoding and the solution space. More precisely, we care only about 10 or 100 entries out of a vector in $\mathbb{R}^{48,588}$, which may be too large of a discrepancy to find any promising solutions. This could cause the search to converge to the first promising candidate solutions and not break out of the local optimum – both subsets of size 10 and 100 rapidly converged onto the 4 largest QTL. While for this work we wanted to evaluate the algorithm with hardset feature numbers, in real world scenarios it is better to co-evolve the number of selected features as an additional parameter in the algorithm.

31.7 Future Work

It is clear that applying domain knowledge is a promising avenue for DE search in this application. Evolutionary computation approaches often can easily accommodate prior knowledge about a problem to enhance their performance. A further opportunity for applying domain knowledge is to consider epistasis. Our method, in its current form, would not effectively find feature subsets with epistatic effects since they are not explicitly modelled by RRBLUP; even though in practice, they are implicitly captured to some extent through the genetic relationships in closely related populations. But more broadly, epistatic effects are non-linear and therefore cannot be captured with ridge regression. However, with minimal effort, a model that can capture non-linear effects—e.g., a small neural network—can be used in the DE fitness function to find feature subsets corresponding to markers that have epistatic interactions with each other.

We believe that alternative AI approaches like DE will become necessary when applying genomic prediction to full sequence data. Genome-wide association studies are well known to not be able to identify many variants that all have small effects [31] and the multiple testing problem with millions of variants will further confound the issue. For genomic prediction, there is very little to gain from sequence data with RRBLUP (or the equivalent GBLUP) as the changes to the genomic relationship matrix are minimal, and consequently the predictions will essentially be the same even if millions of additional SNP are included in the model. Bayesian approaches should be better able to discern effects and it is expected that these methods should have higher accuracy, in the same way as BayesR did with our simulation – but they are currently computationally intractable at the sequence level.

31.8 Conclusion

We have presented a competitive feature selection algorithm for genomic prediction problems. Currently, DE performs competitively with genome-wide methods using a fraction of the number of SNP and seems a promising alternative to current prediction methods. Feature selection with DE, or any other method, has the advantage of giving an interpretable result. With the rise of complex non-linear approaches like deep convolutional and recurrent neural networks becoming popular in bioinformatics and biomedicine, a precise and readable result could lead to a more complete understanding of the complex genotype-phenotype mapping. Differential evolution presents a non-statistical alternative to current state of the art methods with potential to easily apply domain knowledge to this difficult problem.

Acknowledgements We dedicate this chapter to BEACON's director Erik Goodman – advisor, colleague, mentor and friend. This work was supported by the National Science Foundation under Cooperative Agreement No. DBI-0939454 (BEACON 2012); by the Next-Generation BioGreen 21

Program (Project No. PJ01322204), Rural Development Administration, Republic of Korea and by the National Institute of Food and Agriculture (AFRI Project No. 2019-67015-29323).

References

1. Abraham, G., Tye-Din, J.A., Bhalala, O.G., Kowalczyk, A., Zobel, J., Inouye, M.: *Accurate and robust genomic prediction of celiac disease using statistical learning.* PLoS Genetics **10**, e1004137 (2014)
2. Al-Mamum, H.A., Kwan, P., Clark, S., Lee, S.H., Song, K.D., Lee, S.H., Gondro, C.: *Genomic best linear unbiased prediction using differential evolution.* In: Proceedings of the AAABG 21st Conference, pp. 145–148 (2015)
3. Altidor, W., Khoshgoftaar, T.M., Van Hulse, J.: *Robustness of filter-based feature ranking: A case study.* In: Proceedings of the Twenty-Fourth International Florida Artificial Intelligence Research Society (FLAIRS) Conference, pp. 453–458 (2011)
4. Banzhaf, W., Nordin, P., Keller, R., Francone, F.: *Genetic Programming - An Introduction.* Morgan Morgan Kaufmann Publishers Inc. (1998)
5. Bean, J.C.: *Genetic algorithms and random keys for sequencing and optimization.* ORSA Journal on Computing **6**(2), 154–160 (1994)
6. Bhattacharyya, S., Sengupta, A., Chakraborti, T., Konar, A., Tibarewala, D.N.: *Automatic feature selection of motor imagery EEG signals using differential evolution and learning automata.* Medical & Biological Engineering & Computing **52**(2), 131–139 (2014)
7. Biesiada, J., Duch, W.: *A Kolmogorov-Smirnov Correlation-Based Filter for Microarray Data.* In: M. Ishikawa, K. Doya, H. Miyamoto, T. Yamakawa (eds.) Neural Information Processing, pp. 285–294. Springer (2008)
8. Bolón-Canedo, V., Sánchez-Marono, N., Alonso-Betanzos, A., Benítez, J.M., Herrera, F.: *A review of microarray datasets and applied feature selection methods.* Information Sciences **282**, 111–135 (2014)
9. Clark, S.A., Hickey, J.M., van der Werf, J.H.: *Different models of genetic variation and their effect on genomic evaluation.* Genetics Selection Evolution **43**(1), 18 (2011)
10. Das, S., Suganthan, P.: *Differential evolution: A survey of the state-of-the-art.* IEEE Transactions on Evolutionary Computation **15**, 4–31 (2011)
11. De Los Campos, G., Naya, H., Gianola, D., Crossa, J., Legarra, A., Manfredi, E., Weigel, K., Cotes, J.M.: *Predicting quantitative traits with regression models for dense molecular markers and pedigree.* Genetics **182**(1), 375–385 (2009)
12. Ding, C., Peng, H.: *Minimum redundancy feature selection from microarray gene expression data.* Journal of Bioinformatics and Computational Biology **3**(02), 185–205 (2005)
13. Dorigo, M., Maniezzo, V., Colorni, A.: *Ant system: Optimization by a colony of cooperating agents.* IEEE Transactions on Systems, Man, and Cybernetics, Part B (Cybernetics) **26**(1), 29–41 (1996)
14. Eiben, A., Smith, J.E.: *Introduction to Evolutionary Computing.* Springer-Verlag Berlin Heidelberg (2003)
15. Erbe, M., Hayes, B.J., Matukumalli, L.K., Goswami, S., Bowman, P.J., Reich, C.M., Mason, B.A., Goddard, M.E.: *Improving accuracy of genomic predictions within and between dairy cattle breeds with imputed high-density single nucleotide polymorphism panels.* Journal of Dairy Science **95**(7), 4114–4129 (2011)
16. Esquivelzeta-Rabell, C., Al-Mamum, H.A., Lee, S.H., Song, K.D., Gondro, C.: *Evolving to The Best SNP panel for Hanwoo Breed Proportion Estimates.* In: Proceedings of the AAABG 21st Conference, pp. 473–476 (2015)
17. Firpi, H.A., Goodman, E.: *Swarmed feature selection.* In: 33rd Applied Imagery Pattern Recognition Workshop (AIPR'04), pp. 112–118 (2004)

18. Forneris, N.S., Legarra, A., Vitezica, Z.G., Tsuruta, S., Aguilar, I., Misztal, I., Cantet, R.J.C.: *Quality control of genotypes using heritability estimates of gene content at the marker.* Genetics **199**(3), 675–681 (2015)
19. Goddard, M.: *Genomic selection: Prediction of accuracy and maximisation of long term response.* Genetica **136**(2), 245–257 (2009)
20. Goddard, M., Hayes, B., Meuwissen, T.: *Using the genomic relationship matrix to predict the accuracy of genomic selection.* Journal of Animal Breeding and Genetics **128**(6), 409–421 (2011)
21. Goddard, M.E., Hayes, B.J.: *Genomic selection.* Journal of Animal Breeding and Genetics **124**(6), 323–330 (2007)
22. Goldberg, D.: *Genetic Algorithms in Search, Optmization and Machine Learning.* Addison Wesley (1989)
23. Gondro, C.: *Primer to Analysis of Genomic Data Using R.* Springer International Publishing (2015)
24. Habier, D., Fernando, R.L., Dekkers, J.C.M.: *The impact of genetic relationship information on genome-assisted breeding values.* Genetics **177**(4), 2389–2397 (2007)
25. Habier, D., Fernando, R.L., Garrick, D.J.: *Genomic BLUP Decoded: A Look into the Black Box of Genomic Prediction.* Genetics **194**(3), 597–607 (2013)
26. Habier, D., Fernando, R.L., Kizilkaya, K., Garrick, D.J.: *Extension of the Bayesian alphabet for genomic selection.* BMC Bioinformatics **12**(1), 186 (2011)
27. Hayes, B., Bowman, P., Chamberlain, A., Goddard, M.: *Invited review: Genomic selection in dairy cattle: Progress and challenges.* Journal of Dairy Science **92**(2), 433–443 (2009)
28. Hayes, B.J., Pryce, J., Chamberlain, A.J., Bowman, P.J., Goddard, M.E.: *Genetic architecture of complex traits and accuracy of genomic prediction: Coat colour, milk-fat percentage, and type in Holstein cattle as contrasting model traits.* PLoS Genetics **6**, e1001,139 (2010)
29. Heffner, E.L., Jannink, J.-L., Sorrells, M.E.: *Genomic selection accuracy using multifamily prediction models in a wheat breeding program.* The Plant Genome **4**, 65–75 (2011)
30. van Heel, D.A., Franke, L., Hunt, K.A., Gwilliam, R., Zhernakova, A., Inouye, M., Wapenaar, M.C., Barnardo, M.C.N.M., Bethel, G., Holmes, G.K.T., Feighery, C., Jewell, D., Kelleher, D., Kumar, P., Travis, S., Walters, J.R., Sanders, D.S., Howdle, P., Swift, J., Playford, R.J., McLaren, W.M., Mearin, M.L., Mulder, C.J., McManus, R., McGinnis, R., Cardon, L.R., Deloukas, P., Wijmenga, C.: *A genome-wide association study for celiac disease identifies risk variants in the region harboring IL2 and IL21.* Nature Genetics **39**, 827–829 (2007)
31. Hirschhorn, J.N., Daly, M.J.: *Genome-wide association studies for common diseases and complex traits.* Nature Reviews Genetics **6**(2), 95–108 (2005)
32. Holland, J.: *Adaptation in Natural and Artificial Systems: An Introductory Analysis with Applications to Biology, Control and Artificial Intelligence.* University of Michigan Press, Ann Arbor (1975)
33. Islam, S.M., Das, S., Ghosh, S., Roy, S., Suganthan, P.N.: *An adaptive differential evolution algorithm with novel mutation and crossover strategies for global numerical optimization.* IEEE Transactions on Systems, Man, and Cybernetics, Part B (Cybernetics) **42**(2), 482–500 (2012)
34. Khushaba, R.N., Al-Ani, A., Al-Jumaily, A.: *Differential evolution based feature subset selection.* In: 2008 19th International Conference on Pattern Recognition, pp. 1–4 (2008)
35. Khushaba, R.N., Al-Ani, A., AlSukker, A., Al-Jumaily, A.: *A combined ant colony and differential evolution feature selection algorithm.* In: M. Dorigo, M. Birattari, C. Blum, M. Clerc, T. Stützle, A.F.T. Winfield (eds.) Ant Colony Optimization and Swarm Intelligence, pp. 1–12. Springer (2008)
36. Koza, J.R.: *Genetic Programming: On the Programming of Computers by Means of Natural Selection.* MIT Press, Cambridge, MA, USA (1992)
37. Kwong, Q.B., Ong, A.L., Teh, C.K., Chew, F.T., Tammi, M., Mayes, S., Kulaveerasingam, H., Yeoh, S.H., Harikrishna, J.A., Appleton, D.R.: *Genomic Selection in Commercial Perennial Crops: Applicability and Improvement in Oil Palm (Elaeis guineensis Jacq.).* Scientific Reports **7**, 2872 (2017)

38. Luque-Baena, R., Urda, D., Claros, M.G., Franco, L., Jerez, J.: *Robust gene signatures from microarray data using genetic algorithms enriched with biological pathway keywords*. Journal of Biomedical Informatics **49**, 32 – 44 (2014)
39. Luque-Baena, R.M., Urda, D., Subirats, J.L., Franco, L., Jerez, J.M.: *Application of genetic algorithms and constructive neural networks for the analysis of microarray cancer data*. Theor Biol Med Model **11**(Suppl 1), S7–S7 (2014)
40. Meuwissen, T.H.E., Hayes, B.J., Goddard, M.E.: *Prediction of total genetic value using genome-wide dense marker maps*. Genetics **157**(4), 1819–1829 (2001)
41. Moghaddar, N., Swan, A.A., Van Der Werf, J.H.J.: *Comparing genomic prediction accuracy from purebred, crossbred and combined purebred and crossbred reference populations in sheep*. Genetics Selection Evolution **46**, 58 (2014)
42. Nearchou, A.C., Omirou, S.L.: *Differential evolution for sequencing and scheduling optimization*. Journal of Heuristics **12**(6), 395–411 (2006)
43. Patterson, H.D., Thompson, R.: *Recovery of inter-block information when block sizes are unequal*. Biometrika **58**(3), 545–554 (1971)
44. Peng, H., Long, F., Ding, C.: *Feature selection based on mutual information criteria of max-dependency, max-relevance, and min-redundancy*. IEEE Transactions on Pattern Analysis and Machine Intelligence **27**(8), 1226–1238 (2005)
45. Qin, A.K., Suganthan, P.N.: *Self-adaptive differential evolution algorithm for numerical optimization*. In: 2005 IEEE Congress on Evolutionary Computation, vol. 2, pp. 1785–1791 (2005)
46. Raymer, M.L., Punch, W.F., Goodman, E.D., Kuhn, L.A., Jain, A.K.: *Dimensionality reduction using genetic algorithms*. IEEE Transactions on Evolutionary Computation **4**(2), 164–171 (2000)
47. Rechenberg, I.: *Evolutionsstrategie*. Holzmann-Froboog, Stuttgart (1975)
48. Saeys, Y., Inza, I.n., Larrañaga, P.: *A review of feature selection techniques in bioinformatics*. Bioinformatics **23**(19), 2507–2517 (2007)
49. Schwefel, H.P.: *Evolution and Optimum Seeking*. Wiley (1995)
50. Storn, R., Price, K.: *Differential evolution – a simple and efficient heuristic for global optimization over continuous spaces*. Journal of Global Optimization **11**(4), 341–359 (1997)
51. Su, G., Brndum, R., Ma, P., Guldbrandtsen, B., Aamand, G., Lund, M.: *Comparison of genomic predictions using medium-density (54,000) and high-density (777,000) single nucleotide polymorphism marker panels in Nordic Holstein and Red Dairy Cattle populations*. Journal of Dairy Science **95**(8), 4657 – 4665 (2012)
52. Van Raden, P.M.: *Efficient methods to compute genomic predictions*. Journal of Dairy Science **91**(11), 4414–4423 (2008)
53. Visscher, P.M., Wray, N.R., Zhang, Q., Sklar, P., McCarthy, M.I., Brown, M.A., Yang, J.: *10 Years of GWAS Discovery: Biology, Function, and Translation*. The American Journal of Human Genetics **101**(1), 5 – 22 (2017)
54. Whittaker, J.C., Thompson, R., Denham, M.C.: *Marker-assisted selection using ridge regression*. Genetical Research **75**(2), 249–252 (2000)
55. Wientjes, Y.C.J., Veerkamp, R.F., Calus, M.P.L.: *The effect of linkage disequilibrium and family relationships on the reliability of genomic prediction*. Genetics **193**(2), 621–631 (2013)
56. Xavier, A., Xu, S., Muir, W., Rainey, K.: *NAM: Association studies in multiple populations*. Bioinformatics **31**(23), 3862–3864 (2015)
57. Zapata-Valenzuela, J., Whetten, R.W., Neale, D.B., McKeand, S.E., Isik, F.: *Genomic Estimated Breeding Values Using Genomic Relationship Matrices in a Cloned Population of Loblolly Pine*. G3: Genes, Genomes, Genetics **3**, 906–916 (2013)

Part VI
Evolution Education

Chapter 32
Overcoming Classroom Skepticism with Evolution in Action

J. Jeffrey Morris

Abstract Public acceptance of evolution is conspicuously low in the United States. This is especially true in the American Southeast, where the famous "Scopes Monkey Trial" made the religious objections of rural Tennesseans to Darwin's theory famous. In this piece I argue that disbelief in evolution in the Southeast is caused more by social factors than by scientific ones, and I present some efforts that I have successfully used to side-step these social issues in order to effectively teach evolutionary biology to diverse college students in Alabama. I close by calling on other educators to work to defuse the religious, political, and social minefields that separate academic biologists and the public in order to be more effective communicators and teachers of evolution.

In 2015 I accepted a faculty position at the University of Alabama at Birmingham. I was very excited about this opportunity – not only am I a native Southerner, but Alabama is also an under-appreciated reservoir of biological awesomeness. For instance, according to a global biodiversity mapping effort [10], Alabama has more species of fish and trees than any other state in the US. Many of these species are endemic or threatened, marking Alabama as a prime candidate for increased federal land protection [11]. E.O. Wilson – a Harvard evolutionary biologist and Alabama native who maintains close ties to the state – has joined with others to advocate for the creation of a National Park in Alabama's Mobile-Tensaw Delta to protect and showcase this treasure trove of organisms [15]. Couple all that with Alabama's growing biotech infrastructure, its history of technological innovation driven by its role in the US manned spaceflight program, and the fact that it almost never snows down here, and it sounded to me like a pretty swell place to be a professor.

J. Jeffrey Morris
Department of Biology
University of Alabama at Birmingham
e-mail: evolve@uab.edu

© Springer Nature Switzerland AG 2020
W. Banzhaf et al. (eds.), *Evolution in Action: Past, Present and Future*,
Genetic and Evolutionary Computation, https://doi.org/10.1007/978-3-030-39831-6_32

In spite of that, some of my colleagues were surprised that I had accepted the job. In the world of postdocs, it's usually a joyous occasion when somebody gets the golden ticket to a real tenure-track slot. But when I told some colleagues that I was headed to Alabama, their attitude went from unreserved happiness for my success to a sort of confused middle-ground where they weren't sure if they should maybe console me instead. I became aware that some of the other postdocs in my social network wouldn't even countenance *applying* for jobs in the South, despite the brutal competition for limited tenure-track positions in the US. A common refrain about this widespread refusal to work in the South was that nobody wanted to deal with teaching evolutionary biology to a bunch of fundamentalist Christians.

Regardless, I was excited, and jumped into my new position headfirst. I've been fascinated by evolutionary biology since I was a little kid, and I injected it front-line-and-center into every course I developed at UAB. I fully embrace the ideas outlined in *Vision and Change* [3] for improving US biology education: *everything* should go back to the core concepts of biology, and evolution is the most foundational and unifying idea across all biological disciplines. Perhaps it will surprise you to learn, then, that I haven't received any backlash from my (statistically) overwhelmingly Christian students. Certainly my colleagues have reported difficulties, but not me. I think the root of my success – or at least my avoidance of conflict – lies in my focus on contemporary "evolution in action" as opposed to deep-time evolutionary biology. In this article I lay out why I think this technique is an effective way to overcome the stubborn resistance to the acceptance of evolutionary theory in some parts of the world.

32.1 Why Do Southerners Resist Accepting Evolutionary Theory?

70% or more of Southerners refuse to accept the validity of evolutionary theory [1]. This fact has been drilled into the pop culture psyche of the US as far back as the 1950's, when the play (and later movie) *Inherit the Wind* brought a dramatized version of the "Scopes Monkey Trial" into the public eye. *Inherit the Wind* chronicles the show trial of a substitute teacher in rural Tennessee who taught the theory of natural selection in his high school classroom in opposition to a recently-passed law forbidding it. While both sides of this trial were openly Christian, the prosecution favored a much more fundamentalist, literal interpretation of the Bible, argued by the flamboyant orator (and former presidential candidate) William Jennings Bryant. Perhaps as a consequence, the connection between rejection of evolution and fundamentalist Christianity – and particularly Southern fundamentalist Christianity – gained widespread notoriety.

I can verify from personal experience that Southern fundamentalist ministers rail against Darwin and evolution enthusiastically and frequently. However, I am not convinced that rejection of evolution by Southern Christians is actually motivated by religion per se. Rather, I believe it is exacerbated and reinforced by the cultural conflict between the largely rural South and the urban elites that have traditionally dominated both academia and the media. Many writers – including ardently atheistic, hard-leftist academics like Stephen Jay Gould – have acknowledged that there is no *intrinsic* reason why one cannot both believe in God and accept evolution [9], and it is often pointed out that many openly religious people, including prominent members of the Catholic clergy, have endorsed the compatibility of Darwinism and Christian thought [8, 14]. But there might be extrinsic reasons that the two do not rest well together, at least in the South. For example, consider some of these salty quotes from H.L. Mencken's widely-distributed accounts of the Scopes Trial [12] (emphasis mine):

> "The **inferior man's** reasons for hating knowledge are not hard to discern. He hates it because it is complex – because it puts an unbearable burden upon **his meager capacity** for taking in ideas. Thus his search is always for short cuts. All **superstitions** are such short cuts."
>
> "But when the journalists swarmed down, and their dispatches began to show the country and the world exactly how **the obscene buffoonery** appeared to realistic city men, then the yokels began to sweat coldly, and in a few days they were full of terror and indignation. Some of the bolder spirits, indeed, talked gaudily of direct action against the authors of the "libels." But the history of the **Ku Klux** and the American Legion offers overwhelmingly evidence that 100 per cent Americans never fight when the enemy is in strength"

Here we see open hostility directed toward the people who reject evolution. Religion is a one-dimensional joke, an embarrassing anachronism. The people of Dayton, Tennessee, are superstitious yokels, juxtaposed against good realistic city folk like Mencken and his audience. In a grotesque example reminiscent of today's social media-fueled political angst, we even see Fundamentalists compared to the Ku Klux Klan. Imagine being a Tennessean with no strong opinion one way or the other about evolution and reading an article by H.L. Mencken basically calling your neighbors troglodytes for believing in the Bible – is it any wonder you would develop a negative opinion of the topic?

One sees similar – or even more vehement – denunciations of both Christianity and Southerners coming from academic biologists. While Richard Dawkins has authored some of the most accessible and paradigm-shifting books about evolution, he is better known as an anti-religious firebrand; provocatively titled books like *The God Delusion* along with frequent derisive comments on both social and legacy media platforms emphasize a connection between evolutionary biology and antipathy toward religious people. Dawkins isn't alone in this; personal experience (and a quick click-tour around academic Twitter) tells me that visceral contempt toward religion (and the religious South) is common amongst professional biologists.

I believe that this top-down hostility coming from academia and the media contributes to the polarization of US society into "civilized people who 'believe' in evolution" and "Christians who reject evolution". Under tribal conditions like this, people are motivated to signal their allegiance to one group or another, and it is my belief that prominent displays of evolution rejection act as honest signals of group loyalty that help Christians cohere as a group in the face of what they perceive as an existential threat to their way of life. Ironically, Dawkins [6] talked about this kind of thing as a "greenbeard" – some kind of blatantly obvious "tell", like having a long green beard, that lets group members tell each other apart from the rest of the population, allowing the benefits of altruistic behaviors to be faithfully targeted only toward other altruists [4]. Rejection of evolution is a great greenbeard because it is both costly and group-specific. The cost is social, in that a person who rejects evolution is unlikely to be welcome in the elite social groups that fundamentalist Christians perceive as their enemies. Similarly, rejection of evolution is so thoroughly associated with fundamentalist Christianity (at least in the US) that it is hard to imagine it coming from a member of any other social group.

Importantly, if people's rejection of evolution is a greenbeard, this suggests that *no amount of education will affect its prevalence*. Because its value as a greenbeard has nothing to do with the rightness or wrongness of the theory, no amount of logic or evidence will sway peoples' attitudes. Indeed, students who reject evolution generally understand the theory as well as those who accept it (studies summarized in Barnes et al. [2]). If we want to get more people in the South to accept evolution, we don't need to try harder to teach the facts – we need to eliminate the greenbeard value of evolution rejection. One way to do this is to increase the cost of rejecting evolution by showing that evolutionary theory has direct and immediate impacts on students' daily lives. My strategy to accomplish this has been to make evolution real and immediate to students using contemporary examples of evolution in action in my courses.

32.2 Evolution in Action in the Classroom

Most people – including many scientists – think of evolution as a slow process that takes extremely long periods of time to produce observable change. This "deep-time" evolution is relatively easy to reject because it basically has no bearing on day-to-day human existence. Like other technical fields that deal with deep-time questions – e.g. cosmology or geology – whether you believe one explanation or another for the current state of the universe has no *practical* effect on your life. But when the theories that underpin those fields start to be relevant to questions that might make us money or save our lives, all of that changes. For instance, whether or not the continents used to be joined as Pangaea millions of years ago might be viewed as geologic trivia, but the ability to use plate tectonics theory to understand or even predict earthquakes obviously has immediate practical utility.

In my view, evolution is the same way. When evolution courses focus on deep-time, religious greenbearding doesn't carry any risks – for the purposes of day-to-day decision making, it simply doesn't matter whether humans evolved or were created from dust 6000 years ago. But evolution takes place all around us, all the time, and in some cases changes happen so fast you can see the effects not just in a single lifetime, *but in the space of a few months*. In these cases, rejecting evolutionary theory carries a real and immediate risk. In these cases, greenbearding is no longer just a signal; it is a *gamble*, and the rejector risks making costly wrong decisions. Thus, my courses focus on these systems – rapidly evolving populations that have economic, cultural, or health impacts on living human beings.

During my time as a postdoc at the BEACON Center for the Study of Evolution in Action at Michigan State University in the early 2010's, I became familiar with many of these examples of "evolution in action". For example, BEACON researchers use evolutionary theory to craft genetic computer programs that develop unexpected solutions to real-world problems (e.g. [7]). Others use mutating computer viruses that compete over computer resources to study how evolution works in "living" things that aren't made of DNA and protein [4, 13]. My own work dealt with microbial evolution, where gigantic population sizes and short generation times allow bacteria and viruses to adapt to changed environments extremely rapidly – in some cases being able to radically change their metabolism over a few days. On the one hand, microbial evolution towards antibiotic resistance is one of the greatest threats facing our species, but biotechnological innovators are also able to use "directed evolution" to produce biological solutions to many of our most difficult industrial and infrastructure problems. Also, the burgeoning field of microbiome research promises to change how we think about medicine and human health, and understanding the rules by which microbe/animal symbioses work clearly requires an understanding of the rapid evolution going on in the microbial component.

The above examples are all excellent fodder for the classroom. For example, students can download and run the computational evolution software for themselves, letting them get first-hand observations about the effects of random change and selection on functionality in an environment completely devoid of religious or moral judgment [16]. And the microbial stuff is all over the news, from horror stories about MRSA to the glut of pro- and prebiotic supplements hitting the market. Unlike the deep-time evolution of humanity or the evolution of modern birds from dinosaur ancestors, students can see these systems changing under selection in their own lifetimes. Because they are modern and directly observable, these systems don't challenge religious beliefs like deep-time phenomena do. But more importantly, evolution in action has real, concrete effects on the students' lives. Rather than being a costless greenbeard, to resist believing in the evolution of antibiotic resistance, or the power of evolutionary theory to improve engineering design or predict markets, would have a real, measurably negative impact on the student. In other words, by focusing on evolution in action, I've increased the cost of rejection of evolution that (possibly) negates its usefulness as a greenbeard.

32.3 Cop-out or Cultural Competence?

Personally, I'm pretty proud of the fact that I've been able to teach evolution in Alabama for five years without running into any serious bumps. However, a few people have told me that my approach is sort of a cop-out. In his popular blog, evolutionary biologist Jerry Coyne warns people of "accommodationism", the sin of watering down biology in order to keep the Christians at bay [5]. Am I guilty of accomodationism when I tiptoe around the evolution of primates in my course?

Possibly. But I would counter with a couple of points. First, we all know that you can't cover "all the material" in any introductory course. You have to pick and choose examples and lessons that best convey the core concepts of the field. So here we have to ask, what are the core concepts of evolutionary biology? If we believe that "humanity evolved from non-human animals" – a very specific statement – is one of those core concepts, then yes, I'm copping out and depriving my students in an effort to avoid religious resistance. On the other hand, if we believe (as I certainly do) that humans are only one of millions of species, and that the forces resulting in our own evolution are identical to those that produced oak trees, sharks, enteric bacteria, and all the rest, then we would be doing our students a disservice by focusing on a needlessly controversial example when other pedagogically effective options are available.

Second, we ought to ask ourselves what our job is in the biology classroom. As a teacher of evolution, am I supposed to be producing professional evolutionary biologists, or am I supposed to send people out into the world with an understanding of how the forces of natural selection and random drift impact our world? Should I count my success by how many students I convince to give up religion, or by how many people learn from me how to apply evolutionary concepts to real-world situations ranging from the clinic to their own bank accounts? One could argue that maintaining a hard focus on the ancient past actually reduces the ability of students to see the evolution in action all around them.

Last, and I think most importantly, I don't believe that skirting around hot-button issues like human evolution actually prevents students from learning about these things. In fact, I think it might be *more effective* at getting students to approach these kinds of topics. Nobody that walks into my classroom is unaware that the overwhelming majority of scientists believe humans evolved from non-human animals, but many are unaware of the more subtle evolutionary struggles happening around (and inside) them. The fact that the same math unites these contemporary examples with deep-time "macroevolution" can't be overlooked by the student, and perhaps in the fullness of time the puzzle pieces will click together in even the most skeptical mind if you can get it to suspend its disbelief long enough to learn those critical core concepts. By creating an environment where religious students can confront this cognitive dissonance privately, in their own time, I like to believe that I've planted a seed that will continue to work on them over their lifetimes, making it easier to accept evolutionary truths.

Rather than a cop-out, then, I think of my teaching approach as part of the broader attitude of tolerance of diversity that the modern university strives to embrace. Education researchers talk about the importance of "cultural competence" – the process of understanding and openly acknowledging the diverse backgrounds and beliefs of students as a way to create a comfortable classroom environment conducive to learning and collaboration [2]. The idea is often applied to racial and ethnic minorities, but it is certainly also apropos to religious students, who are even more numerous than underrepresented minority students. Just as it is important to acknowledge the history of conflict and bad blood between ethnic groups, we biologists have to own the fact that we haven't been passive players in the evolution of resistance to evolution; in the classrooms where we occupy positions of power and authority, we need to be the ones to extend the olive branch. A recent study showed that a mere *6 minutes* of class time dedicated to defusing religious animosities in an evolution class (e.g., by merely highlighting religious leaders who accept evolution and/or scientists who are Christians) had significant impacts on how much conflict students perceived between their religious beliefs and the course material [17]. While we should never compromise rigorous science to accommodate supernatural beliefs, we should, where possible, give religious students an opportunity to learn and experience our field without unnecessarily bombarding them with anti-religious sentiment. A consistent focus on modern-world evolution in action creates just this opportunity, and it's why I think I'm able to effectively teach evolution in one of the most Christian parts of the United States.

There was a time in the past when people didn't believe in heliocentrism. When the evidence was all abstruse mathematics, or dependent on observations using rare telescopes or other technologies, it was easy for common people to reject the ideas of astronomers in favor of biblical cosmology. However, as the laws of physics were worked out and their universality was demonstrated in countless terrestrial applications, people came to accept them because of their practical utility. The fact that the same mechanics clearly also applied to the motions of the celestial bodies made it inevitable that, given time, people would accept what the scientists of the day already knew. So too in evolutionary biology – focusing on the *practical* benefits of understanding evolution is the route to universal acceptance, not open confrontation with religion. Quite the contrary, the latter is possibly why it has taken so long for people to take the elegant and inevitable truths of evolution to heart.

References

1. Acceptance of evolution by state. Subnormal Numbers (2010). https://subnormalnumbers.blogspot.com/2010/04/acceptance-of-evolution-by-state.html
2. Barnes, M.E., Brownell, S.E., Perez, K.E.: *A Call to Use Cultural Competence When Teaching Evolution to Religious College Students: Introducing Religious Cultural Competence in Evolution Education (ReCCEE)*. CBE—Life Sciences Education **16**, es4 (2017)
3. Brewer, C.A., Smith, D., Bauerle, C., DePass, A., Lynn, D., O'Connor, C., Singer, S., Withers, M., Anderson, C.W., Donovan, S., Drew, S., Ebert-May, D., Gross, L., Hoskins, S.G., Labov, J., Lopatto, D., McClatchey, W., Varma-Nelson, P., Pelaez, N., Poston, M., Tanner, K., Wessner, D., White, H., Wood, W., Wubah, D.: Vision and change in undergraduate biology education: A call to action. In: C.A. Brewer, D. Smith (eds.) Vision and Change. American Association for the Advancement of Science, Washington, DC (2011)
4. Clune, J., Goldsby, H.J., Ofria, C., Pennock, R.T.: *Selective pressures for accurate altruism targeting: evidence from digital evolution for difficult-to-test aspects of inclusive fitness theory.* Proceedings of the Royal Society B: Biological Sciences **278**, 666–674 (2011)
5. Coyne, J.: *Why Evolution Is True.* "Accomodationism", Search term (2018)
6. Dawkins, R.: *The Selfish Gene.* Oxford University Press, Oxford (1976)

7. Dolson, E., Banzhaf, W., Ofria, C.: Applying ecological principles to genetic programming. In: W. Banzhaf, R. Olson, W. Tozier, R. Riolo (eds.) Genetic Programming Theory and Practice XV. Springer, Cham, Switzerland (2018)
8. Funk, C., Alper, B.A.: *Highly religious Americans are less likely than others to see conflict between faith and science.* Pew Research Center (2015)
9. Gould, S.J.: *Rocks of Ages: Science and Religion in the Fullness of Life.* Ballantine, New York, NY (1999)
10. Jenkins, C.N., Pimm, S.L., Joppa, L.N.: *Global patterns of terrestrial vertebrate diversity and conservation.* Proceedings of the National Academy of Sciences **110**, E2602–E2610 (2013)
11. Jenkins, C.N., Van Houtan, K.S., Pimm, S.L., Sexton, J.O.: *US protected lands mismatch biodiversity priorities.* Proceedings of the National Academy of Sciences **112**, 5081–5086 (2015)
12. Mencken, H.: *Coverage of the Scopes Trial.* Baltimore Evening Sun (1925)
13. Ofria, C., Wilke, C.O.: *Avida: A software platform for research in computational evolutionary biology.* Artificial Life **10**, 191–229 (2004)
14. PP: *Pope Pius XII: Encyclica Humani generis.* Vatican City (1950)
15. Raines, B.: *E.O. Wilson on Alabama as the center of American biodiversity and becoming a scientific rock star.* AL.com (2016)
16. Smith, J.J., Johnson, W.R., Lark, A.M., Mead, L.S., Wiser, M.J., Pennock, R.T.: *An Avida-ED digital evolution curriculum for undergraduate biology.* Evolution: Education and Outreach **9**, 1–11 (2016)
17. Truong, J.M., Barnes, M.E., Brownell, S.E.: *Can six minutes of culturally competent evolution education reduce students' level of perceived conflict between evolution and religion?* The American Biology Teacher **80**, 106–115 (2018)

Chapter 33
How to Increase Creativity in Research

Joan E. Strassmann

Abstract Few things are more important to a research career than discovery. A great idea is important and new. It might lead to the founding of a new field, or a new approach to an existing one, and sprout hundreds of additional studies. Where do these ideas come from? The answer to this question lies in the general field of creativity research. Here I discuss how creativity might be enhanced in ten steps. They involve being motivated to be creative, to learn creativity techniques, to interact with others outside your core group, to foster eureka moments, to recognize good new ideas when you see them, to work with others, to encourage creativity in others, to give feedback in a kind way, to defer ownership of ideas, and to learn science improvisation. The references in these sections give a peek into the growing field of creativity studies. Greater attention to and understanding of the sources of creative insight will help lead us to that magic idea that is new and important.

Introduction

What if I told you there was something important to all fields, including all those taught at universities, from business to engineering, from biology to history, from sociology to art? This something we know is important and we talk about all the time. Yet we are mostly unaware that it in itself is a scientific discipline. The results of research in this discipline can help all of us. What I am talking about is creativity, typically defined as the generation, evaluation, and application of new and useful ideas [1]. Actual innovation based on those ideas further requires that they be implemented [5, 22]. As far as I could determine, research into creativity is mostly

Joan E. Strassmann
Washington University in St. Louis, St. Louis, Missouri, USA
and Wissenschaftskolleg zu Berlin
e-mail: strassmann@wustl.edu

done by psychologists and sociologists who are mostly housed in business schools, though Buffalo State University has a center explicitly for creativity.

Another surprising thing about how we approach creativity is that the all important early stages of investigation are often left out of the scientific method. Beginning with the hypothesis leaves out the wondering, the observations, the brainstorming, and other creative work that went before. Even Platt in his famous paper on strong inference has as step one: "1) Devising alternative hypotheses" [20]. Even this has to be preceded by less directed exploration before appropriate hypotheses are likely to come to mind.

This minireview hopes to show how creativity research matters to science. In a way it follows Loehle's guide to creativity in research from nearly 20 years ago, a fabulous paper still important to read [15]. I summarize some of the major findings of this field in the form of advice for practitioners interested in becoming more creative. There are ten points explained, along with supporting and sometimes contradicting research. I am not an expert in this field and so my reading may have led me down rabbit holes that professionals would have avoided. Or my naiveté may have led me to see simplicity where others see only complexity. Ideally all of us should have the privilege of taking courses in creativity taught by experts, like the one taught by Markus Baer at my own institution, or one taught by Jing Zhou at my former institution, Rice University, or the creativity workshops put on by KnowInnovation, http://knowinnovation.com/.

There are three people that I see as formative for the discipline. Perhaps most important and general is Theresa Amabile and her book, *Creativity in context* [3]. Or just read her fantastic piece in the *Harvard Business Review* [4]. In this work, Amabile charts out what stimulates creativity, including expertise, motivation, and creative thinking skills. She goes on to show how to motivate creativity in others. The other two people take a more specific approach. Ronald Burt's book, *Structural holes: the social structure of competition*, talks about the importance of knowing individuals outside your core group and the importance of weak social ties [10]. An earlier book that has generated a lot of workshop techniques in creativity is Alex Osborn's *Applied imagination: principles and procedures of creative thinking* [18]. Seeing who is citing these works in recent years is a good way of getting an entrance into the field. Here are my ten points, distilled from my reading.

33.1 Be Motivated to be Creative

It may seem obvious that having creativity as an actual goal will increase its practice. If you want to be creative, you will be sensitive to the conditions that foster it. You will use creativity as something against which to judge ideas, choosing the one that seems to be more creative over the one that seems less so. But actual creativity can be scary. Why? A creative idea might take you on a path less travelled. That sounds good, but it might lead into an area not so popular, perhaps where funding is less, or invitations are fewer. You might be challenged more. What some might call crackpot creativity would be a form that does not lead to anything, as might happen with ideas that cannot be tested, or with projects that do not connect to what has gone before, so avoid that. What do I mean by connecting to what has gone before? It is simply that even creative ideas come from somewhere and should be connected to the literature, showing how they diverge. Help people see the connections to your new ideas in the fabric of knowledge. Or as Louis Pasteur famously said in 1854: "chance only favors the prepared mind" (Wikiquote.org).

Creativity might challenge leaders in ways that hurt you if not handled carefully. This is actually a big topic in business management research [4, 6, 31]. So remember that creativity is not always comfortable, but creative ideas, set in the context of what has gone before, are worth the risk and effort.

33.2 Separate Idea Generation from Idea Judgement

There is a lot known about explicit teaching of creativity that goes far beyond my favorite point: the importance of separating idea generation from idea judgement [22, 23]. Brainstorming is the name Osborn gave to this process that allowed ideas to be gathered free of judgement as the first step in creative problem solving [18]. Figure 33.1 illustrates an idea board that is the result of brainstorming on the question of how long bird pairs remain together. The approach of creative problem solving has been developed and expanded on particularly at Buffalo State University's International Center for Studies in Creativity [22]. These authors argue that "creativity is a teachable skill, one that makes creative thinking predictable, teachable, repeatable, and accessible for all"[22]). Their model for creativity is called the thinking skills model and has steps for "exploring the vision, formulating challenges, exploring ideas, formulating solutions, exploring acceptance, and formulating a plan."[22]. Each of these steps involves divergent and convergent thinking, separated in time, where divergent thinking is the brainstorming step of writing down all ideas without judging them.

There are other methods and also ways of judging their efficacy. For example, there is deBono's book on the six thinking hats [11]. These hats are like personalities that the student puts on for approaching a problem. These hats help people think from different perspectives. Puccio reviews other approaches to teaching creativity that probably all bear study [22]. For me, the take-away message is to try multiple different ways of solving problems.

John Platt, in a paper entitled *Strong Inference*, that had a huge influence on my own education, argues for a specific kind of creativity in science [20]. He argues for extreme care in designing experiments and explicit consideration of multiple hypotheses. The focus is to be on what the evidence disproves as much as what it supports, following Popper's perspective [21]. Furthermore, Platt talks about the importance of group discussion of ideas. After all, experiments take a long time, so we need to be sure to be doing the very best experiments that test ideas in the simplest, clearest possible ways.

Idea Board

Fig. 33.1: The divergence or idea generation phase of discovery can involve group participants putting their ideas on sticky notes and then posting them on a larger board for later sorting and evaluation.

33.3 Interact with Scientists Outside Your Core Group

It is all too often true that members of a group have the same idea. And why not? After all, they are exposed to the same information and so look at the same problems to solve. One straightforward way to have really novel ideas is to have contact with others outside your core group. For this, it might be outside your discipline, or outside your university. It could mean that people are more likely to have new ideas if they attend a diverse set of scientific meetings rather than sticking loyally to one.

Social scientists call the areas between groups structural holes [9]. Someone with ties to more groups that are different from one's own is at risk for having more good ideas [9]. He argues that even this point is not a new idea, but that it, like other old ideas, got more traction in a new setting, that of his business application to the company. Burt shows in a network drawing two individuals, named Robert and James, with equal numbers of strong and weak connections, but Robert has a strong tie to individuals in two separate groups while all of James's ties are within one group. The area between the groups where information is lacking is what is called a structural hole and it is this that Robert is in a position to bridge. In Figure 33.2 the concept is illustrated with Nancy being more at risk for having new ideas than Michele.

It might be hard to understand the structural holes in your own networks, but they are related to another concept: that of weak ties [32]. Ties to individuals outside your core group are likely to be weaker, because you interact with those people less frequently and share less of a common language. Zhou et al. [32] predicted that people with a middle number of weak ties would be most creative and supported this view with a study of within enterprise ties in a Chinese technology business. The people with whom you have weak ties may be less similar to yourself in a variety of ways. So be extra sensitive to the possibility of new ideas coming from

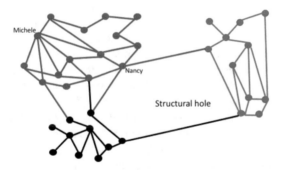

Fig. 33.2: People in different networks that can be university departments, disciplines, or different universities. Those with more ties across groups are more at risk for novel ideas. In this case, Nancy is more likely to develop novel ideas than Michele is.

such interactions. For me, I'll be paying closer attention to some new collaborations with geologists.

33.4 Foster Eureka Moments and Serendipity

We try to work through problems logically, step by step, reading the necessary background material, identifying the gaps and then finding the solution. But sometimes this does not work, even for the most creative of people. Sometimes after worrying about a problem for a long time no solution comes. And then suddenly it does, perhaps after a relaxing nap, or even in a moment of panic. These sudden solutions often come complete, seem perfect, and yet are not worked out in detail. They are called eureka, or aha moments But are these moments really engaging different thought processes? A critical analysis says no. A vivid story illustrates the problem.

An iconic example of a eureka moment involves a fire racing up a grassy hillside and threatening the firefighters dropped in to battle it [14]. The unexpectedly intense fire came too fast to be outrun. Most of the men ran anyway, a couple to a side ridge, the others straight up. When their leader, Wagner Dodge, realized that running was not a survivable option, he had a sudden insight. He set a fire in the grass in front of him, then lay down in the burned area, calling for his team to join him. No one did. He survived as the main fire raced over him, jumping over his tiny area where his little fire had removed the grass fuel. Lying there in the scorched field, Dodge breathed the thin layer of oxygen right on the ground as the firestorm actually picked him up and thrust him down as it passed. His men nearly all perished.

How did Wagner Dodge come up with this strategy that had never before been used? Some argue that it was a eureka moment, entirely different from other problem solving methods [14]. Others argue that it was a logical extension of knowing both that one could survive fires by running into them to burned out areas and that one could fight fires with back fires [26]. Either way, his ability to think at all under such circumstances was remarkable. Proponents of the view that eureka thinking is not really a thing also dissect other apparent sudden insights and find that detailed work preceded them and careful revision followed [26]. There is a debate over whether or not eureka moments are fundamentally different from painstakingly working through a problem [27].

Sometimes insights solve problems other than those you were working on, something called serendipity. It is also of considerable value for creative research [16, 24]. Like other forms of creativity, serendipity comes to prepared minds, can be fostered by triggers, but only if one is able to pay attention to them [16]. It is clear that new ideas build on knowledge, the more the better, and then your brain will sometimes make unexpected connections. So try to welcome sudden insights that solve the problem at hand or another one.

33.5 Recognize Novel and Creative Ideas

There is little point to a good idea that is ignored. When this is brought up, most people think of their own ideas that were not sufficiently recognized. To them I say write a review paper that puts your new ideas in context, but I digress. Just as important to having good ideas is recognizing them when you come across them [3]. This is the convergence step in the creativity process, where the many ideas presented from the divergence step are evaluated. Idea evaluation usually involves taking the idea and evaluating how well it will solve a particular problem [17, 25]. Steele et al. [25] evaluate judging ideas as a skill separate from one's belief in one's own creativity (called creative self-efficacy or CSE). Their experiments with undergraduates had them evaluate new restaurant ideas for college towns and with these experiments the authors showed that "a belief in one's ability to evaluate and revise original ideas"is very important for creativity [25]. They also found this trait correlated with conscientiousness. This is important, because overall, conscientiousness is a trait that is a strong predictor of effective job performance, though not necessarily of original creativity [7]. Thus, some people not confident in their creativity may be creative in this other way, by recognizing the best ideas out of a pool.

33.6 Work with Others

Collaboration and the multi-author papers that result are becoming more and more common in all realms of inquiry [28]. Wuchty and his team discovered that the percentage of papers produced by groups of two or more is increasing, not just in science and engineering, but in social sciences, patents, and to a lesser extent, even in the humanities [28]. Some of the trend toward team work can be attributed to the increasing technical complexity of science and engineering. The trend toward multiple authors could be the result of people being credited with authorship that might once have only been acknowledged.

Not only is teamwork increasing, but teams are producing work that is more cited than that by lone individuals. But what about the truly brilliant studies that are the most cited? Are they not the work of lone genius? Not according to this study, which found teams also at the extreme high end of citations. A team-authored paper in science or engineering was 6.3 times more likely to be cited over a thousand times than was a single-authored paper [28]. There is a case to be made for the raw creativity that comes from bouncing around ideas in a team, since even arts and humanities saw a higher impact of teamwork that is increasing with the years.

But there is also an argument to be made against too much or too early teamwork. There are a lot of dynamics that happen in groups that are not always positive for idea generation. One study tested two idea generating scenarios, one with teamwork from the start and the other where teams were formed after individuals generated ideas [12]. Unlike other such studies, they were most interested in where the best ideas were generated, not just total number, though they also looked at this and at idea evaluation. Their findings were discouraging for advocates of early teamwork, because they found that when people first generated ideas independently, then formed groups, they came up with more ideas, better ideas, and those ideas were better than the groups working in teams the whole time [12]. In these experiments ideas were judged by a separate panel of graduate students who were formally trained apart from this study in business idea valuation. They used business-approved inter-rater reliability ratings [13]. Their experiments involved asking the subjects to either design sports products like footwear for swimming, or gear for golf, or to design things like dorm lighting [12]. Furthermore, they found that when team members built on each other's ideas, the quality of the ideas either stayed the same, or actually diminished. Personal experience argues that this does not hold for longer term scientific collaborations, but it is hard to imagine what the appropriate controlled experiments would be.

Some of the findings of Girotra et al. are discouraging news for advocates of group brainstorming, following Osborne's influential book [18]. Still, Girotra et al. rightly put the focus on both what the very best ideas are, not just the number of ideas generated, and on the ability of the groups to evaluate the ideas. The hybrid model of independent idea generation followed by teamwork and judicious consideration of add-on ideas seems good.

33.7 Encourage Creativity in Others

Enterprises often say they want a creative work force, but then do everything to discourage creativity [4]. Yet the ways to support creativity are known. Perhaps first is to encourage an innate drive to be creative. More systematically, Amabile argues there are six areas that impact creativity: challenge of the project, freedom to solve a problem in one's own way, sufficient resources, a work group with members who bring diverse expertise and background, encouragement, and organizational support [4]. By contrast, things like harsh judgements of new ideas, changing goals and targets, unrealistic deadlines, and unhealthy group politics can kill creativity [4]. One might interpret these directives as encouraging a supervisor to carefully manage the teams she puts together, to establish clear goals and realistic timelines, and then to step aside, leaving the goals and timelines exactly as first stated, but not interfering with how teams work. Hands on, then hands off, except to encourage, or provide requested information.

One can also take a more personal approach to encouraging creativity in others. Research is not a solitary discipline. There is variability among people in both intrinsic creativity and motivation to be creative [3]. Two personality traits in particular have been thought to interact with creativity: openness to experience, and conscientiousness, with the former having a positive and the latter having a negative impact [31]. In experiments using office workers, the first relationship was supported. Conscientiousness, on the other hand, was found to be negative for creativity only when superiors were particularly controlling and tasks were particularly narrowly defined [31]. There is no evidence that you can change the personality of your team, but if you are a supervisor, bear in mind that being too controlling will have a negative impact on the creativity of some of your best workers, those that are naturally conscientious.

Creativity can also be increased in others by increasing contact with creative people who are willing to take the risk of trying new ways of doing things or forging into new areas. This is effective under circumstances where superiors support creativity and offer developmental comments, not controlling ones [30].

33.8 Give Others Honest Feedback on Their Ideas in an Informational not Controlling Way

Any time there is a group, there is likely to be a power structure. It could be employer/employee, teacher/student, older/younger, expert/neophyte. In these circumstances critiques of ideas are likely to flow from the former to the latter category. It has been found that it is not just the content of comments that matters. It is also their form. Evaluators should do their best to diminish the natural rank asymmetries and give their comments in a kind and informational way, rather than a controlling one [30]. This kind of response will be more effective in generating further creativity [29].

33.9 Defer Ownership of Ideas when Asking for Opinions

The strength of working in teams goes against the idea of the individual genius, though it is hard to deny that ideas both originate and are elaborated in individual brains. What happens between those brains that makes collaboration powerful is communication. What exactly that structure is can influence the effectiveness of collaboration. So how do you get the most engagement and collaboration out of others to whom you have brought a new idea? It turns out that if you mark an idea as your own others will give you fewer and less creative and useful answers [8]. In their experiments on this topic, Brown and Baer assigned 230 undergraduates the task of developing a promotion strategy for a new restaurant. The students were told that the proposal they were given was developed by another student, when in fact it was developed by the researchers. It came along with either possessive language to some students, "my proposal, not yours,"and more neutral language to other students, "the proposal."They found that when labeled with the territorial language, the students gave fewer and less creative suggestions [8]. Extending these results to scientific ideas and experiments would suggest that idea ownership is best not emphasized in a collaborative setting, if you expect others to work hard at improving the ideas. This might be because the more tightly ownership of an idea is indicated, the less likely will people think the owner will be willing to change anything.

33.10 Learn Science Communication through Improvisation

If you cannot explain clearly what your ideas are, it is likely that you do not understand them as well as you might. This is because if you cannot explain them you probably cannot sort apart what is new about the idea and what is assumed background knowledge. If someone else does not have that background knowledge, then they will not understand your idea. Steve Pinker calls this the curse of knowledge in a brilliant chapter in his must-read book: *The sense of style* [19]. But knowing about this problem is not the same as solving it.

So, how do you learn how to communicate to anyone? Alan Alda puts his money on Improv as the solution with his book and his workshops [2] (https://www.aldacenter.org/). What exactly does this mean? It means that practicing explaining things that are hard can help you learn to explain your own science more effectively. A lot goes on in a science improv workshop, but the key thing is to learn to explain things that are hard. Imagine telling someone from 500 years ago about a cell phone. Imagine your life depended on it. Think of the arguments you would use. Think of what that person does and does not know. Do this in successive rounds of shorter and shorter time. Then switch it up and do the same sort of thing with your science, first taking 5 minutes, then 3 then 1 to convince a postdoc to join your lab, or a funder to fund it, for example. This is exactly what Aniek Ivens did with us in a workshop she ran http://www.aniek.nyc/academic-improv. Figure 33.3 shows a student explaining her research to another.

Fig. 33.3: Two undergraduates participate in an improv activity, with one explaining her research to the other, after several rounds of other kinds of explanation.

Some might think that improv is more related to creativity in communication than it is to scientific creativity. But communication and creativity are closely tied together. I once heard a student at the end of an improv activity related to her project exclaim that only now did she really understand it. When you can sort your ideas more clearly into what is new and what is foundation, I think creativity will blossom. But this has not, as far as I am aware, been put to the kinds of careful tests discussed elsewhere in this paper.

Conclusion

Creativity is important to success in just about any area because the point of creativity is to find something new and better. It could be a new understanding of the natural world, a new painting, a new theory, a new industrial process, a new way of teaching or learning, and myriad other ways of being new. If creativity is expressed in so many varied areas of human life and human endeavors, how could there be one approach to studying it, or one set of conclusions? This is something worth pondering. It may be that creativity for artists is very different from creativity for entrepreneurs. This is certainly true for many aspects of the creative output. How can we compare the unfinished painting I saw in Patrick Chamberlain's studio (http://kavigupta.com/artist/patrick-chamberlain/) to the sticky plastic credit card case on the back of my smartphone, or to Nancy Moran's discovery of how the symbiosis between aphids and bacteria works? While this is likely to be challenging, the processes have something in common. First, discoveries in both science and art are channeled through the marvel of the human brain. Second, they each represent solutions to problems posed by their creators. So if creativity, at least for now, is always channeled through human brains and always involves solving problems in new ways, then it is not unreasonable to look for commonalities of process.

Another challenge for those looking for ways to improve creativity is how much we can actually learn about real world creativity in the area that matters to us from experiments that take minutes or hours and are done on problems solved by under-graduates who are likely to be getting credit or pay for participation? This frame-work forms much of what we know in many areas of the science of being human, or social sciences, so I am not going to recommend that we discard such studies. But what if challenges that people get an hour or so to solve really do not scale up to those that take months or years? I do not know what to think about this pos-sible challenge, except that in all scientific areas we generalize from experiments that are necessarily simplifications. Their value comes when they are repeated and verified in variable contexts. From the numbers of citations and the huge volume of experiments, I hope that the points I have chosen to emphasize here have some validity.

Is there anything new that might be learned from this paper? Are all the conclu-sions what someone who wanted to be creative might do anyway? Let's go through the ten points. 1. I suppose it should come as no surprise that an important factor in creativity is the motivation to be creative. But some might think that creativity is innate and cannot be learned, whereas anyone might feel that their motivation for creativity could be increased, given the space for it. So the first step towards increas-ing creativity is to value it and look for it. 2. Separating idea generation from idea judgement (divergence before convergence) and creativity training in general can help you discover useful ideas. So much of our training in science is about evaluat-ing ideas. Should we not also have training in how to have good ideas? Should this not be as much a part of our training as scientists as the so-called scientific method is? Divergence before convergence in ideas is somewhat counter-intuitive because it seems messy to keep around ideas you think will not work rather than clearing them away to focus on the good ideas. Keeping as many as possible from the start and not judging until the first phase is over is a key point of brainstorming. I like it because the dark horse idea might be the most novel. It might be an idea you think will not work, but then see a way around it. It also requires a formalization of cre-ativity. Alone, or in a group, you identify which phase you are in and restrict the activities to that phase. I think this one is both important and counter-intuitive. 3. Interacting with others outside your core group as a source of new ideas may also seem counter-intuitive, for why should you get good ideas for solving your problem from someone working on something entirely different? It is the interaction of what you know and what they know that can come to surprisingly creative and workable new ideas. I think this one has been important in my own career as I and others brought social insect thinking to the puzzles of social microbes. If you are open to ideas from those you know less well, you may find something exciting. 4. Another important point in the quest for creativity is how exactly the new idea forms in your brain. Do you just work through the problem, with one logical step leading to the next, or does it sometimes seem like a leap, a eureka moment? Scientists disagree on whether the moments of sudden insight come from completely different brain processes, or from the same processes as more methodical idea processing. But it does seem to be true that ideas or parts of bigger solutions sometimes just come to us and we should welcome such ideas. Giving these ideas space to intrude might be facilitated by putting direct work aside, as with vacations, or by letting our brain refresh itself, as with sleep. 5. Up to now, I have discussed how to foster creativity. But once you have a set of ideas, judging which are the best is a crucial next step, one on which research has also focused. I suppose this one is not particularly insightful, except perhaps making the judgement a conscious process separate from idea generation can help bring the best ideas forward.

The second five points all consider how to stimulate creativity in yourself or others. 6. Working in a team can be fabulous, but it can also squelch great ideas that might come from socially less comfortable group members. Combined with point 2, that idea generation and idea evaluation should be separated, one path through group work could combine individual idea generation with group evaluation. Is this new? Perhaps not, but it can put a focus on how to get ideas from the shyest group members. 7. Encouraging creativity in others may seem like an obviously positive thing to do. But how often do we ask students to simply do the experiments in the grant proposal? How often do we welcome different approaches from others? Like the first point on encouraging creativity in oneself, encouraging creativity in others helps. 8. No matter how much we say we want creativity in others, if we then attack those new ideas or approaches, we will not be encouraging creativity. Yet new ideas are easy and good new ideas are hard. New ideas will still need to be critiqued, but in an informational and not controlling way, not a particularly new thought. 9. Studies mentioned above indicate that people are more likely to put in some effort at idea critiquing if the ideas are presented in a neutral way, with no ownership claimed. This seems to go along with the creativity that comes from an open atmosphere, but the strength of the effect was surprising to me. 10. The final idea for enhancing creativity steps back a bit to the very process of communicating ideas effectively. Move outside your comfort zone to the theater of improv, where you learn to read your audience and truly communicate clearly. When you do this, even you will better understand what you actually have to say.

Besides these ten points, the point of this piece is that there is a literature out there on a topic we all care about. Step outside your group for a bit and set aside some time for reading on creativity. It can only enhance your ability to discover.

Acknowledgements It is my pleasure to dedicate this piece to Erik Goodman on the occasion of his 75th birthday. The BEACON Evolution in Action Center that he leads is an excellent example of an institution that fosters creativity in research. I thank Andy Burnett for introducing me to creativity techniques and for help with references. For quick and detailed help with references in specific areas of the field I thank Markus Baer, Robert Weisberg, and Jing Zhou. I am an amateur intruder into the field and no doubt have made eclectic choices as to approach and inclusion and perhaps I have also made downright mistakes. I thank David Queller, Kay Holekamp and two anonymous referees for helpful comments on an earlier version. I thank the Wissenschaftskolleg zu Berlin for the fellowship that allowed me the time to research and write this piece and the collegiality to acquire ideas outside my group.

References

1. Acar, S., Burnett, C., Cabra, J.: *Ingredients of creativity: Originality and more.* Creativity Research Journal **29**, 133–144 (2017)
2. Alda, A.: *If I Understood You, Would I Have this Look on My Face?: My Adventures in the Art and Science of Relating and Communicating.* Random House Trade Paperbacks (2018)
3. Amabile, T.: *Creativity in context: Update to the social psychology of creativity.* Hachette (UK) (1996)
4. Amabile, T.: *How to kill creativity*, vol. 87. Harvard Business School Publishing, Boston, MA (1998)
5. Anderson, N., Potocnik, K., Zhou, J.: *Innovation and creativity in organizations: A state-of-the-science review, prospective commentary, and guiding framework.* Journal of Management **40**, 1297.–1333 (2014)
6. Baer, M., Brown, G.: *Blind in one eye: How psychological ownership of ideas affects the types of suggestions people adopt.* Organizational Behavior and Human Decision Processes **118**, 60–71 (2012)
7. Barrick M.R., M.M., Judge, T.: *Personality and performance at the beginning of the new millennium: What do we know and where do we go next?* International Journal of Selection and Assessment **9**, 9–30 (2001)
8. Brown, G., Baer, M.: *Protecting the turf: The effect of territorial marking on others' creativity.* Journal of Applied Psychology **100**, 1785 (2015)
9. Burt, R.: *Structural holes and good ideas.* American journal of sociology **110**, 349–399 (2004)
10. Burt, R.: *Structural holes: The social structure of competition.* Harvard university Press (2009)
11. De Bono, E.: *Six thinking hats.* Penguin, UK (2017)
12. Girotra, K., Terwiesch, C., Ulrich, K.: *Idea generation and the quality of the best idea.* Management Science **56**, 591–605 (2010)
13. Gwet, K.: *Inter-rater reliability: dependency on trait prevalence and marginal homogeneity.* Statistical Methods for Inter-Rater Reliability Assessment Series **2**, 9 (2002)
14. Lehrer, J.: *The eureka hunt.* The New Yorker **28**, 40–45 (2008)
15. Loehle, C.: *A guide to increased creativity in research: inspiration or perspiration?* BioScience **40**, 123–129 (1990)
16. McCay-Peet, L., Toms, E.: *Investigating serendipity: How it unfolds and what may influence it.* Journal of the Association for Information Science and Technology **66**, 1463–1476 (2015)
17. Mumford, M., Mobley, M., Reiter-Palmon, R., Uhlman, C., Doares, L.: *Process analytic models of creative capacities.* Creativity Research Journal **4**, 91–122 (1991)
18. Osborn, A.: *Applied imagination: Principles and procedures of creative thinking.* Charles Scribners Sons, New York (1953)
19. Pinker, S.: *The sense of style: The thinking person's guide to writing in the 21st century.* Penguin Books (2015)
20. Platt, J.: *Strong inference.* Science **146**, 347–353 (1964)
21. Popper, K.: *Philosophy of science.* In: C. Mace (ed.) British Philosophy in the Mid-Century. George Allen and Unwin, London (1957)
22. Puccio, G., Cabra, J., Fox, J., Cahen, H.: *Creativity on demand: Historical approaches and future trends.* AI EDAM **24**, 153–159 (2010)
23. Puccio, G., Murdock, M., Mance, M.: *Current developments in creative problem solving for organizations: A focus on thinking skills and styles.* Korean Journal of Thinking and Problem Solving **15**, 43 (2005)
24. Silver, S.: *The prehistory of serendipity, from bacon to walpole.* Isis **106**, 235–256 (2015)
25. Steele, L., Johnson, G., Medeiros, K.: *Looking beyond the generation of creative ideas: Confidence in evaluating ideas predicts creative outcomes.* Personality and Individual Differences **125**, 21–29 (2018)
26. Weisberg, R.: *On the "demystification" of insight: A critique of neuroimaging studies of insight.* Creativity Research Journal **25**, 1–14 (2013)

27. Weisberg, R.: *Toward an integrated theory of insight in problem solving.* Thinking & Reasoning **21**, 5–39 (2015)
28. Wuchty, S., Jones, B., Uzzi, B.: *The increasing dominance of teams in production of knowledge.* Science **316**, 1036–1039 (2007)
29. Zhou, J.: *Feedback valence, feedback style, task autonomy, and achievement orientation: Interactive effects on creative performance.* Journal of Applied Psychology **83**, 261 (1998)
30. Zhou, J.: *When the presence of creative coworkers is related to creativity: Role of supervisor close monitoring, developmental feedback, and creative personality.* Journal of Applied Psychology **88**, 413 (2003)
31. Zhou, J., George, J.: *When job dissatisfaction leads to creativity: Encouraging the expression of voice.* Academy of Management Journal **44**, 682–696 (2001)
32. Zhou, J., Shin, S., Brass, D., Choi, J., Zhang, Z.X.: *Social networks, personal values, and creativity: Evidence for curvilinear and interaction effects.* Journal of Applied Psychology **94**, 1544 (2009)

Chapter 34
Student Learning Across Course Instruction in Genetics and Evolution

Emily G. Weigel, Louise S. Mead and Teresa L. McElhinny

Abstract Genetics and evolution are interconnected topics — evolutionary change requires inheritance and correspondingly, genetic variation is required for selection to have any impact on a population. However, misconceptions and naive ideas of both genetic and evolutionary concepts can fundamentally impact a student's understanding of biology. It is therefore important to understand what information students obtain in various courses at the undergraduate level, and how knowledge of concepts in one course might impact learning in another course. This is particularly important with respect to genetics concepts, as Genetics courses are often a prerequisite to Evolution courses and serve frequently as students' introduction to the basic concepts that underlie evolution. This study compared student performance related to key genetics concepts after taking both Fundamental Genetics and Evolution courses to taking Fundamental Genetics alone and tracked student performance as they progressed through the Genetics-Evolution course sequence. We created a 16-question assessment, developed from published literature on these topics, and administered the survey at three timepoints: the end of Fundamental Genetics, the beginning of Evolution and again at the end of the Evolution course. Our data suggest students do complete Fundamental Genetics with a few misconceptions related to genetic information pertinent to evolution, and that these concepts are varyingly

Emily G. Weigel
Department of Biological Sciences, Georgia Institute of Technology, Atlanta, GA 30332
Beacon Center for the Study of Evolution in Action, Michigan State University, East Lansing MI 48824, USA

Louise S. Mead
BEACON Center for the Study of Evolution in Action, Michigan State University, East Lansing MI 48824, USA
Department of Integrative Biology, Michigan State University, East Lansing, MI 48824

Teresa L. McElhinny
Department of Integrative Biology, Michigan State University, East Lansing, MI 48824

© Springer Nature Switzerland AG 2020   513
W. Banzhaf et al. (eds.), *Evolution in Action: Past, Present and Future*,
Genetic and Evolutionary Computation, https://doi.org/10.1007/978-3-030-39831-6_34

corrected by taking Evolution. This research highlights the advantages of both tracking and comparing students as they progress through a Genetics-to-Evolution course sequence, particularly with respect to how faculty can leverage course sequencing to improve student performance.

34.1 Introduction

In *On the Origin of Species* (1859) [19], Darwin devotes an entire chapter to the Laws of Variation and further explores the importance of variation, and the link between traits in parents and offspring in *The Variation of Animals and Plants Under Domestication* (1868) [20]. Despite a clear role for inheritance in natural selection, Darwin was unable to provide an adequate model [14]. It was not until the discovery of the work of Gregor Mendel that a particulate inheritance hypothesis was accepted [35]. Mendel, however, understood the importance of his work to evolution [35], writing "This seems to be the one final way of finally reaching a solution to a question whose significance for the evolutionary history of organic forms cannot be underestimated [35]."

Although Darwin was able to articulate his theory of natural selection using a faulty model (pangenesis and blended inheritance), it was clear by the 1920's that the application of Mendel's work (Mendelian inheritance) supported evolution by natural selection [14]. In particular, the idea that mutations in genes provide new sources of variation on which selection acts forms the basis for evolutionary change.

The link between evolution and genetics is clearly emphasized in our current approaches to teaching and learning in biology. *Vision and Change in Undergraduate Biology*, the core structural advisement from the American Association for the Advancement of Science (AAAS), lists evolution as a core concept in biology [31]. Specifically, *Vision and Change* recognizes the evolution and diversity of life on Earth as not only critical to understanding biology, but also inherently a function of genetic processes. Although the core foundational idea of evolution originally developed prior to the genetics and genomics era, we know that [genetic] inheritance is an important component to student understanding of evolution [39]. Despite the clear intertwined nature of evolution and inheritance, most biology majors take separate courses in genetics and evolution, and student understanding of these two topics is not evaluated together.

Prior research suggests a positive relationship between evolution acceptance and genetics understanding [36], but the role of genetics instruction in understanding evolutionary processes is unclear despite studies exploring student understanding of genetics [4, 15, 32, 53]. Moreover, student understanding of genetics can include a litany of incorrect ideas, from issues of relationships between genes, chromosomes and cells, to the transmission and display of traits [4, 15, 32]. Particularly resistant incorrect ideas may not change after instruction [53], thus persisting as a student makes their way through the curriculum.

An understanding of evolution, likewise, eludes many students at all levels: K-12 [22, 37]; lower-level undergraduate (both biology majors and non-majors [6, 39, 52]; and upper-level undergraduate [9, 17, 18, 27, 58]. Naïve mental models can plague student understanding as they advance to higher-level courses [5], and as with concepts in genetics, misconceptions can often persist even after instruction [10, 41]. If left uncorrected, these misconceptions continue to manifest themselves as undergraduates progress through the (often linear-pathed) biology curricula.

Recent work has argued that teaching genetics before evolution may increase student understanding of evolution, because the concepts of mutations and alleles can be translated into allele frequency change [34]. Using a common garden experiment, where the order of modules on genetics and evolution were swapped, Mead [34] found that lower-performing students scored better in both genetics and evolution understanding when genetics was taught first. However, the study investigated the order of genetics and evolution in a single class of secondary students. Most post-secondary institutions introduce evolution and genetics in introductory courses that include coverage of the concept that heritable genetic variation is critical for population evolution. However, the need for mastery of these processes for a complete understanding of evolution may not be emphasized and at most institutions, upper-level courses in genetics and evolution are emphasized, and taught separately. Rarely do undergraduates take courses in population or quantitative genetics.

Furthermore, at many institutions, students are expected to take an upper level Genetics course prior to taking an upper level Evolution. In an informal survey conducted via the Evoldir community listserv, of 123 institutions worldwide, 6% list courses in Genetics and Evolution as co-required, 49% list Genetics as a prerequisite to Evolution, 41% have no defined relationship between the courses, and 4% have other. Furthermore, despite its central role in biology, most of the 4% indicated Evolution was not offered as a stand-alone course. Thus, even with evidence suggesting genetics knowledge may improve evolution understanding, roughly only half the institutions in our sample mandate Genetics prior to Evolution, and nearly an equal number make no clear recommendation for the order of these courses and most likely do not emphasize the connection between the two.

Students learn by integrating new knowledge into existing knowledge frameworks [47, 2]. If existing and new knowledge are incompatible, naive ideas can interfere with students' ability to understand new information as they progress through courses, particularly if frameworks are not reorganized or replaced [21, 48, 51]. Curricular frameworks must therefore be carefully crafted such that the requisite knowledge is not only presented, but sufficiently mastered over several experiences, and concepts covered in one course linked to those in follow-up courses [13, 16, 43, 55].

The National Research Council (NRC) advocates for basing curricula on empirical evidence for how students reason and develop competence in the domain [1]. Yet research is lacking to show how genetics knowledge prepares students to conceptualize evolution, specifically whether students can adequately link the underlying genetic characters of a population to its evolution. As Nehm and Ridgway have shown [40], even "emerging expert" students seem to accurately describe natural selection but lack the incorporation of non-adaptive mechanisms, such as genetic drift, present in true expert models of evolution [3, 24, 30, 45, 46].

Because naive ideas can originate prior to explicit instruction, and these ideas tend to persist throughout learning, it is important to understand what information (and misinformation) students obtain as they progress through course sequences. Particularly with respect to Evolution and Genetics, learning should be tracked to determine what information students learn in Genetics, and what is added or replaced by taking Evolution, particularly as Genetics is a mandatory prerequisite at many institutions and introduces many fundamental evolutionary genetics ideas. With the goal of tracking and elucidating student learning of genetic concepts important to understanding evolution, our research explored three questions: 1) Which genetic concepts important for understanding evolution do students understand following Genetics and Evolution coursework, compared to taking a Genetics course alone? 2) For students who take these courses immediately one after the other, does their understanding of the concepts change between the end of Genetics and the beginning of Evolution, and after a full course on Evolution? 3) Do students maintain any specific misconceptions after instruction in either course, and if so, which topics appear to be the most difficult for students?

34.2 Methods

34.2.1 Courses of Focus

This study concerned student performance at a large Midwestern university in relation to enrollment in two courses: Genetics (Fall 2012) and Evolution (Spring 2013). Genetics is a large (300 students) traditional lecture course with 3 exam-based assessments and a cumulative final. Evolution is a considerably smaller course (50-75 students), and is also an upper level course, with both a lecture and a recitation. Both Genetics and Evolution courses require completion of a cell and molecular biology course, and Evolution requires completion of Genetics. Concepts covered in both courses align with expectations outlined by the biology education community [54]. Assessments in Evolution are a combination of exams, quizzes, homework, in-class discussion, and a final exam. The syllabi for each course suggest course coverage includes the genetic basis of evolution, population genetics, and gene expression in explanations of genetic variation. And while not required as a prerequisite for either Genetics or Evolution, many students do complete one semester of an introductory biology course that covers biological diversity and organismal biology, including principles of evolution, transmission genetics, population biology, community structure and ecology.

34.2.2 Study Population

Prior to completing the study's assessments in both Fundamental Genetics and Evolution, students were asked to give consent to participate in this study (x12-1182e). We administered surveys to students enrolled in Fundamental Genetics during the Fall 2012 semester. We then administered the same assessment to students enrolled in Evolution; we will refer to these participants as 'comparison students'. To track a cohort of students as they complete the course sequence (hereafter 'tracked students'), student data were matched across Fundamental Genetics and Evolution; these students took Evolution immediately following Fundamental Genetics ($N = 18$). Students who failed to complete assessments in their entirety were not included in these analyses. Table 34.1 provides demographics for the study population. GPA and ACT scores were self-reported.

Table 34.1: Demographic information of participants for each of the groups compared. The number of female and non-white, non-Hispanic students are shown as percentages. Year in College, GPA, and ACT score are shown as averages.

Criteria	Student Completing Genetics Only	Student Completing Genetics and Evolution	Tracked Students
Female	48.10%	75%	83.3%
Non-white, Non-Hispanic	30.0%	5.0%	0%
Year in College	3.96	3.78	3.61
GPA	3.66	3.74	3.83
ACT Score	26.38	27.89	27.75

34.2.3 Survey Design

Our assessments consisted of a number of survey questions assessing class demographics and prior course performance, followed by a series of sixteen questions assessing student knowledge of course-related topics. Questions were taken directly from the genetics assessment literature, specifically the Genetics Literacy and Assessment Instrument (GLAI) [11], Genetic Drift Inventory (GeDI) [49], and the Genetics Assessment For Core Understanding (GCA) [54], and course textbooks (Pearson's Mastering Biology Sequence; Campbell et al. 2008; see Table 34.2 [12]). Questions were either multiple choice with each distractor tied to a specific misconception, or agree/disagree format coupled with fill-in-the-blank formats to explain why a particular answer was given. Questions were chosen to test basic understanding of the fundamental genetic underpinnings of evolution, span several Bloom levels [7], and correspond to documented misconceptions.

Table 34.2: Question Sources and Concepts Measured in this Assessment.

General Concept/Learning Goal	Item number on current assessment	Original test item
Interpret results from molecular analyses to determine the inheritance patterns and identifies of human genetics that can mutate to cause disease	1	GCA No.19
Extract information about genes, alleles, and gene functions by analyzing the progeny from genetic crosses	2	GCA No.4
Compare different types of mutations and describe how each can affect genes and the corresponding mRNAs and proteins	3	GCA No.5
Most human traits, including diseases, result from the products of multiple genes interacting with environmental variables; examples include height, heart disease, cancer, and bipolar disorder	4	GLAI No.3
Understanding Mendelian patterns of inheritance, and their biological basis, allows probability statements about the occurrence of traits in offspring	5	GLAI No.7
Occasional errors in DNA structure and replication result in genetic variation	6	GLAI No.15
Occasional errors in DNA structure and replication result in genetic variation	7	GLAI No.9
Understanding Mendelian patterns of inheritance, and their biological basis, allows the probability of trait occurrence in offspring to be predictive of parental traits	8	Mastering Biology
Occasional errors in DNA structure and replication result in genetic variation	9	Mastering Biology
Understanding Mendelian patterns of inheritance, and their biological basis, allows probability statements about the occurrence of traits in offspring	10	Mastering Biology
Genetic drift can be a driver of evolution, not just natural selection	11	GEDI
The magnitude of the effect of random sampling error for genetic drift from one generation to the next depends on the population size	12	GEDI
Random sampling error tends to cause a loss of genetic variation within populations, which in turn increases the level of genetic differentiation among populations	13	GEDI
Mutation and genetic drift can contribute to genetic variation in a population	14	GEDI
Standing genetic variation in a population can be acted on by selection	15	GEDI
Genetic drift can create genetic differences between populations over time	16	GEDI

34.2.4 Survey Assessment Administration

The assessment was administered at three curricular timepoints: at the end of Fundamental Genetics (to establish a baseline), at the beginning of Evolution (to determine knowledge retained following Fundamental Genetics and prior to Evolution), and at the end of Evolution (to examine learning and retention of nave ideas). All students consenting to the study completed the assessments within the initial (in the case of Evolution pre-test) and final week of the semester (in the case of the post-Genetics and post-Evolution tests). Assessments at each time point were identical, and students were not given feedback about the accuracy of their responses at any time.

Regardless of the accuracy of their responses, students were incentivized with extra credit for the completion of the assessment at the end of Genetics (N=224; 70.22% participation rate). Because of a reduced class size as students transition into a much smaller Evolution course and the need to keep sample sizes high (N =38; 69.09% participation rate), only students who completed both the pre-and post-tests within Evolution were compensated, although two chose to complete the post-assessment without having completed the pre-assessment.

34.2.5 Survey Assessment Analysis

Prior to scoring, all assessment responses were de-identified, and in the case of the tracked students, matched with prior assessment responses. Individual item performance was compared for both students in the comparison and tracked populations. In addition to scoring answers as correct/incorrect, we further examined student responses for questions that were consistently incorrect. Finally, items that showed improvement across administration of the assessments were noted as possibly the result of additional instruction in evolution.

34.3 Results

34.3.1 Statistical Analyses

Several statistical models were chosen based on the questions asked and statistical agreement with model assumptions; however a significance cut off of $p = 0.05$ was used for all statistical tests. All tests were run in R (R Core Team, 2013, version 3.0.2 [50]).

To assess our instrument, we first conducted a series of Item Response Theory (IRT) tests, and the performance results that follow are reflecting the use of the assessment survey instrument. We partition our performance analyses into two parts:

Part 1 addresses 'comparison' students (comparing those who complete both Genetics and Evolution to those who complete only Genetics); and Part 2 addresses 'tracked' students (following a cohort to document how student knowledge changes as students progress through the course sequence).

Table 34.3: Item performance (expressed as percent of students answering a question correctly) for students who took only Genetics and students who took both Genetics and Evolution. Total average score and standard error for samples are also included.

Question	Students who took only Genetics	Students who took both Genetics and Evolution
1	89.0%	91.7%
2	38.8%	47.2%
3	54.0%	91.7%
4	90.8%	94.4%
5	89.3%	100%
6	68.4%	52.8%
7	73.3%	86.1%
8	87.4%	88.9%
9	54.9%	86.1%
10	58.7%	50.0%
11	54.9%	69.4%
12	65.0%	80.5%
13	75.0%	91.7%
14	87.0%	88.9%
15	86.0%	88.9%
16	85.0%	97.2%
Average Score	12.77	13.05
SE	0.20	0.41

34.3.2 Instrument Analysis

Rasch analyses were conducted on the overall instrument to examine both the difficulty of the items and conceptual independence of the items. Item Characteristic Curve (ICC) plots revealed independence of the items. From a graphical model check plotting the predictive ability of each question, question 5 appeared to deviate from the parameter estimates of the remaining questions. Therefore, an Andersons' likelihood ratio test for goodness of fit with mean-split criterion was performed to determine if this question violated invariance of the model estimates for the instrument; the results supported inclusion of all questions in the instrument for analysis

(*Andersen LR-test, LR-value= 22.642, df= 14, p = 0.066*). Figure 34.1 provides a summary of item difficulty and parameter distribution.

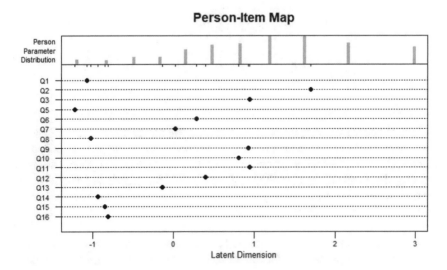

Fig. 34.1: Item difficulties for each of the 16 questions in the assessment. The upper panel displays student performance on the instrument, and the lower panel shows the difficulty of each item. Constructed based on the total number of respondents in the study and, in the case of repeated testing, used only on the first assessment taken by the individual ($N = 262$).

34.3.3 Part 1: Comparing Students between Courses

To compare students taking only Genetics to those taking both Genetics and Evolution, a multiple logistic regression model was conducted with course (either post-Genetics or post-Evolution), gender, year in college, GPA, ACT scores, and race predicting the sum of correct responses. Two-way interaction terms were added to the original model one at a time to examine which combination of predictors yielded the best model based on AIC values. Main effects and interactions, even when not significant, were retained through model selection for the lowest (best) AIC model. When accounting for gender, year in college, GPA, ACT score, and race, students who complete both Genetics and Evolution score higher on our overall assessment than students who have only completed Genetics ($\beta = 0.55310, SE = 0.22135, p = 0.01315$; see Table 34.4). Thus, taking both Genetics and Evolution results in higher scores on genetics concepts compared to taking Genetics alone.

We also examined four questions that appeared to be missed most often. Responses suggested students either had a lack of knowledge, held a misconception, or

Table 34.4: Multivariate regression model predicting the sum of the correct responses.

	β	SE	t	p	$exp(\beta)$
Survey	0.55310	0.22135	2.499	0.01315*	1.738634
Gender	0.15587	0.30993	0.503	0.61550	1.168674
Year in college	-0.12183	0.22119	-0.551	0.58230	0.8852989
GPA	0.215	0.220	0.975	0.331	1.239862
ACT score	0.22203	0.21801	1.018	0.30953	1.248609
Race	-0.94765	0.35912	-2.639	-0.00888**	0.3876509

performed better following the Evolution course. We identified two "lack of knowledge" questions: one concerning translating genotype to phenotype using a Punnett Square (Question 10, Supplemental Material) and one concerning genetic drift as a non-adaptive mechanism of evolution (Question 11, Supplemental Material). Students struggled to translate genotype to phenotype using a Punnett Square. Correct responses were not predicted by Gender, Year, GPA, ACT score, nor race as main effects ($p > 0.11$ in all cases), however, adding the interaction term of Year*ACT to this model shows year in college, ACT score, and the interaction of ACT and year as significant predictors of a correct response (Table 34.5). Similarly, students struggled to identify genetic drift as a non-adaptive mechanism of evolution (Question 11, Supplemental Material). However, increased ACT scores are a significant predictor of a correct response ($\beta = 0.08346, SE = 0.04025, p = 0.0381$), although no other significant variables or interactions were found (all $p > 0.05$).

Table 34.5: Logistic regression model predicting the response to question 10 with interaction terms.

	β	SE	z	p	$exp(\beta)$
Survey	-0.26170	0.20083	-1.303	0.192537	0.7697419
Gender	0.44579	0.28002	1.592	0.111388	1.561723
Year in college	4.72143	1.54968	3.047	0.002314**	112.3288
GPA	0.13764	0.19626	0.701	0.483092	1.147562
ACT score	0.57159	0.17366	3.291	0.000997***	1.771081
Race	-0.44350	0.32182	-1.378	0.168165	0.6417862
Year*ACT	-0.16937	0.05548	-3.053	0.002265**	0.8441965

One question displayed a clear misconception: defining a mutation (Question 2; Supplemental Material). Gender ($\beta = 0.90309, SE = 0.28538, p = 0.001553$) and year in college ($\beta = 0.43755, SE = 0.20584, p = 0.033527$) were significant predictors of a correct response. Substantively, this means that the odds of a correct response were 2.47 times higher for men than for women, and 1.55 times higher for those further along in school than those less far along in school. All two-way interactions were again explored between the predictors, but none turned out to be significant (all $p > 0.05$).

Students who take Evolution after Genetics improved specifically in their knowledge of standing genetic variation (Question 9; Supplemental Material). Students who completed both Evolution and Genetics were 2.38 times more likely to provide correct answers to this question compared to students who only completed Genetics ($\beta = 0.86593, SE = 3.086, p = 0.00203$). Subsequent analysis revealed that two interactions were significant in the best model by AIC: year*ACT and GPA*ACT. When both interaction terms are included in the model, neither emerges as a significant predictor of a correct response (perhaps because both involve the ACT variable and thus introduce high levels of multi-colinearity into the model).

34.3.4 Part 2: Tracking Students across Courses

We were able to track a total of 18 students from the end of Genetics through to the end of Evolution. Table 34.6 provides a summary of the performance on each item for each of the three time points. In addition, to examine student performance as they progressed through the course sequence, repeated measures ANOVA, ANCOVA and Generalized Estimating Equations (GEE) were used. First, we conducted a repeated measures ANOVA using the Greenhouse-Geisser correction for violations of sphericity. In this model, survey (whether post-Genetics, pre-Evolution, and post-Evolution) was our within-subjects factor to test whether a statistically significant change was observed in the sum survey score over time. Note that an ANCOVA was performed to examine whether ACT score influenced survey outcomes in interactions and is reported where appropriate. Figure 34.2 depicts the overall mean change in the sum of correct responses across surveys, which was not found to be significant (Repeated measures ANOVA with Greenhouse-Geisser correction, $F(1.612, 25.788) = .776, p = .445$).

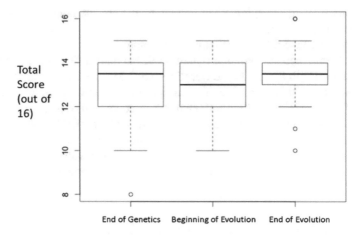

Fig. 34.2: Boxplot of the Sum of Correct Responses Across Time. Central boxes represent values
from the lower to upper quartile (25 to 75 percentile; first and third quartiles), and are
intended to give an approximate 95% confidence interval for differences in the two
datasets. Bold lines represent the median, and extreme values are represented by open
circles.

Finally, the remaining models examined how the odds of getting a correct answer on a given survey question changed over time. Because repeated measures
AN(C)OVA cannot be used for binary dependent variables (such as whether a
question is right or wrong) which violate its assumptions (linearity, constant variance, and normality, etc.) and logistic regression cannot be used due to the non-
independence that exists within subjects, we chose a generalized estimating equations (GEE) model. This model met the model assumptions and converged better than the alternative multilevel model with random effects for subject and time
(which account for within-subjects dependencies). The correlation structure for each
GEE model was set to be "unstructured", meaning that the correlations between
time points were estimated from the data. This is the most relaxed assumption one
can make about the within-subjects correlations and ideal to use in the absence of
specific predictions for changes between consecutive surveys.

The two questions identified above as connected to a lack of knowledge for comparison students were also found to be a challenge for the tracked students. Namely,
students struggled in translating genotype to phenotype and in identifying genetic
drift as a non-adaptive mechanism of evolution. However, among our tracked students, data suggested students taking the sequence contiguously may be better at
defining the term "mutation" (Question 2; Supplemental Material). Of the four incorrect responses, 76% initially stated that mutations must alter the amino acids;
these responses slowly convert to the correct response over time, that a mutation
is any change in the genetic code. However, this improvement was not ultimately
strong enough to be significant ($B = .329, SE = .244, p = .18, OR = 1.39$).

Table 34.6: Overall per item performance on the assessment (as expressed by percent correct per question), as well as the score average and SD of overall assessment for students at 3 curricular timepoints.

Question	End of Genetics	Beginning of Evolution	End of Evolution
1	83.0%	77.8%	94.4%
2	33.0%	44.4%	50.0%
3	67.0%	83.3%	94.4%
4	88.9%	94.4%	100%
5	100%	94.4%	100%
6	77.8%	66.6%	55.6%
7	83.3%	77.7%	88.9%
8	100%	72.2%	88.9%
9	66.7%	83.3%	88.9%
10	55.6%	44.4%	44.4%
11	61.1%	61.1%	66.7%
12	94.0%	83.3%	83.3%
13	94.0%	66.7%	94.4%
14	94.0%	88.9%	100%
15	88.9%	88.9%	94.4%
16	100%	94.4%	100%
Average Score	*12.89*	*12.94*	*13.44*
SE	*0.47*	*0.33*	*0.35*

However, a significant improvement was seen among the tracked students with respect to their understanding of the importance/role of standing genetic variation ($B = .689, SE = .347, p = .047, OR = 1.99$; Question 9; Supplemental Material). Among this group, a GEE model also indicated a significant interaction between survey and ACT scores ($B = .095, SE = .046, p = .041$), where scores tend to increase more across the surveys with increasing ACT score.

34.4 Discussion

Students performed adequately on the overall assessment, averaging about 75-80% correct responses across the assessment. Our results suggest that in general, taking a course in Genetics followed by a course in Evolution can help students retain genetic concepts important to understanding evolution. However, our tracking of a subset of students through this course sequence suggests the specific advantages can be hard to identify.

It is important to note that tracked student scores did not appear to change appreciably from the end of Genetics to the beginning of Evolution; although this is the most ideal case of back-to-back courses, we can confirm students are not immediately 'forgetting' concepts from semester to semester, and it is possible that student ideas on these topics may be formed prior to taking their Genetics course, and their Genetics and Evolution courses may only serve to cement certain ideas, rather than correct and replace them.

Introductory Biology experiences, for example, may cover this material [8, 10, 28, 39, 57, 59]. Students in introductory biology courses do tend to harbor the same naive ideas about evolution, focusing on the organismal and less on the molecular basis of evolution and rarely connecting how phenotypic diversity arises from genetic variation caused by mutation [56, 57]. As experts consistently include heredity and genetic variation in explanations of evolutionary change [40], it appears gains on these topics can occur, but they are modest and likely a result of specific instructional methods [57]. If these specific interventions which lead to gains are missing from one's educational background, regardless of if in introductory biology or the courses since, naive ideas may persist (see [26, 39]).

Indeed, students explanations of evolutionary phenomena frequently do not include genetic concepts like variation and heredity, even after explicit instruction [40, 41, 42, 57]. The inferences necessary to 'see' mutation as a mechanism may be particularly challenging [16, 25], in addition to the inherent difficulties of understanding natural random processes [23, 29, 33]. For this reason, some concepts may be inherently more difficult and deserve more attention in the curriculum.

We observed similar difficulties in our students specifically, defining mutations, recognizing the importance of standing genetic variation, translating genotypes to phenotypes, and identifying genetic drift as a non-adaptive force of evolution. Following an Evolution course, however, students did appear to improve in recognizing the importance of standing genetic variation in populations. The concepts of defining mutations, translating genotypes to phenotypes, and recognizing genetic drift as a force of evolution continued to be problematic even following a course in Evolution. As these courses are often the last of the upper-level required courses for majors, many students may therefore be graduating with an incomplete understanding of the genetics underlying evolution.

Somewhat heartening, however, is the fact that some improvement can happen. Among our tracked students, the data suggest that students may grow better at defining a mutation. Originally, students assumed most commonly that mutations must change amino acids to qualify as a mutation, but gradually students widened their definition to include general genetic changes, such as those to noncoding regions. Grasping that mutations need not have phenotypic effects underscores how modern phylogenies are often constructed and evolutionary relationships are determined. Repeated exposures to the concept of a mutation and phylogeny by integrating evolution into courses may help students to link genetic variation to organismal variation.

It is also possible that there are 'ceiling effects' in the scores attainable on the assessment, as the scores among the tracked students appeared to be consistent, with just the variation narrowing at the end of Evolution as lower-scoring students improved. Repeated exposure could be the mechanism by which under-performing students improve, offering students multiple chances to demonstrate mastery. However, because these questions were generally multiple choice and coded as correct or not, we may be missing progressively more accurate models of evolutionary thinking as the students advance in the curriculum [40, 41], particularly for lower-performers, who may be prone to having multiple areas of improvement. Explicit tests, potentially at more refined intervals with more open-ended responses [10, 38, 39, 44] or oral interviews [41] should be used to address what misconceptions exist and compare gains made across the curriculum in detail.

Longitudinal research on where and how misconceptions arise and change throughout the curriculum (from Introductory Biology on) will guide the development of curricula and make evident where department-level changes need to be made. Our method of tracking students through the curriculum provides a starting point from which instructors can begin to compare and track student performance as they progress through the curriculum as designed, and make modifications if necessary. Because tracking can be difficult to do in practicality, particularly as it can take more time and student schedules and class sizes are variable, we advocate the combination of approaches of tracking students as well as direct comparisons between students in these courses. A better understanding of the gains made, and for whom and when, will help the field in developing how learning progressions intersect in genetics and evolution curricula.

Acknowledgements The authors would like to thank the study participants for their time, Luanna Prevost, Mark Tran, Jeremy Albright, and Amy Lark for assistance with the project, the UGA Biology Education Group for critical comments, and the Future Academic Scholars in Teaching 2012-2013 cohort and mentors for support. This material is based in part upon work supported by the National Science Foundation under Cooperative Agreement No. DBI-0939454. Any opinions, findings, and conclusions or recommendations expressed in this material are those of the author(s) and do not necessarily reflect the views of the National Science Foundation.

References

1. How people learn: Brain, mind, experience, and school: Expanded edition. National Research Council by National Academies Press (2000)
2. Knowing what students know: The science and design of educational assessment. National Research Council by National Academies Press (2001)
3. Ackermann, R.R., Cheverud, J.M.: *Detecting genetic drift versus selection in human evolution.* Proceedings of the National Academy of Sciences **101**(52), 17,946–17,951 (2004)
4. Agorram, B., Clement, P., Selmaoui, S., Khzami, S.E., Chafik, J., Chiadli, A.: *University students' conceptions about the concept of gene: Interest of historical approach.* US-China Education Review, ISSN 1548-6613, USA **7**(2), 9–15 (2010)
5. Alters, B.J., Nelson, C.E.: *Perspective: Teaching evolution in higher education.* Evolution **56**(10), 1891–1901 (2002)
6. Anderson, D.L., Fisher, K.M., Norman, G.J.: *Development and evaluation of the conceptual inventory of natural selection.* Journal of Research in Science Teaching **39**(10), 952–978 (2002)
7. Anderson, L.W., Krathwohl, D.R., Airasian, P.W., Cruikshank, K.A., Mayer, R.E., Pintrich, P.R., Raths, J., Wittrock, M.C.: *A taxonomy for learning, teaching, and assessing: A revision*

of Bloom's taxonomy of educational objectives, abridged edition. White Plains, NY: Longman (2001)

8. Andrews, T., Price, R., Mead, L., McElhinny, T., Thanukos, A., Perez, K., Herreid, C., Terry, D., Lemons, P.: *Biology undergraduates' misconceptions about genetic drift*. CBE - Life Sciences Education **11**(3), 248–259 (2012)

9. Balgopal, M.M., Montplaisir, L.M.: *Meaning making: What reflective essays reveal about biology students' conceptions about natural selection*. Instructional Science **39**(2), 137–169 (2011)

10. Bishop, B.A., Anderson, C.W.: *Student conceptions of natural selection and its role in evolution*. Journal of Research in Science Teaching **27**(5), 415–427 (1990)

11. Bowling, B.V., Acra, E.E., Wang, L., Myers, M.F., Dean, G.E., Markle, G.C., Moskalik, C.L., Huether, C.A.: *Development and evaluation of a genetics literacy assessment instrument for undergraduates*. Genetics **178**(1), 15–22 (2008)

12. Campbell, N.A., Reece, J.B., Urry, L., Cain, M., Wasserman, S., Minorsky, P., Jackson, R.: *Biology*. Pearson Benjamin Cummings, New York (2008)

13. Carey, S.: *The origin of concepts*. Oxford University Press, New York (2009)

14. Charlesworth, B., Charlesworth, D.: *Darwin and genetics*. Genetics **183**(3), 757–766 (2009)

15. Chattopadhyay, A.: *Understanding of genetic information in higher secondary students in northeast india and the implications for genetics education*. Cell Biology Education **4**(1), 97–104 (2005)

16. Chi, M.T., Slotta, J.D., De Leeuw, N.: *From things to processes: A theory of conceptual change for learning science concepts*. Learning and Instruction **4**(1), 27–43 (1994)

17. Dagher, Z.R., BouJaoude, S.: *Scientific views and religious beliefs of college students: The case of biological evolution*. Journal of Research in Science Teaching **34**(5), 429–445 (1997)

18. Dagher, Z.R., Boujaoude, S.: *Students' perceptions of the nature of evolutionary theory*. Science Education **89**(3), 378–391 (2005)

19. Darwin, C.: *On the origin of species by means of natural selection*. John Murray (1859)

20. Darwin, C.: *The variation of animals and plants under domestication*, vol. 2. John Murray (1868)

21. Demastes, S.S., Settlage Jr, J., Good, R.: *Students' conceptions of natural selection and its role in evolution: Cases of replication and comparison*. Journal of Research in Science Teaching **32**(5), 535–550 (1995)

22. Donnelly, L.A., Kazempour, M., Amirshokoohi, A.: *High school students' perceptions of evolution instruction: Acceptance and evolution learning experiences*. Research in Science Education **39**(5), 643–660 (2009)

23. Garvin-Doxas, K., Klymkowsky, M.W.: *Understanding randomness and its impact on student learning: Lessons learned from building the Biology Concept Inventory (BCI)*. CBE - Life Sciences Education **7**(2), 227–233 (2008)

24. Gould, S.J., Lewontin, R.C.: *The spandrels of San Marco and the Panglossian paradigm: A critique of the adaptationist programme*. Proc. R. Soc. Lond. B **205**(1161), 581–598 (1979)

25. Hmelo-Silver, C.E., Marathe, S., Liu, L.: *Fish swim, rocks sit, and lungs breathe: Expert-novice understanding of complex systems*. The Journal of the Learning Sciences **16**(3), 307–331 (2007)

26. Hokayem, H., BouJaoude, S.: *College students' perceptions of the theory of evolution*. Journal of Research in Science Teaching **45**(4), 395–419 (2008)

27. Ingram, E.L., Nelson, C.E.: *Relationship between achievement and students' acceptance of evolution or creation in an upper-level evolution course*. Journal of Research in Science Teaching **43**(1), 7–24 (2006)

28. Kalinowski, S.T., Leonard, M.J., Andrews, T.M.: *Nothing in evolution makes sense except in the light of DNA*. CBE - Life Sciences Education **9**(2), 87–97 (2010)

29. Klymkowsky, M.W., Garvin-Doxas, K.: *Recognizing student misconceptions through Ed's Tools and the Biology Concept Inventory*. PLoS Biology **6**(1), e3 (2008)

30. Lande, R.: *Natural selection and random genetic drift in phenotypic evolution*. Evolution **30**(2), 314–334 (1976)

31. Ledbetter, M.L.S.: *Vision and change in undergraduate biology education: A call to action presentation to Faculty for Undergraduate Neuroscience, July 2011.* Journal of Undergraduate Neuroscience Education **11**(1), A22 (2012)

32. Lewis, J., Wood-Robinson, C.: *Genes, chromosomes, cell division and inheritance - Do students see any relationship?* International Journal of Science Education **22**(2), 177–195 (2000)

33. Mead, L.S., Scott, E.C.: *Problem concepts in evolution part II: Cause and chance.* Evolution: Education and Outreach **3**(2), 261–264 (2010)

34. Mead, R., Hejmadi, M., Hurst, L.D.: *Teaching genetics prior to teaching evolution improves evolution understanding but not acceptance.* PLoS Biology **15**(5), e2002,255 (2017)

35. Mendel, G.J.: *Versuche über Pflanzen-Hybriden? [Experiments Concerning Plant Hybrids]? [1866].* Verhandlungen des naturforschenden Vereines in Brünn (1865)

36. Miller, J.D., Scott, E.C., Okamoto, S., et al.: *Public acceptance of evolution.* Science **313**(5788), 765 (2006)

37. Moore, R., Brooks, D.C., Cotner, S.: *The relation of high school biology courses & students' religious beliefs to college students' knowledge of evolution.* The American Biology Teacher **73**(4), 222–226 (2011)

38. Nehm, R.H., Beggrow, E.P., Opfer, J.E., Ha, M.: *Reasoning about natural selection: Diagnosing contextual competency using the ACORNS instrument.* The American Biology Teacher **74**(2), 92–98 (2012)

39. Nehm, R.H., Reilly, L.: *Biology majors' knowledge and misconceptions of natural selection.* AIBS Bulletin **57**(3), 263–272 (2007)

40. Nehm, R.H., Ridgway, J.: *What do experts and novices "see" in evolutionary problems?* Evolution: Education and Outreach **4**(4), 666–679 (2011)

41. Nehm, R.H., Schonfeld, I.S.: *Measuring knowledge of natural selection: A comparison of the cins, an open-response instrument, and an oral interview.* Journal of Research in Science Teaching **45**(10), 1131–1160 (2008)

42. Nieswandt, M., Bellomo, K.: *Written extended-response questions as classroom assessment tools for meaningful understanding of evolutionary theory.* Journal of Research in Science Teaching **46**(3), 333–356 (2009)

43. Ohlsson, S.: *Resubsumption: A possible mechanism for conceptual change and belief revision.* Educational Psychologist **44**(1), 20–40 (2009)

44. Opfer, J.E., Nehm, R.H., Ha, M.: *Cognitive foundations for science assessment design: Knowing what students know about evolution.* Journal of Research in Science Teaching **49**(6), 744–777 (2012)

45. Orr, H.A.: *Testing natural selection vs. genetic drift in phenotypic evolution using quantitative trait locus data.* Genetics **149**(4), 2099–2104 (1998)

46. Parker, G.A., Smith, J.M.: *Optimality theory in evolutionary biology.* Nature **348**(6296), 27 (1990)

47. Piaget, J.: *Success and understanding.* Harvard University Press, Cambridge, MA. (1978)

48. Posner, G.J., Strike, K.A., Hewson, P.W., Gertzog, W.A.: *Accommodation of a scientific conception: Toward a theory of conceptual change.* Science Education **66**(2), 211–227 (1982)

49. Price, R.M., Andrews, T.C., McElhinny, T.L., Mead, L.S., Abraham, J.K., Thanukos, A., Perez, K.E.: *The genetic drift inventory: A tool for measuring what advanced undergraduates have mastered about genetic drift.* CBE - Life Sciences Education **13**(1), 65–75 (2014)

50. R-Core-Team: *R: A language and environment for statistical computing* (2013)

51. Sinatra, G.M., Brem, S.K., Evans, E.M.: *Changing minds? Implications of conceptual change for teaching and learning about biological evolution.* Evolution: Education and Outreach **1**(2), 189–195 (2008)

52. Sinatra, G.M., Southerland, S.A., McConaughy, F., Demastes, J.W.: *Intentions and beliefs in students' understanding and acceptance of biological evolution.* Journal of Research in Science Teaching **40**(5), 510–528 (2003)

53. Smith, M., Knight, J.: *Using the genetics concept assessment to document persistent conceptual difficulties in undergraduate genetics courses.* Genetics **191**, 21–32 (2012)

54. Smith, M.K., Wood, W.B., Knight, J.K.: *The genetics concept assessment: A new concept inventory for gauging student understanding of genetics.* CBE - Life Sciences Education **7**(4), 422–430 (2008)
55. Smith III, J.P., DiSessa, A.A., Roschelle, J.: *Misconceptions reconceived: A constructivist analysis of knowledge in transition.* The Journal of the Learning Sciences **3**(2), 115–163 (1994)
56. Speth, E.B., Long, T.M., Pennock, R.T., Ebert-May, D.: *Using Avida-ED for teaching and learning about evolution in undergraduate introductory biology courses.* Evolution: Education and Outreach **2**(3), 415–428 (2009)
57. Speth, E.B., Shaw, N., Momsen, J., Reinagel, A., Le, P., Taqieddin, R., Long, T.: *Introductory biology students' conceptual models and explanations of the origin of variation.* CBE - Life Sciences Education **13**(3), 529–539 (2014)
58. Tran, M.V., Weigel, E.G., Richmond, G.: *Analyzing upper level undergraduate knowledge of evolutionary processes: Can class discussions help?* Journal of College Science Teaching **43**(5), 87–97 (2014)
59. White, P.J., Heidemann, M., Loh, M., Smith, J.J.: *Integrative cases for teaching evolution.* Evolution: Education and Outreach **6**(1), 17 (2013)

Supplemental Material

Instrument

1. Polydactyly is an inherited trait that results in extra fingers or toes. In the United States 0.1% of the population exhibits polydactyly. People with polydactyly have the genotype Pp, where P represents the allele that causes polydactyly and p represents the normal allele of this gene. Which of the following is true?

    a) The P allele is more frequent in the US than the p allele.

    b) The P allele is less frequent in the US than the p allele.

    c) The two alleles, P and p are at approximately equal frequencies in the US population.

    d) There is not enough information to answer this question.

2. Suppose that a single DNA base change of an A to a T occurs and is copied during replication. Is this change necessarily a mutation?

    a) Yes, as it is a change in the DNA sequence of an organism.

    b) Yes, but it must be a base change occuring in gametes.

    c) Yes, but it must be a base change occuring in the coding part of a gene.

    d) Yes, but it must be a base change that alters the amino acid sequence of a protein.

    e) Yes, but it must be a base change that alters the appearance of the organism.

3. An isolated population of prairie dogs has longer than average teeth. As a result they can eat more grass with less effort and are better able to survive and reproduce. The mutation(s) that resulted in longer teeth:

    a) allowed the teeth to grow longer over several generations until they reached an optimal length for eating grass.

    b) arose in many members of the population simultaneously and then lead to longer teeth.

    c) happened as a result of chance within the prairie dog population and then lead to longer teeth.

    d) occurred because the prairie dogs needed to be more efficient at eating grass to survive and reproduce.

    e) would only occur in a prairie dog population that eats grass and would not occur in a population that lives on seeds.

4. Adult height in humans is partially determined by our genes. When environmental conditions are held constant, humans have a wide variety of heights (not just short, medium, and tall). Height is probably influenced by:

    a) one gene with two alleles.

    b) a single recessive gene.

    c) a single dominant gene.

    d) several genes and alleles.

    e) only paternal genes.

5. Sometimes a trait seems to disappear in a family and then reappear in later generations. If neither parent has the trait, but some of the offspring do, what would you conclude about the inheritance of the trait?

    a) Both parents are carriers of the recessive form of the gene.

    b) Only one parent has two copies of the recessive form of the gene.

    c) Only one of the parents has a dominant form of the gene.

    d) Only one parent has a copy of the recessive form of the gene.

    e) It is most likely the result of new mutations in each parent.

6. Which of the following is a characteristic of mutations in DNA?

    a) They are usually expressed and result in positive changes for the individual.

    b) They are usually expressed and cause significant problems for the individual.

    c) They occur in the body cells of a parent and are usually passed onto offspring.

    d) They usually occur at very high rates in most genes of all known organisms.

e) They usually result in different versions of a gene within the population.

7. Mutations in DNA occur in the genomes of most organisms, including humans. What is the most important result of these mutations?

a) They produce new genes for the individual.
b) They produce new enzymes for the individual.
c) They provide a source of new cells for the individual.
d) They provide a source of variation for future generations.
e) They produce new chromosomes for future generations.

8. In peas, the round allele is dominant over the wrinkled allele. If a plant with round peas is crossed to a plant with wrinkled peas, all of the resulting plants have round peas. What is the genotype of the parents in this cross?

a) $Rr \times Rr$
b) $RR \times rr$
c) $rr \times rr$
d) $Rr \times rr$
e) $RR \times Rr$

9. Which of the following is a true statement concerning genetic variation?

a) It tends to be reduced by the processes involved when diploid organisms reproduce.
b) It arises in response to changes in the environment in which the organism lives.
c) It is creased by the direct action of natural selection on the population.
d) It must be present in the population before natural selection can act upon it.
e) High average heterozygosity populations predict less genetic variation.

10. Black fur in mice (B) is dominant to brown fur (b). Short tails (T) are dominant to long tails (t). What fraction of the progeny of crosses $BbTt \times BBtt$ will be expected to have black fur and long tails?

a) 3/16
b) 8/16
c) 6/16
d) 9/16
e) 1/16

Please read the following statements to answer questions 11-16.

A small island is home to a unique population of land snails. This population was founded by 10 individuals that floated to the island on a log, and it has been isolated from the large mainland population ever since.

A. Biologists compared the genetic variation of the mainland and island populations a few years after colonization. Please indicate whether a biologist would agree or disagree with each of the following statements.

11. The biologists observed genetic drift but not evolution because the island snails were just as well-suited to their environment as the ones on the mainland.

A. Agree   *B. Disagree*

Why?

_____

12. The biologists observed genetic drift and concluded that the island population had fewer versions of each gene than the mainland population.

A. Agree   *B. Disagree*

Why?

___

B. After forty generations, biologists measured the genetic variation of the island snail population again. They concluded that the population of snails on the island had remained isolated and that genetic drift had occurred. Please indicate whether a biologist would agree or disagree with each of the following statements about the processes that contributed, at least in part, to genetic drift in the population of island snails.
Please indicate whether a biologist would agree or disagree with each of the following statements about what occurred during the forty generations since colonization.
13. The island population experienced random changes in the frequency of certain traits that made them genetically distinct from the mainland population.

A. Agree   *B. Disagree*

Why?

___

14. The island population may have experienced mutation in addition to random changes in the frequency of certain traits.

A. Agree   *B. Disagree*

Why?

___

15. The island population may have adapted to conditions on the island if random genetic change increased survival and reproduction of some individuals.

A. Agree   *B. Disagree*

Why?

___

16. The island and mainland populations will be less similar to each other than they were 100 generations ago.

A. Agree   *B. Disagree*

Why?

___

With what gender do you most identify?
   Male
   Female
   Prefer not to Answer

What year are you in school?

What is your overall GPA?

What semester did you/are you taking Genetics?

What was your approximate SAT or ACT score? (score corrected for ACT)

With what race/ethnicity do you most identify?
American Indian/Alaskan Native
Asian
Black/African American
Native Hawaiin/Other Pacific Islander
White (Not Hispanic or Latino)
White (Hispanic or Latino)
Other (text entry)

Chapter 35
The Evolution of the Scientific Virtues Toolbox Approach to Responsible Conduct of Research Training

Chet McLeskey, Eric Berling, Michael O'Rourke and Robert T. Pennock

Abstract The Scientific Virtues Toolbox is a novel scientific virtue-based workshop model for responsible conduct of research (RCR) training. This paper gives a brief overview of how Pennock's vocational virtue theory, which had previously been delivered in courses, was transformed into discussion-based RCR workshops using the Toolbox structured dialogue method. The interdisciplinary BEACON Center, which combined biologists, computer scientists, and engineers, and which aimed to model a culture of excellence, ethics, and inclusion, proved to be an ideal environment to develop and test this approach to the cultivation of scientific character. The paper describes the guided-dialogue structure of the workshops, the nature of the discussion prompts, the pilot assessments carried out, and how the workshops are now evolving beyond their scientific origin.

Key words: Scientific integrity, scientific virtues, science ethics, responsible conduct of research, RCR training, research integrity, toolbox dialogue initiative.

Chet McLeskey
Center for Interdisciplinarity, Michigan State University, East Lansing, MI, USA

Eric Berling
Center for Interdisciplinarity, Michigan State University, East Lansing, MI, USA

Michael O'Rourke
Center for Interdisciplinarity, Department of Philosophy, AgBioResearch, Michigan State University, East Lansing, MI, USA

Robert T. Pennock
Lyman Briggs College and Departments of Philosophy and Computer Science & Engineering, Michigan State University, East Lansing, MI, USA e-mail: pennock5@msu.edu

© Springer Nature Switzerland AG 2020
W. Banzhaf et al. (eds.), *Evolution in Action: Past, Present and Future*,
Genetic and Evolutionary Computation, https://doi.org/10.1007/978-3-030-39831-6_35

35.1 The History of the Scientific Virtues Toolbox RCR Project

From the beginning, the BEACON Center aimed to excel not only in scientific research, but in all aspects of its activities as a science and technology center. Included in this aim were its ethics commitments. The National Science Foundation responsible conduct of research (RCR) requirement for Science & Technology Center (STC) grants stated:

> The Director shall require that each institution that applies for financial assistance from the Foundation for science and engineering research or education describe in its grant proposal a plan to provide appropriate training and oversight in the responsible and ethical conduct of research to undergraduate students, graduate students, and postdoctoral researchers participating in the proposed research project. [7]

In its strategic plan, BEACON promised to not just meet the letter of the requirement, but to set a model for excellence, including its implementation of this RCR requirement. It stated that goal in the following way:

> Ethics Optimal Outcome. Center participants will understand shared and discipline-specific practices of Responsible Conduct of Research (RCR) and will embody general scientific norms/virtues, including curiosity, honesty, objectivity, integrity, community, and transparency. (BEACON 2016)

Phrasing this goal in virtue-theoretic terms was unusual, but very intentional. The vision was based on a philosophical approach that Pennock had been developing since the late-1990's that looked at science as a methodological practice that had an inherent moral structure characterized by what he called "scientific virtues." This scientific virtue (SV) approach is based in an analysis of excellence inspired by Aristotle, who wrote that virtues are those character traits that lead to flourishing [2]. Aristotle had considered this in terms of general human virtue and human flourishing seen in relation to the human nature and purpose (telos), but if one thinks of it in the context of particular vocational practices rather than human life as a whole, one is able to identify traits that should lead to the flourishing—i.e., the excellence—of that practice. For the scientist, therefore, cultivating scientific virtues should lead to scientific flourishing. BEACON provided an opportunity to move the SV project out of the philosophical armchair and test it in practice.

There was a special challenge in that BEACON's goal also reflected a commitment to developing an ethical culture in a complex interdisciplinary and institutional setting. As an STC, BEACON is an "integrative partnership" that builds "intellectual and physical infrastructure" at the "interfaces" of several disciplines, including biology, electrical engineering, computer science, and philosophy [7]. Its success depended on its ability to integrate knowledge from these different disciplines, and it has worked hard to create a culture of interdisciplinarity that provides the resources and support necessary to achieve this integration. BEACON conducts interdisciplinary scientific, engineering, and educational research; holds a weekly interdisciplinary seminar series and an annual research Congress; and facilitates interdisciplinary courses. These programs are designed to bring representatives from different disciplines together to expand their research networks, catalyze integrative conversations, and create the conditions for successful interdisciplinary collaboration.

Further, BEACON is a consortium of several universities: Michigan State University, the University of Washington, the University of Texas, the University of Idaho, and North Carolina A&T. These very different universities include two land grants, three Research 1 universities, one small research university, and a Historically Black Colleges and Universities (HBCU) institution. The multiple institutions that constitute BEACON form another dimension of difference across which knowledge integration must take place.

Developing a positive culture of research ethics in this context means taking seriously the many differences that exist among BEACON members and integrating them with equity and balance. An ethical approach that privileged, say, life science over computer science would not do, nor would an approach that privileged the experience of larger research universities over smaller universities. The SV approach is well designed to operate in a context of multiplicity like this. Grounded in virtue theory as its foundation, the SV approach highlights what BEACON members have in common as scientific researchers while accommodating differences in emphasis. BEACON members may work at a computer rather than in the field or conduct research along with a substantial amount of teaching rather than run a big lab, but they are in the end scientists who strive to combine curiosity with honesty, persistence, and humility to evidence. By standing on common ground supplied by a conception of excellent science, BEACON members can develop the collective beliefs, values, and habits that constitute a distinctive and authentic culture of research ethics.

A key part of using the SV approach to create such a culture involves creating conditions in which BEACON members can reflect on and compare their own ways of thinking about what it means to be a virtuous scientist. Unless there are collective opportunities for conversation among BEACON members about excellent science, any progress toward a common, positive culture of research ethics would be accidental and unlikely. It is here that the Toolbox approach comes into the picture.

The Toolbox Dialogue Initiative (TDI, formerly the "Toolbox Project") was originally based at the University of Idaho, a BEACON consortium partner, and became involved in BEACON at the proposal stage. TDI is a research and outreach initiative that runs dialogue-based workshops designed to build collaborative and communicative capacity in interdisciplinary teams [4, 5, 8]. Within BEACON, these workshops have been opportunities for participating scientists to have conversations structured around issues of importance to the community.

TDI was originally enlisted by BEACON leadership to serve the consortium in two roles: first, it was to run workshops designed to help build functional interdisciplinary teams, and second, it was to collect data about interdisciplinary performance that could aid in center evaluation. In the words of the proposal, TDI was to "provide training workshops and contribute to assessments to provide not only evaluative information, but more importantly, formative information to improve functionality of the teams."

TDI's initial involvement with BEACON came during the first Congress in 2010, when it ran five Toolbox workshops to help encourage interdisciplinary understanding in the newly funded center. Focusing on scientific research, these workshops were designed to bring to the surface differences in epistemological and ontological assumptions that interdisciplinary researchers might not have otherwise recognized. They were conducted with five groups: the BEACON Executive Committee, the EHRD Steering Committee, and three cross-cutting theme research groups focused on viruses, speciation, and evo/devo. These standard Toolbox workshops ran for 2.5 hours and included a framing talk introducing the participants to the approach and a roughly 90-minute dialogue sandwiched between two periods during which participants filled out the "Scientific Research Toolbox instrument".

The Scientific Research Toolbox instrument is a survey-like instrument designed to stimulate and focus conversation that includes provocative prompts organized into six thematic groups called "modules": Motivation, Methodology, Confirmation, Reality, Values, and Reductionism [8]. The prompts are Likert items associated with a 5-point Likert scale ("Disagree" through "Agree") with "Don't know" and "Not applicable" options. Examples of the prompts include "Scientific research must be hypothesis driven", "Value-neutral scientific research is possible", and "Scientific research need not represent objective reality to be useful."

After scoring these prompts, participants discuss their views of them, noting similarities and differences and developing, coordinating, and even negotiating their alternative perspectives. Participants move through the instrument along a trajectory they determine, transitioning from module to module in a way that tracks their interest. The dialogues are facilitated, but the facilitation approach is light-handed and aimed at making sure the participants visit each of the modules. Rarely do participants discuss all the prompts, and occasionally they will visit some modules only briefly. The point of these dialogues is to enhance understanding about issues that are important to the group, which may only be a smaller subset of the overall number of prompts on offer in the instrument.

As BEACON progressed, the relationship of TDI with BEACON evolved. In 2012, Pennock and O'Rourke applied for and received BEACON seed funding to support the combination of the SV Project with the Toolbox approach in a way that could contribute to the RCR training required of graduate students and post-doctoral scholars in the center. Prior to initial work on the project, O'Rourke moved to MSU, where he joined Pennock, Berling, and McLeskey in adapting the Toolbox approach to the RCR context. Its role was still to provide contexts for structured conversation, but in this case the structure was supplied by the SV approach.

35.2 Adapting the Approaches to the BEACON Context

BEACON provided a unique situation that allowed the SV Project and TDI to combine and evolve in ways that neither might have done on their own. Pennock had previously developed a curriculum to teach science ethics with an SV-based approach for regular undergraduate courses and had taught different versions, with one organized around philosophical theory and a second where the material was presented by way of readings (biographical and autobiographical) of the lives of scientists. The collaboration with TDI yielded a new workshop model designed to deliver RCR training for BEACON participants. The structured dialogue format of the Toolbox provided a framework for adapting the SV-based approach in a new and fruitful way that could be tailored for the needs of graduate students, postdocs, and other researchers. It also accommodated research faculty in face-to-face discussion of scientific values within an ethical framework. The Toolbox approach provided a way to give ethical structure to a conversation in workshop format that was appropriate for researchers who would never have the chance to take a full, formal course on research ethics. Bringing the two approaches together resulted in what we called the Scientific Virtues Toolbox (SVT) workshops.

The SVT workshops are designed with the goal of providing participants with an engaging experience that allows them to meaningfully explore RCR issues through the lens of Pennock's SV theory, which emphasizes the role of character in exemplary scientific practice. These SVT workshops will ultimately serve as one element of a broader virtues-based approach to RCR training and education. This distinct approach, with its emphasis on the Scientific Virtues, augments standard and traditional approaches to RCR training in a way that we believe is likely to strengthen the impact RCR training has on future behavior.

Each SVT workshop module focuses on a particular scientific virtue, such as curiosity. Workshop modules have been created for the following themes: the purpose of science, the purpose of engineering, curiosity, honesty, courage, humility to evidence, meticulousness, perseverance, and skepticism, with plans underway for modules for attentiveness, collaborativeness, and objectivity. A participant's first workshop session involves the Purpose of Science and Curiosity modules, with subsequent sessions involving other modules.

The workshop sessions are designed for small groups and typically run for an hour to an hour and a half. Various aspects of the SVT workshops are similar to standard Toolbox workshops, such as initiation with a preamble that introduces the approach and use of a Toolbox instrument comprising modules containing Likert items (see Figure 1). The preamble for the SVT workshops includes some background on the Scientific Virtues Theory and TDI. After the preamble is delivered, participants are divided into small groups, ideally ranging from 5-10 individuals, and one facilitator is assigned to each group. The facilitator helps administer materials, collect data, and moderate the ensuing discussion as needed. Through this moderation, the facilitator may also play the role of "Socrates in the Room" (see below for more information). Workshop participants then complete the SVT instrument and participate in a 30 to 45-minute dialogue.

During this dialogue portion of the SVT workshops, the prompts structure the discussion by encouraging participants to share and compare their reactions and responses to the prompts. After the discussion, participants then complete the same instrument they completed at the beginning, scoring their reactions to the prompts on the Likert scale. Finally, one to two weeks after the workshop, participants who consented to a follow-up interview are sent a survey that invites them to reflect upon their experiences and offer feedback.

Fig. 35.1: Sample prompts from the Perseverance Toolbox module.

Since beginning our SVT approach in 2013, we have administered over twenty-five SVT workshop sessions, including four or more sessions at each of the annual BEACON Congresses (2013-2018) and several BEACON Friday Seminar sessions. At the request of two professors who had participated in one of our workshops, we have also administered workshop sessions as part of the required curriculum for two graduate-level science courses and an undergraduate engineering course. In total, over 180 unique participants ranging from undergraduate students through senior faculty members have completed an SVT workshop session.

Participants who have already completed their first SVT workshop and discussed the Purpose of Science and Curiosity modules can participate in further workshop sessions with new modules at the BEACON Congress or the occasional BEACON Friday Seminar. These modules for returning participants have collectively had over 80 attendees, though this does not represent 80 unique attendees since a participant may have attended multiple additional modules over the past few years.

Adaptation occurred in the other direction as well. We made several changes in the Toolbox approach to make it work better for RCR training. First, in contrast with standard Toolbox workshops designed to help teams discover things about themselves, these were didactic workshops aimed at imparting ethical lessons to participants about virtuous scientific practice and RCR. These lessons were supported by recognizing commonalities as well as differences across the participant group, and to move past subjective opinions about ethics to a more rigorous account in which shared goals and values point the way to responsible conduct. Second, we structured the dialogues so that the participants talk about all the prompts, dividing time evenly between the modules. Third, because we were interested in making sure that participants discussed all the prompts, we reduced the total number of modules to two, each containing six to eight prompts. Fourth, we changed the nature of the prompts so that they better fulfilled the didactic purpose and helped reveal the connection of values to responsible conduct. And finally, the facilitation style changed so that the workshop leader operates not just as a facilitator, but more like a philosophical guide—a role we came to call "Socrates in the room".

The final point about our change in facilitation style deserves special emphasis. As we noted above, the traditional TDI approach utilizes a more hands-off approach to facilitation, which works well when the goal of the workshop is primarily focused on communication and resolving issues within a research team. With the RCR workshops, however, there is a pedagogical requirement—the participants need to learn something about responsible scientific conduct that will impact their behavior in the future. There are two primary factors at work here one involves what is learned, and the other involves how it is learned. Taking cues from Plato's Socrates, we developed a Socratic role for the facilitator. This represents a move toward more guided discussion than is typical of TDI workshops, in an effort to make sure that participants leave the workshops with a deeper understanding of the nuances of RCR and the dispositions needed to act well when faced with difficult situations.

There are several key characteristics of what is commonly known as the "Socratic Method" or, in ancient Greek scholarship, the *elenchus*, that explain its relevance to our effort. The method itself is largely one of identifying contradictions within an interlocutor's belief set by way of interrogation. In a typical case, the structure involves Socrates asking the interlocutor for something—typically this is the definition of a virtue term (like piety, as we see in Plato's Euthyphro), but could also be the reasons for doing or not doing something. The interlocutor will then offer an answer, A, which will serve as the basis for further questioning. Socrates then proposes that A logically implies consequence X. The interlocutor is supposed to follow the reasoning that Socrates provides and agree that X is a logical consequence of holding A. Socrates then offers another proposition, which could be an alternative definition of the virtue term, an example of something we take to be an instance of the virtue (say, an example of something that is courageous), or something that seems tangential to the conversation but later is shown to be relevant. This second proposition, B, is understood to be a reasonable alternative to A and is then subjected to the same sort of reasoning as A. The result of this second line of questioning is the logical consequence Q.

Thus, we have two definitions, A and B, both of which the interlocutor finds reasonable (if not absolutely true), and two logical consequences, X and Q respectively. Given that the interlocutor agrees with the line of reasoning, Socrates then shows how the consequences X and Q cannot be held simultaneously—i.e. they are contradictory. The goal of this line of inquiry is to bring the person to the point of recognizing their own ignorance of the topic at hand. In the case of virtue, the idea is that people often assume they know what is good and bad, and trust their instincts and intuitions more than they should. The Socratic method is meant to show them that things are more complicated than they thought. Properly executed, the Socratic method can help people who are quite willing to admit that they do not know what morality requires of them in a given situation as well as those who are confident in their moral or practical knowledge. It is a process of bringing the interlocutor to the point of *aporia*—i.e., a kind of confusion that is supposed to be uncomfortable—leading the interlocutor to pursue the truth more fully. Because this is a moral exercise, it is critical that the interlocutor follow the line of reasoning and also say what they truly believe. If people are not sincere in this, they will not get the benefit of the inquiry.

In relation to RCR, we have modified the Socratic method outlined here to press people on issues that are related but not obviously so. How does a proper understanding of courage lead one to being more open to criticism? How does an understanding of curiosity in scientific practice lead one to be less likely to fabricate data? These sorts of connections fall into line with Socratic inquiry. Socrates assumed that no one does what they know to be bad, i.e., no one errs willingly. Given this, the key is to know what sorts of things will hurt you and what will benefit you. This is a teleological value structure, with "the good" or "virtue" being the telos. In the more familiar Aristotelian sense, *eudaimonia* (often defined as happiness' or flourishing') is the telos and virtuous activity is what enables one to be eudaimon.

The form that Socratic facilitation takes in SVT workshops varies from workshop to workshop. In some cases, participants make good points that do not get the uptake they should by the rest of the group. Here, our Socrates can step in to challenge others to take a more serious look at what their fellow interlocutors are saying. In other cases, Socrates may need to moderate the views of participants whose voices dominate the discussion. In any case, the role of Socrates is not to lecture, but to guide, with varying amounts of influence and pressure. Importantly, Plato's Socrates claimed that he was not a teacher, but rather a mid-wife whose job was to bring out the knowledge and understanding that was already there in the minds of his interlocutors as well as determine the bounds of that knowledge in order to understand ourselves better. It is only after we are challenged and forced to accept the limitations of our understanding that we can meaningfully discuss with others the nature of what is good and how to improve ourselves. This is just as true in the context of RCR as it is in everyday life.

When designing the prompts that make up a module, we have several goals. First, prompts are constructed not as survey items with precise wording to elicit consistent, uniform responses, but, rather, as stimuli to guide thought. They are designed to help participants articulate issues about which there may be multiple reasonable responses, thereby stimulating productive reflection and discussion. In some cases, this means that prompts are intentionally ambiguous, so that participants can work through the nuances of context or multiple possible interpretations.

Second, prompts aim at content that is meaningfully relevant and that will ideally help participants come to an appreciation of the focal virtue and the ways in which it may play a role in the practice of science.

Module topics are not created ad hoc but reflect both Pennock's theoretical approach and empirical findings from his national survey of exemplary scientists (Pennock and Miller, in preparation). As noted above, the characteristic scientific virtues arise in relation to science's central, guiding purpose—its *telos*. For that reason, the first module that we have every participant do involves a discussion of the purpose of science. We typically combine this with a module on curiosity, which, together with honesty, stands at the core of science's characteristic values that arise in relation to its telos [11]. For subsequent workshops, participants are given modules that focus on other virtues important to science as identified theoretically and substantiated by Pennock's survey.

We organize the prompts of each module to cover three areas and loosely ordered them so that dialogue participants are more likely to see how virtues relate to conduct. The first couple of prompts in each module explore the concept of some particular virtue in a general sense (e.g., "Being meticulous means paying attention to every detail"). The prompts that follow examine the virtue's role in scientific practice (e.g., "Perseverance in science is indistinguishable from dogmatism"). The final few prompts explicitly connect the virtue to RCR issues (e.g., "A responsible scientist is more skeptical of bold and exciting findings").

This ordered structure changes the nature of the dialogue about RCR, especially in relation to traditional RCR training. The traditional approach often begins with some explicit definition or statement of a particular form of research misconduct, accompanied by examples of misconduct, real or hypothetical, with some history of why the rule came to be made and how it is enforced [13]. Our approach reverses this, beginning with internal values and then showing why responsible conduct flows out of them; the prompts are worded and ordered so that participants come to the discussion of particular RCR issues only after having talked about the values that underlie them. The SV-based approach also allows us to raise broader issues about responsible conduct and ethical culture that go beyond the standard RCR issues of falsification, fabrication, plagiarism and such [10].

Our usual procedure is to devise a set of prompts for a module and then test them in practice at a BEACON seminar or the BEACON Congress. The final version of a module will usually have six prompts, but we typically include eight or nine prompts in a test module so we can determine which work best in live dialogue. We record the dialogues, take notes, get participant feedback, and later discuss how these worked in different sessions. This formative process allows us to pick a subset of prompts that worked best and revise the wording as needed to make them more effective.

35.3 Outcomes

We have already published a few articles about this novel RCR approach [3, 9, 10, 12], and we have another manuscript now in progress that will survey and discuss the RCR literature in light of our virtue-based perspective. Berling et al. [3] reports pilot results from the first few BEACON workshops (e.g. 2013-2015 Congresses), showing that overall respondents (n=44, response rate 67%) believed that a virtues-based approach could contribute to RCR training and preferred our SV approach over traditional RCR training methods.

Since those pilot workshops, we have added to and refined the post-workshop survey questions sent to participants and collected these additional data from eleven workshops groups (Figure 2). These groups have included nine voluntary workshops offered at the 2017 BEACON Congress, 2018 BEACON Congress, and a 2017 BEACON Friday Seminar. These workshops have had a total attendance of 88, with 43 responses to the post-workshop survey (response rate = 48.9%). Also included are a graduate science course and an undergraduate engineering course in which SVT modules were used by the professor as part of the required curriculum, though consent to have deidentified data collected was still voluntary. These courses had a total of 40 students, with 36 responding to the post workshop survey (response rate = 90.0%).

	Workshop	Participants	Responses to Survey	Response Rate
Voluntary BEACON Workshops	2017 BEACON Friday Seminar: Objectivity	11	8	72.7%
	2017 BEACON Congress: Purpose of Science & Curiosity Groups (2)	24	8	33.3%
	2017 BEACON Congress: Meticulousness Groups (2)	18	7	38.9%
	2018 BEACON Congress: Purpose of Science and Curiosity Groups (2)	18	9	50.0%
	2018 BEACON Congress: Skepticism Groups (2)	17	11	64.7%
	TOTAL	**88**	**43**	**48.9%**
Courses	2016 Science Graduate Course	8	6	75.0%
	2017 Engineering Undergraduate Course	32	30	93.8%
	TOTAL	**40**	**36**	**90.0%**

Fig. 35.2: Attendance and post-workshop survey response rates to the workshops conducted since refining the survey sent 1-2 weeks after participation.

The responses to the closed-ended questions from the post workshop survey can be seen in Figure 3. For ease of visualization, the responses to the voluntary BEACON workshops are grouped together as are the responses to the two course curriculum workshops. The agreement column combines "Strongly Agree" and "Agree" responses together, and the disagreement column combines "Strongly Disagree" and "Disagree" responses.

Participants from both the voluntary BEACON workshops and the student courses largely view the workshop's discussion as an open exchange of thoughts and ideas (88.4% (38) and 86.1% (31)). Participants also overwhelmingly reported that the SVT prompts were effective conversation starters (95.3% (41) and 94.4% (34)). Given this, we believe that the prompts and discussion-environment of the SVT workshops can produce efficient, targeted discussion that increases the chances that the ensuing dialogue resonates with participants.

Overall, the respondents from both the voluntary BEACON sessions and the student courses reported enjoying their SVT experience (93.0% (40) and 83.3% (30), respectively). While enjoyment is not a direct measure of effective pedagogy, we

QUESTION	Voluntary BEACON Workshops			Course Curriculum Workshops		
Participants	88			40		
Respondents	43			36		
Response Rate	48.9%			90.0%		
	Agree	**Neutral**	**Disagree**	**Agree**	**Neutral**	**Disagree**
Overall I enjoyed the SV Toolbox Workshop	93.0% (40)	4.7% (2)	2.3% (1)	83.3% (30)	16.7% (6)	0% (0)
The SV Toolbox prompts were effective conversation starters.	95.3% (41)	2.3% (1)	2.3% (1)	94.4% (34)	5.6% (2)	0% (0)
Our conversation was an open exchange of thoughts and ideas.	88.4% (38)	9.3% (4)	2.3% (1)	86.1% (31)	8.3% (3)	0% (0)
Participating in the SV Toolbox workshop is likely to help my professional development.	65.1% (28)	30.2% (13)	4.7% (2)	44.4%** (16**)	38.9%** (14**)	13.9%** (5**)
Since the SV Toolbox workshop, I have thought about the topics raised in the workshop.	60.5% (26)	25.6% (11)	14.0% (6)	63.3%* (19*)	13.3%* (4*)	23.3%* (7*)
Since the SV Toolbox workshop, I have discussed the topics raised in the workshop with others.	48.8% (21)	27.9% (12)	23.3% (10)	50.0% (18)	22.2% (8)	27.8% (10)
My participation in the SV Toolbox workshop contributed to a change in my views of scientific [focal virtue].	30.2% (13)	27.9% (12)	41.9% (18)	26.7%* (8*)	33.3%* (10*)	40.0%* (12*)

Fig. 35.3: Responses to the post-workshop survey sent to participants 1-2 weeks after completing the workshop. Responses with (*) are only responses from the undergraduate engineering course (30 respondents, 93.8% response rate), as these two questions were not asked of the graduate science course students. For the results marked (**), one of the respondents left the field blank, so the responses here only total 35.

believe that enjoyable and engaging RCR training approaches are more likely to be effective than approaches that are not. A comparative test would be required to establish this, of course, but such a test poses a challenge given that assessing the effectiveness of RCR training remains a problem for the field (see Antes et. al [1] and May and Luth [6] for useful discussions on this). We do take it to be encouraging that a majority of our respondents (60.5% (26) and 63.3% (19*)) reported thinking about topics from the workshop after completing it. Nearly half (48.8% (21) and 50.0% (18)) of respondents discussed the topics of the workshop with someone else. These results demonstrate that some participants are mentally engaging with the content of our workshops even after leaving the session. Furthermore, nearly a third of BEACON participants (30.2% (13)) and 26.7% (8*) of students report a change in their views on the focal virtue in light of the workshop discussion.

Adding responses from these workshops to responses from past workshops [3], we see a continuation of the trends first observed. When looking at the first impression from the BEACON workshops, participants overwhelmingly believe that a virtues-based approach can add to the development of RCR, and this trend is also seen in responses from students who received SVT modules as part of their course work (Figure 4). Additionally, BEACON workshop participants prefer a virtues-based approach to RCR training over more traditional approaches (Figure 5). While students from the two courses were also asked this question, their responses are not included since many of them reported not having sufficient experience with traditional methods to make a judgment (e.g. "I have no experience with traditional RCR training, but I expect it would be less interesting and engaging than the toolbox exercise" and "The virtue approach was exceptional for my first RCR experience. As it was my first, I'm not sure what a more traditional approach is, therefore I can't answer.").

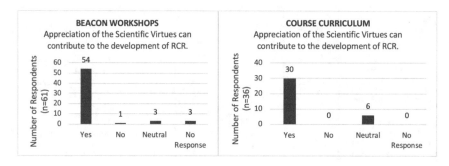

Fig. 35.4: Responses on the left are from BEACON Workshop participants ranging from 2013-present, with each response being a unique individual. In the case of workshop participants who have responded to this survey question multiple times by virtue of having completed multiple workshops, only their first response is included here (so it represents 61 unique respondents). Early workshop participants were asked the open-ended question "Explain whether you think appreciation of the Scientific Virtues can contribute to the development of RCR training," and these responses were coded into "Yes," "No," "Neutral," and "No Response." Later workshops asked participants to rate their response to the statement "Appreciation of the Scientific Virtues can contribute to the development of RCR" on a 5-point Likert scale from "Strongly Disagree" through "Strongly Agree," and in order to combine this data with earlier years "Strongly Disagree" and "Disagree" are treated as "No" responses while "Strongly Agree" and "Agree" responses are treated as "Yes". Responses on the right are from the students of the two courses where SVT modules were a part of their curriculum. Responses were scored on the Likert scale as a reaction to the prompt, and converted here to "Yes" and "No" for comparison purposes.

We find these results promising. While we do not know how effective the SVT approach is in affecting future behavior when faced with RCR issues or whether

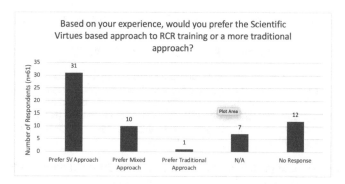

Fig. 35.5: Respondents from BEACON workshop participants ranging from 2013-present, with each response representing a unique individual. In the case of workshop participants who have responded to this survey question multiple times by virtue of having completed multiple workshops, only their first response is included here (so it represents 61 unique respondents). Respondents were asked the open-ended question, and responses were coded to "Prefer SV Approach," "Prefer Mixed Approach," "Prefer Traditional Approach," "Not Applicable," and "No Response."

it is more effective than the traditional approaches, we are optimistic that our approach can meaningfully augment traditional approaches. Given that participants enjoy their workshop experience, believe that the virtue-based approach can contribute to RCR training, and think about and discuss the issues after the workshop is over, we believe that their engagement is likely to have a positive influence in their future behavior. Future work will include the development of additional modules and continued offering of workshops to expand our sample size.

We have also begun to disseminate aspects of the SV Project and the SV Toolbox at national and international professional meetings. For instance, Berling, McLeskey, and Pennock each gave presentations related to this work at the Character and Virtue in the Professions conference held at Univ. of Birmingham in the UK in June 2016 and also at the 5th World Conference on Research Integrity held in Amsterdam, The Netherlands in May 2017. Pennock gave a talk on "Rethinking Science Ethics Training in light of the Scientific Character Virtues" at the National Science Foundation Science & Technology Centers Directors Meeting in Washington DC in August 2016, and gave the invited keynote talk on "Curiosity and the Moral Character of Science" for the Undergraduate Research Summer Institute (URSI) Conference at Vassar College in September of the same year, and led a week-long workshop "A Virtuous Instinct for Truth" at Friends General Conference Gathering in Toledo, Ohio in July 2018. We have presented posters and presentations at various other professional meetings and reported on our efforts at NSF meetings and BEACON site visits.

The reaction at these events, as from our workshop participants, has been very encouraging. To mention just one example, the SVT RCR efforts were called out for special praise by external evaluators in a recent BEACON site review in their official report: "[We] commend their innovative approach to the ethical conduct of research based on the virtues of science.'" Site visitors cited it as an example of an opportunity for continued "major involvement at the national level to engage in leadership related to reforming American science education," noting that:

> There is a strong emphasis on understanding and promoting the nature of science and the value of science in education through the Science Virtues project, which serves as example of how to instill responsible/ethical practices in science education. Simultaneously, the Science Virtues project could influence STEM identity development and be used to help mitigate some of the challenges in recruiting and retaining underrepresented groups in STEM. (Site Visit Report 2016)

Based upon these promising early results and feedback, the project has now begun to evolve beyond its initial development within BEACON.

35.4 Adaptive Radiation of the SVT Approach beyond the Natural Sciences

Although its initial foray concentrated on the sciences, this virtue-centric SVT approach is widely applicable. Pennock saw early on that other vocational types of disciplines, namely medicine and engineering, would be natural next steps in what we now call the Vocational Virtues Project. During the development of the RCR component of the project, we wondered what it might look like to apply our approach to other disciplines within the university. Being a group of philosophers, we naturally thought about the impact we could make in the humanities—an area of the university that has had far less exposure to RCR. Coincidentally, a new mandate took effect at Michigan State University requiring all graduate students to engage in RCR training. This meant areas such as the humanities and the arts would now need to develop and implement an RCR curriculum. The stage was set for us to adapt to a new university environment for which we were now particularly well suited.

In the fall of 2017 we initiated development of an RCR curriculum tailored for the humanities. Using our work on the Scientific Virtues as a model, we began the process of filling in the teleological structure that reflects what it is to be excellent in humanities fields with the dispositions and values that support such excellence. To explore possible overlap between humanities and science in what counts as excellence, we started with the concept of curiosity and the role it plays in fulfilling the purpose of humanistic scholarship. Defining the purpose of the humanities is a tall order, although there may be some parallels to the purpose of science. For instance, a common refrain in science is that its purpose is to discover empirical truths about the world around us, and many humanists see themselves as truth-seekers as well even if not in the same empirical sense as scientists. There are obvious methodological and epistemic differences between science and the humanities (and even among humanities disciplines themselves), however, so we might not expect that the pursuit of truth will look the same for each.

Likewise, the role of curiosity about the world around us (and, in some cases, within us) may not look the same when one examines scholarly activity in the humanities compared to science. The object of this curiosity, the truth being sought, understandably varies from discipline to discipline, and the methodological concerns vary with the epistemic demands of the discipline. For example, what counts as a good explanation in a given domain will directly impact the sorts of methods that are likely to lead to such an explanation. Moreover, some humanists do not see themselves as offering explanations at all but rather as offering interpretations or even wholly new creations. This means that if virtues such as skepticism, humility to evidence, and others are relevant to the humanities, they may manifest quite differently. Other virtues, such as honesty and perseverance, may turn out to be quite similar across disciplines. However, we might also expect to see some divergence in the prominence (if not existence) of virtues, such as objectivity, that are central to science but are optional, or at least less central, for some of the humanities, such as art or creative writing.

The interdisciplinary nature of BEACON afforded us the opportunity to bring together different sorts of scientists and engineers and, when combined with TDI's approach to structured conversations, this led to an appreciation for the challenges that confront us in transitioning to the humanities. For pilot workshops involving humanities graduate students, we adapted the approach to meet the needs of scholars in the humanities broadly construed. There are challenges that require working and re-working the details, such as different manifestations of the virtues, different virtues taking prominent roles, greater differences in terminology and research practices between disciplines, as well as a larger population of potential participants. Our expansion into this new, unexplored environment is a direct outgrowth of the work we started in BEACON.

35.5 Conclusion

Research environments are demanding on a number of fronts. The pressure to produce results quickly, the highly technical language used to communicate challenging concepts, and the difficulties associated with value systems that are often entrenched and unspoken all contribute to the sometimes frantic and often stressful conditions faced by researchers of all stripes. RCR work occupies a special place in this environment, as it is called upon to critique and improve the research environment while also being subject to the same sorts of pressures and time and funding constraints that are being critiqued. BEACON's initial investment of time, money, and most importantly support and buy-in at all levels allowed us to do work that is branching out into other areas of the university and into the wider world. Having an understanding of the difficulties presented in doing RCR work and recognizing the importance of that work led directly to the flowering of an approach that impacts people who have never heard of BEACON, and perhaps never will. BEACON has provided a warm and hospitable environment for this early work in fostering the development of new kind of RCR training rooted in core scientific values. Drawing inspiration from Charles Darwin and the words that concluded the *Origin of Species*, we might hope that "from so simple a beginning" a newly enriched culture of integrity will continue to evolve.

Acknowledgements This material is based upon a collaboration between the Scientific Virtues Project and the Toolbox Dialogue Initiative that would not have been possible without the BEACON Center for the Study of Evolution in Action. The authors would also like to thank James Foster, Robert Heckendorn, Brian Robinson, and MSU's Center for Interdisciplinarity. This work was funded by a grant to Pennock by the National Science Foundation under Cooperative Agreement No. DBI- 0939454, which also supported Berling and McLeskey. O'Rourke's work on this project was supported by the USDA National Institute of Food and Agriculture, Hatch project 10016959. Any opinions, findings, and conclusions or recommendations expressed in this material are those of the author(s) and do not necessarily reflect the views of the funding agency.

References

1. Antes, A., et al.: *Evaluating the effects that existing instruction on responsible conduct of research has on ethical decision making*. Academic Medicine **85**(3), 519–526 (2010)
2. Aristotle: Nicomachean Ethics. In: J. Barnes (ed.) The Complete Works of Aristotle, Translated by Ross Urmson. Princeton University Press, Princeton, NJ (1985)
3. Berling, E., McLeskey, C., O'Rourke, M., Pennock, R.T.: *A new method for a virtue-based responsible conduct of research curriculum: Pilot test results*. Science & Engineering Ethics **25**(3), 899–910 (2019)
4. Eigenbrode, S., O'Rourke, M., Wulfhorst, J.D., Althoff, D.M., Goldberg, C.S., Merrill, K., Morse, W., Nielsen-Pincus, M., Stephens, J., Winowiecki, L., Bosque-Prez, N.A.: *Employing philosophical dialogue in collaborative science*. BioScience **57**, 55–64 (2007)
5. Looney, C., Donovan, S., O'Rourke, M., Crowley, S., Eigenbrode, S.D., Rotschy, L., Bosque-Pérez, N., Wulfhorst, J.D.: Seeing through the eyes of collaborators: Using toolbox workshops to enhance cross-disciplinary communication. In: M. O'Rourke, S. Crowley, S.D. Eigenbrode, J.D. Wulfhorst (eds.) Enhancing Communication and Collaboration in Interdisciplinary Research. Sage Publications, Thousand Oaks, CA (2014)
6. May, D.R., Luth, M.T.: *The effectiveness of ethics education: A quasi-experimental field study*. Science and Engineering Ethics **19**, 545–568 (2013)
7. NSF: Science and technology centers: Integrative partnerships program solicitation. `https://www.nsf.gov/pubs/2014/nsf14600/nsf14600.pdf` (2014)
8. O'Rourke, M., Crowley, S.: *Philosophical intervention and cross-disciplinary science: The story of the toolbox project*. Synthese **190**, 1937–1954 (2013)
9. Pennock, R.T.: *Fostering a culture of scientific integrity: Legalistic vs. scientific virtue-based approaches*. Professional Ethics Report **28**(2), 1–3 (2015)
10. Pennock, R.T.: Beyond research ethics: How scientific virtue theory reframes and extends responsible conduct of research. In: D. Carr (ed.) Cultivating Moral Character and Virtue in Professional Practices, pp. 166–177. Routledge Press (2018)
11. Pennock, R.T.: *An Instinct for Truth: Curiosity and the Moral Character of Science*. The MIT Press, Cambridge, MA (2019)
12. Pennock, R.T., O'Rourke, M.: *Developing a scientific virtue-based approach to science ethics training*. Science & Engineering Ethics **23**(1), 243–262 (2017)
13. Shamoo, A.E., Resnik, D.B.: *Responsible Conduct of Research*. Oxford University Press, New York, NY (2015)

Chapter 36
The Influence of Instructor Technological Pedagogical Content Knowledge on Implementation and Student Affective Outcomes

Amy M. Lark, Gail Richmond and Robert T. Pennock

Abstract To investigate how instructors' technological pedagogical content knowledge (TPACK) influences the way they implement novel educational technologies and how this influences students' affective responses to the technology, we looked at how instructors with varying amounts of TPACK with regard to a specific educational tool—the digital evolution platform Avida-ED—implemented it in their classrooms. We then compared the nature of these implementations to student affective outcomes as measured by a post-implementation student survey. We found that the degree of instructor expertise influences implementation decisions, and that these decisions in turn influence student affect. Effective implementation of new educational technologies requires significant pedagogical knowledge, and warrants opportunities for training and professional development with regard to those technologies.

Key words: Avida-ED; digital evolution; evolution education; student affect; technological pedagogical content knowledge (TPACK)

Amy M. Lark
Department of Cognitive and Learning Sciences, Michigan Tech University, USA e-mail: amlark@mtu.edu

Gail Richmond
Department of Teacher Education, Michigan State University, USA e-mail: gailr@msu.edu

Robert T. Pennock
BEACON Center for the Study of Evolution in Action and Department of Philosophy, Michigan State University, East Lansing, MI, USA e-mail: pennock5@msu.edu

© Springer Nature Switzerland AG 2020   551
W. Banzhaf et al. (eds.), *Evolution in Action: Past, Present and Future*,
Genetic and Evolutionary Computation, https://doi.org/10.1007/978-3-030-39831-6_36

36.1 Introduction

36.1.1 Technology Use in the Classroom

Teacher knowledge and beliefs play key roles in both the adoption and successful implementation of reform-based curricular materials and instructional technologies [9, 14, 34, 35, 46, 48]. However, even teachers whose beliefs are closely aligned with practices advocated by reforms (for example, the Next Generation Science Standards) may struggle with implementing curricular materials if they lack the knowledge of how to do so effectively. In his writings on pedagogical content knowledge (PCK), Shulman referred to this as curricular content knowledge: knowledge of curriculum and materials appropriate for illustrating particular concepts, as well as the knowledge of how to use such materials effectively [43, 44]. In the past decade or so, other researchers have extended Shulman's ideas to include technology, resulting in what has been named *technological pedagogical content knowledge* (or TPACK), a framework for understanding the complex interactions between content, pedagogy, and technology, and how these interactions produce effective and creative uses of technology for teaching [21, 22, 28, 49]. Although "technology" can refer to any tools that might be used as instructional aids, including white boards, overhead projectors, and the like, TPACK usually deals specifically with emergent technologies such as computers, the Internet, social media, and software that have been adopted for educational purposes. Because these technologies change so quickly, and new technologies are constantly being invented, their effective integration into the classroom poses an additional challenge for teachers.

Technological pedagogical content knowledge is a young and rapidly growing field; introduced in 2005 by Koehler and Mishra, it has generated dozens of research articles since that time [20, 49]. However, the degree to which TPACK intersects with research on teacher beliefs has not yet been clearly established, particularly with regard to how teachers' curricular knowledge influences their use of technology in the classroom, and how implementation influences student affective outcomes.

36.1.2 Affective Factors and Influence on Science Learning

From the perspective of most educators, there is much more to teaching than purely cognitive outcomes. Although the primary goal of education is ostensibly to facilitate student conceptual understanding of subject matter, teachers often have related goals that are affective in nature. These may include helping students to appreciate the nature of the discipline under study; to develop practical and critical thinking skills; or to engage with the subject matter, all of which could lead to a lifelong love of the subject and of learning itself. Teachers often put an enormous amount of effort into developing lesson materials and experiences that will be interesting to their students in order to influence not only cognitive but also affective student outcomes.

Student interest has been a prolific area of educational research for decades. The birth of the field is often attributed to the writings of John Dewey, who first explored the topic in his 1913 book *Interest and Effort in Education*. After a 50-year period where very little research was conducted on interest, the field was reinvigorated in the late 1980s, primarily by the work of researchers studying text-based learning [41], and has since expanded to produce valuable insights via hundreds of studies.

In his book, Dewey developed a working definition of interest: "Genuine interest is the accompaniment of the identification, through action, of the self with some object or idea, because of the necessity of that object or idea for the maintenance of a self-initiated activity" [11, p. 14]. Defined as such, interest is connected to ideas such as motivation and self-concept. Indeed, studies have shown that interest can result in increased motivation and achievement in students [3, 5, 29, 41, 45, 47], and may influence student decisions of future educational or vocational pursuits [1, 18]. Interest is also closely associated with several other affective factors, including enjoyment, attention, and self-efficacy, all of which have in turn been associated with learning, achievement, and intent [1, 19, 25, 41], including educational and career aspirations.

Educational researchers working in the field have identified two distinct kinds of interest: individual or personal interest, and situational interest. As defined by Mitchell, "A personal interest refers to an interest that people bring to some environment or context. On the other hand, situational interest refers to an interest that people acquire by participating in an environment or context" [29, p. 425]. Teachers generally have little control over students' individual interests that have been developed over time outside of the classroom. However, they have considerable control over situational interest, and can engender such interest by developing various classroom activities and experiences in which to engage students [5, 29]. Characteristics of activities that create situational interest include active (i.e., hands-on) engagement of students, student choice, opportunities for students to work in social contexts (e.g., groups, cooperative learning), and meaningful content [3, 38, 41, 47]. Students also perceive as interesting activities that elicit enjoyment, provide opportunities for exploration, demand high levels of cognitive engagement, and fill perceived gaps in knowledge [38]. In terms of motivational theories, these kinds of activities also carry value for students as well as an expectation that the students will be successful [4, 5].

Considering the implications of this research for issues of academic success and persistence in science, technology, engineering and math (STEM) fields, a substantial body of work has been devoted to identifying factors that increase student interest and engagement in STEM. In alignment with the findings of disciplinary-based education research, these studies show that pedagogy engaging students in authentic science practices that afford students agency (e.g., by allowing them to pursue questions of interest) has positive influences on student affect [3, 38, 41, 47]. Science teacher pedagogical decisions, in addition to affecting learning outcomes, may therefore indirectly influence student persistence in STEM.

To date, few studies have investigated the influence of instructor familiarity with instructional technologies on student affective outcomes, as they are mediated by implementation decisions. Therefore, the specific purpose of this study was to pursue the following questions: 1) How are instructors' instructional decisions influenced by their TPACK?; and 2) How do these instructional decisions influence student affective outcomes?

36.2 Methods

The study reported here was part of a larger investigation into the influences of a new educational technology on student learning outcomes. The particular technology used—Avida-ED—is a free, open-source evolutionary model system that allows students to observe and engage in scientific practices with evolving populations of digital organisms. The larger study followed a multiple-case approach to characterize the implementations of Avida-ED, and to evaluate the influence of engagement with Avida-ED on student cognitive and affective outcomes.

36.2.1 Instructional Context

The study, conducted in the fall and spring semesters of the 2012-2013 academic year, included eleven instructors in ten different classes (each taken as a case) at eight institutions across the United States. The institutions were different in size, control (i.e., public or private), and location (Figure 36.1). Classes ranged from introductory biology courses (and one Advanced Placement biology course) to senior-level undergraduate seminars in evolution (Figure 36.2).

The technological tool introduced in these courses, Avida-ED, is an educational version of a research platform (Avida), which scientists and engineers use to investigate questions about biological evolution not amenable to studies with organic life forms, as well as to evolve solutions to engineering problems. Avida-ED was designed to allow students to both observe evolution in action and to manipulate various parameters to design experiments in the same evolutionary model system that researchers use. For detailed discussions of the Avida-ED software, see [32].

36.2.2 Characterization of Implementations

To accurately characterize each case, we drew from several data sources including semi-structured instructor interviews and other exchanges (e.g., over email), course syllabi, and other course materials including lecture presentations and assignments. These data were used to characterize the various contexts (institutional, course, and instructor) in which the implementation occurred, and to develop a detailed description or vignette of the implementation itself (included in supplementary materials).

Institution Code	Level	Control	Student Pop'n (est.)	Carnegie Classification (Size and Setting)	Carnegie Classification (Basic)
A	High school	Private	500	N/A	N/A
B	4 year	Private	1,650	S4/HR: Small four-year, highly residential	Bac/A&S: Baccalaureate Colleges-Arts & Sciences
C	4 year	Public	10,500	M4/R: Medium four-year, primarily residential	DRU: Doctoral/Research Universities
D	4 year	Public	30,500	L4/NR: Large four-year, primarily nonresidential	RU/H: Research Universities (high research activity)
E	4 year	Public	34,750	L4/R: Large four-year, primarily residential	RU/VH: Research Universities (very high research activity)
F	4 year	Public	43,000	L4/NR	RU/VH
G	4 year	Public	48,000	L4/R	RU/VH
H	4 year	Public	51,000	L4/NR	RU/VH

Fig. 36.1: List of participating institutions, characterized by size (estimated student population) and Carnegie classification data (if applicable). Rounded values are based on average student enrollments for the 2012-2013 academic year, determined from information made publicly available on institution websites.

Case Code	Class level	Course type	Lecture/ Lab	Major/ Non-Major	N (matched pre/post)
A_APBio	Lower	Advanced Placement Biology (High School)	Combined	N/A	17
B_300Evo	Upper	Evolution	Lecture	Major	9
C_400Evo	Upper	Evolution	Lecture	Major	15
D_100Evo	Lower	Evolution	Lecture	Major	12
E_200Bio_HC	Lower	Biology (Honors College)	Combined	Non-Major	30
F_400Evo	Upper	Evolution	Combined	Major	33
G_100BioLabA	Lower	Biology	Lab	Major	153
G_100BioLabB	Lower	Biology	Lab	Major	234
G_100BioRes	Lower	Biology (Residential College)	Combined	Major	101
H_100CompBio	Lower	Computational Biology	Combined	Major	24

Fig. 36.2: Case summaries. Case codes are designated by institution code (see Figure 36.2) and course level/type. Class levels are designated as Lower (AP, 100-, or 200-level) or Upper (300- or 400-level). Only students taking both the pre- and post-assessment were included in data analyses; therefore, the number of students enrolled in each course may actually be greater than what is reported here.

36.2.3 Instructor TPACK

The eleven instructors who participated in the larger study were selected due to their acquaintance with the investigators, and who volunteered subsequent to invitation by the research team. A key difference between the instructors in the study was their degree of familiarity with Avida-ED. Three instructors were considered expert users; they had worked with Avida-ED to a significant extant and possessed a deep understanding of its theoretical underpinnings, affordances, and limitations. Four instructors were considered experienced; they had used Avida-ED prior to this implementation but their knowledge of the program was not as deep as the experts. The remaining instructors were considered novices, as this was their first time implementing Avida-ED in the classroom, and they possessed very little knowledge about its affordances and limitations. Instructors were assigned "TPACK designations" associated with their level of experience: 1 for novices, 2 for experienced, and 3 for expert Avida-ED users (Figure 36.3). This three-fold classification fit with the levels of instructor experience observed in this study.

All three expert Avida-ED users are members of BEACON, an NSF-funded Science Center for the study of evolution in action. Avida and Avida-ED are flagship products of BEACON research in computational experimental evolution and evolution education. The expert users were familiar with Avida through their participation in BEACON, where work on Avida and Avida-ED is presented frequently. They are familiar with the literature on both programs, in some cases having contributed directly to that literature, and have used both platforms in research and teaching capacities. Indeed, one instructor (YN) had used Avida-ED in her Master's thesis, and the dissertation research of another (TT) had been based entirely on his experiments in Avida. The third instructor (NR) has served as co-principal investigator on several funded research projects involving Avida. These instructors are intimately aware of the programs' capabilities. They understand the substrate neutrality of natural selection as a process [10], and how, therefore, the evolution of digital organisms serves as a model for the evolutionary process, including biological evolution [32, 33].

The four instructors designated as "experienced" had each used Avida-ED in their courses at least twice before. This gave them a good understanding of the capabilities of the software and insight about the kinds of technical issues their students might run into. Unlike the novice users, all four of these instructors had planned on using Avida-ED in their courses, and they had therefore scheduled it in the lesson sequence in a way that made pedagogical sense. Having used Avida-ED before gave the experienced instructors the advantage of foresight. However, although these instructors were familiar with using Avida-ED as an instructional technology, they lacked the deep, theoretical understanding of the expert users.

Although the novice instructors had never used Avida-ED in their courses, they had all heard of the software at some point prior to volunteering to participate in this study (usually in the form of a research presentation). Based on what they knew, they had all decided that Avida-ED aligned well with their objectives for student learning. However, unlike the experienced and expert users, they did not necessarily have the luxury of planning carefully where Avida-ED would best fit within the course sequence, and instead inserted it where it made the most sense to them; in one case, that meant replacing a planned activity with Avida-ED. All of the novice instructors spent some time experimenting with the software prior to introducing it to the students, but because they had never used it before they were unable to anticipate some of the technical issues students experienced. In essence, they were learning alongside their students.

Case Code	Instructor(s)	Position	Familiarity with Avida-ED	TPACK designation
A_APBio	YN	HS Teacher	Expert	3
B_300Evo	NN	Professor	Experienced	2
C_400Evo	TN	Assistant Professor	Novice	1
D_100Evo	JO	Senior Lecturer (tenured)	Novice	1
E_200BioHC	HD	Postdoctoral Fellow	Experienced	2
F_400Evo	NR	Associate Professor	Expert	3
G_100BioLabA	AE	Coordinator	Experienced	2
G_100BioLabB	KA	Visiting Assistant Professor	Experienced	2
G_100BioRes	RY/AR	Postdoctoral Fellows	Novice	1
H_100CompBio	TT	Instructor (non-tenure)	Expert	3

Fig. 36.3: Instructor profiles.

36.2.3.1 Post-implementation Survey of Students

Instructors were asked to administer a short survey to their students subsequent to lessons involving Avida-ED with the goal of eliciting student feedback on their experiences using the tool (Figure 36.4). All survey items were measured on a five-point Likert scale, with the low end of the scale corresponding to negative impressions and the high end to positive or favorable impressions. Items addressed student attentiveness and participation during activities involving Avida-ED, as well as interest and enjoyment, the degree to which students felt that Avida-ED helped increase their understanding of evolution and of the nature of science, the degree to which they felt comfortable discussing the subject of evolution, and their intentions regarding continued use of the program or sharing with others. These data were intended to help gauge student interest and engagement with the program, and to provide useful feedback to the Avida-ED Curriculum Development team for improving the software and associated curriculum materials.

Surveys were returned for seven of the ten cases. Survey data were summarized and shared with instructors prior to exit interviews. To facilitate interpretation of survey responses, the Likert-scale items were converted to numeric scores ranging from 1 (most negative response) to 5 (most positive response), and we calculated mean scores for each item. These numeric values are of course not continuous variables, but they allowed us to create a correlation matrix showing associations between survey items and other variables, including instructor familiarity with Avida-ED and assessment data (Figure 36.5).

Student content knowledge was assessed prior and subsequent to instruction with Avida-ED. Although learning outcomes are not the focus of the current study, we have included these — in the form of average normalized gains — for illustrative purposes only (see Figures 36.4 and 36.5). The assessment outcomes of the larger study are detailed elsewhere [24].

36.3 Results

36.3.1 Implementation of Avida-ED

Qualitative comparison of cases revealed that implementations differed primarily in duration, amount of instructional support provided by the instructor, the degree to which the implementation was student-centered, and the source of lesson materials used (see [23] for detailed descriptions of implementations).

36.3.1.1 Duration

Duration of implementations ranged from one to fifteen weeks, with a median duration of two weeks. It is important to note that although the number of weeks serves as a rough estimate of duration, the actual duration is dependent on the number of hours spent on Avida-ED in class as well as the amount of time students spent working with Avida-ED on their own outside of class. Therefore, the actual duration is difficult to estimate. For example, YN (A_APBio) used Avida-ED in her class for two weeks. That included five days of guided inquiry learning to use the software, each class period lasting 50 minutes, for a total of 250 minutes spent on Avida-ED in class. The students spent an additional week outside of class developing their projects. The students in the introductory biology laboratory at Institution G experienced a similar duration, becoming familiar with Avida-ED over the course of one three-hour lab period and working on group projects outside of class for one week. In contrast, the students in G_100BioRes did not use any class time to work with the software (being given one week to complete a homework assignment outside of class), and no more than 30 minutes of class time were devoted to discussion of Avida-ED.

36.3.1.2 Instructional Supports

The instructors differed with regard to the type and amount of instructional support they provided students. These supports consisted of materials and other resources that instructors incorporated into their lessons to facilitate learning or made available for students to reference as necessary. Several instructors required their students to read introductory materials prior to engaging in activities with Avida-ED; for example, students in A_APBio, E_200Bio_HC, and G_100BioLabA & B were assigned a Discovery magazine article on Avida ("Testing Darwin", [50]). These readings served as an introduction to the idea of digital evolution. Most instructors used class time to introduce students to the software, either in the form of an interactive tutorial or demonstration. TT's students (H_100CompBio) were required to watch an introductory video on Avida-ED, while RY and RA made a similar YouTube video available as a reference to their students as they worked on their assignment outside of class. Other reference materials included the Avida-ED user's manual and the project website. In most of the courses, students worked in groups (only students in C_400Evo, G_100BioRes, and H_100CompBio worked individually). Students in these cases could help one another as they worked on their projects in or out of class.

36.3.1.3 Source of Curricular Materials

Instructors had full control over implementations of Avida-ED in their classrooms. As such, the cases differed in the sources of the lesson materials that were used, and these varied with the degree to which instructors were familiar with Avida-ED. Expert users designed their own materials while novices used existing materials produced by others, and experienced users used some combination of the two (e.g., materials that were produced by modifying existing materials). Instructors who were most familiar with Avida-ED created rich learning environments and challenging activities for their students. As an example, two of the expert users (YN and NR) were able to directly link experiments with digital organisms to biological systems. In her AP biology class, YN first engaged students in an artificial selection experiment with E. coli, and later repeated the experiment with Avida-ED, allowing students to observe the same patterns occurring in parallel contexts. Similarly, NR designed a series of experiments that would allow students to replicate the findings of a classic experiment by Luria and Delbrck [27], using first bacteria, and then Avida-ED.

The expert instructors were able to draw on their knowledge of Avida-ED to create new and challenging experiences for their students. Experienced users, while not designing entirely new instructional materials, were familiar enough with the software to begin adapting existing lesson materials to suit their own objectives. For example, all of the experienced instructors engaged their students in a more or less complete inquiry cycle. Two experienced users, NN and HD, used introductory materials and activities that had been designed by the Avida-ED Curriculum Development team. AE and KA used some of these introductory materials but worked together to create a new set of laboratory exercises based on three common misconceptions associated with evolution. Novice Avida-ED users, in contrast, used materials that had been created by the Avida-ED Curricular Development team, and these were implemented more or less as written.

36.3.1.4 Student Centered-ness

Despite differences in the specific activities used, all of the instructors participating in the study implemented Avida-ED in ways that were, to varying degrees, both aligned with reform recommendations and consistent with its intended use, specifically as a tool for the integration of evolution content and the active, authentic engagement of students in research activities. Many of the instructors endeavored to take full advantage of Avida-ED's strengths by having students pose their own questions and design a study. This sort of open-ended project led to engagement in a wide range of research activities. However, even implementations that were more guided (e.g., A_APBio and D_100Evo) remained well within the range of practices considered learner-centered inquiry [30].

36.3.2 Student Surveys

The number of student respondents for each case is listed in Figure 36.4; response rates were at or near 100% for each of the seven cases for which survey data were collected. With the exception of case G_100BioRes, student Likert survey responses were generally favorable (mean greater than 3; Figure 36.4). Students reported that they were attentive in class and actively participated in exercises involving Avida-ED. They had little difficulty understanding and using the software. Most students were interested in Avida-ED and enjoyed using it. Many students, particularly those in A_APBio and H_100CompBio, reported that Avida-ED helped them to better understand both evolution and the nature of science, and that they felt much more comfortable discussing the topic of evolution. However, most students were not particularly interested in continuing their experimentation with Avida-ED or in sharing the program with their families and friends. A correlation matrix (Figure 36.5) shows the Pearson correlation coefficients between different survey items, as well as instructor TPACK designation (based on familiarity with Avida-ED) and student normalized gains from the content assessment. There were several strong associations between various affective factors. For example, students who reported that working with Avida-ED increased their understanding of the nature of science ("Increase Science") also tended to report an increased understanding of evolution ("Increase Evolution"; $r = 0.97; p < 0.01$). Students who reported increases in their understanding of both science and evolution tended to feel more comfortable ("Comfort") discussing the topic of evolution ($r = 0.97$ and 0.90, respectively; $p < 0.01$).

Several of the affective measures were also significantly and positively associated with instructor TPACK. Notably, students reported higher levels of interest ("Interest") and enjoyment ("Enjoy") in the lesson activities with increasing instructor knowledge of Avida-ED ("Instructor TPACK designation"; $r = 0.85$ and 0.89 respectively; $p < 0.01$). Student perception of increased understanding of evolution was significantly associated with instructor TPACK as well ($r = 0.91; p < 0.01$). Other significant, positive associations with instructor TPACK included student at-

Case Code	A_AP-Bio	B_300 Evo	E_200 Bio_HC	G_100 BioLabA	G_100 BioLabB	G_100 BioRes	H_100 CompBio
N	19	9	31	154	268	112	28
Your general level of attentiveness during lessons involving Avida-ED	3.53	3.44	3.26	3.56	3.40	2.67	3.61
Your general level of participation during lessons involving Avida-ED	4.63	4.11	4.23	4.69	4.52	2.93	4.14
The relative ease with which you were able to understand Avida-ED	3.95	3.56	3.13	4.07	3.89	3.53	4.11
The relative ease with which you were able to use Avida-ED	3.68	3.44	3.45	4.18	3.99	2.73	3.93
Your overall interest in Avida-ED	3.47	3.44	2.87	2.92	2.61	2.01	3.36
Your overall enjoyment of Avida-ED	3.89	3.44	3.42	3.53	3.25	2.46	3.57
Avida-ED significantly increased my understanding of evolutionary processes	4.53	3.11	3.92	3.80	3.65	2.66	4.25
Avida-ED significantly increased my understanding of the process of science	3.95	2.78	3.75	3.49	3.42	2.70	3.96
I feel much more comfortable discussing the topic of evolution	3.84	2.56	3.83	3.57	3.57	2.76	4.14
I will continue to experiment with Avida-ED in my free time	2.05	2.78	1.42	2.05	1.85	1.73	2.96
I have or will share Avida-ED with friends or family	2.37	2.44	1.79	2.08	1.99	1.75	3.00
Instructor TPACK designation	3	2	2	2	2	1	3
Content gain (g-avg)	17%	-10%	18%	1%	2%	20%	9%

Fig. 36.4: Summary of survey data for each of seven cases. Survey items were based on a 5-point Likert scale (1= low; 5 = high); item mean response is reported for each case. Survey response rate was at or near 100% for each of the seven cases for which survey data were collected.

tentiveness during instruction with Avida-ED ("Attentive"; $r = 0.83; p < 0.05$), student increase in understanding of science ($r = 0.83; p < 0.05$), student comfort with discussing the topic of evolution ($r = 0.72; p < 0.05$), student participation ("Participate"; $r = 0.67; p < 0.05$), and student intention to share Avida-ED with others ("Share"; $r = 0.78; p < 0.05$). Interestingly, neither instructor TPACK nor any of the affective factors were associated significantly with student learning outcomes. For a complete listing of associations, see Figure 36.5.

36.4 Discussion

Instructor technological pedagogical content knowledge, specifically with regard to Avida-ED, seems to be a function of familiarity with the tool. Expert users were simultaneously more comfortable with the tool, and they were able to create their own curricular materials; these materials included learning contexts that were richer

	Attentive	Participate	Under-stand	Use	Interest	Enjoy	Increase Evolution	Increase Science	Comfort	Continue	Share	Instructor TPACK desig-nation
Participate	0.89											
Under-stand	0.57	0.43										
Use	0.89	0.89	0.67									
Interest	0.85	0.64	0.31	0.51								
Enjoy	0.92	0.88	0.38	0.72	0.90							
Increase Evolution	0.76	0.75	0.43	0.67	0.66	0.86						
Increase Science	0.64	0.65	0.36	0.63	0.50	0.74	0.97					
Comfort	0.55	0.56	0.37	0.62	0.33	0.59	0.90	0.97				
Continue	0.52	0.11	0.52	0.29	0.64	0.35	0.13	0.03	-0.04			
Share	0.67	0.28	0.60	0.43	0.76	0.57	0.48	0.40	0.33	0.92		
Instructor TPACK designa-tion	0.83	0.67	0.53	0.63	0.85	0.89	0.91	0.83	0.72	0.50	0.78	
Content g-avg	-0.50	-0.37	-0.26	-0.45	-0.39	-0.24	0.19	0.32	0.37	-0.58	-0.33	-0.05

| $p < 0.05$ |
| $p < 0.01$ |

Fig. 36.5: Correlation matrix showing associations between survey items (column and row headings), instructor familiarity with Avida-ED, and assessment data (average normalized student gains in content and acceptance scores). Cells shaded in yellow are significant at the 0.05 level while cells shaded in orange are significant at the 0.01 level.

and more meaningful than those of less experienced instructors. Thus, it appears that in this study instructor TPACK had a direct influence on instructor implementation decisions: instructors with greater TPACK were afforded much more flexibility and creativity in their implementations.

Koehler and Mishra [21] argue that particular technologies have specific affordances and constraints. These constraints are either inherent to the technology or imposed externally by the user, a condition known as "functional fixedness"; that is, the manner in which the ideas we hold about an object's function can inhibit our ability to use the object for a different, creative function. Functional fixedness often stands in the way of creative uses of a technology, and as such makes it difficult for teachers to imagine how that technology might be useful as a tool for teaching and learning. In the case of Avida, functional fixedness may lead an instructor to believe that it is useful only as a simulation in a computational context, and may preclude that instructor from understanding how it can be used as a dynamic model of biological evolution.

An example of functional fixedness in this study arose when instructors who were not as familiar with Avida expressed frustration that the program was not more closely aligned with their preferred model system. Avidians are asexual, and so genetic recombination during sexual reproduction cannot account for variation in the population. Avida cannot be used to model the Central Dogma of molecular biology [7], because the Avidian genome consists of computer code that directly accounts for phenotype — phenotypic expression is not mediated by transcription, translation, or protein synthesis as in biological organisms with genomes composed of nucleic acids. In addition, Avidians are haploid; they are essentially prokaryotic organisms, which may pose a challenge for inexperienced instructors who are fixed in thinking just in terms of organisms that are diploid and eukaryotic. It can be difficult for such instructors to appreciate the value of using a digital system to model "real life." Here is one such example, from a post-implementation interview with instructors RY and AR, who co-taught an introductory biology lecture (G_100BioRes):

> RY: "I wonder what the cost/benefit analysis would be for spending all of that time on [learning to use] Avida-ED. Because [Avidians] are asexual — well, because they are so weird. They're just these letters. And [students] have to make this connection, and I wonder if it takes so much for us to"
> AR: "There's a lot of conceptual jumps that they have to make to apply the Avida system."
> RY: "We really want them to explain what's happening in the cases we use. What's happening in elephants, or what's happening in snakes. What's happening in real life. There's a couple of things we need to deal with there, and I wonder if they get caught on — I wonder if we spent more time, how easy it would be for them to move from genes and alleles and As, Ts, Cs, and Gs and proteins to the Avidians. And since the Avidians don't have that gene to protein to phenotype step that is really emphasized [in our course] then it is not the model that we are using There are no proteins, and the little colored balls are not nucleotides, and they are prokaryotes, which we don't emphasize prokaryotes in our class. That's just us."
> AR: "I think that's just biology in general."
> RY: "If this were the cell and molecular class and we were looking at transcription and translation, then maybe we would spend more time on prokaryotes. But when you're teaching an organismal class, you just don't use prokaryotes."

The issue motivating RY's comments seems to be concern over the universality of Avida as a model system. A perennial challenge of using model organisms in science courses is that it can be difficult for students to see the general, broad patterns that the models are meant to illustrate; instead, students can get bogged down by the idiosyncrasies of particular organisms or systems [15, 16]. This difficulty is not limited to Avidians, but also occurs when Drosophila, E. coli, or other model organisms are the focus. To avoid this "forest for the trees" phenomenon, it is important for instructors to connect model organisms to other systems that exhibit the same fundamental patterns. This is not simply a matter of increasing their relevancy for students, although it helps achieve that as well. The main point is to help them to understand how model-based reasoning works in science—models allow scientists to study phenomena in a manner that transcends specific examples, ultimately allowing them to make broader generalizations. Helping students understand this basic form of inductive inference—drawing general conclusions from relevantly similar instances—is essential for understanding the nature of scientific reasoning. This inferential power is the intended purpose of Avida as a research tool and Avida-ED as a teaching tool: to serve not as a one-to-one digital representation of an organic prokaryotic organism, but instead as a general model of universal evolutionary mechanisms. Rather than seeing unfamiliar features as a problem or a reason to avoid using some model organism, instructors with greater experience and expertise can use this as an opportunity to get students to think about how model systems work in science generally—as a way of abstracting and generalizing the causal features of interest—by using the specific model system as an illustration of this ubiquitous mode of scientific reasoning. Expert users of Avida-ED understood this, and as a result were able to use the tool accordingly, drawing parallels between biological and digital evolution on a level that is appropriate to the concepts being learned and that are common to both systems. The less experienced users, on the other hand, may fall into the same trap as their students and get caught up in the specifics (or rather, the differences) of the digital system to the extent that they struggle to see how it can be generalized to biological systems.

Another notable difference between expert and less experienced users has to do with the curricular materials they used in their implementations of Avida-ED. All three experts designed their own materials, while less experienced users tended to use materials that had been developed by someone else (e.g., the Avida-ED Curriculum Development team); this was true for all of the novice users. This is significant, because not only did the less experienced users have to deal with navigating unfamiliar software, they also had to contend with using unfamiliar lesson materials. In terms of pedagogical knowledge, this left the less experienced users, and the novices especially, doubly disadvantaged.

The findings presented here make a strong case for the importance of professional development opportunities aimed at preparing teachers to effectively implement reform-based curricular materials in the classroom [34, 48], especially those dependent on emerging technologies. Such professional development will only become more essential as the use of technology in teaching increases in popularity. In the case of Avida-ED, one way around the issue of functional fixedness is to direct instructors to the extensive literature on Avida and to provide professional development experiences that involve experimenting with Avida-ED. This work is already being undertaken at BEACON by the Avida-ED Curriculum Development team (e.g. see the project website http://avida-ed.msu.edu). As instructors become more experienced with using the program, they should also be encouraged to design their own curricular materials for use with Avida-ED, just as the expert users have in this study. As Koehler and Mishra [21] point out:

> The teacher, Dewey argued, is not merely the creator of the curriculum, but is a part of it: teachers are curriculum designers. The idea of teachers as curriculum designers is based on an awareness of the fact that implementation decisions lie primarily in the hands of particular teachers in particular classrooms. Teachers are active participants in any implementation or instructional reform we seek to achieve, and thus require a certain degree of autonomy and power in making pedagogical decisions. (p. 21; emphasis in the original)

In order to achieve the goals advanced by reform initiatives such as Vision & Change and the Next Generation Science Standards, teachers at all levels of schooling (K-16) must first possess an appropriate attitude toward change, one that embraces reform-oriented pedagogical practices [8, 35], and secondly must be provided with opportunities to improve their (technological) pedagogical content knowledge by engaging in professional development activities (reviewed in [17, 34, 36, 40, 48].

36.4.1 Student Affective Response

Although course learning goals are most commonly focused on subject matter competencies, it is also important to cultivate student interest and engagement in the classroom. Interest has been linked to motivation, self-efficacy, achievement, and intent [1, 4, 5, 18, 19, 25, 26, 29, 41, 45]. Instructors who create and maintain situational interest in the classroom — by engaging students in cognitively challenging, hands-on, meaningful inquiry-based activities, and providing opportunities for choice and social interaction — may influence student personal interests and inspire students to persist in a field [29, 37]. Student affect in response to instructional approaches in science education could therefore be critical for retention of STEM students and increasing the number of graduates and workers in STEM fields, both of which have been cited as crucial for the economic future of the country [31].

Avida-ED, as a virtual lab space intended for the observation and study of evolution in action, is well positioned to produce situational interest in the classroom. From the assessment data collected in this study, we found significant increases in student normalized gains [24]. These results suggest Avida-ED is a promising tool for teaching evolution in ways consistent with reform recommendations. In addition, survey data provide evidence that Avida-ED elicits a favorable affective response from students. Therefore, Avida-ED may be useful not only as a tool for facilitating student learning about evolution, but also for increasing student interest in evolution specifically and science more generally.

Associations between student responses to survey items were consistent with the literature on interest. Student interest was tied very closely to enjoyment and attentiveness, which were also associated with participation and ease of use, all positive interactions that are associated with increased motivation and learning [1, 4, 5, 25]. In addition, the degree to which students perceived increases in their understanding of evolution and of the nature of science were very strongly correlated, and each of these factors was related to how comfortable students felt when discussing evolution. The students' perceptions of what they learned and their confidence in discussing these ideas may be linked to their self-efficacy, or the situation-specific belief that one can succeed in a given domain [2]. Self-efficacy has been shown to be important for predicting an individual's success and persistence within a domain [2, 13, 25, 45]. Student self-efficacy can influence academic performance, and can be positively influenced by certain instructional strategies, particularly those that are student-centered [13, 25, 26]. Instructors who engage students in these activities have an influence on student learning and on their self-efficacy. This is also linked to students' interest in science and confidence to do science: "Attitudinal and affective variables such as self-concept, confidence in learning mathematics and science, mathematics/science interest and motivation, and self-efficacy have emerged as salient predictors of achievement in mathematics and science. These factors also predict mathematics and science avoidance on the part of students, which affects long-term achievement and career aspirations in the mathematics/science fields" [45, p. 324].

Several studies in the research literature on interest have found a positive effect of interest and related factors, such as engagement, motivation, and self-efficacy, on student learning and achievement [41, 42]. The data from this study, however, do not support a necessary association between student affective and cognitive outcomes. In particular, no measures of affect reported by students were significantly correlated with normalized learning gains—even students who didn't feel positively about Avida-ED still learned by using it. There were only weak positive associations between student perceived learning of evolution and actual gains in content and acceptance scores ($r = 0.19$ and 0.57, respectively), though neither was statistically significant; this suggests that students may have difficulty predicting their own learning gains, particularly if their experience with Avida-ED was not positive. Negative emotion, such as frustration and anxiety, felt during an instructional task can greatly influence student interest and engagement [1, 3, 25, 45]. Unlike in most of the cases, students in G_100BioRes reported a predominantly negative experience with Avida-ED, having the lowest average for most survey items. However, their learning gains were very similar to other lower-division courses. Indeed, when removing this case from the analysis, the correlation between average normalized student gain in content score and student perceptions of what they learned becomes significant ($n = 6, r = 0.85, p < 0.05$), suggesting that students who had positive experiences with Avida-ED were much more likely to indicate that it had a positive influence on their learning.

Despite this interesting lack of association between affective and cognitive outcomes in this study, there were several strong correlations between various affective factors and instructor TPACK. In particular, students of more experienced instructors reported significantly greater levels of interest and enjoyment, as well as a greater perceived increase in understanding of evolution. Notably, the one case in this analysis that was taught by novices, G_100BioRes, was also the only course in which students reported a largely negative experience using Avida-ED. Although the specific reasons for this are unknown, one could speculate, based on the nature of the implementation, that students were frustrated by being given a challenging assignment with Avida-ED, receiving very little introduction to the software and very little instructional support, and having little time to complete the assignment — recall that the entire implementation lasted only one week, and most of this time the students were working independently outside of class. The instructors may simply have lacked sufficient pedagogical knowledge of Avida-ED to support the students as they worked on the assignment. Research has shown that instructors can significantly influence student situational interest by providing adequate background knowledge needed for completing a task [41] and possessing a high degree of content knowledge [37]; both of these factors appear to have been lacking in the case of G_100BioRes.

36.4.2 *Implications for Science Education*

The outcomes of this study have provided valuable insights for the successful classroom implementation of reform- and research-based tools like Avida-ED, and potentially for science education in general. Drawing from the cases in this study, one may propose the following set of best practices for successful implementation of educational technology.

First, instructors must develop an understanding of the way specific instances (e.g. model organisms) allow scientists to infer broader conclusions through model-based reasoning. It is critical that instructors and their students do not become too caught up in the specifics of a particular model system, especially when the goal is for students to understand the target phenomena broadly at a level that transcends individual contexts [15, 16]. Related to this point, it is good practice to include multiple contexts for illustrating phenomena [6]. Instructors who do not fully understand how the evolutionary mechanism works across what might seem to be very different contexts may not immediately recognize the applicability of a system like Avida. Evolution is a substrate neutral process that will occur spontaneously in any system that possesses variation, inheritance, and selection [10]. In this sense, digital and biological evolution are merely two different contexts or instances in which to observe the general causal mechanism of evolution. Selecting a single context to the exclusion of all others may interfere with students' ability to generalize across contexts, particularly if the chosen context is not universally representative. It is therefore problematic when instructors do not see the value of model organisms such as bacteria and Avidians. For such instructors, haploid, asexual organisms are so far removed from their preferred multicellular, diploid, sexually-reproducing examples that they fail to articulate the more generalized definition of evolution that is inclusive of prokaryotes. In so doing, they are missing the majority of life forms on the planet and the universality of the evolutionary processes that produced them. This can have unintended negative consequences on student learning.

A second point that is obvious but essential for TPACK is that instructors must be familiar with teaching tools, including the limits and affordances of a tool. From a practical standpoint, instructors need to be technically familiar with the tool in order to anticipate or troubleshoot student difficulties. Beyond that, instructors need to be able to assess the extent to which a particular tool aligns with their objectives, and to understand a tool's capabilities so that it can be used to its full potential. All of these issues serve to underscore the importance of pedagogical content knowledge, and particularly technological pedagogical content knowledge. To increase levels of TPACK, it is advisable that instructors engage in professional development activities around the curricular materials they choose to use. For Avida-ED, this is especially true in light of the students' more negative affective responses when instructor TPACK was low. As noted, in contrast to other studies on student interest, the association between student affect and learning outcomes was not significant— using Avida-ED still helped them learn evolutionary concepts even if they did not feel positively about the tool. There was, however, a strong relationship between student affective response and instructor familiarity with Avida-ED, suggesting that instructor TPACK does influence students' enjoyment and interest—important aspects of learning that should not be neglected.

Finally, to avoid negatively affecting student enjoyment and interest, it is important to provide students with sufficient time and instructional support. The duration of an implementation matters [12, 39], and students should be given ample time to become familiar with a tool and to complete all associated tasks. In this study, two weeks seemed to be enough time for most students to learn how to use Avida-ED and engage in a more or less full inquiry cycle. Allowing for less time may have had a negative impact on student affect in the case of G_100BioRes. Also, it is advisable to provide students with a demonstration or tutorial — a series of simple tasks designed to familiarize students with the tool — allowing them to ease into it, progressing from simple to more complex concepts and practices. Students may also appreciate the provision of supportive materials such as supplementary articles, videos, or user's manuals. Instructor TPACK does matter, and it is reasonable to hypothesize that instructors who are less confident about their own understanding of a tool or some new instructional technology may be less able to model its appropriate use. This deserves further investigation. Again, although the current study did not find a significant effect of instructional supports on student learning outcomes, there is more to STEM education than mere content knowledge. We did find evidence of a significant association of instructional support with student affect, which has been shown to influence student interest and intention, giving another good reason to pay attention to instructor TPACK as part of professional development.

Acknowledgements This material is based in part upon work supported by the National Science Foundation under Cooperative Agreement No. DBI- 0939454 as part of BEACON's Avida-ED Curriculum Development and Assessment Study to Pennock. Any opinions, findings, and conclusions or recommendations expressed in this material are those of the author(s) and do not necessarily reflect the views of the funding agency.

References

1. Ainley, M., Ainley, J.: *Student engagement with science in early adolescence: The contribution of enjoyment to students' continuing interest in learning about science.* Contemporary Educational Psychology **36**(1), 4–12 (2011)
2. Bandura, A.: *Perceived self-efficacy in cognitive development and functioning.* Educational Psychologist **28**(2), 117–148 (1993)
3. Bergin, D.: *Influences on classroom interest.* Educational Psychologist **34**(2), 87–98 (1999)
4. Brophy, J.: *Conceptualizing student motivation.* Educational Psychologist **18**(3), 200–215 (1983)
5. Brophy, J.: *Toward a model of the value aspects of motivation in education: Developing appreciation for particular learning domains and activities.* Educational Psychologist **34**(2), 75–85 (1999)
6. Brown, D.E.: *Using examples and analogies to remediate misconceptions in physics: Factors influencing conceptual change.* Journal of Research in Science Teaching **29**(1), 17–34 (1992)
7. Crick, F.: *Central dogma of molecular biology.* Nature **227**(5258), 561–563 (1970)
8. Cronin-Jones, L.L.: *Science teacher beliefs and their influence on curriculum implementation: Two case studies.* Journal of Research in Science Teaching **28**(3), 235–250 (1991)

9. Czerniak, C.M., Lumpe, A.T.: *Relationship between teacher beliefs and science education reform*. Journal of Science Teacher Education **7**(4), 247–266 (1996)
10. Dennett, D.C.: *Darwin's dangerous idea: Evolution and the meanings of life*. Simon & Schuster, New York (1995)
11. Dewey, J.: *Interest and effort in education*. Riverside Press, Cambridge (1913)
12. Durlak, J.A., DuPre, E.P.: *Implementation matters: A review of research on the influence of implementation on program outcomes and the factors affecting implementation*. American Journal of Community Psychology **41**, 327–350 (2008)
13. Fencl, H., Scheel, K.: *Engaging students: An examination of the effects of teaching strategies on self-efficacy and course climate in a nonmajors physics course*. Journal of College Science Teaching **35**(1), 20–24 (2005)
14. Gess-Newsome, J., Southerland, S.A., Johnston, A., Woodbury, S.: *Educational reform, personal practical theories, and dissatisfaction: The anatomy of change in college science teaching*. American Educational Research Journal **40**(3), 731–767 (2003)
15. Grosslight, L., Unger, C., Jay, E., Smith, C.L.: *Understanding models and their use in science: Conceptions of middle and high school students and experts*. Journal of Research in Science Teaching **28**(9), 799–822 (1991)
16. Harrison, A.G., Treagust, D.F.: *A typology of school science models*. International Journal of Biology Education **22**(9), 1011–1026 (2000)
17. Hew, K.F., Brush, T.: *Integrating technology into K-12 teaching and learning: Current knowledge gaps and recommendations for future research*. Educational Technology Research and Development **55**(3), 223 – 252 (2007)
18. Hidi, S., Renninger, K.A.: *The four-phase model of interest development*. Educational Psychologist **41**(2), 111–127 (2006)
19. Hidi, S., Renninger, K.A., Krapp, A.: Interest, a motivational variable that combines affective and cognitive functioning. In: D.Y. Dai, R.J. Sternberg (eds.) Motivation, Emotion, and Cognition: Integrative perspectives on intellectual functioning and development, pp. 89–115. Lawrence Erlbaum Associates, Inc., New Jersey (2004)
20. Koehler, M.J., Mishra, P.: *What happens when teachers design educational technology? The development of technological pedagogical content knowledge*. Journal of Educational Computing Research **32**(2), 131–152 (2005)
21. Koehler, M.J., Mishra, P.: Introducing TPCK. In: AACTE Committee on Innovation and Technology (ed.) Handbook of Technological Pedagogical Content Knowledge (TPCK) for Educators, pp. 3–29. Routledge, New York (2008)
22. Koehler, M.J., Mishra, P., Kereluik, K., Shin, T.S., Graham, C.R.: The technological pedagogical content knowledge framework. In: J.M. Spector, M.D. Merrill, J. Elen, M.J. Bishop (eds.) Handbook of Research on Educational Communications and Technology (Fourth ed.), pp. 101–111. Springer, New York (2014)
23. Lark, A.: Teaching and learning with digital evolution: Factors influencing implementation and student outcomes. Ph.d. dissertation, Michigan State University (2014)
24. Lark, A., Richmond, G., Mead, L.S., Smith, J.J., Pennock, R.T.: *Exploring the relationship between experiences with digital evolution and students' scientific understanding and acceptance of evolution*. The American Biology Teacher **80**(2), 74–86 (2018)
25. Linnenbrink, E.A., Pintrich, P.R.: *The role of self-efficacy beliefs in student engagement and learning in the classroom*. Reading & Writing Quarterly **19**(2), 119–137 (2003)
26. Linnenbrink-Garcia, L., Pugh, K.J., Koskey, K.L.K., Stewart, V.C.: *Developing conceptual understanding of natural selection: The role of interest, efficacy, and basic prior knowledge*. The Journal of Experimental Education **80**(1), 45–68 (2012)
27. Luria, S.E., Delbrück, M.: *Mutations of bacteria from virus sensitivity to virus resistance*. Genetics **28**(6), 491–511 (1943)
28. Mishra, P., Koehler, M.J.: *Technological pedagogical content knowledge: A framework for teacher knowledge*. Teachers College Record **108**(6), 1017–1054 (2006)
29. Mitchell, M.: *Situational interest: Its multifaceted structure in the secondary mathematics classroom*. Journal of Educational Psychology **85**(3), 424–436 (1993)

30. National Research Council: *Inquiry and the National Science Education Standards: A guide for teaching and learning*. National Academies Press, Washington, D.C. (2000)

31. Olson, S., Riordan, D.G.: Engage to excel: Producing one million additional college graduates with degrees in science, technology, engineering, and mathematics. Report to the President. Executive Office of the President, Washington, D.C.

32. Pennock, R.T.: *Learning evolution and the nature of science using evolutionary computing and artificial life*. McGill Journal of Education **42**(2), 211–224 (2007)

33. Pennock, R.T.: *Models, simulations, instantiations, and evidence: The case of digital evolution*. Journal of Experimental & Theoretical Artificial Intelligence **19**(1), 29–42 (2007)

34. Powell, J.C., Anderson, R.D.: *Changing teachers' practice: Curriculum materials and science education reform in the USA*. Studies in Science Education **37**, 107–136 (2002)

35. Roehrig, G.H., Kruse, R.A.: *The role of teachers' beliefs and knowledge in the adoption of a reform-based curriculum*. School Science & Mathematics **105**(8), 412–422 (2005)

36. Roehrig, G.H., Luft, J.A.: *Constraints experienced by beginning secondary science teachers in implementing scientific inquiry lessons*. International Journal of Science Education **26**(1), 3–24 (2004)

37. Rotgans, J.I., Schmidt, H.G.: *The role of teachers in facilitating situational interest in an active-learning classroom*. Teaching and Teacher Education **27**(1), 37–42 (2011)

38. Rotgans, J.I., Schmidt, H.G.: *Situational interest and learning: Thirst for knowledge*. Learning and Instruction **32**, 37–50 (2014)

39. Sadler, T.D., Burgin, S., McKinney, L., Ponjuan, L.: *Learning science through research apprenticeships: A critical review of the literature*. Journal of Research in Science Teaching **47**(3), 235–256 (2010)

40. Schneider, R.M., Krajcik, J., Blumenfeld, P.: *Enacting reform-based science materials: The range of teacher enactments in reform classrooms*. Journal of Research in Science Teaching **42**(3), 283 – 312 (2005)

41. Schraw, G., Flowerday, T., Lehman, S.: *Increasing situational interest in the classroom*. Educational Psychology Review **13**(3), 211–224 (2001)

42. Schraw, G., Lehman, S.: *Situational interest: A review of the literature and directions for future research*. Educational Psychology Review **13**(1), 23–52 (2001)

43. Shulman, L.S.: *Those who understand: Knowledge growth in teaching*. Educational Researcher **15**(2), 4–14 (1986)

44. Shulman, L.S.: *Knowledge and teaching: Foundations of the new reform*. Harvard Educational Review **57**, 1–22 (1987)

45. Singh, K., Granville, M., Dika, S.: *Mathematics and science achievement: Effects of motivation, interest, and academic engagement*. The Journal of Educational Research **95**(6), 323–332 (2002)

46. Southerland, S.A., Gess-Newsome, J., Johnston, A.: *Portraying science in the classroom: The manifestation of scientists' beliefs in classroom practice*. Journal of Research in Science Teaching **40**(7), 669–691 (2003)

47. Subramaniam, P.R.: *Motivational effects of interest on student engagement and learning in physical education: A review*. International Journal of Physical Education **46**(2), 11–19 (2009)

48. Sunal, D.W., et al.: *Teaching science in higher education: Faculty professional development and barriers to change*. School Science and Mathematics **101**(5), 246–257 (2001)

49. Voogt, J., Fisser, P., Pareja Roblin, N., Tondeur, J., van Braak, J.: *Technological pedagogical content knowledge - a review of the literature*. Journal of Computer Assisted Learning **29**(2), 109–121. (2013)

50. Zimmer, C.: *Testing Darwin*. Discovery Magazine (2005, February)

Chapter 37
Exploring Evolution in Action in the Classroom, through Human Genetic Diversity and Patterns

Michael Wiser

Abstract This chapter outlines the difficulty and importance of tackling human evolution in the classroom, and the different demands this places on the instructor. It discusses lessons for teaching evolution in action, specifically in the context of human evolution and population genetics. These lessons include explicit hands-on activities for the students in analyzing and interpreting human genetic data, and addressing the information flow from scientific journal articles to popular presentation in the media. This content was implemented in a course designed for graduate students though with the occasional advanced undergraduate student with background in STEM fields other than biology. The chapter finally reflect on what the students gain from these lessons, and what limitations remain.

37.1 Evolution in Action

Evolution in Action encompasses a lot of different things. For many of us in our research, one of the primary points is that evolution is not solely a historical process. It's not just that current species and populations are the result of evolutionary forces, but that they are subject to those same forces now. In addition to the fact that things have evolved, they currently are evolving, and will continue to evolve in the future that is, unless they go extinct.

In this article, I first outline the difficulty and importance in tackling human evolution in the classroom, and the different demands this places on the instructor. I discuss lessons I've developed for teaching evolution in action, specifically in the context of human evolution and population genetics. These lessons include explicit hands-on activities for the students in analyzing and interpreting human genetic data, and addressing the information flow from scientific journal article to popular

Michael Wiser

Beacon Center for the Study of Evolution in Action, Michigan State University, East Lansing, MI, USA e-mail: mwiser@msu.edu

Fig. 37.1: A popular style of depicting human evolution. Image source: Flikr user Patriziasoliani,
        CC BY-NC 2.0 licensing.

presentation in the media. I have implemented this content repeatedly in a course designed for graduate students though with the occasional advanced undergraduate student who have backgrounds in STEM fields other than biology. I then reflect on what the students gain from these lessons, and what limitations remain.

The reality of current evolution is not always immediately apparent to students. Popular images of evolution, such as the one in Figure 37.1, typically present a clear linear path of adaptation. This may reinforce misconceptions, such as that humans evolved from chimpanzees, rather than that humans and chimpanzees have a recent common ancestor that was neither a chimp nor a human. Particularly for students without a strong background in biology, many aspects of evolution as it applies to human populations can be confusing and non-intuitive. Further, evidence shows both that many students are more interested in topics of biology as they relate to humans [4, 7], and that resistance to evolutionary thinking can be stronger in students when they are presented with evolution in the context of humans than in the context of other organisms [6]. How, then, can we resolve these competing demands on instruction?

One option, of course, would be to steer clear of discussions of human evolution. The data supporting the idea that students become more engaged in topics of biology as those topics are explicitly related to humans come overwhelmingly from studies of undergraduate students in biology courses [4, 7]. Many of those students, of course, are intending careers in health professions, and thus the connection to humans is obvious. When students are less interested in careers in health or life sciences, though, such connections are less obviously useful. It would be very easy to argue that the core evolutionary principles we want these students to internalize don't require reference to human evolution, and that further our students would have little reason to be particularly interested in human evolution.

Yet in some senses, I feel avoiding discussions of human evolution would be an abdication of my job. For many of the students I teach (see Curriculum context and student population, below), this is their first biology class since high school, and almost certainly their first time directly grappling with evolutionary concepts. Whether these students choose to move to industry or stay in academia after graduation, many of them may well be the only person in their department or company with training in evolutionary biology. If we can address and resolve misconceptions about human evolution in this group, they will be able to answer questions and correct misunderstandings of their future colleagues and/or students.

A second option would be to rely on some high-quality educational resources produced by others. Education is no less a collaborative effort than science is, and we do not need to constantly reinvent the wheel. There are some fantastic videos and interactive activities that are freely available from such reputable sources as the Howard Hughes Medical Institute that deal directly with human evolution, which provide data on what the likely selective pressures are, and why different populations have evolved different adaptations. Indeed, I make use of several of these in my course, dealing with topics such as sickle cell disease and its relationship to malarial resistance [3], selection for darker pigments near the equator and their role in reducing the incidence of birth defects by protecting folate in the blood and simultaneous selection for lighter pigments nearer the poles and their role in preventing vitamin D deficiency [1], and selection for lactase persistence among pastoralists [2]. These resources, though, have typically been developed explicitly for a high school or undergraduate audience, and thus there is more I can do to engage my specific student population.

Therefore, I've taken a different approach. I address human evolution directly in an evolution in action framing, using both some of the previously mentioned resources, and some that I've developed specifically for this course. Notably, students gain some experience working with genotypic data, looking at single nucleotide polymorphisms (SNPs) correlated with human genetic traits. They each research a trait of interest, and learn first-hand how information changes between scientific journal article, responsible presentation of that information to a general audience, and what one can find in a general news article. And in working with these data, they get a much stronger sense of what it means when we say that the overwhelming majority of human genetic diversity is found within populations, rather than between them. Beyond this, they also have a chance to learn what sort of information can be gleaned from personal genomics, and what (currently) cannot, and how to evaluate evidence in relationship to claims of human genetic divergence.

37.2 Curriculum Context and Student Population

The Evolutionary Biology for Non-Life Scientists (EBNLS) course was developed to meet the needs of students in the BEACON Center, a National Science Foundation Science and Technology Center focused on studying evolution in action. The majority of the faculty in this center are biologists, studying evolution in a variety of biological systems, both in the field and in the lab. A sizeable minority, though, are computer scientists and engineers, using evolution as a tool to solve complex engineering problems. Students who have a background in mathematics, computer science, engineering, or physics, but who do not have a background in biology, take the EBNLS course. Students with a background in biology, but not in computer science, take an alternate course of Introduction to Computational Methods in Biology. In a later semester, students from both courses jointly take a multidisciplinary, project-based course where they engage in collaborative research on evolutionary questions in a computational setting. The intended target audience is overwhelmingly graduate students in computer science and engineering, though my co-instructors and I occasionally have advanced undergraduates enrol as well, most often when they're working in the lab of one of the BEACON computer scientists or engineers.

In this context, we have several major learning outcomes we want for the students in the EBNLS course. We want students to 1) understand the major evolutionary forces (selection, drift, mutation, recombination, migration, etc.) which act on populations; 2) develop and be proficient with the vocabulary to effectively communicate with evolutionary biologists on topics of mutual interest; and 3) gain sufficient background information to be able to take advanced courses in evolution, such as those offered through the Ecology, Evolutionary Biology, and Behavior program (EEBB) at Michigan State University. The fact that this course is not designed to substitute for other graduate course work in evolutionary biology allows us significant flexibility in choosing which topics to cover, and at what depth.

37.3 Context of Personal Genomics within the EBNLS Course

The general structure of the EBNLS course was designed years ago, in work led by Dr. Louise Mead, and with significant contributions from (now) Drs. Caroline Turner, Emily Weigel, and, more recently, me. We have arranged this course in three modules: the chemical basis of life; mechanisms of evolutionary change; and evolutionary change in the context of genomes. The personal genomics and human evolution lessons now comprise a large fraction of this third module, which I've taken the lead on adapting. Because of its placement near the end of the term, students have already had lessons on basic mechanisms of information flow within biological molecules, population genetics, molecular phylogenetics, and evolutionary analysis. This unit, then, enables students to apply the newly-learned evolutionary concepts to humans, as well as visualize actual genetic data. I also use this unit as an opportunity for students to gain practical experience with how scientific information is presented in a variety of venues, and to see how what is written in a scientific journal gets translated into news articles or summaries of findings written for a general public.

37.4 Student Experience

Students initially encounter the ideas of the personal genomics module on the first day of the term, when we discuss the course outline and themes, and the expectations for both the students at the instructor. I explain that in the final module of the course we will discuss genome evolution with particular attention paid to human evolution and its ongoing, rather than purely historical, nature. I explain that due to funding from BEACON we have kits from the company 23andMe for their use, if they want them. No one has to use the kit if they don't want to; if a student does not want to use the kit, I can give them my login information so they can still access all the information they need for their project in this unit. Further, no students will be asked to provide any of their genetic data to us during the context of the class. If they wish to make their data available as part of the class dataset after the semester is over, we will give them instructions then on how to do so. The data they analyze comes from a set of people who have volunteered their anonymized data for classroom purposes.

I also explain that while there are interesting things they could learn, they should be aware of potential drawbacks. First, it is possible that they will discover that they have a higher risk for certain genetic conditions than is typical. However, biology is not destiny. Such testing can show increased or decreased likelihoods, but it doesn't provide guarantees of what will or will not happen, and students should contact a genetic counselor if they are concerned by a result. I also show them how the traits that have the potential to be medically sensitive are not shown immediately, but that users need to go through several pages of informed consent about what the test results might mean in order to see them. Secondly, if a close relative of theirs has also used the service, it is possible that they may find that their genetic relationship is different than previously assumed. To allay this concern, I discuss how for many people kin relationships are based on interactions rather than shared blood; adopted children are no less of family members than are genetic children. I further demonstrate how users can adjust their settings to either allow or prevent others from contacting them through the service, or even seeing their information. Finally, I note that it is theoretically possible the information could be leaked in a data breach from the company. However, as users are not required to provide any personal information other than an email address, they can take steps to minimize this threat if they so choose. Only after discussing these potential drawbacks do students get their kits.

At the beginning of the third module, approximately two months later, we take half of a class period to walk through the type of information that can be learned from these sorts of personal genomics tests. I log into my own results to demonstrate, and encourage the students to bring their laptops, and either look at these things in their own results, or in the temporary log in I give them to see my results. We start our discussion with the topic of ancestry: how these companies assign it, what their reference populations are, and what biases exist in their data (such as that the demographics of their customer base do not closely match global human demographics, so certain populations are drastically over-represented, and others are under-represented). We talk about how one can trace ancestry differently for the autosomes (most of the human chromosomes), and the sex chromosomes, X and Y, and how differences in inheritance patterns among these can reveal different population histories. We further discuss how the information from the mitochondria shows only what has happened along the matriline from the student to their mother, to her mother, to her mother, stretching back into the distant past of humanity and how this can differ from other parts of the same genome. And we briefly touch on a topic that will come up in a later class about how racial categories are primarily socially constructed rather than matters of biology, and how this has led to the whole history of scientific racism, which we take the time to debunk in a later lesson.

At this point, we also take a moment to discuss how the sampling of any population is a snapshot of one particular point in time. Humans have shaped the planet in many different ways, and one of those relates to the speed at which we now move about the globe. We discuss how reference populations for ancestry can be established mostly depending on recorded histories of all of an individual's genealogical ancestors being from one particular region for at least a certain number of generations but how these reference populations can change over time, due either to large population movements or adaptation to specific environmental conditions. Which regions could even theoretically have different human population groups associated with them is influenced both by the distant past of migrations and population mixing, and the greatly increased transportation of the past few hundred years.

As we move beyond ancestry, we discuss the different types of traits assessed, reinforcing earlier content about the different between simple Mendelian traits, polygenic traits, and complex traits that depend on both multiple genes and environmental conditions. We talk about the difference between causal genetic variants those that directly lead to a genetic condition, through changes in the gene product or how it is regulated and linked genetic variants, which are statistically associated with a trait but don't directly contribute to it. We also tie this in to other lessons on human alleles under current selection (such as lactase persistence), past selection (such as variants that reduced susceptibility to past plagues), or both (such as sickle cell trait, and its resulting resistance to malaria).

The primary student work with personal genomics comes in the form of a project. Each student picks one of the traits that is not medically sensitive; recent examples include whether a person has wet or dry earwax, and whether a person can detect the scent of metabolites from asparagus in their urine. For this project, they start with the company's description of the trait. Next, they read all of the scientific papers the company cites for this trait, so they can determine what evidence there is that this trait is linked to this genotype. Then they have to search for a news article about the trait or condition, and read that as well. I provide all of the students with the genotypes of the people who have provided their data for the classroom dataset in one file; in another, I provide high-level ancestry from these same individuals. This data has been collected from volunteers who are part of the BEACON community, including some former class members who contributed their data after they completed the course, but not while enrolled. Finally, they have to create a class presentation about the trait: what we know about it and how we know this, what is left to determine, and how the information about this changes as one moves from scientific journal article to a summary provided by a company to what shows up in the popular press. The students need to try their best to figure out which genotypes correspond with which high-level ancestry, and explain their reasoning.

37.5 Reflection

In my view, the personal genomics unit in this course offers several major advantages. It serves as a route for making certain content more immediately relevant for the student population. This allows me to teach the material first, and then relate it back to their research interests even when that relationship is only apparent after the students have learned the material. The data available to the students allows them to grapple with what it means that most human genetic diversity is found within populations. Material relating to human genetics also showcases stark differences between the careful findings in research publications and the more sensational headlines of news reports, allowing students to think a little more critically when they next encounter a headline about how, for example, redheads are going to go extinct.

There are, of course, some limitations as well. Due to the low enrolment numbers in the EBNLS course, I do not have sufficient statistical power to make meaningful comparisons of pre- and post-instruction student responses to commonly accepted assessment instruments. As such, I cannot currently show whether the personal genomics lessons have resulted in important learning gains. Conversely, the small class size and discussion format of it allow for in depth informal assessment of the students, based on the questions they ask, their responses to questions posed, and their presentations.

One of the key instructional challenges in this student population is relating content in the basic science to examples they fundamentally care about. In some sections of the course, this is easy: students readily grasp the importance of population genetics and the impacts of population size, selection strength, mutation rate, etc. to the genetic and evolutionary algorithms they are likely to use in their engineering research. Other topics, such as genome-wide association studies, or the mapping of phenotypic traits to genome locations, can be much more abstract for a population of engineers unlikely to ever need to implement these sorts of studies. However, the underlying concepts, such as how we can determine which sections of a complex code are necessary for a function, directly relate to research in evolutionary computation. Relating the impact of individual genetic changes to human evolution and, even better, showing how students can use this information to assess potential impacts from their own genetic variants can do a lot to make the content more meaningful to the students. Once the students have engaged with the material, making the underlying conceptual connections is much easier.

The discussion of human evolution also allows us to potentially short-circuit a common difficulty in evolutionary thought among those relatively new to the topic: the problem of pan-adaptationist thought [5]. It is easy to come up with stories about why certain traits would be selectively advantageous; it is much harder to provide evidence that they are. Because humans are such a young species, and because our large population size is a very recent development in terms of genetic timescales, there are a number of interesting traits found in some but not all people and where we wouldn't expect the benefits to be limited to only certain geographic regions. When students ask about the adaptive significance of a trait, this provides an opportunity to brainstorm what might be adaptive about it, what other explanations there could be for this trait, and what evidence we would need to be able to make a reasonable claim that the trait is adaptive.

Arguably the most important instructional content in this unit concerns how genetic diversity is distributed through human populations. Many students enter this course having heard the true statement that race is a social construct rather than a biological reality. But most of them have very little idea what that means. Discussions of selective pressure on traits that visibly differ between populations in the world, such as pigmentation, or height and breadth, and how these pressures are extremely dependent on the environment highlight that these aren't traits where one physical type is actually better than another. Instead, the relative costs and benefits depend on the environment. Spending time discussing migration patterns, and how human genetic diversity does not support the idea of biological races of humans mapping strictly to continental boundaries, does a lot to put social and biological concepts of races into context. The assignment where students need to try to match SNP genotypes with broad-scale ancestry patterns takes this a significant step further. I fully expect none of the students to manage to assign the genotypes to the correct ancestry, but this is the primary point of having them try. No matter how often we explain verbally that the vast majority of human diversity is found within populations, not between populations, students still typically think they'll be able to do this until they have a direct experience failing to do so. Since I've implemented this activity, every single student has stated afterwards they'd be unable to assign genotypes to ancestry, sometimes adding that might be able to do so for exceptionally rare genotypes only ever seen in one population, but not for typical traits.

Human evolution also offers an excellent entry point for discussions about science journalism, and the difference between the information in a scientific paper and what gets conveyed to a broader public audience. When I first proposed adding this element to the class, I was concerned that faculty members would consider it a poor use of scarce class time. Instead, I've received nearly universally positive feedback both from the BEACON faculty and from colleagues in the biology education community that this is a valuable addition to the curriculum. Even more importantly, a substantial fraction of the students have told me that they previously didn't realize what a disconnect there often is between cautiously worded scientific findings and enthusiastic press coverage, or why certain science journalists are routinely referenced as being more reliable sources than others. Providing students with a hands-on experience of why one might be skeptical of popular press coverage of scientific findings does more to advance notions of fair skepticism than just telling them to be skeptical of what they read.

This curricular unit would have been difficult to develop and implement without the support of a center like BEACON, where the administration has been happy to invest resources into an educational opportunity for their students. Because of their support, I've been able to develop a set of genetic data that my future students and I will be able to use even after funding for the center ends. Through this unit, a small part of BEACON's legacy can be reaching a population of scientifically-literate people who otherwise would have virtually no exposure to evolution, and particularly to questions of human evolution.

Acknowledgements I would like to acknowledge Erik Goodman for approving the funding for the personal genomics unit in this course. Louise Mead, Caroline Turner, and Emily Weigel were instrumental in developing previous versions of this course, and Louise in particular was involved in the first run of this personal genomics curriculum.

References

1. The Biology of Skin Color. HHMI BioInteractive. (available at https://www.hhmi.org/biointer active/biology-skin-color#)
2. The Making of the Fittest: Got Lactase? The Co-evolution of Genes and Culture. HHMI BioInteractive. (available at https://www.hhmi.org/biointeractive/making-fittest-got-lactase-co-evolution-genes-and-culture)
3. The Making of the Fittest: Natural Selection in Humans. HHMI BioInteractive. (available at https://www.hhmi.org/biointeractive/making-fittest-natural-selection-humans)
4. Chamany, K., Allen, D., Tanner, K.: *Making Biology Learning Relevant to Students: Integrating People, History, and Context into College Biology Teaching.* CBE Life Sci. Educ. **7**, 267–278 (2008)
5. Gould, S.J., Lewontin, R.C.: *The spandrels of San Marco and the Panglossian paradigm: a critique of the adaptationist programme.* Proc R Soc Lond B **205**, 581–598 (1979)
6. Pobiner, B.: *Accepting, understanding, teaching, and learning (human) evolution: Obstacles and opportunities.* Am. J. Phys. Anthropol. **159**, 232–274 (2016)
7. Weber, K.S., et al.: *Personal microbiome analysis improves student engagement and interest in Immunology, Molecular Biology, and Genomics undergraduate courses.* PLOS ONE **13**, e0193,696 (2018)

Part VII
The Evolution of Erik Goodman

Chapter 38
Academic Biography of Erik D. Goodman

Erik D. Goodman

Abstract This contribution summarizes the academic biography of the founding director of the NSF-funded BEACON Center for the Study of Evolution in Action, Dr. Erik D. Goodman, in a first person narrative.

Pre-college Education

I was fortunate to receive an excellent education during my childhood years, both in public schools and from my family members. My grade school teachers were very supportive, mostly forgiving my talkative nature and even giving me special opportunities to do things that kept my interest. From third grade on, I was an avid reader, mostly of science fiction, and I guess as a result, also a good speller. The only thing that endeared me to my fourth grade teacher during the dreadful year I spent in Winston-Salem, North Carolina while my Dad was doing a year as a postdoc, was my performance in the school spelling bee, in which I beat all but one seventh grader. Otherwise, to the teacher and the kids in the class, I was just a Yankee with a smart mouth, and that was not a good thing. I was delighted when my Dad accepted a faculty position in Zoology at Michigan State University, and I entered Red Cedar Elementary School in East Lansing. I had two excellent teachers there—a Mrs. Hunnicutt for fifth grade and Mrs. Harner for sixth. Two of my fellow students and I even got to do some special projects with the principal, after being pretty bored with the regular curriculum.

In my pre-college days, I was also a keen student of genetics, as my Dad was always eager to tell me about his human genetics research. I also loved linguistics, which my great uncle, Anders Orbeck, taught me after I finished mowing his lawn every few weeks while I lived in East Lansing. Uncle Anders was a professor of linguistics in the English Department at MSU, specializing in Old Norse. I credit these familial influences with keeping me interested in both science and language, a

W. Banzhaf et al. (eds.), *Evolution in Action: Past, Present and Future*,
Genetic and Evolutionary Computation, https://doi.org/10.1007/978-3-030-39831-6_38

combination that has brought me a great deal of enjoyment throughout my life and has also facilitated much of my international involvement.

My high school was one of the best in the country, even though located in Winston-Salem, NC. It boasted more National Merit Scholarship semi-finalists than all but a few other schools in the country, and offered four years of Latin, two years of Greek, in addition to German, French and Spanish. I took two years of Latin, two years worth of German (packed into one year), and two years of French. I was a science/math geek, so had biology, chemistry, and in my senior year, physics and AP chemistry (the only AP science course offered there at that time). I had some wonderful teachers, and thoroughly enjoyed my high school years, although I was way too "geeky" to be part of the elite social circle. But at Reynolds High, that didn't isolate me socially in school, as I had good relationships with many of the "in crowd" people during school hours, just not after school, when I spent my time with my geekier friends. But we had a great time, especially after I learned to drive and bought an old "junker" car to transport us around.

I earned good grades in high school and did well on the National Merit and SAT tests, which earned me a full ride scholarship to Duke University, which I almost took. But I had been invited to Michigan State University to take their Alumni Distinguished Scholarship test, and had traveled to East Lansing for that in February, 1962. I was delighted when they offered me an ADS scholarship, which would pay all my college expenses, and I couldn't wait to get back to Michigan State and out of the South. So in September, 1962, I began my college education, which certainly took me in directions I had never anticipated.

College Days

My first quarter, I took the "usual" science major courses—organic chemistry, physics, calculus (but a special proof-oriented course MSU offered to its major scholarship students), Russian 101, and also a formal philosophy course called "Principles of Right Reason," which was really a course in semiotic, taught by Henry Leonard, a past president of the American Philosophical Association. I became "hooked" on formal philosophy, and Leonard offered me a chance to do a one-on-one reading course with him in the winter quarter, which I eagerly accepted. From there, I took many other philosophy courses—several in logic plus others in philosophy of language, philosophy of science, philosophy of probability, etc. I also took a few other courses leading toward a philosophy major, including ethics, which, contrary to my expectation, has turned out to be handy when I was called on to teach engineering ethics. I continued studying Russian for three years.

Then the unexpected happened: I enrolled in a computer programming course, and a new phase of my life began. I devoured computer science as rapidly as I could absorb it, although it wasn't yet a formal major at MSU or most other universities. So I ended up graduating with a math major, because that went well with all the formal logic and computer science courses I had taken, as well as the math courses.

But my interest in computer science was more intense and motivating than any other subject had ever been for me, and my academic passion was found!

My social development grew by leaps and bounds in college, and I became a resident advisor ("RA") in Wilson Hall, where I lived for four years. My college days were greatly enriched by my involvement in both student government and in the residence halls program, in which MSU was a national leader. I became very interested in how students received academic advising, and conducted a university-wide study of the process, under the auspices of AUSG, the All-University Student Government. I also initiated a program that lasted many years and was replicated across the campus. It was called "South Campus Major Night," and brought faculty members from about 100 departments to Wilson Hall for one evening each year to answer questions about their departments' majors on an informal basis, for the seven thousand residents of the dorms in the South Campus complex.

Master's Program

I began my Master's program during my senior year, and stayed at MSU for that degree. The first year, I was supported as a Graduate Advisor in the newly opened Hubbard Hall, and the second year, I was a research assistant for Dr. Herman Koenig, a pioneering systems scientist and Chair of the Department of Electrical Engineering and Systems Science. That assistantship turned out to be important to me. I did state modeling work on the undergraduate education system of the university, writing a FORTRAN program for that purpose. That prepared me for teaching state modeling and computer simulation to biologists, which I did for the first decade after completing my doctorate.

During my Master's program at MSU, the Computer Science Department was not yet established, so I majored in Systems Science, while taking courses in both disciplines. I specialized in automata theory, and planned to do my Ph.D. work at the University of Iowa with an automata theorist there. But Carl Page (father of Google founder Larry Page) convinced me to go instead to his alma mater, the University of Michigan, and in August of 1968, I began my study there, in the Department of Computer and Communication Sciences, one of the first computer science departments in the country.

Ph.D. Program

Computer and Communication Sciences at Michigan was a very competitive program, and almost all of the students in my incoming Ph.D. class (about a dozen of us) studied together every week in preparation for the Ph.D. Qualifying Examination. All of us who studied together passed the exam, and I believe the two who had not studied with us did not. The qualifying exam had three parts: formal systems (like math), natural systems (biology, language), and artificial systems (including topics commonly taught as "computer science"). The CCS Department was much more multidisciplinary and forward-looking than most computer science programs then or now, including coursework in complex adaptive systems, natural language processing, linguistics and speech production, neuroanatomy, information theory, and cell biology.

My second big discovery of a field that fascinated me was when I took John Holland's two complex adaptive systems courses. I immediately became fascinated with the idea and power of algorithms based on evolutionary principles, although the term "genetic algorithm" was not coined until about five years after I left Michigan. But Holland had already fleshed out the essence of the GA and proved a number of important theorems.

I did my doctoral thesis research in John Holland's Logic of Computers Group, University of Michigan, 1969-71. My thesis work included the first use of a genetic algorithm to solve a "real-world" problem—i.e., one for which the answer was unknown and actually wanted—as opposed to "toy" benchmark problems. My work began from a computer simulation initiated by Dr. Roger Weinberg, a Ph.D. in Biochemistry and fellow (but second-) Ph.D. student in the Logic of Computers Group. We modeled the growth of an E. coli cell, using fourteen generalized "pools" of cellular constituents (think cell membrane, cell wall, ATP, ADP, MRNA, TRNA, pools like that). I added a more sophisticated replicon (Helmstetter-Cooper) and capability to grow on multiple media, producing a more robust model capable of simulating shifts among growth media, enabling the use of more data to indirectly parameterize the model based on its behavior. Based on what I had learned in Holland's two courses covering what we'd now call genetic algorithms and complex systems, I developed a genetic algorithm using integer-mapped real representations of over forty rate constants, most of which were involved in calculating nonlinear enzyme activities for pathways between selected pairs of the fourteen "pools" based on a model of allosteric modification of enzyme activity. We did not know about epigenetics or gene regulatory networks at that point, so having context-sensitive rate coefficients was our way of approximating such effects, which we knew had to be present somehow. The integers were uniformly mapped to logarithms of real number rate constants, to allow search of exponentially distributed rate constants across many orders of magnitude. The GA used integer-truncated Gaussian mutation and two-point crossover. This huge search was possible only because it could be run almost continuously on Logic's IBM 1800 computer, being checkpointed whenever another user wanted the machine for another purpose, but then restarted immediately afterward. It ran cumulatively for over half a year. When I joined the faculty at Michigan State University in fall, 1971, I had access to another IBM 1800, where I finished the runs needed for my dissertation. After completing the defense in January, 1972, I published an article about a cellular-space embedding of the model, in the journal Biomedical Computing, based on my study of cellular automata under Prof. Arthur Burks (a patentee on the early ENIAC computer), also of the Logic of Computers Group. However, my initial attempt to publish about the GA process I used to evolve the model's parameters was not accepted, and the pressures of assistant professorship discouraged me from pursuing it further. At that time, this material was very difficult to publish both because there was no GA community that could have reviewed it knowledgeably, and because the paper could only describe the results of a single, year-long run, which could not practically be replicated in order to draw any statistically valid conclusions. However, the GA did produce a solution that generated good agreement with changes in doubling times resulting from shifts among several growth media (e.g., changing environmental conditions) that approximated those seen for E. coli. That enabled me to begin to look at modeling colony formation.

An Aside on Music: An Important Parallel Journey

It is an adage that people who are mathematically inclined also have an aptitude for music, and for me, that interest in music was expressed from my earliest years. My mother sang and taught me many songs when I was very young, and I could pick out harmonies to Christmas carols when I was still in grade school. I started playing violin in the school music program in East Lansing in fifth grade, and my parents soon enrolled me in private lessons. Later, as a student at Walter French Junior High School in Lansing, Michigan in eighth grade, I was concertmaster of the orchestra, and also did a few performances with a buddy, Phil Carr, for things like PTA meetings, with both of us singing and playing fiddle. These violin lessons were the only classical music training I ever received everything else was learned "by ear." But I had to give up violin when I got to high school, because my paper route kept me out of the school orchestra, which practiced after school. My best friend in high school, Paul Licker, taught me some chords on the guitar, and by my freshman year in college, I was happily playing and singing folk songs—Joan Baez, Bob Dylan, and many others. In fact, for two summers (after my sophomore and junior years), I led "hootenannies" for the summer orientation students when they came to campus for a week before entering MSU as freshmen. Those were great summers, and the protest songs were very popular!

Although I never noticed bluegrass music while living in North Carolina, I was introduced to an excellent band in Ann Arbor during grad school, and that changed my life, too. I immediately began learning to play bluegrass banjo, and as soon as I became a faculty member at MSU, joined with some other faculty members to form a bluegrass band, the Bluegrass Extension Service. I later also played with the RFD Boys, the group in Ann Arbor that introduced me to bluegrass and also introduced me to my wife, Cheryl. So I owe that band quite a debt! The Bluegrass Extension Service played over 1000 gigs in local bars and folk and bluegrass festivals over the 25 years from 1972-1997!

Teaching and Collaborative Research in Systems Ecology

In 1972, MSU's large NSF RANN grant required that many biologists—particularly ecologists—be able to work with systems scientists to develop mathematical models and computer simulations of biological systems, at many levels. When I was hired in Fall, 1971, before completing my Ph.D., I was asked by my mentor and the RANN grant PI, Prof. Herman Koenig, to develop a series of courses to train systems scientists and biologists to work together to develop such models and computer simulations. Dr. Patricia Helma (now Werner) and I developed two parallel courses, followed by a third course in which the graduates from the first two were grouped to work together on developing a computer simulation associated with the research of one of the biology graduate students. In the first course, "Systems Concepts for Biologists," I taught the principles of state modeling of dynamic systems, with special emphasis on developing the students' intuition for the qualitative behavior of differential equations. I then taught several numerical techniques for solving the differential equations constituting a state model. I taught them to write FORTRAN code to simulate such systems, also introducing probability distributions and stochastic modeling, including generation by computer of discrete and continuous random variables with arbitrary distributions. We studied classical population models as differential equations (logistic growth, predator-prey systems, etc.). Meanwhile, Pat taught the engineering students principles of population biology, including mathematical models and concrete examples. In the spring, we formed 10-12 teams combining the classes, in Ecosystem Analysis, Design and Management. Teams generated over 100 models during a ten-year period. Many teams produced computer simulations that ended up as parts of the doctoral research of one of the biology graduate students, and some led to independent publication of the modeling work. This three-course sequence brought MSU recognition as a leading place to study the newly emerging field of systems ecology, and more than twenty faculty members from many universities came on leave to MSU to participate in this course sequence. The three-course sequence became the model on which we based BEACON's flagship courses almost 40 years later.

Early Research Modeling Ecosystem Dynamics (Using a Genetic Algorithm)

At Michigan State University, I continued to use genetic algorithms for parameterizing compartment models for nutrient and pesticide transport in aquatic and terrestrial environments, 1973-81, under Koenig's grants from NSF, and later my own from EPA. The first EPA project looked at fate of pesticides in aquatic environments, including an artificial stream on the grounds of the Monticello Nuclear Power Station in Minnesota. Later EPA support was to model the fate and effects of the organophosphate pesticide Guthion (azinphosmethyl) in experimental apple orchards. To parameterize the compartment models we developed from the data we collected, we used a genetic algorithm, but did not publish this evolutionary computation work because of a turn in world politics. The GA work had been done by my Iranian Ph.D. student, Mehrdad Tabatabai, but after completing essentially all the work, he had to depart the U.S. suddenly one night, without a degree or a trace, because of the overthrow of the Shah of Iran.

Early CAD/CAM/CAE and ALife Research in the Case Center, and the Birth of the GARAGe

I became Director of the Case Center for Computer-Aided Engineering and Manufacturing in 1983, which included not only a CAD/CAM/CAE research mission but also responsibility for all computer facilities of the College of Engineering. The center had a marvelous staff of graduate student consultants, professional systems analysts, and undergrad staffers and programmers. One of these students, Adrian Sannier, and I did early work in the area of artificial life, using a genetic algorithm to evolve agents that roamed a rich landscape interacting with it and with each other, and they evolved some very clever and unexpected strategies for managing and using the limited resources present. Our Case Center team also did CAD/CAM/CAE research projects for many companies—including GM, Ford, Chrysler, Motor Wheel—and CAD software developers (SDRC, CIMLINC). General Dynamics also sponsored CAD/CAM algorithm development, under the leadership of Mel Barlow (a vice-president and M.E. alumnus) and Russ Owen (a Case Center graduate who started as a design engineer at General Dynamics and rose to a vice presidency at Computer Sciences Corporation). Mel Barlow went on to lead the Case Center's Endowment Fund Campaign, so he and I visited many corporate headquarters across the country in building a multi-million dollar endowment for the Case Center. The center's "golden years," when our mechanical engineering students were highly sought after by CAD developers and corporate users alike, lasted into the 1990s, during which time I did research in computer graphics algorithms, numerically controlled machining toolpath verification robotic spray simulation, lost foam pattern molding for cylinder heads, and many other similar projects. I joined the CAD/CAM Technical Committee of the American Institute of Aeronautics and Astronautics (AIAA), at one point chairing the TC's Research and Future Directions Subcommittee, and the TC gave me their Outstanding Contributions Award in 1990.

In 1993, the computer services division of the Case Center became a freestanding division, the Division of Engineering Computing Services, and the research arm continued under my direction as the Case Center, becoming the home of both the Genetic Algorithms Research and Applications Group (GARAGe), co-directed with Bill Punch, and MSU's Manufacturing Research Consortium, which I directed from 1993-2002, and which also sponsored a number of projects applying evolutionary computation to solve industrial problems.

The early 1980s also saw Ron Rosenberg and I collaborating on development of a systems dynamics software package called SYSKIT, under sponsorship of publisher McGraw-Hill. We developed a very capable package, but it was completely "scooped" by the introduction of the Macintosh computer, just as our PC-based software was entering the market.

Initiating International Research Collaborations

After studying the Chinese language at MSU beginning in 1987 and spending summer, 1988 in Shanghai as a Study Abroad student, I began planning a sabbatical leave in Beijing, which I took with my family, Cheryl and 3-year-old David, in 1993-4. There, with the help of graduate student Wang Gang, of Beijing University of Aeronautics and Astronautics, I wrote the first version of the parallel genetic algorithm package GALOPPS (Genetic Algorithm Optimized for Portability and Parallelism) that I continued developing for many years and used as the starting point for the commercial software package I later developed in Red Cedar Technology, the company I co-founded with Ron Averill in 1999. In fact, I still find the heterogeneous island parallelism provided by GALOPPS useful in my optimization work today.

My collaborations in China, and additional collaborations in Russia that arose from our Dean of Engineering's hosting of Russian visitors, led me to create the International Technology Incubator within the Case Center. I brought several Russian engineering faculty members to the U.S. frequently to consult for American companies, which also gave me the opportunity to fly them all around the Midwest in my airplane for our joint consulting visits, while being reimbursed by the university for the flight expenses. Those were certainly fun times for me! I also participated in many conferences in Russia and China throughout the 1990s, and have continued the annual trips to China to this day. These visits resulted in a number of evolutionary computation consortia being formed between the Case Center and universities in both China and Russia. At one point, after several years of NSF grants to develop ways to improve the effectiveness of globally distributed engineering design teams, I nearly made a dramatic career change, founding a joint for-profit and not-for-profit operation to employ Russian engineers in their "closed" nuclear cities as off-site consultants to U.S. companies, under the U.S. Department of Energy's Nuclear Cities Initiative. But that was killed by U.S. politics, and the Russian consortia ceased operating after things became "chillier" for Russian/American collaboration at the end of the 1990s.

Red Cedar Technology, Inc.

After M.E. Prof. Ron Averill and I tackled a design problem to remove weight and improve crashworthiness of an automotive component for General Motors, GM urged us to form a company. The result was Red Cedar Technology, in which I worked as Vice President for Technology from 1999 until BEACON's founding in 2010. The company started as a consulting services firm, but soon wrote and marketed a design software package we called HEEDS, the Hierarchical Evolutionary Engineering Design System. I enjoyed writing the search/optimization part of the code, called SHERPA, very much, spending summers and my consulting days happily coding away at Red Cedar headquarters. I resigned after BEACON was founded, and in 2013 the company was sold to CD Adapco, another engineering software firm. In turn, that company and its Red Cedar division were sold to Siemens PLM, the industry giant, in 2016. I take great pleasure today in seeing the software we wrote back then still being at the core of a package now used by countless companies worldwide.

The Founding of GECCO and SIGEVO

At the '95 International Conference on Genetic Algorithms (ICGA '95), I was selected to be the General Chair of the 1997 ICGA, to be held at the Kellogg Center at MSU. At and around ICGA 97, a number of leaders involved in ICGA began discussions with other leaders in EC about possible unification of conferences in the field, in order to build the size and reputation of the field, especially for the benefit of new junior faculty members. Key participants included David Goldberg, John Koza (organizer of the Genetic Programming Conferences), Darrell Whitley, Wolfgang Banzhaf, Hans-Paul Schwefel (co-inventor of evolution strategies and organizer of the Parallel Problem Solving from Nature Conferences—PPSN), and myself. Representatives from another smaller group dealing with evolutionary programming chose not to affiliate, allying instead with the IEEE Neural Net Society, but the remaining parties all joined to form the new International Society for Genetic and Evolutionary Computation, which sponsored the first GECCO (Genetic and Evolutionary Computation Conference) in Orlando in 1999, with 600 attendees—the largest for any conference in our field up to that time. I was General Chair of GECCO 2001, the third GECCO, in San Francisco.

I was elected Chair of the International Society for Genetic and Evolutionary Computation (ISGEC) in 2003, succeeding Dave Goldberg. Over the next three years, I worked intensively with Goldberg, Koza, Banzhaf, Whitley and others to seek ways to enhance the visibility of the field and to improve the recognition it provides to new junior faculty. In 2004, we began the process of affiliation with ACM as a new Special Interest Group, the SIG for Genetic and Evolutionary Computation— SIGEVO. The move was designed to make our society more recognizable as an important part of the mainline professional organization in computer science, even though it came at the cost of loss of considerable independence for the society. I became the founding Chair of SIGEVO in 2005, continuing in that role until 2007. I was re-elected to SIGEVO's Executive Committee twice, and continued to serve on the Business Committees for GECCOs through 2010. I am still on the Executive Committee, as are fellow BEACONites Kalyanmoy Deb and Wolfgang Banzhaf, and we all participate actively in its activities.

Mechatronic Systems Design—"Bond Graph/ Genetic Programming"

My musical and administrative collaborator Ron Rosenberg is an expert on bond graphs and had developed a commercial bond graph software analysis package, EN-PORT. We decided to try using bond graphs with EC to evolve novel designs for dynamic systems, in a genetic programming framework. The resulting systems could then be instantiated using many alternative combinations of electrical, mechanical or other components—whichever were most appropriate for a particular application. We got funding from the NSF, and recruited postdoc Kisung Seo and grad students Zhun Fan and Jianjun Hu for the project. This BG/GP method was used for design of new mechatronic systems and also for redesign of existing systems, in what we might now call "Genetic Improvement" of dynamic systems. We redesigned pneumatic railroad braking systems and improved on early designs of "Selectric" typewriters, among others. Because these systems required that both the topology and parameters of a system be evolved at once, the team developed "Structure Fitness Sharing (SFS)," a concept allowing new topologies to remain in a population long enough on average that appropriate parameters could be evolved before the topology was discarded. Such a process today would be called a multi-level search algorithm. Because this de novo synthesis required very long searches, Jianjun Hu developed a new model for sustainable search, allowing continual injection of new random genomes into the population but not forcing them to compete with more highly evolved individuals until their fitnesses are in similar ranges. This approach, called Hierarchical Fair Competition (HFC), was extended to allow it to evolve its fitness thresholds and even into a continuous form in which the fitness distribution itself is sampled for breeding and fitness evaluation purposes without subdividing the population explicitly. HFC and its offspring, Adaptive HFC (AHFC), Quick HFC (QHFC), and Continuous HFC (CHFC) were used by a number of authors during the ten years after their introduction, and the papers were cited over 700 times, but eventually, most of the community moved toward Greg Hornby's Age-Layered Population Structure (ALPS), which appeared soon after HFC was published, to address sustainability of search.

After the NSF grant expired, Zhun Fan joined the faculty of the Technical University of Denmark (DTU) and our collaboration continued. Prof. Fan, now directing the Guangdong Provincial Key Laboratory of Digital Signal and Image Processing of Shantou University, and BEACON have now established a joint research center on Evolutionary Intelligence and Robotics, and continue our collaborations through my annual visits to China and students from his lab coming to BEACON to work with me for one or two years. He has hosted me for many visits to Shantou University and other universities in Guangdong Province, also introducing me to others who have become collaborators on evolutionary multi-objective optimization. I enjoy very much working with the students and faculty that he and his colleagues have sent for stays in BEACON.

Evolution of Control Algorithms for "Green" Greenhouses

In 2006, I began collaborating with Prof. Lihong Xu, Dean of the School of Modern Agriculture Science at Tongji University. The topic of our collaboration has been greenhouse control, and we seek to evolve controllers with superior performance, in terms of crop yield and/or environmental impact, to the "standard" controllers that maintain a fixed internal greenhouse temperature/ humidity regardless of external conditions. Prof. Xu has visited MSU each year since then and has helped to guide the Ph.D. work of Jos Llera and the doctoral and postdoc work of Prakarn Unachak. He has also sent eight excellent Ph.D. students and several postdocs and junior faculty members as visiting scholars, each working with me in BEACON for one- or two-year periods. Prof. Xu has a large and excellent team of researchers working in his lab at Tongji University, and his projects have important places in China's Five-Year Plans. They work not only on greenhouse management, automation and control, but also on related problems such as fish farming. The complexity of these problems has taken us in many directions, including development of multi-objective optimization algorithms, machine vision methods for visualizing plant growth and food status on fish farms, and many others, both practical and theoretical. MSU Ph.D. student Jos Llera is nearing completion of his degree on this problem, and we hope to see his work tested in practice in a greenhouse in China. MSU Prof. Erik Runkle, Dept. of Horticulture, does extension work with Michigan greenhouse operators and has been an invaluable source of guidance for our greenhouse modeling and control work.

Addressing Land Use Issues Using Evolutionary Multi-objective Optimization

Soon after Dr. Kalyanmoy Deb was recruited as the Koenig Endowed Chair, we began working with Dr. Oliver Chikumbo, a control theorist and forestry researcher originally from Zimbabwe, who received his university education in Australia and was working in New Zealand. He had created a model of agricultural land use in the drainage basin of Lake Rotorua, New Zealand. With Prof. Deb, we developed not only the tools to find good solutions in this huge many-objective search space, but also the multi-criterion decision making methodology for providing it to the stake-holders in the community and getting their agreement on how to proceed. Although the Maori elders liked this approach very much, it ran into political problems with the younger Maori farmers, and we had to abandon it. But that was not before we were awarded the Wiley Practice Prize by the Society for Multi-Criteria Decision Making. Subsequently, Prof. Deb and I began collaborating with a team of land use researchers at ETH in Zurich and others from the University of Bonn in Germany. Graduate student Jonas Schwaab came to work with us for a year in BEACON, and we are continuing to address land use problems with him and his teammates in Germany and Switzerland.

BEACON Center: Surpassing All Expectations

In the early 2000s, Rich Lenski and Charles Ofria approached me with an idea to apply for a Science and Technology Center to study evolution in action. They were already collaborating in study of evolutionary mechanisms in both bacteria (in Lenski's Long-Term Evolution Experiment, or LTEE, involving 12 parallel cultures of E. coli transferred to new medium each day) and in "digital organisms," using Ofria's Avida platform of self-reproducing computer programs. They thought I could bring to the center both evolutionary computation (genetic algorithms and the like, based on modeling of biological evolution) and my management skills, since I had directed centers and industrial consortia for more than 20 years. The evolutionary computation would also bring to the center a strong capability for knowledge transfer to industry, which is expected of Science and Technology Centers. Rob Pennock was another co-PI, bringing in deep experience in the battle about teaching evolution in schools, among other things.

I thought this was an absolutely spectacular idea, and despite my earlier resolve never again to direct anything but my own research, quickly said yes to what I saw as an irresistible opportunity. We recruited an excellent team, selected partner institutions including the University of Michigan and the University of Puerto Rico Mayaguez, and prepared the proposal, under the title "Center for the Study of Evolution in Action," or CSEA. However, the proposal was not even selected for a site visit, so CSEA never came into existence.

However, in 2008, another call for STC's was announced, and we quickly decided to try again. We added Kay Holekamp as the fifth co-PI, bringing in her experience with evolution in action among spotted hyena in Kenya, adding a charismatic organism to the research areas of the PI's. This time, we chose as partners NCAT, where I was familiar with Gerry Dozier's GA work and Rich knew Joe Graves, an evolutionary biologist; University of Idaho, where I knew James Foster, who was active in both evolutionary computation and evolutionary biology; University of Texas Austin, where I knew computer scientist Risto Miikkulainen; and University of Washington, where Rich and Charles knew Ben Kerr, an outstanding young evolutionary biologist.

We structured BEACON such that, aside from the central operations staff, all of the research, education and diversity support funds would be allocated annually based on BEACON-peer-reviewed "budget requests." This was an innovative structure for an STC, and we feared that it might be attacked by reviewers as "trying to set up a mini-NSF." However, that was exactly what we wanted to do, for the particular research focus of evolution in action and for our five-university partnership. This time, the reviews were very positive, and the site visit went extremely well. I was astonished when MSU told us that they were proceeding with remodeling space in the BPS Building for BEACON, well before the decision to fund BEACON was announced by NSF. Selection of BEACON as an STC was announced in early February, 2010. I remember well that on February 8, when government offices were closed because of a severe snowstorm in Washington, D.C., I got a call from the person coordinating the 2010 class of STC's, Dr. Joan Frye. She knew that we'd expected to hear that week whether or not we were funded, and she didn't want us to have to wait through the weekend, so she called me with the good news. I was thrilled beyond words, of course, and will always be grateful to her for not forcing us to wait.

BEACON's model would never have worked had the co-PI's been the least bit selfish about how the funding was to be used. In most STC's, the PI's decide how to divide most of the money among their labs, in order to achieve their center's mission. But in BEACON, the mission is not tied to the labs of the PI's; rather, the co-PI's and I go through exactly the same process for getting funded any projects we might come up with as do any other BEACON members. Some of our projects are funded, and some are not. From the beginning, I have turned to the co-PI's for advice about issues that arise, sometimes weekly, and for help in policy formulation, even if those policies are eventually taken to our larger Executive Committee for consideration. Rich, Kay, Charles and Rob have been unfailingly helpful in guiding BEACON. In June, 2010, we were already moving into the completed BEACON Headquarters. We created the Managing Director, Diversity Director, Education Director and Business Manager positions in very short order, and proceeded to make spectacular (and very lucky) hires of Dr. Danielle Whittaker, Dr. Judi Brown Clarke, Dr. Louise Mead, and Ms. Connie James. Much of the success of BEACON has been attributable to the skills and dedication of these four and the other BEACON staff who were later added. I'd like to think that it was my highly tuned ability to spot top talent that helped us to make these hiring decisions, but I only wish that were true I have to attribute a lot of it to luck that these perfectly suited people appeared in the pools of applicants.

We allocated the first year's startup projects in March, 2010, using a clunky set of cobbled-together spreadsheets that I assembled with Brian Baer's help. We held the first BEACON Congress in August, 2010. Meantime, NSF had not yet completed the paperwork to actually fund the new STC's even in early September, and in fact, started the other STC's in September or later. But because BEACON was already in full swing, they agreed to backdate our starting date to August 1, 2010.

Science and Technology Centers are required by NSF to have an External Advisory Committee, and BEACON appointed an outstanding group of scientists and engineers to that body. But that group is external, and advisory in nature, and I also wanted to get regular and detailed feedback on BEACON's operation to help to guide in its development. So at the very beginning of BEACON, we also launched a team of assessment experts to do annual study of BEACON's dynamics, surveying BEACONites to identify things that could be improved. Under the leadership of Drs. Patty Farrell-Cole and Marilyn Amey, they and their students surveyed BEACONites and quickly discovered a number of problems. We needed better communication with our members, and made our BEACON-wide seminars a weekly event. We expanded the BEACON Congress, bringing in more people from our partner institutions, and decided to hold it every year, rather than as originally proposed. We initiated the BEACON Blog, under which ongoing research was highlighted with one or more posts every week on the center's front page, keeping it continually new and worth visiting. We created a special training program for the grad students and postdocs who would be supervising summer research interns and apprentices, after learning that in the first year, the undergrads felt like an unwelcome burden on the senior researchers. That program paid immediate dividends, increasing the quality of the experience for the undergrads and helping the grads/postdocs to see it as a valuable part of their training for a future faculty role.

BEACON Research

BEACON's research has been carried out in about four hundred "seed" projects so far, so it is not possible to summarize them here. The interested reader can consult each year's annual report on the public website, in which Dr. Danielle Whittaker, BEACON's Managing Director, has summarized each project in a paragraph or two. A very readable review of most of the projects is also available by reading back through the several hundred BEACON Blog entries. Suffice it to say that over 1,300 papers about BEACON-sponsored or BEACON-enabled research have been published already, including in journals such as Science, Nature, Proceedings of the Royal Society, PLOS One, Proceedings of the National Academy of Sciences, Evolutionary Computation, IEEE Proceedings, and many other top places. BEACON research is helping to address many of the important challenges to evolutionary biology, including evolution of cooperation, multicellularity, speciation, gene regulation, epigenetics, etc. BEACON researchers attack these questions in the domains of biological organisms and of artificial organisms, vastly increasing their capability to formulate and conduct experiments that illuminate the underlying mechanisms. Still other BEACON researchers are pushing the frontiers of developing new, more powerful evolutionary algorithms and applying them to solving real-world problems that have formerly been impractical to solve. This work can be found through BEACON's Research web page, and I couldn't begin to represent it well here.

Of particular note is the hiring of two endowed chair professors into BEACON's evolutionary computation research team. I recruited Prof. Kalyanmoy Deb to BEACON from IIT Kanpur to fill the Herman and Janet Koenig Endowed Chair (yes, the same Herman Koenig who was my advisor and later hired me on MSU's faculty). Then Dr. John Koza generously endowed a new chair in BEACON, the John R. Koza Endowed Chair in Genetic Programming, and we were able to recruit Prof. Wolfgang Banzhaf to that position. These two greatly enhanced the stature of BEACON's evolutionary computation team, and I have had the great pleasure of collaborating with them individually and together on many projects already, including on topics like multi-level optimization of logistics, automated classification of tariff codes, and evolution of architectures for deep learning networks.

BEACON Education and Outreach

Many of BEACON's initial graduate students were already at one of the BEACON universities before the center was created. So there was a backlog of students ready to take BEACON's newly created trio of courses: Evolutionary Biology for non-Life Scientists, Introduction to Computational Methods in Biology, and Multidisciplinary Approaches to the Study of Evolution. They were structured such that students from the BEACON partner universities could participate remotely, if they did not have equivalent courses already available locally or did not create them. Our engineers and computer scientists took the fall course in evolutionary biology and the evolutionary biologists took the fall course in computational methods. Then the two groups took the spring Multidisciplinary Approaches course, dividing into teams and creating computational models to experiment with. After a few years, the backlog of students was eliminated and the course enrollments decreased, and we are now opening them up more broadly to students across engineering and biology, including combining one of the courses with a similar one in the Department of Computational Math, Science and Engineering.

Many BEACONites proposed and carried out projects to create educational tools and curriculum to improve teaching of evolution, under the able leadership of Dr. Louise "Weezie" Mead. One of BEACON's shining points has been the extension of many of these projects to serve the whole country, which is unusual in STC-developed education projects. These include the founding of the Data Carpentry Workshops, now run nationally as a not-for-profit organization with funding from the Moore Foundation and under the leadership of former BEACONite Dr. Tracy Teal. Another is the Data Nuggets project, which now includes over sixty sets of data and accompanying descriptions that are valuable for teaching students how to analyze real scientific data, and are distributed freely on the web. A third is the Avida-ED software, which provides free access to a first-class, web-based research tool for study of evolution of digital organisms and provides lesson plans for teachers to help them use it effectively in a biology class. Many Avida-Ed workshops are offered each year to train teachers in its use. A fourth is the set of computer games continuing to be developed by BEACON's Idaho partners, led by Terry Soule. Darwin's Demons, the first of these games, features opponents that continually evolve, based on the player's actions, providing a fun way to see evolution in action. All of these BEACON products are creating impacts nationwide. BEACON staff and faculty also assume leadership roles in many professional organizations, including those for biology teachers, helping to assure the prominence of evolution education in our curricula.

BEACON Diversity

While the National Science Foundation has strong expectations of its STC's that they develop diversity within their centers, BEACON was determined from the outset that we would set ambitious goals and then actually surpass them. In the proposal stage, BEACON was able to draw on the unmatched skills of Prof. Percy Pierre, arguably the most impactful leader of graduate student diversity in the country. He helped to shape and start the graduate fellowship program of the Sloan Foundation, and in later years, to administer that program at Michigan State University. We set ambitious goals in the proposal, to exceed NSF's national norms in all categories and all levels. Percy and I put together the diversity program outlined in the proposal. As soon as BEACON was funded, we hired Dr. Judi Brown Clarke, who became the prime mover behind our diversity programs, as BEACON's Diversity Director.

Because Percy and I had already worked together before BEACON, we had a level of trust and mutual confidence that enabled us to set goals and then achieve them, with the understanding that we would cooperate and share resources to enable that. As a result, BEACON was soon able to exceed NSF's national norms for underrepresented minority graduate students, including racial/ethnic minorities in all fields, females in engineering/computer sciences, and persons with disabilities. Through Judi Brown Clarke, we then targeted underrepresented minorities and females in the undergraduate, postdoctoral and faculty ranks. Achieving those goals required creation of several new programs. For undergraduates, Dr. Brown Clarke created and operated a huge summer program, with part being a traditional REU (Research Experience for Undergraduates) program, drawing many participants from BEACON's partner NCAT and other HBCU's. But we wanted to have something for younger students not yet ready for the REU experience, so she created a second program, the Research Apprenticeships Program, which in one summer prepared students for a subsequent summer as an REU participant. We then added a "bridge" between those two, the BEACON Luminaries Program, which provides academic-year support for ten hours/week of research back at the home institution, between the apprentice and REU summers or between REU and grad school. All of this is aimed at providing these students the skills and confidence to know they can succeed in graduate school. BEACON has been hosting 60-80 students in BEACON labs each summer for many years. Much of the funding has been arranged from other sources, so that BEACON pays only about 40% of the total bills, enabling support for such a large number of students. But all of these students benefit from the oversight and training provided and arranged by Dr. Brown Clarke. That training included her oversight of the postdocs and grads working with the students, assuring that they also benefited from their supervisory experience.

The early years saw BEACON not yet meeting its goals regarding diversity of postdocs and faculty. We therefore created a BEACON Distinguished Postdoc Program, under which faculty members could nominate postdocs working on evolution in action to be supported at BEACON for two years. This program brought excellent postdocs to BEACON and put us over the top regarding postdoc diversity in both gender and race/ethnicity. For faculty diversity, and to assist with future undergrad and graduate student diversity, BEACON recruited a few outstanding faculty members outside the BEACON institutions to be brought into BEACON as Faculty Affiliates. These faculty members received some initial research funds, and then became eligible for competition for all other project funds in the future, just as if they were at a BEACON institution. Once again, this program allowed us to exceed the national norms for faculty in all categories.

Dr. Brown Clarke also launched BEACON's program to support persons with disabilities (or "this-abilities," as they are called here), and we now exceed these national norms by several fold.

Pierre, Brown Clarke and Goodman are now working on capturing the processes that have helped BEACON to achieve this notable success. Of course, part of the answer is the particular talents and dedication of Pierre and Brown Clarke, which cannot be replicated elsewhere. But we hope to capture other elements that can be used elsewhere in trying to achieve ambitious diversity goals.

Summary of Goodman's Intellectual Contributions

1. Demonstrated possibility to use genetic algorithm to solve a real-world problem, during thesis research, 1970-72.
2. Taught principles of state modeling and numerical simulation of ecological/biological systems to 400 biologists and 100 engineers, 1972-1981.
3. Led CAD/CAM research in the Case Center for Computer-Aided Engineering, 1983-2002: a. Developed the surface fitting system used by Chrysler to translate their proprietary Tchebyshev sculptured surface part definition database into industry-standard Non-Uniform Rational B-Splines (NURBS). b. Worked on CAD geometry used in manufacturing problems with many students, culminating in a series of contributions to numerically controlled toolpath verification and simulation of high-precision robotic spray application, both of which were applied in industry.
4. Did early work introducing "island" parallelism in evolutionary computing.

   a. Ph.D. student Adrian Sannier used fine-grain parallelism (with reproduction triggered by behavior, influenced by others' behaviors, but not requiring synchronization/generations) in his ALife and load-balancing work, published in the mid 1980s. This use of "island" parallelism was inspired by D.S. Wilson's MSU (Zoology) thesis work in the 1970s demonstrating that parallel populations could evolve behaviors not evolvable in single well-mixed populations.

   b. Developed an island parallelism GA toolkit, GALOPPS (the Genetic ALgorithm Optimized for Portability and Parallelism), 1994-2000, and supported it for the community for six years.

   c. Used GALOPPS in combinatorial optimization for scheduling/packing/nesting/knapsack/protein folding/docking problems during the 1990s, with many students and visitors in GARAGe.

   d. Extended GALOPPS, introducing heterogeneous parallel genetic algorithms allowing different fitness functions and problem representations in different islands, with translation of representations during migration.

   e. Specialized the heterogeneous parallel genetic algorithm to create "iiGA", the injection island (AKA island injection) genetic algorithm; distributed it as part of GALOPPS.

   f. Used iiGA with Ron Averill and David Eby to design radically different and superior lower compartment rails (parts of chassis) for cars. That led GM to ask us to form a company, which became Red Cedar Technology.

5. Redeveloped the ECE Senior Capstone Design Course to include industrially sponsored design projects and training in project management, team collaboration, and six-sigma methodology; taught it 2002-2010.

6. In the Genetic Programming/Bond Graph project funded by NSF, Goodman and bond graph expert Ron Rosenberg worked with postdoc Kisung Seo and Ph.D. students Zhun Fan and Jianjun Hu to create a GP system that created novel designs in the form of bond graphs. These bond graphs could be translated into designs for dynamic systems including mechanical, electrical, hydraulic, pneumatic, thermal, or other types of components. This method was used not only for design of new mechatronic systems, but also for redesign of existing systems, in what we might now call "Genetic Improvement" of dynamic systems, rather than computer programs. It was used to redesign pneumatic railroad brakes, improve on early designs of "Selectric" typewriters, and other systems. Because these systems required that both the topology and parameters of a system be evolved at once, the team developed "Structure Fitness Sharing (SFS)," a concept that allowed new topologies to remain in a population long enough on average that appropriate parameters could be evolved before a topology was discarded. Such a process today would be called a multi-level search algorithm. Because this de novo synthesis required very long searches, Jianjun Hu developed a new model for sustainable search, a process that allows continual injection of new random genomes into the population and does not force them to compete with more highly evolved individuals until their fitnesses are in similar ranges. This approach, called Hierarchical Fair Competition (HFC), was eventually extended by the team to allow it to evolve its fitness thresholds and even into a continuous form in which the fitness distribution itself was sampled for breeding and fitness evaluation purposes without subdividing the population ex-

plicitly. HFC and its offspring, Adaptive HFC (AHFC), Quick HFC (QHFC), and Continuous HFC (CHFC) were used by many authors during the ten years after their introduction, and the papers were cited over 700 times, but eventually, more in the community moved toward Greg Hornby's Age-Layered Population Structure (ALPS), which appeared soon after HFC was published, to address sustainability of search.

7. With Ron Averill, co-founded Red Cedar Technology, Inc. (originally known as "Applied Computational Design Associates"), 1999. The optimization technology was originally based on GALOPPS, Goodman quickly created a much more sophisticated, powerful and parameter-free blackbox optimizer and named it SHERPA, the optimization/search portion of the HEEDS design automation package. Because it was held as a trade secret, the SHERPA algorithm has never been published or described in detail, but has continued to win head-to-head blackbox optimization "bakeoffs" against its commercial and academic rivals. Goodman gave up his VP Technology role at Red Cedar in 2010, when NSF announced the funding of BEACON. In 2013, Red Cedar was acquired by CD Adapco, which in turn was acquired by Siemens PLM in 2016. Most of the Red Cedar staff members continue development and marketing of HEEDS and SHERPA under the Siemens banner today, and licensing, primarily for access to SHERPA, brings in millions of dollars each year—but to Siemens, not to its creators.

8. Co-founded, with Prof. Jennifer Olson of Media & Information, a Study Abroad program and associated specialization (now a minor) in Information and Communication Technology for Development, beginning in 2008. In the ensuing 10 years, the program has grown to serve five schools and we have taken over 70 students in total to Mto wa Mbu, Tanzania, for a month of service learning in the schools there. We continue to update the electrical systems, Internet access, and computing resources each year. The program received initial funding from Lenovo, then five years of support from the Office of the Provost, but now relies entirely on donations for its continuation. The advances in the schools, although they started slowly, have now been dramatic.

9. With Prof. Kalyanmoy Deb and others, I developed methodologies for addressing land use problems, first in New Zealand and later in Switzerland.

10. My primary contributions to the BEACON proposal (and, after funding, to the center) were (i) my organizational skills, which helped us to establish the center's structure and operations—very different from those of most previous Science and Technology Centers, (ii) my experience in training interdisciplinary teams, which was the basis for BEACON's three-course transdisciplinary graduate training sequence, and (iii) my experience in tapping evolutionary principles from nature to improve evolutionary computation algorithms and their use to address industrial problems, contributing strongly to BEACON's knowledge transfer to industry.

Professional Contributions

1. General Co-chair, Eleventh International Computer Graphics Conference, 1987, Detroit (sponsored by Society of Automotive Engineers and Engineering Society of Detroit)
2. American Institute of Aeronautics and Astronautics CAD/CAM Technical Committee Outstanding Contributions Award, 1990
3. Organizer of the first evolutionary computation conference in Russia, EvCA'96, at the Presidium of the Russian Academy of Sciences, under sponsorship of the International Society for Genetic Algorithms. Vladimir Uskov co-organized.
4. General Chair of the last International Conference on Genetic Algorithms, MSU, 1997; collaborated in its combining with the Genetic Programming Conference to become GECCO, the Genetic and Evolutionary Computation Conference, in 1999.
5. General Chair of GECCO in San Francisco, 2001
6. Chair of International Society for Genetic and Evolutionary Computation, 2001-2004
7. Founding Chair of ACM SIGEVO (Special Interest Group on Genetic and Evolutionary Computation), 2005
8. Organizer and Co-chair of a special, ACM-SIGEVO-sponsored EC conference in China, the GEC Summit, in Shanghai, 2009

A more extensive autobiography is available at www.egr.msu.edu/~goodman/ ProfessionalAndPersonalAutobiography.pdf and a list of publications can be found at Google Scholar under "ED Goodman".

Index

Printed in the United States
by Baker & Taylor Publisher Services